Holistic Approaches to Teacher Development:

Leadership, Pedagogical Practices, and Cognitive Insights

Mehmet Tufan Yalçın
Çankırı Karatekin University, Turkey

Alper Uslukaya
Çankırı Karatekin University, Turkey

Bahaddin Demirdiş
Çankırı Karatekin University, Turkey

IGI Global
Scientific Publishing
Publishing Tomorrow's Research Today

Vice President of Editorial	Melissa Wagner
Director of Acquisitions	Mikaela Felty
Director of Book Development	Jocelynn Hessler
Production Manager	Mike Brehm
Cover Design	Phillip Shickler

Published in the United States of America by
IGI Global Scientific Publishing
701 East Chocolate Avenue
Hershey, PA, 17033, USA
Tel: 717-533-8845
Fax: 717-533-7115
Website: https://www.igi-global.com E-mail: cust@igi-global.com

Library of Congress Cataloging-in-Publication Data

Names: Yalçın, Mehmet, 1985- editor. | Uslukaya, Alper, 1981- editor. | Demirdis, Bahaddin, 1991- editor.
Title: Holistic approaches to teacher development : leadership, pedagogical practices, and cognitive insights / edited by Mehmet Yalçin, Alper Uslukaya, Bahaddin Demirdis.
Description: Hershey, PA : IGI Global Scientific Publishing, [2025] | Includes bibliographical references and index. | Summary: "The objective of the book is to provide an integrative framework that enhances the professional growth of teachers by connecting leadership strategies, innovative pedagogical practices, and cognitive insights"-- Provided by publisher.
Identifiers: LCCN 2025000148 (print) | LCCN 2025000149 (ebook) | ISBN 9798337304724 (hardcover) | ISBN 9798337304731 (paperback) | ISBN 9798337304748 (ebook)
Subjects: LCSH: Teachers--In-service training--Case studies. | Teachers--Training of--Case studies.
Classification: LCC LB1731 .H585 2025 (print) | LCC LB1731 (ebook) | DDC 370.71/5--dc23/eng/20250304
LC record available at https://lccn.loc.gov/2025000148
LC ebook record available at https://lccn.loc.gov/2025000149

British Cataloguing in Publication Data
A Cataloguing in Publication record for this book is available from the British Library.

Table of Contents

Preface x

Chapter 1
Teacher Leadership and AI Literacy: An Analysis of Teachers' AI Literacy Levels 1
Ömür Çoban, Karamanoglu Mehmetbey University, Turkey
Mehmet Durnalı, Bülent Ecevit University, Turkey

Chapter 2
Artificial Intelligence in Education and Changing Teacher Roles 25
Ramazan Atasoy, Harran University, Turkey

Chapter 3
Building Smarter Educators: Merging Cognitive and Emotional Intelligence in Teacher Development 47
Vijendra Nath Pathak, Parul University, India
Elizabeth Kardze Annan, School of Social Science and Language, Lovely Professional University, India

Chapter 4
Teachers' Perceptions of Affective Skill Development and Associated Barriers: Strategies and Policy Implications in Secondary Education 69
Alper Uslukaya, Çankırı Karatekin University, Turkey
Mehmet Tufan Yalçın, Çankırı Karatekin University, Turkey

Chapter 5
Shaping Inclusive Mindsets Through Teacher Education and Personal Experience in Türkiye 97
Bahaddin Demirdis, Çankırı Karatekin University, Turkey
Abdullah Çiftçi, Akdeniz University, Turkey

Chapter 6
The Role of Assistant Principals in Fostering Teacher Collaboration and Professional Growth: A Pathway to Effective School Leadership 121
Henderson Lewis, Louisiana State University, USA
Stacy-Ann T. Campbell, Louisiana State University, USA
Saleria S. Shaffer, Louisiana State University, USA
Veysel Altunel, Louisiana State University, USA

Chapter 7
Play, Lead, Teach: Tapping Into the Power of Sports for Teacher Development and Classroom Innovation 147
Amitava Pal, ICFAI University, Jharkhand, India
Pallavi Kumari, ICFAI University, Jharkhand, India
Preksha Yadav, Krupanidhi College of Management, India

Chapter 8
Developing Mathematical Competencies in Teachers: A Holistic Perspective on Effective Teaching and Teacher Well-Being 171
Rocío Rodríguez-Padín, Universidade da Coruña, Spain

Chapter 9
Teaching With Adaptability: Cognitive and Pedagogical Strategies for Geriatric Nursing 205
Tiago Manuel Horta Reis da Silva, King's College London, UK

Chapter 10
Investigating the Relationship Between Teacher Stress and Classroom Performance: A Well-Being Perspective 235
A. K. Monisha, Jain University, India
Sandhya Reddy S., Bishop Cotton Women's Christian College, India
Shree S. Kavya, Bishop Cotton Women's Christian College, India
M. Swetha, Bishop Cotton Women's Christian College, India

Chapter 11
Spiritual Wellbeing and Self-Efficacy as Catalysts for Culturally Responsive Teaching in Diverse Classrooms 261
Vijendra Nath Pathak, Parul Institute of Liberal Arts, Parul University, India
Mamta, School of Social Science and Language, Lovely Professional University, India

Chapter 12
Digital Pathways to Excellence for Bridging Gaps and Building Competencies: The Role of Educational Technology in Transforming Teacher Professional Development 287
Seema Yadav, Bhopal School of Social Sciences, India

Compilation of References 309

About the Contributors 359

Index 363

Detailed Table of Contents

Preface x

Chapter 1
Teacher Leadership and AI Literacy: An Analysis of Teachers' AI Literacy Levels 1
Ömür Çoban, Karamanoglu Mehmetbey University, Turkey
Mehmet Durnalı, Bülent Ecevit University, Turkey

This section emphasises the importance of teacher leadership and determine their artificial intelligence literacy level. Teachers should have a mission that enables students in their classrooms to realise themselves, transform education and training processes according to the requirements of the age group in question, inspire and guide their colleagues in terms of their professional development, and strive for the development of the school and society. In other words, the teacher leader should optimise their potential for the advancement of their students, colleagues, school and society by enhancing their knowledge, abilities and experience. The objective is to enhance teachers' awareness of the potential applications of artificial intelligence in their instructional practices and the broader education and training processes, within the context of Education 4.0. Furthermore, it will assist teachers in making more informed decisions regarding the ethical and beneficial use of artificial intelligence.

Chapter 2
Artificial Intelligence in Education and Changing Teacher Roles 25
Ramazan Atasoy, Harran University, Turkey

This book chapter aims to present a holistic and innovative perspective on AI in education and its impact on evolving teacher roles. It seeks to raise awareness of AI-enabled teaching practices while offering a transformative framework to enhance teachers' professional development and promote sustainable education. Additionally, this chapter provides both theoretical and practical insights into the shifting needs of educators as they navigate the transition from today's complex educational landscape to AI-integrated learning environments, which have the potential to shape future educational paradigms. In line with these objectives, the chapter explores various perspectives on integrating AI to support teacher development, reframe teacher leadership, and equip school administrators and policymakers with a comprehensive understanding of AI's opportunities and challenges in education. Furthermore, it examines how AI enhances education, fosters professional development, and improves student outcomes while addressing potential risks and ethical considerations.

Chapter 3
Building Smarter Educators: Merging Cognitive and Emotional Intelligence in Teacher Development 47
Vijendra Nath Pathak, Parul University, India
Elizabeth Kardze Annan, School of Social Science and Language, Lovely Professional University, India

The chapter, "Building Smarter Educators: Merging Cognitive and Emotional Intelligence in Teacher Development," emphasises the integration of cognitive and emotional intelligence into teacher training. It highlights how these competencies help teachers recognise and address students' needs, create effective lesson plans, mentor and inspire students, manage conflicts, and cope with professional challenges, leading to reduced burnout, job satisfaction, and student success. Strategies discussed include mentoring, collaborative enquiry, mindfulness, SEL programmes, peer coaching, reflective practices, and leveraging technology. Challenges like resource constraints, cultural differences, and balancing personal and professional growth are addressed with solutions such as continuous evaluation, stress management, and fostering supportive environments. The chapter underscores the need for emotionally and cognitively intelligent teachers to foster student-centred learning and enhance academic and professional outcomes.

Chapter 4
Teachers' Perceptions of Affective Skill Development and Associated Barriers: Strategies and Policy Implications in Secondary Education 69
Alper Uslukaya, Çankırı Karatekin University, Turkey
Mehmet Tufan Yalçın, Çankırı Karatekin University, Turkey

This chapter explores secondary school teachers' perspectives on the development of affective skills, including emotional regulation, empathy, and social communication, in classroom settings. Adopting a phenomenological framework, the study utilized semi-structured interviews with 18 secondary school teachers in Türkiye to examine their perceptions of affective skill development, the challenges they encounter, and the strategies they employ. The findings indicate that while teachers recognize the importance of affective competencies, they face systemic challenges such as rigid curricular structures, the dominance of standardized testing, and limited institutional support, which hinder the consistent implementation of affective education. Furthermore, teachers predominantly rely on informal, intuitive methods to assess affective growth, underscoring the need for structured assessment frameworks and ongoing professional development. The chapter concludes by offering policy recommendations to integrate affective education into formal curricula and enhance institutional resources.

Chapter 5
Shaping Inclusive Mindsets Through Teacher Education and Personal Experience in Türkiye 97
Bahaddin Demirdis, Çankırı Karatekin University, Turkey
Abdullah Çiftçi, Akdeniz University, Turkey

This chapter examines the attitudes of teacher candidates in Türkiye towards inclusive education, focusing on the influence of teacher training pathways and personal demographic factors. Drawing on data from 369 participants enrolled in either education faculties or pedagogical formation programmes, the study explores how variables such as gender, age, and prior experience with individuals with special educational needs (SEN) shape inclusive mindsets. The findings reveal significant differences based on gender and familial experience with SEN, while age and type of training programme show no statistically significant impact. Using a validated attitudes scale, this chapter highlights the critical role of experiential learning and targeted professional development in shaping positive perceptions of inclusive education. The chapter concludes with implications for teacher education policy and practices, emphasizing the need for comprehensive, equity-focused training that bridges theory and practice.

Chapter 6
The Role of Assistant Principals in Fostering Teacher Collaboration and Professional Growth: A Pathway to Effective School Leadership 121
Henderson Lewis, Louisiana State University, USA
Stacy-Ann T. Campbell, Louisiana State University, USA
Saleria S. Shaffer, Louisiana State University, USA
Veysel Altunel, Louisiana State University, USA

This chapter examines the evolving role of assistant principals (APs) as instructional leaders and their impact on fostering teacher collaboration and professional growth. The purpose of this chapter is to explore how APs perceive their roles and the challenges they face in balancing instructional leadership and administrative duties. Drawing on theoretical frameworks such as sensemaking and social constructivism, the chapter highlights strategies APs employ to support teachers, build collaborative cultures, and enhance professional learning communities. It also underscores the critical role of APs in shaping school culture, improving teaching practices, and promoting student achievement. Using qualitative data collected through interviews with assistant principals in various school districts, this chapter identifies key competencies, challenges, and recommendations for maximizing the AP's role as a pathway to effective school leadership. Practical implications for professional development, mentorship, and policy considerations are also discussed.

Chapter 7
Play, Lead, Teach: Tapping Into the Power of Sports for Teacher Development and Classroom Innovation 147
Amitava Pal, ICFAI University, Jharkhand, India
Pallavi Kumari, ICFAI University, Jharkhand, India
Preksha Yadav, Krupanidhi College of Management, India

This chapter explores the transformative potential of sports-based approaches in teacher development, focusing on how physical activities and game-based learning foster leadership, cognitive, and emotional skills in educators. By examining global case studies, it highlights the role of sports in enhancing competencies such as decision-making, collaboration, resilience, and emotional intelligence. The chapter offers practical strategies for integrating sports into teaching methods to improve classroom dynamics, promote inclusivity, and encourage cross-disciplinary learning. It also discusses scalable implementation and the use of digital tools like AI to support teacher development, presenting sports as a key platform for shaping the future of teaching.

Chapter 8
Developing Mathematical Competencies in Teachers: A Holistic Perspective on Effective Teaching and Teacher Well-Being 171
Rocío Rodríguez-Padín, Universidade da Coruña, Spain

This chapter explores the development of mathematical competencies in teachers through a holistic lens, emphasizing the integration of pedagogical knowledge, socio-emotional support, and innovative teaching practices. It highlights the need for adaptive and creative approaches that address diverse student needs and foster critical thinking, problem-solving, and engagement. Central to the discussion is the role of leadership in mathematics teaching, where educators act as facilitators of inquiry and mentors within their professional communities. The chapter also examines challenges such as cognitive demands, equity in education, and classroom management, offering strategies to promote teacher well-being and resilience. Key insights include the use of manipulatives, interdisciplinary connections, and reflective practices to enhance teaching effectiveness. By prioritizing professional development and addressing institutional barriers, the chapter advocates for sustainable and transformative mathematics teaching, benefiting both educators and students.

Chapter 9
Teaching With Adaptability: Cognitive and Pedagogical Strategies for Geriatric Nursing 205
Tiago Manuel Horta Reis da Silva, King's College London, UK

This chapter explores the role of adaptability in teaching geriatric nursing, emphasising cognitive and pedagogical strategies that cater to diverse learning needs. It delves into the cognitive processes involved in learning complex geriatric care concepts and the pedagogical methods that enhance adaptability in teaching. The chapter discusses case studies and evidence-based approaches to integrating flexible teaching strategies, promoting critical thinking and problem-solving skills among nursing students. It also addresses the challenges and solutions in implementing adaptable teaching practices in geriatric nursing education. The chapter concludes with recommendations for educators to develop adaptable teaching frameworks that foster comprehensive learning in geriatric care.

Chapter 10
Investigating the Relationship Between Teacher Stress and Classroom Performance: A Well-Being Perspective 235
A. K. Monisha, Jain University, India
Sandhya Reddy S., Bishop Cotton Women's Christian College, India
Shree S. Kavya, Bishop Cotton Women's Christian College, India
M. Swetha, Bishop Cotton Women's Christian College, India

Abstract: Teaching profession is highly demanding, often exposing the teachers to high stress which is affecting their well-being and classroom effectiveness. The study explores the relationship between teacher stress and classroom performance, emphasizing the mediating role of teacher's well-being. Through thematic analysis of teacher's semi structured interviews, key stressors such as workload, administrative pressures and student behaviour were identified. Findings indicated that these stressors negatively influence teaching quality engagement, and student outcomes. The Study also highlights the effectiveness of well-being programs including mindfulness practices, emotional support systems and work-life balance strategies to decrease stress and enhance teacher's performance. Based on the findings, it is recommended that institutions to prioritize well-being programs to foster a healthier and resilient teaching environment. Implementing holistic interventions can support teachers' mental health while also contributing to improved classroom dynamics and student success.

Chapter 11
Spiritual Wellbeing and Self-Efficacy as Catalysts for Culturally Responsive Teaching in Diverse Classrooms 261
Vijendra Nath Pathak, Parul Institute of Liberal Arts, Parul University, India
Mamta, School of Social Science and Language, Lovely Professional University, India

Integrating spiritual wellbeing and self-efficacy equips educators with the emotional resilience and confidence needed to navigate diverse classrooms effectively. Spiritual wellbeing fosters empathy, patience, and emotional balance, enabling teachers to manage stress, form meaningful connections, and create inclusive environments where students feel valued and motivated. Self-efficacy enhances teachers' belief in their ability to positively impact student learning, driving reflective practices, adaptability, and innovative strategies. Together, these elements empower educators to address diverse needs, foster equity, and create compassionate, adaptive learning environments, benefiting both students and the broader educational system.

Chapter 12
Digital Pathways to Excellence for Bridging Gaps and Building Competencies: The Role of Educational Technology in Transforming Teacher Professional Development................................. 287
Seema Yadav, Bhopal School of Social Sciences, India

This chapter explores the integration of educational technology into Continuing Professional Development (CPD) for educators, highlighting the transformative potential of advanced digital tools and strategies. It examines how technologies such as Artificial Intelligence (AI), Virtual Reality (VR), Augmented Reality (AR), and microlearning can create personalized, adaptive, and immersive learning experiences, enhancing the effectiveness of CPD programs. It emphasizes the importance of peer-to-peer learning, mentorship, and global collaboration in promoting knowledge-sharing and professional development. Challenges such as ensuring digital accessibility, addressing attitudes towards technology, and providing ongoing support are also explored. The chapter concludes by stressing the need for inclusive, equitable CPD programs that prioritize teacher agency and hands-on learning, ensuring that educators are equipped with the skills and confidence to navigate the evolving digital landscape and provide high-quality, student-centered education.

Compilation of References 309

About the Contributors 359

Index 363

Preface

INTRODUCTION

Education is undergoing a profound transformation in the face of rapidly evolving pedagogical, cognitive, and technological developments. The acceleration of change in global education systems has redefined the expectations from teachers, requiring them to adapt swiftly and skillfully. As the roles and responsibilities of educators continue to expand, the need for holistic and evidence-based approaches to teacher development becomes increasingly essential. Teachers are now expected to be more than content experts; they are mentors, facilitators, leaders, and lifelong learners. They must navigate complex learning environments, respond to diverse student needs, and integrate new technologies into their practice.

This volume, *Holistic Approaches to Teacher Development: Leadership, Pedagogical Practices, and Cognitive Insights*, addresses these complexities by bringing together a wide range of interdisciplinary perspectives that place the teacher at the center of educational change. By combining theory, practice, and lived experiences, the book presents a rich array of insights into how teachers can grow personally and professionally in response to changing educational demands.

Teaching today takes place in a landscape marked by rapid innovation, increased student diversity, growing mental health concerns, and new forms of digital engagement. In such a climate, educators are called upon to rethink their roles, reflect critically on their practices, and engage in continual professional development. This book responds to these global realities by offering fresh insights and strategies that support educators in thriving amidst complexity. Whether through the lens of emotional intelligence, AI literacy, or culturally responsive teaching, each chapter contributes to building an educator who is not only skilled but also resilient, ethical, and reflective.

Purpose and Vision

The purpose of this book is to bridge the gap between theoretical perspectives and practical strategies in teacher development. By integrating research findings, case studies, and interdisciplinary insights, this volume seeks to:

- Identify essential competencies such as AI literacy, emotional intelligence, and cognitive adaptability
- Showcase leadership practices that support collaboration and professional growth
- Highlight inclusive and culturally responsive approaches in diverse teaching contexts
- Promote well-being, resilience, and lifelong learning as foundations of sustainable teacher development

Coverage of Topics

The chapters in this book are organized into thematic areas that reflect the multifaceted nature of modern teacher development:

- **Foundational Skills for the Future**: Emotional intelligence, self-efficacy, and resilience as essential qualities for effective teaching
- **Technology Integration**: Exploring the impact of AI literacy and digital tools on evolving teacher roles and classroom practices
- **Innovative Pedagogies**: Cognitive and affective strategies that promote adaptability, collaboration, and reflective teaching
- **Equity and Inclusion**: Addressing cultural responsiveness, inclusive mindsets, and spiritual well-being in diverse educational settings
- **Professional Growth and Well-Being**: Investigating the role of leadership, stress management, and supportive environments in sustaining teacher motivation and excellence

Target Audience

This book is intended for a diverse range of stakeholders in education. It is particularly useful for:

- **Teacher educators and mentors** seeking innovative frameworks for professional development
- **Practicing teachers** who want to enhance their skills, reflect on their practice, and stay current with evolving educational demands
- **Educational leaders and school administrators** interested in fostering supportive learning environments and promoting teacher growth
- **Graduate students and researchers** exploring contemporary issues in pedagogy, leadership, and cognitive-emotional development
- **Policy makers and curriculum developers** aiming to design effective teacher education programs that integrate technology, inclusivity, and mental well-being.

OVERVIEW OF THE CHAPTERS

This section provides an overview of the chapters in *Holistic Approaches to Teacher Development: Leadership, Pedagogical Practices, and Cognitive Insights*, summarizing their core themes, approaches, contributions, and implications for advancing teacher education and professional growth. Each chapter offers unique perspectives on the evolving role of teachers, highlighting both the challenges and opportunities presented by contemporary educational demands. The contributions span a range of topics including AI literacy, emotional intelligence, inclusive practices, leadership development, and cognitive strategies. Together, these chapters reflect a shared commitment to supporting educators as they navigate the complexities of 21st-century classrooms. The insights offered serve not only as academic reflections but also as practical tools for shaping the future of teacher development.

Chapter 1: Teacher Leadership and AI Literacy: An Analysis of Teachers' AI Literacy Levels

This chapter explores the intersection of teacher leadership and artificial intelligence literacy by examining how well educators understand and utilize AI technologies in their professional roles.

Approach: Adopts a quantitative method to assess teachers' levels of awareness, usage, evaluation, and ethical understanding of AI through a survey conducted across a broad sample.
Contribution: Demonstrates how AI literacy can empower teachers to enhance their leadership roles, support innovation in the classroom, and guide peers in adopting new technologies.
Implications: Suggests that improving teachers' AI literacy may contribute to more effective educational practices, promote ethical awareness, and encourage stronger engagement with digital tools.
Chapter 2: Artificial Intelligence in Education and Changing Teacher Roles

This chapter discusses how artificial intelligence is transforming educational environments and redefining the traditional roles and responsibilities of teachers.

Approach: Explores the implications of AI integration in instructional processes, administrative tasks, personalized learning, and content creation through thematic analysis.
Contribution: Highlights the shifting expectations placed on educators, emphasizing their evolving roles as facilitators, ethical leaders, and mentors in AI-supported learning ecosystems.
Implications: Suggests the urgent need for teacher training in AI literacy, ethical awareness, and digital pedagogy to ensure responsible, effective use of AI in education and prevent growing inequalities.
Chapter 3: Building Smarter Educators: Merging Cognitive and Emotional Intelligence in Teacher Development

This chapter explores how integrating cognitive and emotional intelligence in teacher development can lead to more effective, adaptive, and resilient educators.

Approach: Combines theoretical discussion with practical strategies, highlighting how emotional and cognitive competencies can be fostered through mentoring, reflective practice, and professional development programs.
Contribution: Emphasizes the need for teachers to be both intellectually capable and emotionally attuned in order to manage classroom dynamics, support student well-being, and enhance learning outcomes.
Implications: Suggests that teacher training programs should adopt a holistic framework that develops both emotional awareness and cognitive skills to prepare educators for the complex demands of 21st-century classrooms.
Chapter 4: Teachers' Perceptions of Affective Skill Development and Associated Barriers, Strategies and Policy Implications

This chapter investigates teachers' perceptions of affective skill development in secondary education, focusing on how they define, support, and evaluate students' emotional competencies.

Approach: Employs a phenomenological research design through in-depth interviews with secondary school teachers to explore lived experiences and classroom strategies related to affective learning.

Contribution: Provides insights into how teachers conceptualize emotional growth, implement classroom practices like group work and empathy-based activities, and navigate systemic challenges such as curriculum pressure and lack of institutional support.
Implications: Highlights the need for practical training, curriculum reform, and policy-level support to better integrate affective education into formal schooling and empower teachers as key agents in students' emotional development.
Chapter 5: Shaping Inclusive Mindsets Through Teacher Education and Personal Experience in Türkiye

This chapter investigates the attitudes of teacher candidates in Türkiye toward inclusive education and how these attitudes are shaped by demographic and experiential factors.

Approach: Analyzes survey data from pre-service teachers in different training programs, examining the influence of gender, age, and personal experience with individuals with special educational needs.
Contribution: Reveals that personal experiences and gender significantly impact attitudes toward inclusion, while age and program type show no effect.
Implications: Emphasizes the importance of experiential learning and equity-focused teacher education programs to cultivate inclusive mindsets.
Chapter 6: The Role of Assistant Principals in Fostering Teacher Collaboration and Professional Growth

This chapter explores how assistant principals (APs) contribute to instructional leadership and support teacher collaboration and growth.

Approach: Uses qualitative interviews with APs to examine how they balance administrative responsibilities with leadership in professional learning communities.
Contribution: Identifies strategies and competencies APs use to foster collaboration, shape school culture, and improve teaching effectiveness.
Implications: Offers insights into AP training, mentorship, and policy changes to enhance their leadership role in schools.
Chapter 7: Play, Lead, Teach: Tapping into the Power of Sports for Teacher Development and Classroom Innovation

This chapter examines how sports-based learning can transform teacher development by enhancing leadership, emotional intelligence, and classroom innovation.

Approach: Draws on international case studies and practical strategies that integrate physical activity and gamified learning into teaching practices.
Contribution: Shows how sports promote collaboration, resilience, and inclusive teaching across disciplines.
Implications: Suggests incorporating sports and digital tools into teacher training to improve engagement, well-being, and instructional effectiveness.
Chapter 8: Developing Mathematical Competencies in Teachers: A Holistic Perspective on Effective Teaching and Teacher Well-being

This chapter addresses how to develop mathematics teaching skills through a holistic lens that integrates pedagogy, emotional support, and innovation.

Approach: Discusses reflective practice, leadership roles, and the use of interdisciplinary methods to support effective and equitable mathematics instruction.
Contribution: Highlights the importance of teacher well-being, adaptive teaching strategies, and socio-emotional learning in math education.
Implications: Recommends sustained professional development and institutional support to empower math educators and improve outcomes.
Chapter 9: Teaching with Adaptability: Cognitive and Pedagogical Strategies for Geriatric Nursing

This chapter examines the importance of adaptability in geriatric nursing education, focusing on how cognitive and pedagogical strategies enhance teaching effectiveness.

Approach: Uses case studies and evidence-based methods to illustrate flexible teaching practices that support critical thinking and problem-solving in complex care contexts.
Contribution: Highlights how adaptable instruction addresses diverse learner needs and promotes deeper understanding of geriatric care.
Implications: Recommends designing responsive teaching frameworks that prepare nursing educators to deliver effective, student-centered instruction in dynamic healthcare environments.
Chapter 10: Investigating the Relationship Between Teacher Stress and Classroom Performance: A Well-Being Perspective

This chapter explores how teacher stress impacts classroom performance, emphasizing the importance of well-being in sustaining teaching effectiveness.

Approach: Analyzes qualitative interview data to identify common stressors and their influence on engagement, instructional quality, and student outcomes.
Contribution: Demonstrates how stress undermines classroom dynamics and highlights strategies like mindfulness and work-life balance to mitigate its effects.
Implications: Advocates for institutional policies that prioritize teacher well-being to foster resilience and improve student learning experiences.
Chapter 11: Spiritual Wellbeing and Self-Efficacy as Catalysts for Culturally Responsive Teaching in Diverse Classrooms

This chapter investigates how spiritual wellbeing and self-efficacy support culturally responsive teaching in diverse learning environments.

Approach: Theoretically explores the emotional and psychological foundations of effective teaching through the lens of spiritual balance and personal belief in competence.
Contribution: Shows how inner resilience and confidence enhance empathy, adaptability, and student-centered teaching in multicultural settings.
Implications: Encourages teacher education programs to integrate reflective practices that nurture both spiritual awareness and professional agency.

Chapter 12: Digital Pathways to Excellence for Bridging Gaps and Building Competencies

This chapter focuses on the role of educational technology in transforming teacher professional development through innovative and inclusive digital tools.

Approach: Examines the use of AI, VR, AR, and microlearning in CPD to support personalized and immersive learning experiences for educators.
Contribution: Highlights how digital strategies enhance teacher competencies, encourage collaboration, and support continuous, relevant professional growth.
Implications: Calls for equitable, tech-integrated CPD programs that empower teachers to navigate the digital age with confidence and adaptability.

FINAL REFLECTIONS AND IMPACT ON THE FIELD

This book contributes to the field of teacher education by emphasizing the need for a holistic, interdisciplinary approach to professional development. Rather than treating teaching as a purely technical process, it highlights the emotional, cognitive, cultural, and ethical dimensions of being an educator in the 21st century. The integration of themes such as AI literacy, emotional intelligence, inclusive teaching, and leadership reflects the complex skill set today's teachers require. By presenting practical strategies alongside theoretical frameworks, the volume supports educators in navigating change with confidence. It serves as both a reflective tool and a practical guide for developing resilient, adaptive, and future-ready teachers.

Each chapter in the book brings a unique perspective on what it means to teach and lead in modern educational contexts. From sports-based development to spiritual well-being, from affective skills to mathematical competencies, the diversity of content reflects the broad scope of teacher responsibilities today. This variety not only enriches academic discourse but also provides actionable insights for practitioners. The chapters encourage educators to reflect on their roles, challenge their assumptions, and explore new ways of growing both personally and professionally. Through this, the book fosters a deeper understanding of teaching as a dynamic and evolving practice.

Ultimately, this volume aims to inspire systemic change in how we support, train, and empower educators. It argues for the centrality of teacher well-being, lifelong learning, and ethical awareness in creating equitable and innovative learning environments. The book also contributes to global conversations around educational reform by addressing pressing challenges such as digital transformation, diversity, and teacher burnout. In doing so, it positions itself as a relevant and necessary resource for anyone involved in shaping the future of education. Its contribution lies in bridging knowledge and practice, and in championing the human side of teaching in an increasingly complex world.

Chapter 1
Teacher Leadership and AI Literacy:
An Analysis of Teachers' AI Literacy Levels

Ömür Çoban
https://orcid.org/0000-0002-4702-4152
Karamanoglu Mehmetbey University, Turkey

Mehmet Durnalı
https://orcid.org/0000-0002-1318-9362
Bülent Ecevit University, Turkey

ABSTRACT

This section emphasises the importance of teacher leadership and determine their artificial intelligence literacy level. Teachers should have a mission that enables students in their classrooms to realise themselves, transform education and training processes according to the requirements of the age group in question, inspire and guide their colleagues in terms of their professional development, and strive for the development of the school and society. In other words, the teacher leader should optimise their potential for the advancement of their students, colleagues, school and society by enhancing their knowledge, abilities and experience. The objective is to enhance teachers' awareness of the potential applications of artificial intelligence in their instructional practices and the broader education and training processes, within the context of Education 4.0. Furthermore, it will assist teachers in making more informed decisions regarding the ethical and beneficial use of artificial intelligence.

INTRODUCTION

Artificial intelligence (AI) is defined as a computing technology that is capable of imitating human cognitive abilities. A review of the literature reveals that AI is defined as a system that can perform human abilities such as perception, reasoning, comprehension, interpretation, generalisation, inference and learning, and can successfully complete many tasks simultaneously (Gondal, 2018). The advent of ChatGPT in late 2022 brought AI to the fore on a global scale. This AI model is capable of mimicking human capabilities, including the generation of text, images, video, music, and software code, as well as

DOI: 10.4018/979-8-3373-0472-4.ch001

answering questions in a manner akin to a personal assistant. The number of AI models is increasing on a daily basis, with new examples including ChatGPT, Alpaca, Bard, Chatsonic, Copilot, Ernie, Hugging Chat, Jasper, Llama: Open Assistant, Tongyi Qianwen, YouChat, and others. As technology advances, the number of applications utilising artificial intelligence continues to grow, with a notable impact on the ease and convenience of people's lives. The use of AI algorithms is becoming increasingly prevalent across a range of fields, including automatic language translation, autonomous vehicles, health diagnosis, sales forecasting and personal assistants (Komalavalli et al., 2020). The pervasive use of AI across a multitude of domains, including healthcare, industry, education, finance, and transportation, underscores the imperative for individuals to cultivate themselves in this field and acquire the requisite competencies to adapt to the innovations ushered in by AI (Erdoğan & Kırılmaz, 2020; Wang et al., 2023).

A review of the literature reveals that the concept of literacy is defined in ways that extend beyond the traditional skills of reading and writing. The ability to use these skills in daily life is also a key aspect of literacy (Güneş, 2019). From this perspective, artificial intelligence literacy (AIL) encompasses a comprehensive range of competencies that extend beyond mere technological expertise, encompassing the social and ethical dimensions of AI. This concept facilitates comprehension of AI technologies, evaluation of the prospective consequences of these technologies, and conscious utilisation of AI-supported products. In light of the growing integration of AI technologies and the impact of these technologies on human life, it can be argued that AI is of paramount importance (Stembert et al., 2019; Su, 2018). It is hypothesised that the effective utilisation of AI technologies can have a considerable impact on individuals' attitudes towards these technologies and their daily usage habits. It is therefore of great importance to measure and develop AIL in order to ensure that all humanity benefits from AI technologies in the most efficient way (Luo et al., 2019; Metelskaia et al., 2018). It is similarly important to emphasise the relationship between the teaching of professional skills and AIL. Teachers are of critical importance in ensuring the effective and efficient utilisation of AI technologies in the field of education. Teachers who possess the requisite literacy in AI are well-positioned to facilitate the acquisition of these skills by integrating the relevant technologies into the educational process. In this context, it is of great importance for teachers to develop their AIL skills, both for their own professional development and for their students to be better prepared for the future (Kandlhofer et al., 2016). The objective of this study is to ascertain the levels of AIL among teacher leaders. The objective is to enhance teachers' awareness of the potential applications of AI in their instructional practices and the broader education and training processes, within the context of Education 4.0. Furthermore, it will assist teachers in making more informed decisions regarding the ethical and beneficial use of AI.

TEACHER LEADERSHIP

One of the most notable works in the field of teacher leadership is Katzenmeyer and Moller's (2013) publication, *Awakening the Sleeping Giant*. In this book, the authors define the teacher leader as an effective educator and trainer in the classroom, a school developer and goal-setter, an innovative practitioner incorporating new methods and practices in their lessons, and a colleague who shares these with others. In the definition, Danielson (2006) emphasises that teacher leaders facilitate students' personal development in the classroom, transform education and training processes in accordance with the requirements of the age, inspire and guide their colleagues in their professional development, and strive for the advancement of the school and society. As evidenced by the aforementioned definitions, a teacher leader is engaged in the process of acquiring new knowledge, developing enhanced skills, and expanding their experience base. They are also committed to optimising their capacity for the advancement of their students, colleagues, school, and society at large. In terms of the means by which this is achieved, the teacher leader typically does so by monitoring and engaging with contemporary developments in classroom teaching practices, developing and implementing projects, organising social, cultural and sporting activities, and participating in decision-making processes and activities aimed at the advancement of the school.

A review of the history of teacher leadership reveals that it has evolved into its current form as a result of three main currents. In the initial phase, teachers served as the primary implementers and observers of curricula and educational policies developed by policymakers. During this period, teachers assumed roles within the intermediate administrative structure, acting in a mentorship capacity. During this period, which extended until the 1980s, teachers were not actively involved in the processes of curriculum development and educational policymaking. Teacher leaders were intermediate decision-makers who only observed the processes and provided mentoring in the field (Smylie et al., 2002). In other words, teacher leaders were not the developers or planners of innovative policies related to education and training in this period, but only its representatives and observers in the field.

The second mainstream of teacher leadership encompasses the period between 1980 and 1990. During this period, teacher leaders began to engage in decision-making processes and contributed to efforts to enhance the quality of education and training in collaboration with their colleagues. This second stream was based on school-based development, and the roles of teachers as programme developers, team leaders and trainers of trainers came to the fore (Silva et al., 2000). As evidenced by the second stream, teacher leaders have become increasingly autonomous, particularly in the areas of curriculum development, enhancing classroom instruction, and collaborating with colleagues.

The third stream of teacher leadership is still in operation today. This stream posits that the teacher leader is a natural leader situated at the nexus of educational and training activities, rather than being an ancillary figure within a managerial structure. Primarily, the teacher leader is a classroom leader who facilitates the advancement of students within the classroom setting and consistently enhances their performance. Moreover, although not formally designated as such, the teacher leader is a leader who, in collaboration with their colleagues, provides students with the 21st-century skills necessary to build a future society and overcome structural limitations such as time, space, and resources (Harris & Muijs, 2003). The principal objectives of these leaders are to enhance the efficacy of the educational institution and to equip their students with the skills necessary to thrive in the future.

In regard to the role of the teacher as a leader in the educational setting, Harris (2005) identified four key elements. (i) Master of curriculum development, (ii) Project writer and school improvement team leader, (iii) Mentor and mentor of less experienced teachers, (iv) Researcher with strong ties to the classroom. In these roles, the teacher is primarily a transformative leader. The teacher's mastery of curriculum development enables them to foster student development through innovative classroom approaches, thereby ensuring the effectiveness of the school by reflecting these approaches in the curriculum.

The characteristics of teacher leaders are also elucidated in the existing literature. Upon examination of these characteristics, it becomes evident that teacher leaders are receptive to professional and individual learning and development, inquiry, investigation, accountability, initiative, and action, exhibiting a high degree of self-assurance (Çoban, 2021; Demir, 2020). Danielson (2006) conducted an analysis of the characteristics of teacher leaders, classifying them under two headings: (i) skills and (ii) dispositions. In terms of skills, the following are identified: participation in the decision-making process, action based on data and evidence, collaboration in alignment with educational objectives, and support for the professional development of colleagues, students, and the school. The disposition-related characteristics include open-mindedness, humility, creativity and inquisitiveness, perseverance, and initiative.

In examining the impediments to teacher leadership, the initial challenge is the assumption that the teacher's role, accountability, and scope of influence are confined to the classroom. Consequently, teachers are only able to engage in school management to a limited extent and are therefore unable to demonstrate the behaviours associated with teacher leadership. The second perception is that school development studies and field practitioners generally perceive the school administrator as the leader and teachers as followers of the leader. Consequently, the management processes and decision-making mechanisms within the school are under the control of the school principal, who holds the sole authority in the management of the institution. To avert this, it is imperative to establish legal frameworks that confer authority upon teachers to assume leadership roles and participate in the governance of educational institutions (York-Barr & Duke, 2004). Another obstacle highlighted in the literature is that teacher leadership can result in increased workload for teachers. As previously noted in the literature on teacher leadership, teachers who assume leadership roles must dedicate time, energy, and effort to the professional development of their colleagues and the advancement of their schools, in addition to fulfilling their responsibilities in the classroom.

Such circumstances may result in adverse outcomes, including stress, time constraints and role conflict. In such circumstances, the teacher may typically opt to assume the teaching role in preference to the leadership role. The culture and climate of the educational establishment are also significant factors in the context of teacher leadership. In an educational establishment with a closed climate, the teacher leader will be unable to demonstrate the behaviours associated with teacher leadership and will encounter obstacles due to a lack of support. Conversely, teachers are more likely to display leadership behaviours in educational establishments with an open climate and a democratic culture (Demir, 2020). A further obstacle that can be identified is colleague jealousy. A teacher leader who endeavours to facilitate the professional development of their colleagues may encounter resentment from those colleagues. Approaches such as "Who is he/she to teach me something" or "I know better than him/her" may impede the teacher leader's ability to exhibit leadership behaviours (Çoban, 2021).

ARTIFICIAL INTELLIGENCE

Technological developments have resulted in the replacement of human labour with technological tools. Artificial Intelligence (AI) is one such technological innovation developed to assist humans in various fields. AI can be defined as a system that creates intelligent machines and computer programmes capable of performing tasks requiring human intelligence and imitating human behaviour (Ergen, 2019; McCarthy, 2007). These tasks are performed using external data, such as big data. The prevalence of AI as a primary reference point in sectors such as manufacturing, healthcare, and supply chains is indicative of its transformative impact on various domains. The capacity of AI to execute tasks that are beyond the capabilities of humans has led to numerous applications, resulting in enhanced performance and productivity (Fathima Anjila, 2021).

The conceptual foundations of artificial intelligence (AI) trace back to before 1956. In 1936, mathematician Alan Turing introduced a mathematical model, which laid the theoretical groundwork for the ideal computer. Subsequently, in 1943, neurophysiologists W. McCulloch and W. Pitts developed the first neural network model, known as the M-P model (Zhang & Zhang, 1999). In 1949, Donald Hebb introduced the 'Hebb learning rule,' an unsupervised learning principle capable of extracting statistical features from training sets and classifying data based on similarities (Kuriscak et al., 2015). This rule represents one of the earliest concepts of machine learning (ML), bearing a close resemblance to human cognitive processes. Turing's work, "Computing Machinery and Intelligence", posed the famous question: "Can machines think?". This paper became one of the foundational texts in AI research, introducing the notion of the "Turing Test" as a measure of a machine's ability to exhibit intelligent behaviour indistinguishable from that of a human (Turing, 1950). In 1952, IBM scientist Arthur Samuel created a checkers programme that could learn implicit patterns from the current game state and use them to guide subsequent moves (Samuel, 2000).

In 1956, the term "artificial intelligence" was coined by John McCarthy during the Dartmouth Summer Research Project on Artificial Intelligence. Since then, AI research has achieved numerous significant milestones, including advancements in machine learning, theorem proving, pattern recognition, problem solving, expert systems, and natural language processing (McCarthy, 1995). In 1957, American psychologist Frank Rosenblatt introduced the 'Perceptron' model, which became a cornerstone of artificial intelligence. This model is based on the creation of a system that employs 'neurons' for pattern recognition, significantly influencing key areas such as machine learning capabilities and the subsequent development of neural networks and connectionist models. Rosenblatt's pioneering work on the perceptron continues to be a central topic in introductory artificial intelligence courses to this day (Samuel, 2000).

The development of AI has experienced periods of stagnation known as "AI winter" (Crevier, 1993). One reason for these setbacks was the limitations of computational power and the failure of early AI systems to live up to the lofty promises made by researchers. Early approaches such as symbolic AI, which focused on programming explicit rules to mimic human thought, faced difficulties in dealing with the complexity and variability of real-world data (Minsky, 1988). For instance, during the Cold War, the US Defence Establishment initiated plans for an automatic Russian-English translation programme, yet progress on the programme was sluggish between 1967 and 1976. This failure in machine translation had a detrimental effect on the development of the entire field of artificial intelligence. This decline in progress was further compounded by several reports, including the 1966 'Language and Machines' report by the Automatic Language Processing Advisory Committee (ALPAC), which exposed the limitations of machine translation (Jiang, et al., 2022). The 1969 report by Minsky and Paper, asserting that the Perceptrons were constrained in their ability to address non-linear problems and were ineffective for more complex issues, further dampened optimism (Mehrling, 1999). Furthermore, the 1973 report by the British Scientific Research Council (SRC), which highlighted that artificial intelligence was lagging behind expectations, also contributed to a slowdown in AI development (Arnold et al., 1989). Another factor that impeded positive progress in the field of artificial intelligence during the 1980s was the inability of experts to recognise the limitations of AI when confronted with highly complex tasks and systems, such as medical diagnostics and vision systems (Jiang, et al., 2022; McCarthy, 1995). During this period, there was a tendency among experts to overlook the potential of artificial intelligence, favouring instead the use of general-purpose computers for problem-solving. This shift in focus further delayed the advancement of AI, as researchers opted for more conventional approaches to tackling intricate issues.

AI saw a major resurgence in the 1990s, driven by improvements in machine learning techniques, better algorithms, and more powerful computing hardware. The development of statistical approaches, particularly in the field of machine learning, allowed AI systems to learn from data rather than relying solely on pre-programmed rules (Jordan & Mitchell, 2015). In the 21st century, AI has truly entered the mainstream, with advancements in areas like natural language processing, computer vision, and autonomous systems. Landmark achievements such as IBM's Deep Blue defeating world chess champion Garry Kasparov in 1997 (Campbell et al., 2002) and the rise of deep learning in the 2010s have demonstrated AI's immense potential. AI technologies are now integrated into daily life, from virtual assistants like Siri and Alexa to autonomous vehicles and medical diagnostic systems.

AI has a broad range of applications in education, with the majority of these applications aiming to enhance teaching, learning and administrative processes (Chen et al., 2022). One of the most significant areas where AI is being utilised is in personalised learning, which refers to the use of technology to tailor educational experiences to meet the unique needs, preferences and learning styles of individual students (Hashim et al., 2022). AI algorithms analyse data from students' interactions with digital platforms, identifying patterns in their learning behaviours and adapting the content to suit their needs (Baker & Inventado, 2016). AI-powered tutoring systems, for instance Carnegie Learning and Duolingo, utilise machine learning algorithms to assess students' comprehension and provide personalised exercises to facilitate progress at a pace that suits the individual student. These systems are designed to offer immediate feedback, enabling students to receive guidance as they work through problems, which is a crucial component of effective learning (Unal & Unal, 2024). Furthermore, AI has the capacity to provide real-time assessments, thereby enabling educators to continuously monitor students' progress and intervene when necessary (Unal & Unal, 2024).

Another significant application of AI in education pertains to the domain of automating administrative tasks. Educational institutions frequently encounter challenges in the management of voluminous data, including student records, attendance records, and academic grades. The integration of AI has the potential to markedly streamline these processes by automating data management, scheduling, and grading (Durnalı & Çoban, 2025). For instance, AI systems are being employed to grade assignments and examinations, particularly in subjects such as mathematics and the sciences, where the answers are objective in nature. This, in turn, reduces the workload of educators and allows them to focus on more complex tasks, such as facilitating discussions or providing one-on-one support.

The utilisation of artificial intelligence (AI) in the realm of intelligent content development, encompassing digital textbooks and curriculum guides, is a notable development. These resources are frequently driven by AI algorithms that adapt to the individual learner's needs, thereby ensuring a dynamic and engaging learning experience. What is more, AI tools such as chatbots and virtual assistants are being employed to enhance communication between students and educators. Specifically, chatbots are capable of providing students with answers to frequently asked questions, while virtual assistants can offer guidance on assignments, deadlines, and course materials (Engreini et al., 2025).

The potential of AI to enhance accessibility in education is significant, particularly for students with disabilities. The utilisation of AI-powered tools, such as speech recognition software, text-to-speech applications, and adaptive learning platforms, can provide substantial support. For instance, students with visual impairments can employ screen readers to access text-based content, while students with learning disabilities can benefit from tools that simplify complex texts or provide step-by-step guidance. Furthermore, AI has the potential to address the challenges faced by students in remote or underserved areas by offering affordable, high-quality learning experiences through online platforms and mobile applications (Zdravkova, et al., 2022).

Another significant benefit of AI in education is its capacity to provide real-time feedback. Conventional education frequently relies on periodic assessments to evaluate student performance, which can be delayed and does not offer immediate insights (Pham & Sampson, 2022). Conversely, AI-driven systems can offer instantaneous feedback, enabling students to identify areas requiring improvement and implement corrective actions promptly (Ahmad et al., 2023). This continuous feedback loop is considered essential for cultivating a growth mindset and promoting lifelong learning (Abulibdeh et al., 2024).

Notwithstanding the considerable potential of AI in the field of education, there are significant challenges and ethical concerns that must be addressed. A primary issue is that of data privacy and the security of students. AI systems require substantial amounts of data to function effectively, including personal information about students' academic performance, behaviour, and learning preferences. The collection and storage of such data give rise to concerns regarding its protection and the individuals who have access to it. It is, therefore, incumbent upon educational institutions to ensure that they comply with data protection laws and implement robust security measures to safeguard students' privacy (Nguyen et al., 2023).

The role of educators is also being called into question as AI becomes more prevalent in the classroom. While AI can support teachers by automating administrative tasks and providing real-time feedback, it cannot replace the human element of teaching (Roll & Wylie, 2016). Educators play a crucial role in fostering social and emotional development, building relationships with students, and providing mentorship. It is vital that AI complements teachers, rather than replacing them, thereby ensuring that human interaction remains at the core of the educational experience (Gocen & Aydemir, 2020).

The potential of artificial intelligence (AI) in education is a subject of considerable interest, with great promise being attributed to its future development. As AI technologies continue to evolve, there is a growing expectation that they will be utilised to create increasingly sophisticated and nuanced applications within the classroom setting (Huang et al., 2021). A particularly exciting prospect is the use of AI to create immersive learning environments through technologies such as virtual reality (VR) and augmented reality (AR). The integration of AI into these technologies has the potential to facilitate highly interactive and engaging learning experiences, allowing students to explore complex concepts in ways that were previously impossible (Gong, 2021).

Furthermore, there is a growing body of evidence to suggest that AI is set to assume an increasingly prominent role in the domain of teacher training and professional development. The integration of AI-powered systems has the potential to provide educators with personalised feedback on their pedagogical practices, thereby facilitating the refinement of their skills and the enhancement of their effectiveness in the classroom environment (Kandlhofer et al., 2016). Additionally, as AI continues to improve, it may also assist in the development of new pedagogical approaches, shifting the focus of education from traditional, teacher-centred methods to more collaborative, student-centred models.

Artificial Intelligence Literacy

As AI continues to permeate various sectors of society, the need for AI literacy (AIL) has become an increasingly vital subject of discussion. The term AIL is often employed to refer to the knowledge and skills necessary for individuals to comprehend the fundamental concepts of AI, its applications, and its societal implications (Wang et al., 2023). AI literacy is generally understood to denote the ability to comprehend the functionality of AI systems, to engage critically with such systems, and to understand their implications for society (Sperling et al., 2024). The concept extends beyond the confines of technical proficiency, which is defined as the ability to use AI tools, to include critical thinking about the ethical, social, and economic implications of AI (Ayanwale et al., 2024). As AI technologies become more pervasive in everyday life, there is an increasing need for a population that can engage meaningfully with these technologies, whether through understanding their benefits, risks, or ethical dimensions. Wang and friends (2023) expressed four sub-dimensions: awareness, usage, evaluation and ethics.

A fundamental aspect of AIL encompasses the possession of a rudimentary comprehension of the operational mechanisms of AI systems and this is called awareness. This encompasses a grasp of pivotal AI concepts such as machine learning, neural networks, natural language processing, and data-driven decision-making. It is imperative to discern the underlying principles that underpin AI technologies, including the distinction between supervised and unsupervised learning, the function of algorithms, and the significance of training data. While a comprehensive technical expertise is not a prerequisite for AIL, it is essential for an individual to possess the capacity to comprehend the foundational principles that underpin the design, implementation, and application of AI systems in diverse contexts. This essential foundational knowledge is indispensable for effective interaction with AI systems, whether in a professional or a personal context (Ma & Chen, 2024).

Another critical dimension of AIL is the ability to apply AI knowledge in everyday contexts and participate effectively in the digital world and this is called usage. This includes understanding how to interact with AI-based products and services, such as virtual assistants, recommendation systems and smart devices. It also entails the ability to protect personal information in an AI-driven environment and the awareness of the potential risks associated with AI, such as data privacy concerns. It is vital for individuals to be equipped with the capacity to discern between the adoption and avoidance of specific AI technologies, guided by their personal predilections and ethical considerations. Additionally, the concept of digital citizenship in the context of AIL encompasses the comprehension of the extensive ramifications of AI on society and the engagement in informed discourse concerning its role in shaping public policy, economic structures, and cultural mores (Kırksekiz, et al., 2024). In light of the rapid advancements in the field of AI, a pivotal aspect of AIL pertains to the capacity for lifelong learning and adaptability. As AI technologies undergo continuous evolution, individuals are compelled to maintain a commitment to ongoing updates of their knowledge and skill sets, in order to remain abreast of novel developments. Consequently, AIL should not be regarded as a static repository of knowledge, but rather as an ongoing process that encourages individuals to remain informed about emergent trends and innovations in AI. This adaptability is of critical importance in a world where AI's capabilities and applications are subject to constant change, requiring individuals to be flexible and open to learning about new AI tools, systems (Nong et al., 2024).

The assessment dimension of AI literacy pertains to the ability to critically evaluate both the technical functionality and the ethical ramifications of AI systems. This process involves a thorough analysis of the design, implementation, and broader societal consequences of AI technologies (Kırksekiz, et al., 2024). An effective evaluation of AIL should include the ability to identify biases within algorithms, assess the fairness of AI decision-making processes, and examine the transparency of AI models (Holmes et al., 2019). Moreover, it is crucial to evaluate the societal impacts of AI, particularly in relation to privacy, autonomy, and employment. Evaluating AI literacy also necessitates an understanding of the potential risks and unintended consequences that may arise from AI deployment, including issues such as discrimination, surveillance, and the erosion of human agency. Thus, AI literacy assessment extends beyond mere technical proficiency, incorporating the individual's capacity to critically and ethically engage with AI technologies, fostering responsible usage and informed decision-making.

A fundamental aspect of AI literacy involves an understanding of the ethical implications of AI technologies, as AI systems become more capable and raise significant ethical questions regarding privacy, autonomy, and fairness. For example, AI-powered surveillance technologies have sparked debates about individual privacy and the potential for invasive monitoring, while AI-based decision-making in areas such as criminal justice has raised concerns about fairness and potential bias (Wang et al., 2023). AI literacy is defined by an awareness of these ethical challenges and the capacity for thoughtful reflection on the broader social consequences of AI deployment. It is essential to understand the ethical responsibilities of AI developers and users, and to recognise the significance of establishing ethical guidelines and regulatory frameworks for AI technologies.

Teacher Leadership and Artificial Intelligence

The advent of AI in education has prompted the question of whether AI will supplant the role of teachers. Indeed, if educators merely enter and exit the classroom and relay the information presented in textbooks, the answer to this question would be affirmative: AI will replace teachers. Nevertheless, given that contemporary educators adopt innovative pedagogical strategies in the classroom with the aim of fostering the holistic development of their students, encompassing academic, emotional, physical, and social growth, the answer to this question is likely to be negative, indicating that the role of the teacher cannot be entirely replaced by AI. As has been observed in the existing literature on the subject, effective teaching involves more than the transfer of knowledge; it requires the development of critical thinking skills, emotional intelligence, social interaction and the ability to respond to the diverse needs of students (Çoban & Gün, 2024). These aspects of education, more than AI, are required by teachers because teachers possess distinctive capabilities to engage, motivate and inspire students (Guan et al., 2020). Furthermore, the formation of strong interpersonal relationships between teachers and students is of paramount importance for the efficacy of the learning process (Shen et al., 2020). These interpersonal relationships play a vital role in fostering a positive and supportive learning environment, which is almost impossible for AI systems to do. Teachers have the capacity to provide the empathy, understanding and guidance necessary to ensure students' development and psychological well-being (Bertolin & Da Rin, 2020).

Rather than perceiving AI as a potential threat, leading educators should leverage its capabilities to enhance the efficacy of educational and training processes within their classrooms and facilitate their professional growth. Indeed, AI has already begun to be employed in numerous domains within the field of education, including learning, teaching, assessment, and educational management. AI provides students with personalised and adaptive learning experiences, as well as learning opportunities that are tailored to their individual needs (Durnalı & Çoban, 2025; Ramnarain et al., 2024). Furthermore, the system assists educators in monitoring and reporting on students' learning processes, making predictions about student development, and providing support for assessment and evaluation (Chiu et al., 2023; Guan et al., 2020; Hwang et al., 2020). Nevertheless, it is anticipated that AI-supported educational systems will also play a pivotal role in identifying at-risk students and intervening in a timely manner (Tsai et al., 2020).

In their qualitative study, Ghamrawi et al. (2024) posited that AI has the potential to enhance teacher leadership. AI can facilitate this assistance by enabling educators to automate administrative tasks, such as preparing reports, interpreting data to make data-driven decisions, and providing instructional leadership in the classroom. This includes monitoring and guiding students' learning and suggesting innovative learning methods and strategies in the classroom. Furthermore, the incorporation of AI in educational institutions offers a valuable chance for educators to expand their leadership responsibilities beyond the confines of their classrooms. The system offers educators the chance to assume the roles of mentors, coaches, and resource providers, while leveraging AI as a valuable tool to assist and direct their colleagues. Furthermore, this can also serve to alleviate the time and workload pressures typically experienced by lead teachers. Nevertheless, the use of AI can also serve to mitigate the negative effects of professional jealousy and envy, particularly in the context of professional development and the provision of mentorship to colleagues. Teacher leaders may utilise AI to ascertain the most efficacious resources for the particular requirements of their colleagues and students (Çoban & Gün, 2024).

A further significant finding of this research is the identification of the competencies that should be possessed by those in leadership roles within the education sector in the context of the advent of AI. These competencies are classified into five categories: technological literacy, adaptability and continuous learning, collaborative and coaching skills, data-driven decision-making and human-centred approaches. Technological literacy in teacher leadership represents a crucial competence, underscoring the necessity for educators to comprehend and adeptly utilise AI-enabled tools, data analytics, and educational technologies. The capacity for adaptability and continuous learning is a key competence for educators, encompassing the ability to embrace new technologies, pedagogical approaches, and innovative classroom practices. It also entails a willingness to engage in ongoing professional development and to effectively integrate AI tools and resources into teaching methodologies. The ability to collaborate and provide coaching is crucial for leveraging AI in a manner that fosters a collaborative culture among colleagues, facilitates peer learning, and enables one to serve as an instructional coach. The competence of teachers to analyse and interpret data generated by AI systems is an essential component of their ability to make data-driven decisions. This encompasses the capacity to determine teaching strategies, personalise learning experiences for their students and make evidence-based decisions in and out of the classroom. In conclusion, the theme of human-centred approaches underscores the necessity for educators to prioritise positive relationships and social-emotional development in order to maintain and develop their leadership functions in the age of AI.

In light of the latest research on teacher leadership, there has been a notable shift in focus towards defining the optimal approach to teacher leadership practices within educational institutions. In this process of developing leadership skills, teachers should engage primarily with activities that facilitate the growth of their personal, social, and professional competencies. For example, they should follow organisations that are engaged in the professional development of teachers, such as the Teacher Informatics Network (TIN) and the Teacher Academy Foundation (TAV). Furthermore, Massive Open Online Courses (MOOCs) provide numerous opportunities for professional development training for teacher leaders. Furthermore, teacher leaders must possess a comprehensive understanding of their students' interests and expectations. They should also be afforded the opportunity to examine their students' home environments through the utilisation of home visits. It is incumbent upon them to collaborate with their colleagues in the development of activities, course materials and the utilisation of digital tools such as Web 2.0, with a view to conducting lessons in a more efficacious manner. It is incumbent upon them to assume responsibility within the decision-making and school development teams of the administrative body. It is also recommended that teachers participate in the organisation of social, cultural and sporting activities for students. It is incumbent upon them to enhance their expertise in areas such as project writing, robotics, and coding, thereby enriching the educational experience of their students and the institution at large (Çoban & Gün, 2024).

The potential impact of AI on education in Türkiye has yet to be fully acknowledged, particularly in terms of its potential benefits and challenges for students and educators. Nevertheless, it seems probable that AI will become more prevalent in our society. Indeed, it can be stated that a considerable number of students and teachers engage with the network, albeit at a basic level. It is therefore possible that any technological errors or accidents related to AI may prove to be extremely difficult to rectify. Students who lack proficiency in questioning and critical thinking are more susceptible to errors when utilising AI-supported applications. Those who are unable to question the rationale behind AI-supported decision-making and the accuracy and ethical standards of the information provided are at greater risk. In light of these considerations, it is imperative that policy makers and educational institutions adopt a more proactive stance with regard to AI awareness and AI education.

This study was conducted in accordance with the parameters set forth by Wang and friends' (2023) AI literacy framework. Wang et al.'s AIL scale comprises four sub-dimensions: awareness, usage, evaluation and ethics. With regard to the awareness sub-dimension, the teacher is conversant with the various AI tools and is aware of the extent to which AI is utilised in a given application. In the usage sub-dimension, the teacher is able to utilise AI in an efficacious manner in both their daily lives and in the processes of education and training. In the evaluation sub-dimension, the teacher is able to discern the limitations of the AI product in problem solving and those of their own abilities. Nevertheless, the user is able to select the most appropriate solutions offered by AI and discard those that are unsuitable. With regard to the final sub-dimension, ethics, the teacher evinces a high degree of sensitivity with respect to the security and confidentiality of information and compliance with ethical principles in the context of AI (Erdoğan & Ekşioğlu, 2024).

The objective of this study was to ascertain the level of AIL among teachers. To this end, the following research questions (RQ) were primarily investigated:

RQ 1. What is the level of AIL among teachers?
RQ 2. Are there significant differences between teacher related variables (gender, seniority, current working time at the school, the level of technology usage) and the level of AIL?

METHOD

Research Design

This quantitative research employs a survey model, which is recommended for studies that aim to describe a past or present situation as it exists. As outlined by Büyüköztürk et al. (2024), the general survey model involves conducting a survey on the entire universe or a sample drawn from it to make a general assessment of the universe, which comprises a vast number of elements.

Sampling

The population of the study consists of 1 million 168 thousand 896 preschool, primary, secondary and high school teachers working in public schools in Türkiye. In the study, it was calculated that 386 teachers should be reached with .95 reliability and .05 margin of error calculations. Considering the time, cost and data security in the study, convenience sampling from non-random sampling methods was used and 402 teachers were reached in this context. Personal information about the teachers participating in the study is given in Table 1.

Table 1. Teachers' personal information

Gender	**n**	**%**
Women	284	69.3
Men	126	30.7
Seniority	n	%
1-5	48	11.7
6-10	81	19.8
11-15	85	20.7
16-20	85	20.7
21+	111	27.1
Current working time at the school	n	%
1-5	165	40.5
6-10	106	25.9
11-15	55	13.4
16	83	20.2
Level of technology usage	n	%
low	36	2.9
medium	254	67.8
High	120	29.3
Total	410	100.0

Data Collection Tools

The data collection tool in this study consists of two parts. The first part includes personal information, and the second part includes the Artificial Intelligence Literacy Scale.

Personal Information Form

This section includes questions about the gender, age, professional seniority, working time in the institution and technology usage levels of the teachers participating in the study.

Artificial Intelligence Literacy Scale (AILS)

Artificial Intelligence Scale (AILS) was developed by Wang et al. in 2023. The scale was adapted into Turkish by Erdoğan and Ekşioğlu (2024). The scale consists of awareness (n=3), usage (n=3), evaluation (n=3) and ethics (n=3) dimensions. Sample items of the scale from each sub-dimension respectively are as follows: "I can identify the artificial intelligence technology used in the applications and products I use"; "I can use artificial intelligence applications or products skilfully to help me in my daily work", "I can evaluate the capabilities and limitations of an artificial intelligence application or product after using it"; "I always follow ethical principles when using artificial intelligence applications or products". Erdoğan and Ekşioğlu (2024) found the Cronbach's alpha (α) value of the scale as .85 for the awareness dimension, .95 for the usage dimension, .91 for the evaluation dimension, .95 for the ethical dimension and .98 for the whole scale. The CFA results are as follows: $\chi2/df)$= 3.10 RMSEA = .079, CFI= .95; TLI= .94; SRMR= .033.

According to the results of the reliability test carried out in this study, Cronbach's alpha (α) value was calculated as .89 in the scale in general and as awareness (α= .73); usage (α= .85); evaluation (α= .90) and ethics (α= .85) in the dimensions. When the results are analysed, it is seen that the scale dimensions are reliable since they are greater than .70 (Tabachnick & Fidell, 2003). In the validity study conducted using the research data, Confirmatory Factor Analysis (CFA) was conducted to see whether the AILS was validated or not. Accordingly, χ^2=133.841, df=48, χ^2/df= 2.788, RMSEA = .066, RMR=.043 GFI= .95, AGFI= .92, NFI= .95. According to these values, the measurement model has excellent fit values (Hu & Bentler, 1999). Hu and Bentler (1999) state that a χ^2/df value less than 2 indicates perfect fit, and a value between 2 and 5 indicates good fit. For RMSEA and SRMR values, it is underlined that being equal to or lower than .05 is an indicator of excellent fit and being between .05 and .08 is an indicator of good fit. The researchers characterise CFI, TLI, NFI, AGFI values greater than .95 as perfect fit and between .95 and .90 as good fit.

Data Collection Process and Analysis

In the process of collecting the data, firstly, ethics committee permission dated 21.11.2024 and numbered 307 decision numbered from Karmanoğlu Mehmetbey University for research permission. In addition, the necessary permissions were obtained from the researchers who adapted the scale to be used in data collection to Turkish in order to use the scales. Then, google forms were delivered to pre-school, primary, secondary and high school teachers working in Türkiye in the 2024-2025 academic year and data were obtained on the basis of volunteerism.

For the normality distribution of the obtained data, skewness and kurtosis values were examined with the help of SPSS 25.0 programme. As a result of the examination, the skewness and kurtosis values of the AILS were found as -.325 and 1.004, respectively. When the data were analysed, it was seen that the values were between -1.5 and +1.5 and showed normal distribution (Tabachnick & Fidell, 2003). In addition, preliminary analysis results such as extreme value, homogeneity, linearity and multicollinearity problems of the data were examined, and it was determined that the data structure was suitable for the analyses. To determine the AI literacy level of the teachers, arithmetic mean and standard deviation was calculated. The intervals on Likert-type scales are typically calculated by dividing the difference between the minimum and maximum values of the scale by the number of categories (Levin et al., 2010; Pimentel, 2019). In this approach, the cut-off points for the 5-point Likert scale used to assess AIL are as follows: 1.00–1.79 indicates 'strongly disagree',1.80–2.59 corresponds to 'disagree',2.60–3.39 represents 'somewhat agree',3.40–4.19 signifies 'agree', and 4.20–5.00 reflects 'strongly agree'. Then, independent sample t-test (according to gender) and one-way analysis of variance (according to seniority, working time in the current school and technology usage level) were performed to see whether there was a significant difference according to the teacher-related variables.

RESULTS

In this section, we analysed the research questions: "What is the level of AIL among teachers?" and "Are there significant differences between teacher related variables (gender, seniority, current working time at the school, the level of technology usage) and the level of AIL?" Table 2 indicates the level of artificial intelligence literacy among teachers.

Table 2. Level of artificial intelligence literacy among teachers.

Variables	Mean (M)	Std. Deviation
Awareness	3.75	.77
Usage	3.44	.86
Evaluation	3.60	.83
Ethic	4.04	.65
AI literacy (total)	3.70	.63

According to the findings. the mean scores of teachers' AIL indicated the following levels: awareness (M=3.75; 'agree'). usage (M=3.44; 'agree). evaluation (M= 3.60 'agree'. ethic (M= 4.04) and AI literacy total (M=4.04; 'agree'). The findings indicate that teachers primarily focus on the ethical implications of artificial intelligence when developing their AI literacy. The findings also demonstrate that teachers exhibit a relatively low level of attention to the usage of artificial intelligence in literacy. Table 3 gives the independent sample t-test results according to teachers' genders.

Table 3. Independent sample t-test results according to teachers' genders.

Sub-dimensions	Gender	n	$\bar{x}$	Ss	t	*p*
Awareness	Female	284	3.68	.75	-2.51	.01*
	Male	126	3.89	.81		
Usage	Female	284	3.41	.84	-1.84	.27
	Male	126	3.51	.89		
Evaluation	Female	284	3.58	.77	-.55	.59
	Male	126	3.63	.94		
Ethic	Female	284	4.06	.60	1.26	.21
	Male	126	3.98	.75		

*p<.05

As illustrated in Table 3, there is a notable discrepancy in the level of awareness exhibited by male and female teachers with regard to their in artificial intelligence literacy ($t_{(408)}$= 2.51, $p<.05$). We can posit that male teachers (X= 3.89) demonstrate a greater degree of awareness with regard to their proficiency in the field of artificial intelligence than their female counterparts (X= 3.68). However, there is no statistically significant difference between teachers' gender and the other subcategories of AIL, usage

($t_{(408)}$= 1.84, p>.05); evaluation ($t_{(408)}$= .55, p>.05) and ethic ($t_{(408)}$= 1.26, p>.05). In Table 4, we can see ANOVA results with application sub-dimensions of AILS according to teachers' seniority.

Table 4. ANOVA results with application sub-dimensions of AILS according to teachers' seniority.

Sub-dimensions	Seniority	n	$\bar{x}$	Ss	F	p
Awareness	Up to 5 years	48	3.93	.70	.98	.42
	6-10 years	81	3.79	.75		
	11-15 years	85	3.67	.83		
	16-20 years	85	3.72	.77		
	21+ years	111	3.71	.77		
Usage	Up to 5 years	48	3.64	.81	1.06	.37
	6-10 years	81	3.48	.95		
	11-15 years	85	3.34	.87		
	16-20 years	85	3.40	.84		
	21+ years	111	3.42	.80		
Evaluation	Up to 5 years	48	3.88	.73	2.22	.06
	6-10 years	81	3.66	.91		
	11-15 years	85	3.59	.81		
	16-20 years	85	3.56	.77		
	21+ years	111	3.47	.85		
Ethic	Up to 5 years	48	4.13	.64	2.88	.02*
	6-10 years	81	4.22	.52		
	11-15 years	85	3.92	.75		
	16-20 years	85	4.00	.67		
	21+ years	111	3.98	.61		

*p<.05

According to Table 4, there is no statistically significant difference between the seniority groups and the subcategories of AILS; awareness ($F_{(4,405)}$= .98, p>.05), usage ($F_{(4,405)}$= 1.06, p>.05), and evaluation ($F_{(4,405)}$= 2.22, p>.05) except for ethic sub-dimension ($F_{(4,405)}$= 2.88, p<.05). We can express that teachers with six to ten years of seniority (X= 4.22) demonstrate a greater proclivity to address ethical concerns pertaining to artificial intelligence literacy than their counterparts with eleven to fifteen years of experience (X= 3.92). As is seen in Table 5, ANOVA results with application sub-dimensions of AILS according to teachers' level of technology usage is given.

Table 5. ANOVA results with application sub-dimensions of AILS according to teachers' level of technology usage.

Sub-dimensions	Level of technology usage	n	$\bar{x}$	Ss	F	*p*
Awareness	Low	36	3.02	.75	79.21	.00*
	Medium	254	3.57	.66		
	High	120	4.43	.61		
Usage	Low	36	2.84	.85	66.40	.00*
	Medium	254	3.22	.75		
	High	120	4.08	.69		
Evaluation	Low	36	3.05	.86	41.72	.00*
	Medium	254	3.44	.75		
	High	120	4.11	.83		
Ethic	Low	36	3.83	.92	7.73	.00*
	Medium	254	3.98	.60		
	High	120	4.22	.65		

*p<.05

According to Table 5, there is a statistically significant difference between teachers' level of technology usage and the subcategories of AILS; awareness ($F_{(2,407)}$ = 79.21, p<.05), usage ($F_{(2,407)}$= 66.40, p<.05), evaluation ($F_{(2,407)}$= 41.72, p<.05) and ethic ($F_{(2,407)}$= 7.73, p<.05). In light of the aforementioned findings, we can posit that the level of AI literacy among teachers who demonstrate a high degree of technological proficiency is superior to that observed among their counterparts who exhibit a medium or low level of technology utilisation.

DISCUSSION

Our first finding provides the mean scores of teachers' AIL and ethical issues are vital for teachers though awareness, usage, evaluation of AI is relatively lower than ethical issues. In the meta-analysis study on the use of artificial intelligence literacy in teaching, Sperling et al. (2024) asserted that the theoretical, practical, and ethical aspects of teachers' actions in their classrooms and teaching processes related to artificial intelligence literacy remain inadequately elucidated, thereby creating a significant gap in this field of study. Additionally, in the study, Ayanwale et al. (2024) highlight the significance of integrating technology in a manner that aligns with ethical principles, particularly in the context of artificial intelligence literacy and usage. In the studies conducted by Wang et al. (2023) and Lérias et al. (2024), the participants demonstrated a relatively high level of awareness regarding artificial intelligence, with ethical principles ranking as the second most important factor. This may be attributed to the fact that Wang et al. (2023) conducted this study with individuals from a diverse range of age groups and professions. Additionally, Lérias and friends (2024) conducted their study in a university. As the study was conducted exclusively with teachers, who are professionally inclined to prioritize ethical values, ethical principles may have been regarded as relatively more significant. Teachers are not only users of artificial intelligence themselves, but also endeavour to ensure its appropriate use by students in the target group. This indicates that teachers place a greater emphasis on ethical principles. In a way that confirms this, Banaz and Demirel (2024) assessed the level of artificial intelligence literacy among teachers. Their item-based analysis revealed that ethical principles were the most highly rated.

Our second key finding reveals male teachers demonstrate a greater degree of awareness with regard to their proficiency in the field of artificial intelligence than their female counterparts. In a study conducted by Zhang et al. (2023), female pre-service teachers exhibited a greater degree of anxiety regarding the utilisation of AI compared to their male counterparts. Additionally, the male participants demonstrated a heightened awareness of AI applications. To them, male pre-service teachers demonstrated superior awareness, utilisation and adaptation of artificial intelligence applications in educational processes when compared to their female counterparts. However, the results of this study are inconsistent with those of Mart and Kaya (2024), who did not identify a significant difference in artificial intelligence literacy according to gender. Conversely, Banaz and Demirel (2024) found that female teachers demonstrated a higher level of awareness regarding artificial intelligence than their male counterparts. The discrepancy between the findings of the two studies may be attributed to the fact that Banaz and Demirel (2024) examined the overall AI literacy score, rather than investigating whether there were notable discrepancies between the sub-dimensions of the AI literacy scale and gender. The present study revealed that male teachers exhibited a higher level of awareness regarding the utilisation of artificial intelligence in comparison to their female counterparts. This discrepancy may be attributed to the fact that male teachers tend to exhibit higher levels of technology literacy than their female counterparts. It is notable that Avcı and Candan's study (2023) also indicated a discrepancy in the awareness of Industry 4.0 between male and female teachers. Specifically, male teachers demonstrated a higher level of awareness than their female counterparts.

The third finding of the study reveals that teachers with six to ten years of seniority demonstrate a greater proclivity to address ethical concerns pertaining to artificial intelligence literacy than their counterparts with eleven to fifteen years of experience. Karataş et al. (2024) posit that less experienced teachers tend to eschew artificial intelligence applications in favour of more traditional methods in order to deliver lessons with greater substance. This is often attributed to factors such as curriculum pressure and a sense of unease about meeting subject completion targets. The researchers highlight that as teachers gain experience, they become more adept at utilising artificial intelligence applications. However, teachers with very high seniority have difficulty in integrating artificial intelligence applications into their lessons and education processes because they are inadequate in the use of technology. Hsu et al. (2023) state that teachers with high seniority are worried about using artificial intelligence because they are not in touch with technology and are insufficient in technology, and they avoid using these applications in their classrooms.

Our last finding points out that the level of AI literacy among teachers who demonstrate a high degree of technological proficiency is superior to that observed among their counterparts who exhibit a medium or low level of technology utilisation. Chiu et al. (2024) posited that there is a robust correlation between proficiency in artificial intelligence and the aptitude for leveraging technology. As expressed by Chiu et al. (2024) the capacity to utilise technology encompasses not only the ability to operate technological instruments, but also the capacity to communicate through them. This entails ensuring the responsible and salutary use of technology, which necessitates the consideration of ethical values. Furthermore, they explained that the concept of artificial intelligence literacy encompasses not only an understanding of technology, but also the ability to utilise it. Furthermore, Al-Abdullatif's study (2024) revealed a correlation between teachers with high levels of AI literacy and those who utilise AI applications and technology to a greater extent. From this perspective, teachers can enhance their utilisation of technology by developing their AI proficiency, or conversely, they can augment their AI literacy by integrating technology into their pedagogical practices. As evidenced by these studies, the relationship between these two variables is cyclical.

CONCLUSION

The development of teachers' AI literacy will assist them in overcoming the obstacles to teacher leadership, including workload, time constraints, and a lack of sufficient educational resources for professional development. Furthermore, it will enhance the quality of classroom practices and the efficacy of educational and training processes. In particular, the judicious use of artificial intelligence applications in accordance with ethical principles will ensure that teachers, their students, and themselves are safeguarded from the potentially harmful consequences of such technology. Most importantly, teachers' use of artificial intelligence will enable them to build stronger bridges with their students who are digital citizens. The more teachers use artificial intelligence applications, the more artificial intelligence literate they become. Similarly, the more AI literate they become, the more their use of AI will increase.

Recommendations

Researchers may devise a training programme with the objective of enhancing the level of artificial intelligence literacy among teachers. Following the implementation of the programme, a quasi-experimental study may be conducted to evaluate its effectiveness. Teachers may elect to pursue training in artificial intelligence literacy as a means of advancing their professional development and enhancing their proficiency in the utilisation of artificial intelligence applications. Decision makers should plan training programmes for teachers, students and school administrators with the aim of developing their literacy in artificial intelligence, their ability to use artificial intelligence applications and their understanding of the ethical principles.

REFERENCES

Abulibdeh, A., Zaidan, E., & Abulibdeh, R. (2024). Navigating the confluence of artificial intelligence and education for sustainable development in the era of industry 4.0: Challenges, opportunities, and ethical dimensions. *Journal of Cleaner Production*, *437*, 140527. DOI: 10.1016/j.jclepro.2023.140527

Ahmad, K., Iqbal, W., El-Hassan, A., Qadir, J., Benhaddou, D., Ayyash, M., & Al-Fuqaha, A. (2023). Data-driven artificial intelligence in education: A comprehensive review. *IEEE Transactions on Learning Technologies*, *17*, 12–31. DOI: 10.1109/TLT.2023.3314610

Al-Abdullatif, A. M. (2024). Modeling teachers' acceptance of generative artificial intelligence use in higher education: The role of ai literacy, intelligent tpack, and perceived trust. *Education Sciences*, *14*(11), 1209. DOI: 10.3390/educsci14111209

Arnold, E. (1989). Artificial Intelligence in the UK: The Alvey Intelligent Knowledge-Based Systems (IKBS) Programme. *AI Communications*, *2*(1), 18–23. DOI: 10.3233/AIC-1989-2102

Avcı, Ü., & Candan, Ö. (2023). Öğretmenlerin bilgi teknolojileri okuryazarlık düzeylerine göre endüstri 4.0 farkındalıklarının incelenmesi [Investigation of teachers' awareness of industry 4.0 according to their information technology literacy levels]. *Necmettin Erbakan Üniversitesi Ereğli Eğitim Fakültesi Dergisi*, *5*(1), 160–178.

Ayanwale, M. A., Adelana, O. P., Molefi, R. R., Adeeko, O., & Ishola, A. M. (2024). Examining artificial intelligence literacy among pre-service teachers for future classrooms. *Computers and Education Open*, *6*(100179), 1–15. DOI: 10.1016/j.caeo.2024.100179

Baker, R. S., & Inventado, P. S. (2016). Educational data mining and learning analytics. In Rupp, A. A., & Leighton, J. P. (Eds.), *The Cambridge handbook of the learning sciences* (Vol. 2, pp. 589–607). John Wiley.

Banaz, E., & Demirel, O. (2024). Türkçe öğretmen adaylarının yapay zekâ okuryazarlıklarının farklı değişkenlere göre incelenmesi [Investigation of Turkish pre-service teachers' artificial intelligence literacy according to different variables]. *Dokuz Eylül Üniversitesi Buca Eğitim Fakültesi Dergisi*, (60), 1516–1529. DOI: 10.53444/deubefd.1461048

Bertolin, M., & Da Rin, D. (2020). Chatbots for K-12 education: A review of recent advances. Opportunities. and Challenges. *Education Sciences*, *10*(7), 181–191.

Büyüköztürk, Ş., Kılıç-Çakmak, E., Akgün, Ö. E., Karadeniz, Ş., & Demirel, F. (2024). *Bilimsel araştırma yöntemleri* [Scientific research methods]. Pegem Akademi.

Campbell, M., Hoane, A. J.Jr, & Hsu, F. (2002). Deep Blue. *Artificial Intelligence*, *134*(1-2), 57–83. DOI: 10.1016/S0004-3702(01)00129-1

Chen, X., Zou, D., Xie, H., Cheng, G., & Liu, C. (2022). Two decades of artificial intelligence in education. *Journal of Educational Technology & Society*, *25*(1), 28–47.

Chiu, T. K., Ahmad, Z., Ismailov, M., & Sanusi, I. T. (2024). What are artificial intelligence literacy and competency? A comprehensive framework to support them. *Computers and Education Open*, *6*(100171), 1–9. DOI: 10.1016/j.caeo.2024.100171

Chiu, T. K., Xia, Q., Zhou, X., Chai, C. S., & Cheng, M. (2023). Systematic literature review on opportunities. challenges. and future research recommendations of artificial intelligence in education. *Computers and Education: Artificial Intelligence*, *4*, 100–118. DOI: 10.1016/j.caeai.2022.100118

Çoban, Ö. (2021). Liderlik ve sorumluluk. [Leadership and responsibility] In Karataş, K. (Ed.), *Eğitim ve 21.yy. becerileri* [[Education and 21st-*century skills]*]. (pp. 258–276). Nobel.

Çoban, Ö., & Gün, F. (2024). Eğitimde mükemmelliğe giden yolda öğretmen liderliği. [Teacher leadership on the road to excellence in education] In Atasoy, R., Çoban, Ö., & Yalçın, M. T. (Eds.), *Eğitimde mükemmellik haritası: Liderlik, beceriler ve başarı* [*Map of excellence in education: leadership, skills and success*]. (pp. 45–60). Eğitim.

Crevier, D. (1993). *AI: The tumultuous search for artificial intelligence*. Basic Books.

Danielson, C. (2006). *Teacher leadership that strengthens professional practice*. Association for Supervision and Curriculum Development.

Demir, K. (2020). Öğretmen liderliği. [Teacher leadership] In Yılmaz, K. (Ed.), *Liderlik: Kuram. araştırma. uygulama* [*[Leadership: Theory, research and implication*]]. (pp. 558–570). Pegem.

Durnalı, M., & Çoban, Ö. (2025). Artificial intelligence in higher education management: Innovations. opportunities and challenges. In González Lezcano, R. A., & Orakçı, Ş. (Eds.), *Innovation trends and educational technology in higher education* (pp. 501–516). Information Age Publishing.

Engreini, S., Amini, R., & Alfana, Y. D. (2025). Effectiveness of using artificial intelligence chatbox integrated case method on science learning outcomes of basic education students. *Jurnal Penelitian Pendidikan IPA*, *11*(1), 1206–1214.

Erdoğan, M., & Kırılmaz, H. (2020). Hasta merkezlilik ve hasta merkezli bakım [Patient centerization and patient-centered care]. *İnsan ve İnsan, 7*(24), 97-126.

Erdoğan, T. E., & Ekşioğlu, S. (2024). Yapay Zekâ Okuryazarlığı Ölçeği'nin Türkçeye uyarlanması [Adaptation of Artificial Intelligence Literacy Scale into Turkish]. *Türk Eğitim Bilimleri Dergisi.*, *22*(2), 1196–1211.

Ergen, M. (2019). What is artificial intelligence? Technical considerations and future perception. *The Anatolian Journal of Cardiology*, *22*(2), 5–7. DOI: 10.14744/AnatolJCardiol.2019.79091 PMID: 31670719

Fathima Anjila, P. K. (2021). Artificial intelligence. In J. Karthikeyan, T.S. Hie, & Ng. Y. Jin (Eds.). *Learning outcomes of classroom research.* (pp. 65-73). L'Ordine Nuovo Pub.

Ghamrawi, N., Shal, T., & Ghamrawi, N. A. (2024). Exploring the impact of AI on teacher leadership: Regressing or expanding? *Education and Information Technologies*, *29*(7), 8415–8433. DOI: 10.1007/s10639-023-12174-w

Gocen, A., & Aydemir, F. (2020). Artificial intelligence in education and schools. *Research on Education and Media*, *12*(1), 13–21. DOI: 10.2478/rem-2020-0003

Gondal, K. M. (2018). Artificial intelligence and educational leadership. *King Edward Medicine Üniversity Annuals*, *24*(4), 1–2.

Gong, Y. (2021). Application of virtual reality teaching method and artificial intelligence technology in digital media art creation. *Ecological Informatics*, *63*, 101304. DOI: 10.1016/j.ecoinf.2021.101304

Guan, C., Mou, J., & Jiang, Z. (2020). Artificial intelligence innovation in education: A twenty-year data-driven historical analysis. *International Journal of Innovation Studies*, *4*(4), 134–147. DOI: 10.1016/j.ijis.2020.09.001

Güneş, F. (2019). Okuryazarlık yaklaşımları. *The Journal of Limitless Education and Research.*, *4*(3), 224–246.

Harris, A. (2005). Teacher leadership: More than just a feel-good factor. *Leadership and Policy in Schools*, *4*(3), 201–219. DOI: 10.1080/15700760500244777

Harris, A., & Muijs, D. (2003). *Teacher leadership: Principles and practices*. National College for School Leadership.

Hashim, S., Omar, M. K., Ab Jalil, H., & Sharef, N. M. (2022). Trends on technologies and artificial intelligence in education for personalized learning: Systematic literature. *Journal of Academic Research in Progressive Education and Development*, *12*(1), 884–903. DOI: 10.6007/IJARPED/v11-i1/12230

Holmes, W., Bialik, M., & Fadel, C. (2019). *Artificial intelligence in education: Promises and implications for teaching and learning*. Centre for Curriculum Redesign.

Hsu, T. C., Hsu, T. P., & Lin, Y. T. (2023). The artificial intelligence learning anxiety and self-efficacy of in-service teachers taking AI training courses. *2023 International Conference on Artificial Intelligence and Education (ICAIE)*, Kobe, Japan, pp. 97-101. DOI: 10.1109/ICAIE56796.2023.00034

Hu, L., & Bentler, P. M. (1999). Cutoff criteria for fit indexes in covariance structure analysis: Conventional criteria versus new alternatives. *Structural Equation Modeling*, *6*(1), 1–55. DOI: 10.1080/10705519909540118

Huang, J., Saleh, S., & Liu, Y. (2021). A review on artificial intelligence in education. *Academic Journal of Interdisciplinary Studies*, *10*(3), 206–217. DOI: 10.36941/ajis-2021-0077

Hwang, G. J., Xie, H., Wah, B. W., & Gašević, D. (2020). Vision. challenges. roles and research issues of Artificial Intelligence in Education. *Computers and Education: Artificial Intelligence*, *1*(100001), 1–5. DOI: 10.1016/j.caeai.2020.100001

Jiang, Y., Li, X., Luo, H., Yin, S., & Kaynak, O. (2022). Quo vadis artificial intelligence? *Discover Artificial Intelligence*, *2*(1), 4. DOI: 10.1007/s44163-022-00022-8

Jordan, M. I., & Mitchell, T. M. (2015). Machine learning: Trends, perspectives, and prospects. *Science*, *349*(6245), 255–260. DOI: 10.1126/science.aaa8415 PMID: 26185243

Kandlhofer, M., Steinbauer, G., Hirschmugl-Gaisch, S., & Huber, P. (2016). *Artificial intelligence and computer science in education: From kindergarten to university*. In 2016 IEEE Frontiers in Education Conference (FIE) (pp. 1-9). NY: IEEE. DOI: 10.1109/FIE.2016.7757570

Karataş, F., Eriçok, B., & Tanrikulu, L. (2024). Reshaping curriculum adaptation in the age of artificial intelligence: Mapping teachers' AI-driven curriculum adaptation patterns. *British Educational Research Journal, 1*(0), 1–27.

Katzenmeyer, M., & Moller, G. (2013*). Uyuyan devi uyandırmak [Awaking the sleeping giant]*. Servet Özdemir (Trans. Ed.). Nobel.

Kırksekiz, A., Yıldız, M., Kıyıcı, M., & Yıldız, M. (2024). Adaptation of artificial intelligence literacy scale: Latent profile analysis. *Sakarya University Journal of Education, 14*(3), 581–596. DOI: 10.19126/suje.1479294

Komalavalli, K., Hemalatha, R., & Dhanalakshmi, S. (2020). A survey of artificial intelligence in smart phones and its applications among the students of higher education in and around Chennai City. *Shanlax International Journal of Education, 8*(3), 89–95. DOI: 10.34293/education.v8i3.2379

Kuriscak, E., Marsalek, P., Stroffek, J., & Toth, P. G. (2015). Biological context of Hebb learning in artificial neural networks, a review. *Neurocomputing, 152*, 27–35. DOI: 10.1016/j.neucom.2014.11.022

Lérias, E., Guerra, C., & Ferreira, P. (2024). Literacy in artificial intelligence as a challenge for teaching in higher education: A case study At Portalegre Polytechnic University. *Information (Basel), 15*(4), 205. DOI: 10.3390/info15040205

Levin, J. A., Fox, J. A., & Forde, D. R. (2010). *Elementary statistics in social research*. Allyn & Bacon Pearson.

Luo, X., Tong, S., Fang, Z., & Qu, Z. (2019). Machines versus humans: the impact of AI chatbot disclosure on customer purchases. *Nanyang Business School Research Paper, 20*(33), 1-30.

Ma, S., & Chen, Z. (2024). The development and validation of the artificial intelligence literacy scale for Chinese college students (AILS-CCS). *IEEE Access : Practical Innovations, Open Solutions, 12*, 146419–146429. DOI: 10.1109/ACCESS.2024.3468378

Mart, M., & Kaya, G. (2024). Okul öncesi öğretmen adaylarının yapay zekâya yönelik tutumları ve yapay zekâ okur yazarlığı arasındaki ilişkinin incelenmesi [Investigating the relationship between pre-service preschool teachers' attitudes towards artificial intelligence and artificial intelligence literacy]. *Edutech Research, 2*(1), 91–109.

McCarthy, J. (2007). From here to human-level AI. *Artificial Intelligence, 171*(18), 1174–1182. DOI: 10.1016/j.artint.2007.10.009

McCarthy, J., Minsky, M. L., Rochester, N., & Shannon, C. E. (1995). A proposal for the Dartmouth summer research project on artifcial intelligence. *AI Magazine*.

Mehrling, P. (1999). The vision of Hyman P. Minsky. *Journal of Economic Behavior & Organization, 39*(2), 129–158. DOI: 10.1016/S0167-2681(99)00029-3

Metelskaia, I., Ignatyeva, O., Denef, S., & Samsonowa, T. (2018). A business model template for AI solutions. *Proceedings of the international conference on intelligent science and technology* (pp. 35-41). NY: Association for Computing Machinery. DOI: 10.1145/3233740.3233750

Minsky, M. (1988). *The society of mind*. Simon and Schuster.

Nguyen, A., Ngo, H. N., Hong, Y., Dang, B., & Nguyen, B. P. T. (2023). Ethical principles for artificial intelligence in education. *Education and Information Technologies*, *28*(4), 4221–4241. DOI: 10.1007/s10639-022-11316-w PMID: 36254344

Nong, Y., Cui, J., He, Y., Zhang, P., & Zhang, T. (2024). Development and validation of an AI Literacy Scale. *Journal of Artificial Intelligence Research*, *1*(1), 17–26. DOI: 10.70891/JAIR.2024.100029

Pham, S. T., & Sampson, P. M. (2022). The development of artificial intelligence in education: A review in context. *Journal of Computer Assisted Learning*, *38*(5), 1408–1421. DOI: 10.1111/jcal.12687

Pimentel, J. L. (2019). Some biases in Likert scaling usage and its correction. [IJSBAR]. *International Journal of Sciences, Basic and Applied Research*, *45*(1), 183–191.

Ramnarain, U., Ogegbo, A. A., Penn, M., Ojetunde, S., & Mdlalose, N. (2024). Pre-service science teachers' intention to use generative artificial intelligence in inquiry-based teaching. *Journal of Science Education and Technology*, •••, 1–14. DOI: 10.1007/s10956-024-10159-z

Roll, I., & Wylie, R. (2016). Evolution and revolution in artificial intelligence in education. *International Journal of Artificial Intelligence in Education*, *26*(2), 582–599. DOI: 10.1007/s40593-016-0110-3

Samuel, A.L. (2000). Some studies in machine learning using the game of checkers. *IBM J Res Dev.44(*1.2), 206–226.

Shen, J., Wu, H., Reeves, P., Zheng, Y., Ryan, L., & Anderson, D. (2020). The association between teacher leadership and student achievement: A meta-analysis. *Educational Research Review*, *31*(100357), 1–19. DOI: 10.1016/j.edurev.2020.100357

Silva, D. Y., Gilbert, B., & Nolan, J. (2000). Sliding the doors: Locking and unlocking the possibilities for teacher leadership. *Teachers College Record*, *102*(4), 779–804. DOI: 10.1111/0161-4681.00077

Smylie, M. A., Conley, S., & Marks, H. M. (2002). Exploring new approaches to teacher leadership for school improvement. *Yearbook of the National Society for the Study of Education*, *101*(1), 162–188. DOI: 10.1111/j.1744-7984.2002.tb00008.x

Sperling, K., Stenberg, C. J., McGrath, C., Åkerfeldt, A., Heintz, F., & Stenliden, L. (2024). In search of artificial intelligence (AI) literacy in teacher education: A scoping review. *Computers and Education Open*, *6*(100169), 1–13. DOI: 10.1016/j.caeo.2024.100169

Stembert, N., & Harbers, M. (2019). Accounting for the human when designing with AI: Challenges identified. CHI'19-Extended Abstracts. Glasgow. Scotland.

Su, G. (2018). Unemployment in the AI Age. AI Matters 3(4). 35–43. Tarafdar. M., Beath. C. M., & Ross. J. W. (2019). Using AI to enhance business operations. *MIT Sloan Management Review*, *60*(4), 37–44.

Tabachnick, B. G., & Fidell, L. S. (2003). *Using multivariate statistics*. Pearson.

Tsai, S. C., Chen, C. H., Shiao, Y. T., Ciou, J. S., & Wu, T. N. (2020). Precision education with statistical learning and deep learning: A case study in Taiwan. *International Journal of Educational Technology in Higher Education, 17*(1), 1–13. DOI: 10.1186/s41239-020-00186-2

Turing, A. M. (1950). Computing machinery and intelligence. *Mind, 59*(236), 433–460. DOI: 10.1093/mind/LIX.236.433

Unal, A., & Unal, Z. (2024). Evaluating the integration of artificial intelligence (AI) in K-12 education. *Journal of Interactive Learning Research, 35*(3), 353–387.

Wang, B., Rau, P. L. P., & Yuan, T. (2023). Measuring user competence in using artificial intelligence: validity and reliability of artificial intelligence literacy scale. *Behaviour & Information Technology, 42*(9), 1324–1337. DOI: 10.1080/0144929X.2022.2072768

York-Barr, J., & Duke, K. (2004). What do we know about teacher leadership? Findings from two decades of scholarship. *Review of Educational Research, 74*(3), 255–316. DOI: 10.3102/00346543074003255

Zdravkova, K., Krasniqi, V., Dalipi, F., & Ferati, M. (2022). Cutting-edge communication and learning assistive technologies for disabled children: An artificial intelligence perspective. *Frontiers in Artificial Intelligence, 5*, 970430. DOI: 10.3389/frai.2022.970430 PMID: 36388402

Zhang, C., Schießl, J., Plößl, L., Hofmann, F., & Gläser-Zikuda, M. (2023). Acceptance of artificial intelligence among pre-service teachers: A multigroup analysis. *International Journal of Educational Technology in Higher Education, 20*(49), 1–22. DOI: 10.1186/s41239-023-00420-7

Zhang, L., & Zhang, B. A. (1999). Geometrical representation of McCulloch-Pitts neural model and its applications. *IEEE Transactions on Neural Networks, 10*(4), 925–949. DOI: 10.1109/72.774263 PMID: 18252589

Chapter 2
Artificial Intelligence in Education and Changing Teacher Roles

Ramazan Atasoy
https://orcid.org/0000-0002-9198-074X
Harran University, Turkey

ABSTRACT

This book chapter aims to present a holistic and innovative perspective on AI in education and its impact on evolving teacher roles. It seeks to raise awareness of AI-enabled teaching practices while offering a transformative framework to enhance teachers' professional development and promote sustainable education. Additionally, this chapter provides both theoretical and practical insights into the shifting needs of educators as they navigate the transition from today's complex educational landscape to AI-integrated learning environments, which have the potential to shape future educational paradigms. In line with these objectives, the chapter explores various perspectives on integrating AI to support teacher development, reframe teacher leadership, and equip school administrators and policymakers with a comprehensive understanding of AI's opportunities and challenges in education. Furthermore, it examines how AI enhances education, fosters professional development, and improves student outcomes while addressing potential risks and ethical considerations.

INTRODUCTION

We are in a new paradigm shift in which artificial intelligence (AI)-enabled technologies are beginning to dominate all disciplines. The implications of this paradigm shift cover a huge and very broad spectrum, ranging from predictions that AI could bring about the end of humanity itself (Barrat, 2023, p. 39) to the unbridled enthusiasm that captivates the masses. Rapid advancements in contemporary information and communication technologies, transcending the limits of future expectations, have significantly accelerated the convergence of artificial intelligence and educational technologies. As a result, these innovations are increasingly shaping and transforming all aspects of school organizations. The backbone of this change is AI-enabled instructional practices applications in the classroom. In this context, AI-enabled educational applications add dynamism to the teaching process, make student learning more comprehensible, rekindle

DOI: 10.4018/979-8-3373-0472-4.ch002

the desire to learn, and provide tremendous convenience in accessing, processing, and making decisions about information. Nevertheless, over time, as in other fields, AI-supported applications and robots in educational institutions are gradually surpassing human capabilities. As this progression continues, our reliance on these technologies is steadily increasing. As time goes on, we are becoming increasingly attached to them (Kaplan, 2015; Schwab, 2017). Furthermore, an increasing number of people are devoting their entire lives to the development of mechanical life, which has resulted in our becoming dependent on and partly enslaved to machines. In other words, there is a growing tendency towards automation and robotization in all areas of life, including the economy, education, social interaction and entertainment.

A new world order is emerging in which the distinction between humans and machines is less defined and the boundary between humanity and technology is shrinking. This paradigm shift will have a profound impact on our lives (Brynjolfsson, & McAfee, 2014). It will transform the way we live, our learning methods and techniques, our expectations and perceptions of education, and particularly the school, which is currently confined within four walls and is regarded as a relic of the industrial society. It is therefore being challenged.

The literature on artificial intelligence in education (AIE) offers insights into innovative AI-based application praxis in teachers' instructional practices and the potential risks and ethical issues in educational processes when experimenting with them, as well as strategies for minimizing these risks. In this context, the deployment of large-scale AI applications in information processing and knowledge production has the potential to exert a significant influence on the field of education. This is due to the fact that these applications are characterized by the capacity to engage with higher-order thinking skills that are fundamental to the processes of human learning. In a context where AI tools are being utilized to automate certain basic levels of knowledge and product generation, it falls upon schools and the relevant educational policymakers to conduct a detailed analysis of the specific content and methods through which students are engaged in the learning process.

This study is significant in that it demonstrates how AI challenges the established practice of rote learning in learning and teaching processes. Furthermore, AI has the potential to facilitate personalized and inclusive education, which has been a prominent area of focus for educational researchers in recent years. The aspiration of many education systems is to provide all students with personalized support and inclusive learning opportunities, enabling them to build stronger connections with teachers for better guidance and to have their achievements appropriately recognized and assessed. Another important dimension of this study is its focus on the changing role of school organizations and teachers in the responsible and conscious adoption, training and monitoring of the use and limitations of AIE, despite its remarkable and compelling promises.

The integration of AI into education offers numerous potential benefits for both individual learners and educational institutions. These advantages encompass innovative learning opportunities, personalized education, intelligent tutoring systems, comprehensive learning assessments, data-driven school management, and strategies for preventing student dropouts. Notably, in the context of evolving teacher roles, AI-driven tools signal a transformative shift. They have the potential to alleviate teachers' workloads, provide educational and pedagogical insights into student performance, facilitate the adoption of innovative teaching practices within the classroom, and deliver supplementary support when needed (Baker & Smith, 2019). The potential applications of AI in education are extensive and transformative. AI offers opportunities to equip teachers with a more comprehensive understanding of how to integrate learning opportunities tailored to individual students. It also supports the advancement of more functional and inclusive educational practices, the implementation of continuous feedback mechanisms in assessment processes, and the design of personalized learning pathways that account for students' readiness to learn. Moreover, education-focused smart learning analytics demonstrate considerable potential in identifying at-risk students, enabling timely interventions, and developing innovative content and teaching strategies specifically designed for both disadvantaged and gifted learners. The insights generated from big data and smart learning analytics empower educators to improve teaching quality and optimize student learning outcomes (Yang et al., 2021).

This new digital literacy, which has attracted great interest around the world, faces the dilemma of a lack of theoretical frameworks and planning by researchers and policymakers, and limited practical experience among teachers. In many education systems, AI applications are not yet on the radar of existential opportunities and threats for students and teachers. However, AI will likely become more widespread in school organizations. In fact, many students and teachers are actively expanding its use, even if only at a basic level. At this stage, technological errors or accidents caused by AI may be extremely difficult to rectify. It seems that students with weak questioning and critical thinking skills are more likely to make mistakes in AI-based applications. The danger is greater for individuals who cannot question why an AI-powered application is making decisions and how accurate and ethical the information it is providing is. In this context, policymakers and schools need to be more proactive about AI awareness and training. In this context, we consider the study's focus on the changing roles and responsibilities of teachers in AI to be valuable in terms of school outcomes.

AI IN EDUCATION

Artificial Intelligence in Education (AIE) refers to the use of artificial intelligence technologies in educational processes to improve, personalize and optimize educational processes. This approach finds a wide range of applications from teaching methods to student assessment, from the development of learning materials to school management processes. Kamalov et al. (2023) conceptualize artificial intelligence as a hallmark of a transformative phase, heralding a new technological epoch across education and various other domains. This paradigm shift is catalyzing the development and implementation of cutting-edge pedagogical tools and methodologies. These innovations, designed to enhance both teaching efficacy and learning outcomes, are undergoing iterative testing and refinement across diverse educational settings and levels. For instance, Bearman and Ajjawi (2023) emphasize the role of AI in personalizing learning pathways, while Markauskaite et al. (2022) highlight its potential to foster collaborative, data-driven teaching practices that address individual learner needs.

AIE has emerged as a disruptive technology with the potential to revolutionize specific educational efforts, such as facilitating hybrid intelligence approaches and/or developing written compositions (Järvelä et al., 2023). The widespread adoption of the internet has accelerated the transition to web-based and online intelligent education systems. The primary driver behind this shift has been the significant advancements in computing technologies, including increased processing power, sophisticated algorithms, and the development of network infrastructures such as the internet and the World Wide Web. Additionally, specialized software and task-oriented programs have diversified the applications of computers in education. For example, educational robots have been utilized to enhance STEM learning, while adaptive chatbots provide personalized learning support by analyzing student interactions in real-time. Studies demonstrate that these technologies not only improve efficiency in administrative tasks but also enhance student engagement and learning outcomes (Belpaeme et al.,2018; Chen et al., 2020;

Holmes et al., 2019). This evolution culminated in the integration of embedded computing systems and complementary technologies, enabling the deployment of humanoid robots and web-based chatbots in educational settings. AIE offers innovative solutions to the education sector with algorithms and data analytics tools that mimic human-like learning and problem-solving abilities. These tools support school administrations and teachers in fulfilling their roles independently or collaboratively.

Balderson (2023) presented the results of a study involving over 5,000 Swedish university students, which will serve as the foundation for the AIE requirement. The study revealed that 95% of participants were familiar with AI, 56% expressed an intention to incorporate it into their studies, and 35% reported frequent usage. Among those who used AI, 68% indicated that it had a positive impact on their performance. However, nearly half of the instructors reported utilizing productive AI in their lesson planning.

AI has had a substantial impact on the reduction of paperwork and workload for trainers, particularly in terms of administrative functions, thereby enabling trainers to focus on their core tasks, namely teaching and the dissemination of content and materials in accordance with the prevailing curriculum.Intelligent tutoring systems have been shown to offer a wide range of functionalities that enable instructors to perform various administrative tasks, including grading and providing feedback (Rus et al., 2013). Other programmes that make use of AI, such as Deeply, Grammarly, Ecree, PaperRater and Turnitin, also provide instructors with functions to perform different tasks, including translation, plagiarism checking, grading and providing feedback to students on areas for improvement. Table 1 provides a comprehensive guide on how artificial intelligence (AI) can be utilized in both education and academic research. AI is revolutionizing education and research by making learning personalized, writing efficient, research insightful, and assessments automated. These tools save time, enhance understanding, and improve outcomes for students and researchers alike. Table 1 demonstrates how AI can revolutionize both education and academic research by providing personalized, efficient, and innovative solutions. research.

Table 1. Artificial intelligence tools used in education and functions

AI Tools	Functions
AdaptiveU	AI-based personalized education platform
Aleks	Individualized learning programs
Carnegie Learning AI	Customized recommendations for math and language education
Century Tech	Personalized learning pathways
ChatGPT	Q&A, text generation, lesson delivery, homework assistance
ChatPDF	Understanding and summarizing PDF articles
ClassPoint	Interactive teaching tools for PowerPoint
Claude AI	Academic writing and analysis support
Cognii	AI-powered analysis of open-ended responses
Connected Papers	Visualizing connections between academic papers
Copy.ai	Research text generation and editing
Coursera AI	Personalized course recommendations
DeepL	Advanced translation and language learning support
DeepSeek	Q&A, text generation, lesson delivery, homework assistance

continued on following page

Table 1. Continued

AI Tools	Functions
DreamBox	Personalized math education
Duolingo AI	AI-powered language learning
edX AI	Student performance analysis and feedback
Elicit	AI-powered academic paper summarization and research assistance
Ello	AI-powered English speaking practice
EndNote	Bibliography and citation management
Explainpaper	Analyzing and explaining academic papers
Grammarly	Grammar and spell-checking, academic writing support
IBM Watson Discovery	Large-scale data and academic content analysis
IBM Watson Education	Predicting student success through big data analysis
Jenni AI	AI-powered suggestions for academic writing
Khan Academy AI	Personalized education and guidance
KnowBot	AI-powered learning assistant
Knowji	AI-powered vocabulary learning app
Lingvist	Language education adapting to individual learning pace
Litmaps	Literature review and academic connection mapping
MATHia	AI-powered math teaching
Mendeley AI	Reference management and academic paper discovery
Nuance Dragon Speech	Speech-to-text for note-taking and learning
Otter.ai	Transcribing lecture recordings into text
Paperpile	Reference management and academic resource organization
Pearson AI	Personalized educational solutions
Prodigy	Math education through educational games
Querium	Smart tutoring for STEM subjects
QuillBot	Paraphrasing, summarizing, and improving academic writing
Quillionz	Automated question generation and exam preparation
Quizizz AI	Automated quiz and test creation
Quizlet AI	Memory enhancement with smart flashcards
Research Rabbit	Discovering relevant academic papers
Scispace	AI-powered academic paper formatting and writing assistant
Scite.ai	Analyzing source reliability and citation contexts
ScribeSense	Digitizing paper-based exams automatically
Semantic Scholar	AI-powered academic paper search engine
Smart Sparrow	AI-powered adaptive learning environments
Socratic by Google	Visual and textual learning support for students
SpeechTexter	Voice-command note-taking and content creation
Thinkster Math	AI-powered math teaching

continued on following page

Table 1. Continued

AI Tools	Functions
TLDR Papers	Summarizing papers and highlighting key points
Trinka AI	Academic writing checking and improvement
Zotero AI	Academic reference management and source recommendations

To provide a comprehensive overview of the application areas of AIE within educational environments, the following key themes will be analyzed: personalized learning, automated administrative tasks, intelligent tutoring systems, content creation and enhancement, and data-driven insights.

PERSONALIZED LEARNING

The convergence and/or integration of AI into education has emerged as a transformative catalyst, reshaping conventional teaching methodologies and broadening the scope of personalized learning. Central to this shift is the concept of personalized learning, powered by sophisticated AI algorithms. Personalized learning focuses on creating educational experiences that match the unique needs of each student. Using AI systems, these programs can assess how well a student is doing and adjust the lessons to fit their progress. For example, the system might repeat a concept if a student struggles or move forward if they excel, ensuring that each learner progresses at their own pace. At its core, personalized learning adopts a student-centered approach by tailoring instructional materials, pacing, and assessments to each learner's unique needs. This diverges from the traditional, one-size-fits-all education model, offering a more flexible and adaptive framework. AI-powered personalized learning platforms leverage machine learning (ML) algorithms to evaluate students' competencies and deliver tailored assessments and insights. As students interact with the platform, the AI continually refines its understanding of their strengths and weaknesses, providing a personalized learning experience that improves over time (Zohuri and Mossavar-Rahmani, 2024). Key principles of personalized learning include flexibility, adaptability, and an emphasis on student agency, creating a dynamic and responsive learning environment that empowers students to take an active role in their education. Unlike the uniformity of traditional education models, this approach tailors learning experiences to reflect the unique skills, preferences, and progressions of individual learners.

The role of AI tools in educational settings is undeniably significant, especially when evaluated within the framework of traditional pedagogical methods. While conventional teaching approaches are widely used, they often risk diminishing students' enthusiasm, curiosity, and intrinsic motivation—factors essential for fostering meaningful and effective learning. In contrast, AI-powered educational systems provide a more dynamic and engaging alternative by adapting to individual learners' unique preferences, cognitive styles, and learning speeds. This adaptability not only enhances motivation and productivity but also facilitates the delivery of personalized learning experiences, thereby improving educational outcomes.

A key advantage of personalized learning is its ability to ensure that each student receives the support and instruction necessary to reach their full potential. This approach not only allows advanced students to be challenged at an appropriate level, but also provides struggling students with the opportunity to catch up. Furthermore, AI has the capability to analyze past student performance to identify areas of difficulty and deliver tailored support to address these specific challenges. One prominent example is the integration of AI-driven digital game-based learning, which offers transformative and engaging opportunities for students. Such systems enable real-time adjustments to the learning process by analyzing data on learner performance, allowing for personalized and immersive exploration (Gaurav et al., 2022). Moreover, the use of AI in feedback mechanisms improves data accuracy and balances the learning flow, ensuring alignment with individual learner profiles. In parallel, virtual reality (VR) and three-dimensional (3-D) technologies have been shown to provide students with experiential learning opportunities, further enriching the educational process (Gaurav et al., 2022).

Studies by Miao et al. (2021) and Tan (2023) underscore its potential to enhance student engagement and improve outcomes by dynamically addressing personal learning needs. Similarly, Bahroun et al. (2023) highlight AI's capacity to refine instructional strategies, create adaptive learning pathways, and drive innovation through data-centric educational practices. Additionally, Baltezarević and Baltezarević (2024) underscore the transformative potential of AI in shaping the future of education. They highlight its ability to deliver highly personalized learning experiences, promote student autonomy, and enhance comprehension. Their research also emphasizes that AI can streamline administrative tasks for educators while creating innovative pedagogical opportunities. Students, as reflected in their findings, view AI as a critical enabler of personalized learning, helping them identify their abilities and creativity. They also recognize AI's potential to automate progress monitoring, foster greater interactivity through innovative teaching approaches, and address individual challenges by analyzing past performance and providing targeted support.

AI-powered software, including tools such as Knowji, Knewton Alta, Khan Academy, DreamBox Learning, ALEKS (Assessment and Learning in Knowledge Spaces), Coursera, edX, Duolingo, Socratic by Google, ChatGPT (OpenAI), Classcraft, Gradescope, zSpace, Quizlet, Kahoot!, and Cognii, exemplifies the integration of technology to enhance education. These tools facilitate a more student-centered approach by personalizing learning experiences, continuously monitoring students' progress, and providing instant feedback to optimize learning outcomes. Moreover, they offer in-depth evaluations to address individual needs, ultimately making educational processes more effective and efficient.

With the daily increase in the number of artificial intelligence-based systems in education, it is crucial to address the pedagogical considerations in the design and development of these technologies. Emphasis should be placed on how these systems can effectively promote self-regulated and active learning for individual students. Furthermore, careful attention must be given to their integration into the curriculum to support and enhance personalized learning experiences.

AUTOMATED ADMINISTRATIVE TASKS

The integration of AI into the educational workforce is reshaping the landscape of teaching and learning, particularly through the adoption of advanced tools like ChatGPT and Gradescope for both administrative and instructional tasks. This technological shift marks a redefinition of educators' roles, steering their focus toward higher-order teaching responsibilities such as nurturing critical thinking, providing ethical and moral guidance, and offering emotional support to students. In this sense, AI serves as a valuable partner for educators, fostering mentorship roles, promising efficiency and innovation while preserving the essential human elements of teaching. By automating repetitive tasks—such as grading, attendance tracking, and report generation—AI has the potential to substantially reduce the administrative burden on teachers, allowing them to invest more time and energy in student-centered activities.

INTELLIGENT TUTORING SYSTEMS

Intelligent Tutoring Systems (ITS) are advanced, AI-powered platforms designed to deliver personalized, real-time educational support, often outside traditional classroom settings. By leveraging sophisticated algorithms, these systems provide targeted feedback and instruction, analyze student performance, adapt content to individual needs, and promote active learning. An ITS typically consists of four key components: the Domain Model, Student Model, Tutoring Model, and User Interface. The Domain Model serves as a repository of subject knowledge and outlines the problem-solving framework. The Student Model tracks a learner's progress, knowledge gaps, misconceptions, and preferences, enabling adaptive and personalized instruction. The Tutoring Model utilizes data from the Student Model to determine instructional strategies, such as task selection and the timing of feedback. Lastly, the User Interface facilitates interaction between the learner and the system, offering an accessible, user-friendly experience (Bourdeau & Grandbastien, 2010; Nwana, 1990; VanLehn, 2011).

Studies suggest that ITS improve student engagement and learning outcomes more effectively than traditional methods (Graesser et al., 2018). Unlike conventional "one-size-fits-all" approaches—which often fail to adequately challenge advanced learners or support those struggling—ITS dynamically monitor a student's progress, identify knowledge gaps, and adjust instruction in real time. They are especially effective in STEM education, where they provide immediate feedback, enhance engagement, and enable students to learn at their own pace (VanLehn, 2011). Despite their advantages, ITS face challenges such as dependency on robust student models, computational complexity, and limited classroom integration due to insufficient evaluation studies or inadequate teacher training (Sedlmeier, 2001). Adaptive learning platforms using advanced techniques like fuzzy logic offer solutions by dynamically adjusting content to suit user knowledge and preferences (Goodkovsky, 2001). However, real-world adoption remains constrained by infrastructure, scalability issues, and institutional barriers. Expanding ITS capabilities to include verbal reasoning and flexible, non-linear learning paths may help overcome these limitations (Lesgold, 1992).

ITS can seamlessly integrate with classroom technologies such as Learning Management Systems (LMS), interactive whiteboards, and virtual classrooms, creating a cohesive learning ecosystem. For example, ITS can export performance data to LMS platforms, enabling teachers to monitor progress and tailor their instruction. Integration with tools like Google Classroom allows ITS to embed adaptive exercises directly into a teacher's curriculum, enhancing both engagement and blended learning. Features such as shared dashboards further facilitate collaboration between teachers and students, supporting innovative teaching practices. By leveraging APIs and cloud-based technologies, ITS can complement traditional classroom methods while advancing educational innovation (Kay et al., 2019).

Several ITS platforms demonstrate the transformative potential of these systems in education. MATHia (by Carnegie Learning) customizes lessons to address individual strengths and weaknesses, particularly in mathematics and language learning. It provides step-by-step guidance and allows students to progress at their own pace, improving comprehension and retention. ALEKS (Assessment and Learning in Knowledge Spaces) specializes in math and science, dynamically adapting to a student's knowledge level to close learning gaps effectively. AutoTutor, powered by natural language processing, offers conversational tutoring that mimics human interaction, making it highly effective in subjects like science and reading comprehension. Lastly, the Generalized Intelligent Framework for Tutoring provides a flexible platform for developing ITS applications, focusing on scalability and customization. These platforms exemplify the diversity and adaptability of ITS in addressing a wide range of educational needs.

Content Creation and Enhancement

AI is increasingly being utilized in education to create and enhance learning content. The field of artificial intelligence is dynamic and constantly undergoing technological advances. Educational institutions need to stay abreast of these developments to utilize the latest tools, course content, and methodological techniques for the benefit of students. This requires a commitment to ongoing content development and adaptation of educational strategies. This approach offers numerous advantages, such as personalizing materials for students, making learning more interactive, and reducing teachers' workloads (Holmes et al., 2019).

AI's role in education is often analyzed through the lens of learning theories and pedagogical practices. For example, constructivist learning theory highlights active engagement and experiential learning as critical to student success. AI tools support these principles by offering interactive, customized experiences that empower students to take ownership of their education. By combining cutting-edge technology with educational theory, AI tools not only transform how students learn but also redefine the role of educators, creating opportunities for more personalized and impactful teaching practices. As these innovations continue to evolve, their potential to revolutionize education becomes even more apparent.

A frequently identified challenge for contemporary educators is the overwhelming number of responsibilities they are required to manage on a daily basis. Teachers often report a sense of being overwhelmed, stemming from an excess of demands, distractions, and a lack of direction and support. Creating AI-powered course content can help teachers in this regard. In this sense, with its easy-to-use interfaces and widespread availability of Generative AI has raised the expectation that it will impact education.

Thanks to the ability of advanced models such as ChatGPT and Deepseek to rapidly generate plausible texts across diverse topics, educators now have greater potential to create more dynamic lesson content and design innovative classroom activities. Recent studies (Jones et al., 2023) have highlighted ChatGPT's capacity to support teachers in developing exercises, examinations, and scenarios for student evaluation. For example, tools such as EduWeaver, which assists in the creation of interactive learning materials, and TeacherMatic, a platform for generating customized lesson plans, exemplify how artificial intelligence is transforming educational practices. Additionally, platforms like Blackboard's AI Design Assistant, Course AI, and Coursera AI offer further support by personalizing and streamlining curriculum development. This integration of artificial intelligence systems into education signifies a shift toward more tailored and adaptive teaching methods, enabling teachers to customize content to meet individual student needs (Smith et al., 2024). Such advancements hold significant promise for enhancing learning experiences and improving the overall quality of education (Liu et al., 2023). Tools like Pearson's AI-driven learning aids in Pearson+ Channels automate repetitive tasks and create customized content for students, allowing educators to focus on more meaningful interactions in the classroom. The GoGuardian Instant Pear Deck, IXL learning, Amplify offers a solution to this issue by enabling educators to upload curriculum resources and target specific standards, with the AI system capable of generating a core lesson in approximately one minute. This reduction in time allows for greater personalisation and integration of life experiences into the lesson. Quizlet uses AI to generate tailored study tools, enabling learners to work at their own pace with materials suited to their specific needs. Duolingo, a popular language-learning platform, analyzes users' proficiency to deliver personalized exercises, creating an efficient and effective learning process. StudyFetch leverages AI to help students organize their notes into flashcards and quizzes while offering 24/7 tutoring support. Tools like Teachable and Thinkific enhance student engagement with auto-generated quizzes, making it easier for students to reinforce and evaluate their understanding of key topics. Microsoft has demonstrated that AI can produce content that supports all activities through its Azure AI Chat technology. However, it appears that the technology can currently be regarded as assistive technology, capable of supporting a variety of roles. Nevertheless, it is evident that greater control mechanisms will be necessary in content production, as any output must be evaluated for integrity, accuracy and precision.

Data-Driven Insights

The utilisation of AI-powered analytics empowers educators to discern learning discrepancies by meticulously analysing patterns inherent in student performance data. These insights allow for the development of tailored interventions for students struggling with specific concepts while also predicting academic outcomes, thereby supporting proactive and data-driven mechanisms for student success. Research indicates that predictive analytics can improve retention rates and enhance learning experiences by delivering real-time feedback to both students and instructors.

Platforms such as Canvas and Moodle, which function as Learning Management Systems (LMS), offer detailed analytics on student engagement, academic progress, and areas requiring improvement. These platforms visualize data in actionable formats, enabling educators and administrators to make better-informed decisions. For instance, LMS tools can flag students at risk of underperforming based on their activity patterns, offering instructors an opportunity to intervene early. This predictive capability helps prevent potential setbacks while fostering a supportive learning environment.

Data-driven insights have transformative implications for educational practices, particularly in three critical areas: personalized learning, data-informed pedagogy, and institutional efficiency. AI facilitates the creation of adaptive learning paths, ensuring that educational content aligns with the unique needs and abilities of individual students. Teachers, equipped with empirical data rather than relying solely on intuition, can refine their instructional strategies to better support student learning. Similarly, administrators can optimize resource allocation by identifying courses or topics requiring additional attention, thereby maximizing institutional efficiency.

AI-enabled applications provide continuous feedback during assessments, offering valuable insights into students' individual learning paces and readiness levels. These insights support the integration of personalized learning opportunities, helping students develop their skills and knowledge at their own pace. Furthermore, intelligent learning analytics—a prominent application of AI in education—enable educators to identify at-risk students and implement timely interventions. As Yang et al. (2021) highlight, data derived from big data analytics and intelligent systems can guide teachers in improving both instructional quality and students' learning outcomes. The following are five ways these tools empower educators to enhance their effectiveness:

Figure 1. Improving teaching quality with smart learning analytics

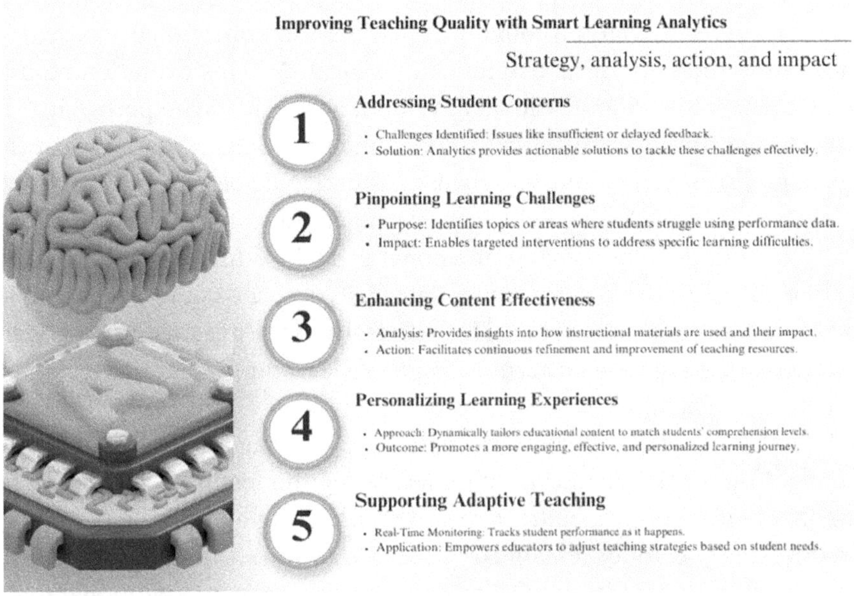

Note: Figure 1 was created by the author using the Canva platform.

These applications of intelligent learning analytics not only enhance teaching practices but also promote the personal and professional development of educators. By using data-driven insights, teachers can refine their teaching strategies and ensure that instructional content evolves alongside students' needs.

Intelligent learning analytics empower students to take ownership of their educational journeys. These tools enable students to explore their own learning pathways, monitor their progress, and compare their performance with peers. Moreover, by identifying and addressing learning gaps, educators can ensure that students receive the support they need to succeed. For example, teachers can provide personalized reports on what students know—or have yet to master—thus bridging learning gaps and enhancing overall understanding.

The integration of intelligent analytics in education facilitates personalized and self-paced learning, enabling students to develop their skills and knowledge according to their unique needs. These systems also offer students actionable feedback, helping them understand how they are progressing and what they need to do to achieve their academic goals.

In conclusion, the integration of AI and data analytics in education not only supports personalized learning tailored to individual student needs but also enhances the overall efficiency of educational institutions. By fostering data-informed decision-making, these technologies hold the potential to reshape the future of education. Intelligent learning analytics are a strategic tool for improving both student success and institutional performance, ensuring that education systems can meet the evolving demands of a data-driven world.

ARTIFICIAL INTELLIGENCE IN EDUCATION: REDEFINING THE ROLES OF TEACHERS

The rapid advancements in AI are profoundly reshaping the educational landscape, prompting a transformative shift in the roles and responsibilities of educators. This evolution requires teachers to actively invest in their personal and professional development, acquiring expertise in AI-driven teaching methodologies and tools to stay relevant in this dynamic environment. This transformation extends beyond technological progress; it represents a broader cultural and sociological shift that impacts individuals across all levels of society. As these changes take place, educators must not only adopt emerging technologies but also commit to continuous professional growth. While AI offers significant potential to revolutionize personalized learning, automation, data analytics, and content creation, it also provides an opportunity to redefine the educator's role. This redefinition includes facilitating learning, offering personalized and individualized support, fostering continuous learning, integrating technology effectively, and exemplifying ethical leadership.

In the age of AI-driven education, teachers are increasingly taking on the role of facilitators, guiding students in navigating AI-powered platforms and interpreting the feedback these systems provide. As a result, their focus has shifted toward cultivating critical thinking and problem-solving skills—areas where human expertise and creativity remain irreplaceable. To meet the diverse needs of students in today's rapidly evolving educational landscape, teachers must go beyond traditional roles and embrace the opportunities offered by AI. By adopting the role of a mentor and integrating AI applications, they can transform learning into a more dynamic and personalized experience.

Tools like adaptive learning platforms, which adjust content to individual performance, and collaborative AI systems, which encourage teamwork and idea-sharing, enable teachers to tailor lessons to individual paces, manage cognitive load, and inspire students through personalized guidance. These innovations empower students to monitor their own progress, regulate their learning, and achieve meaningful, long-lasting success. Yet, in this AI-enhanced classroom, it is the teacher's empathy, motivation, and encouragement that truly breathe life into learning—qualities no machine can replicate. By combining the power of AI with their unique human touch, teachers can lead education into a future that is as innovative as it is deeply personal.

The effective use of AI tools in educational and instructional contexts necessitates the presence of skilled and professionally competent teachers. Achieving targeted learning outcomes requires precise commands, as well as the critical evaluation and filtering of generated content. Therefore, investing in teachers' professional development in AI technologies is essential. Moreover, the potential misuse of AI outputs without the application of inquiry, analysis, and critical thinking processes poses significant risks, particularly for students. Combinatorial problems, which are difficult to address through traditional methods, can be effectively solved using the intuitive methods provided by AI. For instance, classrooms equipped with sensors can analyze students' emotional states and contribute to classroom management. However, the over-controlling application of such technologies may disrupt the natural flow of learning processes. Thus, the integration of AI into educational settings must be carefully examined within the framework of ethical principles.

In educational settings, AI-powered adaptive technologies that enable the customization of learning content to meet individual student needs are significantly supporting teachers by enhancing teaching effectiveness, the retention of learning, and overall educational quality. For instance, AI-based platforms such as Course Hero, Gradescope AI, DeepTutor, CENTURY Tech, MEBI, and Carnegie Learning facilitate personalized learning experiences, thereby easing the workload for both teachers and students. Similarly, assessment tools like ZipGrade, Kahoot, and ExamSoft are part of the AI-driven solutions that contribute to improving teaching quality. A more specific example is the CENTURY platform, which analyzes students' strengths, weaknesses, knowledge gaps, and learning needs through formative assessment data and diagnostic tests, providing a personalized learning pathway. AI helps create the most suitable learning scaffold for students based on these assessments, allowing them to take ownership of their learning and explore areas for development. In addition, AI offers teachers data-driven insights into class and student progress, thereby reducing their workload and enabling more effective individualized support (Baker & Smith, 2019). Furthermore, online and hybrid learning platforms have introduced greater flexibility by making education independent of time and place. AI applications, such as plagiarism detection, grammar analysis, and content evaluation, have transformed teachers' ability to use digital tools in their professional development. Today, AI-powered tools are capable of detecting plagiarism with greater accuracy, automatically correcting grammar mistakes, and analyzing student performance with increased precision. These innovations have revolutionized educational processes, enhancing academic efficiency and accessibility (Durnalı & Çoban, 2025).

AIE assists educators in creating more personalized, flexible, inclusive, and engaging innovative teaching content. These AI-powered applications and platforms can provide teachers and students with data not only about what is being taught and learned but also about how learning occurs and the motivations behind it. Enhancing motivational well-being during the learning process can increase students' willingness to develop more advanced knowledge and skills. This atmosphere of well-being among students may, in turn, support teachers in designing more sophisticated learning environments.

The developments we once could not even imagine are now beyond the limits of our current conceptualization. Although the level of technological advancement varies across national education systems, we are witnessing a transition from simple tools like basic computers and Web 2.0 technologies to independent robots and collaborative robots (cobots)—machines that work alongside humans to assist in their tasks. This transition incorporates adaptive learning systems, intelligent tutoring systems, and other computer-embedded technologies that enhance both administrative processes and learning quality, including smart classrooms and educational cobots. These advancements are likely to have profound implications for education policy, and it appears we will increasingly debate their effects in the near future. Indeed, in the not-too-distant future, AI-supported and possibly humanoid robots are expected to interact directly with both teachers and students. Countries such as Japan, Switzerland, Singapore, and South Korea are already exploring how near-human robots can be effectively used in education.

As humanoid robots become more prevalent in schools in the coming years, the teaching process will evolve into a different dimension. This should not be interpreted as the complete replacement of teachers. If designed ethically and with well-defined roles, educational cobots can take on social roles in classrooms, akin to teaching assistants. They might act as "shadow teachers," capable of understanding the language of students and teachers, reading facial expressions and gestures, and recognizing emotions to collaborate effectively. However, it is important to acknowledge potential concerns about classroom management and autonomy. Evaluating this shift from the perspective of students born into a digital age and in the context of personalized learning and inclusivity would be more appropriate. For the digital-age generation, the integration of cobots into classrooms could serve as a highly motivating and engaging tool for learning. Imagine a classroom where multiple cobots interact with both teachers and students. For instance, a cobot could observe that a group of students in the far-right corner of the room is struggling to set up a scientific experiment. It might then provide feedback to the teacher or offer direct assistance to the students. Such an AI-powered system embodied in a cobot could join the group, assess the experimental setup, and provide personalized guidance or tutoring to facilitate learning.

Schools are inherently heterogeneous, complex, and dynamic institutions shaped by students from diverse socioeconomic backgrounds. This structural and functional diversity often manifests in significant disparities in school performance and quality, particularly in less developed or developing nations. Additionally, individual students differ markedly in their learning styles, pace, and openness to various domains of intelligence. Within this multifaceted context, both across schools and within classrooms, teachers can leverage AI-powered robots to tailor and differentiate their instructional approaches. These technologies enable students to engage with tasks at varying speeds and skill levels, thereby fostering an adaptive and inclusive learning environment. While such innovations may currently seem out of reach for many educational systems, they are poised to become standard practice in developed nations in the near future. When integrated with intelligent learning systems, collaborative robots (cobots) can actively monitor students, identify those requiring additional support, provide targeted interventions, enhance student engagement, and respond to individual questions. These technological advancements hold significant potential to improve learning outcomes for all students within a classroom setting. Moreover, these tools can substantially reduce teachers' workloads by automating routine tasks, thereby enabling educators to focus on implementing innovative teaching strategies, providing personalized guidance when needed, and gaining valuable insights into student learning processes (Baker & Smith, 2019). By delegating administrative and repetitive responsibilities to AI, teachers can dedicate more time to delivering individualized support, particularly to students who require additional motivation or assistance. However, the effective integration of AI tools into the classroom necessitates a balanced approach, ensuring that educators avoid over-reliance on technology. As a result, teachers remain pivotal in guiding students through the ethical dimensions of AI, including issues of bias, privacy, and digital literacy. The adoption of AI in education also requires teachers to engage in continuous professional development, equipping themselves with the skills to utilize AI tools effectively and understand their broader pedagogical implications.

To adapt to the rapidly evolving digital age, educators must prioritize enhancing students' AI literacy and awareness. Notably, evidence from China highlights the significant role AI courses play in fostering students' foundational competencies (Huang, 2021). Moreover, these courses not only bolster skill proficiency and cognitive competence but also promote self-directed learning, collaborative teamwork, and interactive engagement among students. This underscores the need for policymakers in education to prioritize cultivating AI awareness, particularly among teachers, school administrators, and students. AI awareness entails empowering students to create content that allows them to express themselves effectively within digital environments and resources. It also involves developing a holistic perspective on information, enabling them to synthesize new knowledge and systems effectively. Such awareness equips students to actively engage with the learning opportunities of the digital age, adapt to future work environments, and assert their roles as proactive participants in contemporary society.

Research underscores the rapid adoption and diverse applications of AI as a teaching tool and pedagogical resource among educators. While AI holds immense potential as a critical digital competence in the modern era—capturing significant global attention—it also faces substantial challenges. Chief among these are the lack of a well-defined theoretical framework, inadequate strategic planning by researchers and policymakers, and limited hands-on experience among educators. As Huang (2021) highlights, although several AI-related educational initiatives have been launched in countries such as China,

Singapore, Japan, and the United States, stark regional disparities persist. Furthermore, many of these initiatives rely on content that is often outdated, overly repetitive, and lacking in innovative approaches.

The integration of AI into teaching practices has profoundly reshaped the educational landscape, significantly improving the quality and efficiency of educational processes. For instance, school administrators frequently utilize AI-driven applications to streamline administrative tasks, while educators employ these tools to optimize grading, review student assignments, and prepare enriched instructional materials. However, the advent of AI-powered tools like ChatGPT and DeepSeak R1 has sparked concerns about their potential to undermine the originality of student work and complicate assessment processes. These apprehensions have led many educators to prefer traditional assessment methods, as recent research suggests (Rudolph et al., 2023). These developments highlight the critical role of policymakers in shaping the future of AI in education. To maximize its benefits, it is essential to integrate traditional pedagogical knowledge with innovative AI-supported teaching practices. Such an approach can foster learning environments that promote deep, meaningful, and lasting understanding for students, ensuring that AI serves as a tool to enhance, rather than detract from, the educational experience.

Teachers in overcrowded classrooms often struggle to fully grasp the complex dynamics of student engagement and understanding. Key processes, such as cognitive activation and content comprehension, are frequently overlooked due to time and resource constraints. Student learning, shaped by cognitive, emotional, and social factors, is inherently multifaceted. To address these challenges, educational technologies are evolving, with robotics emerging as a promising tool to enrich classroom experiences. Future educational cobots (collaborative robots) hold significant potential to support teachers by monitoring student progress and providing actionable feedback. These cobots, equipped with advanced features like facial and voice recognition, could interact with students in meaningful ways. For instance, a cobot might move through the classroom during project-based learning, offering real-time guidance. Specialized tools for subjects like mathematics or music could allow these robots to provide hands-on demonstrations, fostering deeper comprehension compared to traditional video-based instruction. Research highlights the promise of robotics in education. Robotics kits have proven effective for teaching STEM subjects (Hussain et al., 2024), while social robots have enhanced vocabulary and language acquisition (Tanaka & Matsuzoe, 2012). As these technologies advance, their integration into classrooms offers exciting possibilities, but thoughtful implementation will be key to ensuring they complement—and never replace—the vital role of human teachers.

One significant factor driving the shift in teacher roles is the preference of students—raised in a technologically driven world—for computer-assisted learning materials and, notably, interactions with robots. These students increasingly rely on AI tools as research assistants to support their academic work (Meade et al., 2023). The integration of AI as an educational assistant enables a personalized and optimized learning experience tailored to individual needs, which can significantly enhance students' willingness to learn. This willingness—a critical indicator of engagement and a fundamental source of motivation—is often diminished in traditional educational settings. For example, research shows that social robots, such as Robovie, can actively foster scientific curiosity by encouraging students to ask thoughtful questions and explore topics in depth (Tulsulkar et al., 2021). Traditional teaching practices often struggle to resonate with students immersed in digital technologies, making it harder to sustain their motivation and engagement. Integrating collaborative robots (cobots) into classroom settings provides an innovative solution, as these tools can capture students' attention, foster intrinsic motivation, and maintain their willingness to engage. For instance, cobots have been successfully employed in science classrooms to facilitate group experiments and guide students through complex problem-solving activities.

Another impactful strategy involves the use of AI tools capable of real-time adaptation based on student profiles, allowing instructors to address individual strengths and weaknesses dynamically. Additionally, educators should explore the potential of digital game-based learning environments, as well as immersive technologies like virtual reality (VR) and 3D modeling applications. Research has shown that students can learn subjects more effectively and with greater enjoyment when interactive games are integrated into the educational process. For instance, platforms such as Classcraft, Socratic or Duolingo exemplify the potential of gamified learning environments by fostering meaningful interaction between teachers and students (Johnson et al., 2021). These tools provide dynamic, interactive content that can transform both instructional and independent learning processes. For example, VR applications have been used in history lessons to transport students to ancient civilizations, fostering an emotional connection to the subject matter while enhancing retention (Gaurav et al., 2022). Similarly, 3D modeling has enabled students in STEM fields to visualize and interact with complex systems, such as molecular structures or engineering designs, in ways that traditional methods cannot achieve. By incorporating these cutting-edge technologies, teachers can create a learning environment that not only engages students but also prepares them for the demands of an increasingly digital world.

EQUITY IN EDUCATION WITH AI

The growing integration of artificial intelligence (AI) in education has significantly enhanced the skills of students, teachers, and school administrators, while simultaneously intensifying discussions about ethical concerns. Central to these concerns are issues surrounding the security of educational data and the protection of personal information. AI technologies now enable the transfer of detailed feedback—ranging from students' academic performance to their personal, emotional, and social contexts—into advanced systems such as humanoid robots. While this development appears promising—for instance, envisioning a teacher equipped with tools that assess students' emotional states and recommend tailored interaction strategies—it also poses substantial risks. AI-enabled tools, sensors, and humanoid robots integrated into smart classrooms generate real-time data streams from numerous sensors, providing extensive insights into student behavior. Although such systems have the potential to encourage teachers to adopt diverse instructional methods, they raise critical concerns regarding the misuse of data and threats to personal information security.

Another major criticism focuses on the perception that many AI-driven educational technologies reduce teachers to passive participants in opaque *black box* systems (Timms, 2016). This dynamic risks exacerbating power imbalances and perpetuating inequalities in education. For example, while AI systems may suggest alternative materials for classroom activities, questions arise regarding the appropriateness of these recommendations—particularly in multicultural classrooms or under-resourced schools. Would algorithm-generated suggestions adequately account for the unique needs and contexts of such environments? These uncertainties highlight the importance of critically evaluating AI's role in decision-making processes within education. Furthermore, there is concern that AI systems could inadvertently constrain teachers' professional agency and creativity. For instance, a teacher might deliberately design a sequence of simpler exercises to build students' confidence for motivational or pedagogical purposes. In contrast, AI systems, driven by algorithmic logic, might prioritize strictly didactic objectives, thereby limiting such flexibility. This restriction could negatively impact teachers' professional growth while disrupting the natural dynamics of classroom management. Over time, excessive reliance on AI in teaching may contribute to a decline in essential teacher skills (Berendt et al., 2020). Critics also emphasize the risks associated with inadequate AI literacy among teachers and students. Overreliance on automated decision-making systems could lead to significant educational and pedagogical challenges. Without thorough evaluation, schools may inadvertently adopt flawed AI tools under the assumption that they deliver superior insights or outcomes (Jones et al., 2020). Such missteps could undermine the effectiveness of educational practices and further complicate the integration of AI in learning environments.

Potential Risks in AIE

The integration of AI in education raises significant concerns, particularly regarding data privacy, security, and its impact on the roles of educators. AI systems, which collect sensitive data from students and teachers, are vulnerable to misuse and cyberattacks, necessitating the establishment of clear legal and ethical guidelines. Additionally, the unequal access to AI tools can deepen educational disparities, further exacerbating the digital divide. AI algorithms, if not carefully monitored, may reinforce biases based on gender, race, or socioeconomic status, resulting in discrimination and unfair treatment. Over-reliance on AI could also diminish essential human interaction, limiting opportunities for socialization and impeding the development of critical thinking skills. This shift could also erode the unique value that teachers provide in terms of mentorship, emotional support, and adaptability to students' individual needs, potentially leading to a standardized approach to education that overlooks the diverse capabilities and creativity of students. Moreover, the increasing automation of educational tasks raises concerns about the future role of teachers and job security. As AI-driven platforms take over lesson delivery, assessment, and personalized learning, the human connection in education may weaken, with AI systems failing to address the emotional and psychological needs of students. In addition, the surveillance-based evaluation systems that some AI models employ may infringe upon students' rights and freedoms, creating ethical dilemmas. Technical malfunctions or errors within AI systems can disrupt learning, and an overdependence on technology could present long-term challenges, both for individuals and broader societal structures. While AI has the potential to enhance certain aspects of education, it must complement, rather than replace, the irreplaceable role of teachers in fostering critical thinking, creativity, motivation, and holistic student development.

Future Direction

The rapid evolution of AI in the digital era continues to expand the boundaries of innovation across diverse fields. One of its emerging and transformative subsets, AIE is an evolving domain that holds significant promise yet remains fraught with complexities. Despite its potential, our understanding of AIE is still developing, particularly regarding its opportunities and the challenges it poses (Heffernan & Heffernan, 2014). Although current applications of AIE are in their nascent stages, they are steadily approaching maturity. However, this progress is accompanied by ethical dilemmas and skepticism from educators and school administrators, who often resist integrating novel practices into traditional educational systems. On one hand, AIE offers remarkable potential to foster creativity, independent thinking, and critical reasoning by transforming conventional learning opportunities (Richardson & Milovidov, 2019). On the other hand, it has been criticized for stifling originality and perpetuating a mechanized, "robotized" approach to education. The duality of these perspectives underscores the importance of critically evaluating the role of AIE in the educational landscape. Notably, research indicates that educators who are innovative and collaborative tend to embrace AIE more readily, integrating it into their teaching practices to diversify instructional methods. This growing interest is reflected in an increasing body of scholarly work exploring AIE's implications for teaching and learning. As educational management researchers and forward-thinking education systems pioneer new approaches, valuable insights are emerging about how AI might be seamlessly integrated into future classrooms.

In the near future, classrooms may incorporate robots capable of automating routine tasks, such as navigating between desks, gathering student feedback, and managing educational tools. While these intelligent systems could revolutionize classroom dynamics, their potential impact on the roles of teachers and schools continues to spark heated debate among educators. It is essential for institutions to define clear parameters for AIE applications, distinguishing those that enhance learning from those that require additional oversight or raise ethical concerns. For traditional education systems that view teachers primarily as transmitters of knowledge, the rise of AIE may seem daunting. The era of education centered solely on the possession and dissemination of information is rapidly fading, as AI now provides unparalleled access to data and its presentation. Within this shifting paradigm, teachers must adapt by embracing innovative practices that highlight their uniquely human abilities—empathy, emotional connection, and mentorship—which no machine can replicate. Additionally, educators are expected to take on broader roles, such as guiding students in developing digital literacy and navigating the risks of hyper-digital environments, including the phenomenon of *digital obesity*.

Looking ahead, AIE is poised to complement, rather than replace, educators. By automating administrative and repetitive tasks, AIE could free teachers to focus on cultivating creativity, emotional intelligence, and higher-order thinking skills in their students. This anticipated paradigm shift emphasizes collaboration, where technology and humanity work in harmony to reimagine education for the future. As we navigate this transformation, the human element in teaching will remain indispensable, ensuring that education continues to inspire, connect, and empower.

Let's Reflect on...

- Can AI replace teachers, or should it be seen as a complementary tool to their roles?
- What new skills will teachers need with the integration of AI into education processes?
- How does the teacher's role in addressing students' emotional and social needs differ from the capabilities of AI?
- Can AI-supported tools reduce inequalities in education, or do they risk creating new forms of inequity?
- How does teachers' collaboration with AI affect the student-teacher relationship?

The increasing integration of artificial intelligence into education has sparked complex reflections on the evolving roles of teachers and the ethical, emotional, and social dimensions of learning. As AI systems become more capable of automating routine tasks, providing real-time feedback, and delivering personalized instruction, the teacher's role is not diminished but rather transformed. Educators are no longer mere transmitters of knowledge but facilitators of critical thinking, ethical reasoning, and emotional intelligence—domains where AI remains limited. The reflective questions presented underscore a fundamental tension: while AI can optimize learning processes, it cannot replicate the nuanced human relationships, moral judgment, or empathetic connections that are essential to meaningful education. Teachers must now navigate this dual landscape—leveraging AI to enhance efficiency and engagement while safeguarding the human elements that foster trust, curiosity, and intrinsic motivation. This transition demands new competencies, including AI literacy, digital pedagogy, and a robust ethical framework. Moreover, it calls for inclusive strategies that ensure AI tools do not exacerbate existing inequalities or marginalize vulnerable learners. Teachers' collaboration with AI should thus be intentional, critically informed, and student-centred. Rather than viewing AI as a threat, educators can position it as a powerful partner—provided its use is guided by pedagogical purpose and human values. In this evolving educational paradigm, it is the teacher's ability to think reflectively, act ethically, and connect authentically that will remain the most valuable force in shaping the future of learning.

CONCLUSION

The discourse around AI swings between two extremes: it is either celebrated as humanity's rising star or feared as a harbinger of its decline (Barrat, 2023, p. 39). AI captivates audiences with its promise, even as its ethical dilemmas and unpredictable advancements raise concerns. The educational sector similarly grapples with this duality. On one side are the opportunities AI offers, such as enhanced personalization and efficiency; on the other are challenges like digital security and the growing reliance on technology, sometimes referred to as digital obesity. Central to this transformative change is how the role of teachers will be redefined.

AI's influence on educational organizations is growing rapidly. Research suggests that a cautious and ethically grounded classroom environment, supported by robust security protocols, can enhance teaching and learning. The day when educational robots assist teachers in schools seems imminent. In response, educators are expected to adopt innovative instructional practices that foster cognitive engagement and adapt to the needs of students born into a world of technology. As traditional teaching methods are questioned, interactive and AI-supported instructional practices seem poised to become a necessary step forward for educators.

AI-assisted systems reduce teachers' workloads, streamline administrative tasks, and provide personalized support that fosters dynamic and inclusive learning environments. However, these advancements also reshape teachers' responsibilities and professional development needs. Teachers must receive training in AI literacy and the critical use of digital technologies. They also have a new responsibility: raising students' awareness about AI's development and its potential risks. Educational research must further evaluate the compatibility of AI applications with educational goals and ensure they align with ethical and moral standards. National education systems, too, must develop policies that establish legal and pedagogical foundations for robust digital learning environments. As AI transforms education, stakeholders must prioritize not only the opportunities it offers but also the ethical and practical frameworks necessary for its responsible implementation. Only then can AI truly enhance the learning landscape for sustainable education.

REFERENCES

Bahroun, Z., Anane, C., Ahmed, V., & Zacca, A. (2023). Transforming education: A comprehensive review of generative artificial intelligence in educational settings through bibliometric and content analysis. *Sustainability (Basel)*, *15*(17), 12983. DOI: 10.3390/su151712983

Baker, T., & Smith, L. (2019). *Educ-AI-tion Rebooted? Exploring the future of artificial intelligence in schools and colleges*. https://media.nesta.org.uk/documents/Future_of_AI_and_education_v5_WEB.pdf

Balderson, K. (2023). AI In Education statistics you should know. https://mspoweruser.com/ai-in-education-statistics

Baltezarević, R., & Baltezarević, I. (2024). Students' attitudes on the role of artificial intelligence (AI) in personalized learning. *International Journal of Cognitive Research in Science* [IJCRSEE]. *Engineering and Education*, *12*(2), 387–397. DOI: 10.23947/2334-8496-2024-12-2-387-397

Barrat, J. (2023). *Our final invention: Artificial intelligence and the end of the human era*. Hachette UK.

Bearman, M., & Ajjawi, R. (2023). Learning to work with the black box: Pedagogy for a world with artificial intelligence. *British Journal of Educational Technology*, *54*(5), 1160–1173. DOI: 10.1111/bjet.13337

Belpaeme, T., Kennedy, J., Ramachandran, A., Scassellati, B., & Tanaka, F. (2018). Social robots for education: A review. *Science Robotics*, *3*(21), eaat5954. Advance online publication. DOI: 10.1126/scirobotics.aat5954 PMID: 33141719

Berendt, B., Littlejohn, A., & Blakemore, M. (2020). AI in education: Learner choice and fundamental rights. *Learning, Media and Technology*, *45*(3), 312–324. DOI: 10.1080/17439884.2020.1786399

Bourdeau, J., & Grandbastien, M. (2010). Modeling tutoring knowledge. In *Advances in intelligent tutoring systems* (pp. 123–143). Springer Berlin Heidelberg. DOI: 10.1007/978-3-642-14363-2_7

Brynjolfsson, E., & McAfee, A. (2014). *The second machine age: Work, progress, and prosperity in a time of brilliant technologies*. WW Norton & company.

Chen, X., Xie, H., Zou, D., & Hwang, G. J. (2020). Application and theory gaps during the rise of artificial intelligence in education. *Computers and Education: Artificial Intelligence*, *1*, 100002. DOI: 10.1016/j.caeai.2020.100002

Durnalı, M., & Çoban, Ö. (2024). Artificial intelligence in higher education management: Innovations, opportunities and challenges. In González Lezcano, R. A., & Orakçı, Ş. (Eds.), *Innovation trends and educational technology in higher education* (pp. 501–516). Information Age Publishing.

Gaurav, D., Kaushik, Y., Supraja, S., Yadav, M., Gupta, M. P., & Chaturvedi, M. (2022). Empirical study of adaptive serious games in enhancing learning outcome. *International Journal of Serious Games*, *9*(2), 2. Advance online publication. DOI: 10.17083/ijsg.v9i2.486

Heffernan, N. T., & Heffernan, C. L. (2014). The ASSISTments ecosystem: Building a platform that brings scientists and teachers together for minimally invasive research on human learning and teaching. *International Journal of Artificial Intelligence in Education, 24*(4), 470–497. DOI: 10.1007/s40593-014-0024-x

Holmes, W., Bialik, M., & Fadel, C. (2019). *Artificial intelligence in education: Promises and implications for teaching and learning*. Center for Curriculum Redesign.

Huang, X. (2021). Aims for cultivating students' key competencies based on artificial intelligence education in China. *Education and Information Technologies, 26*(5), 5127–5147. DOI: 10.1007/s10639-021-10530-2

Hussain, S., Lindh, J., & Shukur, G. (2024). The effect of LEGO training on pupils' school performance in mathematics, problem solving ability and attitude: Swedish data. *Journal of Educational Technology & Society, 9*(3), 182–194.

Järvelä, S., Nguyen, A., & Hadwin, A. (2023). Human and artificial intelligence collaboration for socially shared regulation in learning. *British Journal of Educational Technology, 54*(5), 1057–1076. DOI: 10.1111/bjet.13325

Jones, K. M. L., Rubel, A., & LeClere, E. (2020). A matter of trust: Higher education institutions as information fiduciaries in an age of educational data mining and learning analytics. *Journal of the Association for Information Science and Technology, 71*(10), 1227–1241. DOI: 10.1002/asi.24327

Kamalov, F., Calonge, D. S., & Gurrib, I. (2023). New era of Artificial intelligence in education: Towards a sustainable multifaceted revolution. *Sustainability (Basel), 15*(16), 1–27. DOI: 10.3390/su151612451

Kaplan, J. (2015). *Humans need not apply: A guide to wealth and work in the age of artificial intelligence*. Yale University Press.

Liu, M., Ren, Y., Nyagoga, L. M., Stonier, F., Wu, Z., & Yu, L. (2023). Future of education in the era of generative artificial intelligence: Consensus among Chinese scholars on applications of ChatGPT in schools. *Future in Educational Research, 1*(1), 72–101. DOI: 10.1002/fer3.10

Markauskaite, L., Marrone, R., Poquet, O., Knight, S., Martinez-Maldonado, R., Howard, S., Tondeur, J., De Laat, M., Buckingham Shum, S., Gašević, D., & Siemens, G. (2022). Rethinking the entwinement between artificial intelligence and human learning: What capabilities do learners need for a world with AI? *Computers and Education: Artificial Intelligence, 3*, 100056. DOI: 10.1016/j.caeai.2022.100056

Meade, M., Ogundamisi, R., & Tiratelli, M. (2023). Stay calm and educate. *ChangeMaker-funded project investigating SRI students' perspectives on generative AI.*https://www.ucl.ac.uk/teachinglearning/sites/teaching_learning/files/student_perspectives_on_ai_report.docx

Miao, F., Holmes, W., Huang, R., & Zhang, H. (2021). *AI and education: A guidance for policymakers.* UNESCO Publishing.

Nwana, H. S. (1990). Intelligent tutoring systems: An overview. *Artificial Intelligence Review, 4*(4), 251–277. DOI: 10.1007/BF00168958

Richardson, J., & Milovidov, E. (with Europarat). (2019). *Digital citizenship education handbook: Being online, well-being online, rights online.* Council of Europe.

Rudolph, J., Tan, S., & Tan, S. (2023). ChatGPT: Bullshit spewer or the end of traditional assessments in higher education?. *Journal of applied learning and teaching, 6*(1), 342-363.

Rus, V., D'Mello, S., Hu, X., & Graesser, A. (2013). Recent advances in conversational intelligent tutoring systems. *AI Magazine, 34*(3), 42–54. DOI: 10.1609/aimag.v34i3.2485

Schwab, K. (2017). *The Fourth Industrial Revolution.* Crown Business.

Tanaka, F., & Matsuzoe, S. (2012). Children teach a care-receiving robot to promote their learning: Field experiments in a classroom for vocabulary learning. *Journal of Human-Robot Interaction, 1*(1), 78–95. DOI: 10.5898/JHRI.1.1.Tanaka

Timms, M. J. (2016). Letting artificial intelligence in education out of the box: Educational cobots and smart classrooms. *International Journal of Artificial Intelligence in Education, 26*(2), 701–712. DOI: 10.1007/s40593-016-0095-y

Tulsulkar, G., Mishra, N., Thalmann, N. M., Lim, H. E., Lee, M. P., & Cheng, S. K. (2021). Can a humanoid social robot stimulate the interactivity of cognitively impaired elderly? A thorough study based on computer vision methods. *The Visual Computer, 37*(12), 3019–3038. DOI: 10.1007/s00371-021-02242-y PMID: 34345091

VanLehn, K. (2011). The relative effectiveness of human tutoring, intelligent tutoring systems, and other tutoring systems. *Educational Psychologist, 46*(4), 197–221. DOI: 10.1080/00461520.2011.611369

Yang, S. J. H., Ogata, H., Matsui, T., & Chen, N.-S. (2021). Human-centered artificial intelligence in education: Seeing the invisible through the visible. *Computers and Education: Artificial Intelligence, 2*, 100008. DOI: 10.1016/j.caeai.2021.100008

Zohuri, B., & Mossavar-Rahmani, F. (2024). Revolutionizing education: The dynamic synergy of personalized learning and artificial intelligence. *International Journal of Advanced Engineering and Management Research, 9*(1), 143–153. DOI: 10.51505/ijaemr.2024.9111

Chapter 3
Building Smarter Educators:
Merging Cognitive and Emotional Intelligence in Teacher Development

Vijendra Nath Pathak
https://orcid.org/0000-0003-0731-7003
Parul University, India

Elizabeth Kardze Annan
https://orcid.org/0009-0004-7935-8431
School of Social Science and Language, Lovely Professional University, India

ABSTRACT

The chapter, "Building Smarter Educators: Merging Cognitive and Emotional Intelligence in Teacher Development," emphasises the integration of cognitive and emotional intelligence into teacher training. It highlights how these competencies help teachers recognise and address students' needs, create effective lesson plans, mentor and inspire students, manage conflicts, and cope with professional challenges, leading to reduced burnout, job satisfaction, and student success. Strategies discussed include mentoring, collaborative enquiry, mindfulness, SEL programmes, peer coaching, reflective practices, and leveraging technology. Challenges like resource constraints, cultural differences, and balancing personal and professional growth are addressed with solutions such as continuous evaluation, stress management, and fostering supportive environments. The chapter underscores the need for emotionally and cognitively intelligent teachers to foster student-centred learning and enhance academic and professional outcomes.

INTRODUCTION

A teacher or an educator is a trained professional who teaches or educates pupils usually in a school setting such as in school or university. Teachers play a significant role in students' learning and achievements (Bardeck & Klassan, 2020), whether in academia, personal, or professional life (Chehi,2012). The role of teachers is no longer confined to only developing the cognitive aspect of learners but now includes developing students' emotional, character, and personality aspects (Boyd,2005). According to Chechi (2012, Pathak, 2024), in this modern century, students need to be emotionally intelligent and flexible learners proficient in literacy, numeral, communication, and self-management, so teachers must

DOI: 10.4018/979-8-3373-0472-4.ch003

possess specific attributes that will equip them to teach emotional and social skills. This calls for teachers who are self-aware, empathetic, and able to deal with distressing situations (Saxena & Sexena, 2012). Also, teachers should be knowledgeable, effective communicators, have a well-organized pedagogy, motivational and inspirational skills, cordial character, and good classroom management skills. This points to the importance of teachers possessing both high cognitive and emotional intelligence levels. According to Siddique et al. (2020), teaching is a demanding field requiring teachers to possess specific skills and competencies to ensure an effective teaching process. They went on to mention that the overall development (such as the physical, cognitive, emotional, and intellectual) of students is dependent on the teaching process. Cognitive intelligence is the mental ability of a person to acquire and understand knowledge or information to reason and judge well. While emotional intelligence, on the other hand, refers to the ability of a person to understand and manage the emotions of both oneself and that of others. A teacher should be able to recognize and understand the learning motivation and behaviors of students which could enable the teacher to create an ideal learning environment fostering social inclusion and interaction, active engagement, and motivation to learn (Jha & Singh, 2012; Siddique et al.,2020), which could improve the learning attitudes of students and their academic performance. According to Guskey (2002), teacher development is essential to promote highly effective teaching and students' achievements, and the growing need for teachers' continuous learning and upgrading their professional skills by gaining more knowledge and getting abreast with current trends in teaching (Avdor,2008). Research shows that teacher development can be achieved through teacher training programs and teachers' professional development programs.

According to Learning Matters (n.d), a teacher training program is an online or offline program designed to equip teachers with abilities and pedagogy skills to help them teach, manage, and connect with students, which would result in positive students learning and achievements. These programs help teachers to grow professionally as they gain the opportunities to learn and adapt new ways, methods, strategies, skills, and tools, which helps to produce self-efficacious and motivated teachers. Also, it equips teachers with the capacity to recognize and understand students which helps them to implement effective measures to improve the learning process of students.

Agarwal (2024) defined Teachers' Professional Development (TPD) as a continuous practice of reflection, learning, and action process designed to equip teachers with the knowledge, refine their skills, and adapt modern teaching methods that meet the ever-evolving needs of students. The benefits of Teachers' Professional Development are; It promotes growth, challenge, and skill development for teachers as they get the opportunities to learn and acquire knowledge and skills needed to navigate the new educational policies, new technological advancements, and diverse student body that has changed the education field. This helps teachers integrate technology into the classroom, help implement new curriculum standards, and address the diverse learning styles and behaviors of students. It also equips teachers to deal with challenges effectively, collaborate with colleagues, and witness students' improvement in learning which contributes to job satisfaction and fulfillment, a new-found purpose with their profession, and less burnout as they are able to deal with stress. Teachers' Professional Development programs provide teachers with research evidence-based instructional strategies to enhance student learning. These strategies can develop students' critical thinking abilities, foster an in-depth understanding of study materials, and improve academic outcomes. The program fosters a collaborative learning environment for teachers, where they interact, connect, share, and learn from each other's expertise and experiences. This practice can also be incorporated in the classroom, where teachers guide students to work in teams, which creates a conducive environment for learning and support. Professional Development for teachers helps build confident and self-efficacious teachers who believe in their abilities to handle demanding situations, implement innovative approaches, and inspire students to embrace risks and challenges, which could lead to a positive learning environment.

This calls for the need for teachers to possess the ability to face cognitive and emotional challenges effectively (Day et al., 2007) because their profession is a highly social and emotional field (Hargreaves, 1998; Sutton and Wheatley, 2003 as cited in Dolev & Leshan, 2016). It is important to integrate cognitive and emotional intelligence in teacher development to build more brilliant educators for the benefit of students and the progress of the education sector.

UNDERSTANDING COGNITIVE INTELLIGENCE AND ITS ROLE IN TEACHER DEVELOPMENT

Intelligence refers to the ability to learn and acquire knowledge or information and use it to solve problems, adapt to one's environment, and think rationally or critically.

According to Cambridge Dictionary (n.d), intelligence is the capacity to learn, understand, judge, or give reasonable opinions. The multiple intelligences theory by Howard Gardner draws our attention to the fact that intelligence is not a single factor but consists of multiple factors that should be considered. His theory proposed nine intelligences which are:

- Logical-mathematical intelligence: This refers to the ability to solve mathematical and logical problems. It helps develop and solve numerical equations and abstract problems.
- Linguistic intelligence: It deals with spoken and written language abilities. Hence, it helps to learn languages quickly.
- Visual-spatial intelligence: This type of intelligence deals with visualising, recognising, and manipulating patterns in space.
- Body-kinesthetic intelligence: This is a type of intelligence that involves the ability to utilise one's entire or parts of the body (body-mind coordination) to solve problems.
- Musical intelligence: This type helps to identify, create, and use musical patterns such as pitches, rhythm, tones, and tunes.
- Intrapersonal intelligence: This intelligence helps to be aware and understand oneself better by appreciating one's strengths and weaknesses.
- Interpersonal intelligence is understanding human dynamics, such as people's emotions, motivations, and ambitions.
- Naturalistic intelligence refers to the ability to recognise and understand nature and the environment. Existential intelligence: It is the capacity to ponder on and answer deep questions about human existence and things beyond human understanding (example, the presence of a supreme being).

Inmaculada et al. (2021) referred to cognitive intelligence as general mental ability or general intelligence, defined as the ability to learn a task, subject, or skill precisely and quickly under optical instruction. This is similar to Spearman's two-factor theory's general factor (g-factor). The g-factor in this theory refers to the central or single factor that enables a person to perform tasks in various areas. One's ability to perform a mathematical task well is expected to perform equally well in a logical reasoning task.

Lungu (2019) defined cognitive intelligence as the ability of a person to learn and process information, knowledge or data and also be able to adapt to social and cultural situations.

Theories of Cognitive Intelligence

Vygotsky Sociocultural Theory (1978 as cited in McLeod, 2024): The theory posits that social and cultural factors influence cognitive development. One's social interaction shows the level of development of mental abilities (for example, the language and reasoning abilities of the individual). McLeod (2024) defined cognitive development as a social myalgia process where a person's values, beliefs, and problem-solving abilities are acquired through social interaction with more knowledgeable members of society, including parents, teachers, or elders who can provide guidance and modelling. Vygotsky's theory proposes a collaborative and cooperative learning relationship between children and their teachers or peers. With this, he proposed two educational strategies: Zone of Proximal Development (ZPD), which is synonymous with scaffolding (it was coined by Wood, Bruner, & Ross in 1976 as cited in McLeod). ZPD or Scaffolding is when teachers create and provide support structures for students to help them master specific skills. The second strategy is reciprocal teaching (where the teacher and students take turns to lead discussions or teaching using methods such as summarizing and clarifying subject matters). Vygotsky mentioned that cognitive development does not centre around developing basic mental abilities such as attention and memory but involves language and communication, logical reasoning, problem-solving, planning, self-regulation, and meta-cognition (as cited in McLeod,2024).

Cattell-Horn-Carroll (CHC) Theory of Cognitive Abilities (Flanagan & Dixon, 2014): According to (Flanagan & Dixon, 2014), this theory is the most comprehensive and empirically supported psychometric theory of the structure of cognitive abilities to date. Raymond Cattell first developed the theory in the 1940s when he proposed the Fluid-Crystallized (Gf-Gc) theory. Fluid intelligence refers to an individual's inductive and deductive reasoning abilities, usually influenced by in-born attributes or through learning from social interaction with one's environment. In contrast, crystallised intelligence is the acquired knowledge abilities of an individual, which are influenced by socialisation (Cattell, 1957, 1971). According to Horn (1968 as cited in Flanagan & Dixon, 2014), later in 1965, John Horn contributed to the theory by introducing the 8-factor model which includes *Visual perception or processing* (Gv), *Short-term memory or Short-term acquisition and retrieval* (SAR or Gsm), *Long-term storage and retrieval* (Glr), *Speed of processing* (Gs), *Auditory processing ability* (Ga), *Reaction time and Decision speed* (Gt), *Quantitative factor* (Gs), and *Broad reading-writing factor* (Grw), Then, John Carroll expanded on the theory by developing the Three-Stratum Theory (Carroll, 1997). Stratum I was termed *broad abilities* (this refers to individualised fundamental traits that impact our behaviours in specific domains; these skills vary in their emphasis on procedures, subject matter, and their way of accountability). This stratum includes fluid and crystallized intelligence and approximately 70 other cognitive abilities. Stratum II or *narrow abilities* (they represent the specific abilities that reflect the influence of learning and experience or the particular strategies adopted by an individual for maximum performance). Finally, Stratum III, which he termed the general factor is similar to Spearman's, which states that a single cognitive ability is responsible for higher-order cognitive processes such as meta cognition and learning (Carroll, 1997).

Information Processing Theory (Cuncic, 2023): This theory discusses the flow of information in the human mind, focusing on the mental processes involved in acquiring, storing, and processing knowledge. According to the theory, the human mind operates like a computer information processor. The model developed for this theory proposes that cognitive processes occur in four stages which are the encoding stage, storage stage, retrieval stage, and transformational stage.

The Role of Cognitive Intelligence in Teacher Development

Using the taxonomy for teaching, learning, and assessment (2021), a revised version of Bloom's Taxonomy, also known as the Taxonomy of Educational Objectives, proposed six major categories of educational objectives: knowledge, comprehension, application, analysis, synthesis, and evaluation (Bloom et al., 1956, as cited in Armstrong, 2010).

Knowledge: It "involves the recall of specifics and universals, the recall of methods and processes, or the recall of a pattern, structure, or setting."

Comprehension: It "refers to a type of understanding or apprehension such that the individual knows what is being communicated and can use the material or idea being communicated without necessarily relating to other material or seeing its fullest implications."

Application refers to the "use of abstractions in particular and concrete situations".

Analysis: It is the "breakdown of a communication into its constituent elements or parts such that the relative hierarchy of ideas is made clear and/or the relations between ideas expressed are made explicit".

Synthesis: It is the "putting together of elements and parts to form a whole".

Evaluation: It involves the "judgements about the value of material and methods for given purposes". (Taxonomy of Educational Objectives. Handbook One, pp. 201-207 (Armstrong, 2010)

According to Armstrong (2010), the revised taxonomy, developed in 2001 by a group of researchers, redefined Bloom's original framework by introducing a hierarchy of action-oriented cognitive processes. These processes are described as mental operations through which individuals engage with and apply knowledge. The six categories in this hierarchy are: Remember, Understand, Apply, Analyze, Evaluate, and Create. "Remember" involves recognizing and recalling previously learned information. "Understand" refers to the ability to interpret, exemplify, classify, summarize, infer, compare, and explain ideas or concepts. "Apply" encompasses executing or implementing procedures in various contexts. "Analyze" includes differentiating, organizing, and attributing knowledge to understand relationships among parts. "Evaluate" refers to making judgments based on criteria and standards through checking and critiquing. Finally, "Create" involves generating new ideas, planning processes, and producing original outputs or products based on synthesized knowledge.

In addition to this taxonomy of cognitive processes, a separate classification was developed to describe the types of knowledge utilized in cognition. These knowledge categories serve as the foundation upon which the cognitive processes operate. "Factual Knowledge" includes basic elements such as terminology and specific details relevant to a subject. "Conceptual Knowledge" involves understanding classifications, categories, theories, models, and structures that organize information. "Procedural Knowledge" encompasses the know-how required to perform tasks, including subject-specific methods, techniques, and criteria for applying procedures appropriately. Finally, "Metacognitive Knowledge" refers to strategic thinking, knowledge of cognitive tasks, self-awareness, and the ability to monitor and regulate one's own learning processes. Together, these categories illustrate how diverse cognitive operations and types of knowledge interact in educational settings, offering a structured approach to curriculum design, teaching strategies, and assessment.

These knowledge distinctions show why some teachers are better at explaining concepts and theories than others, while others are best at using practical examples to explain concepts. The teaching profession is a complex and intellectually demanding field that requires teachers to possess certain cognitive abilities. Additionally, the educational strategies proposed by Vygotsky in his sociocultural theory will be used to describe the role of cognitive intelligence in teacher development: the zone of proximal development or Scaffolding and residual teaching.

Zone of proximal development or scaffolding: The zone of proximal development is the area where a child receives the required tools, instructions or guidance, which aids the child in developing specific skills that they can perform independently after a while. It can also improve the child's mental functioning (Vygotsky, 1978). Zone of proximal development (ZPD) is the difference between actual and potential learning, which further suggests that collaborative and cooperative learning with teachers could help model certain behaviours and guide students' performance. Scaffolding is a structure erected to support workers reaching certain heights of a building during construction, which is later removed after work. The same principle applies in this situation, where teachers structure personalized activities to support students in reaching their potential learning, passing through the zone of proximal development. The support can vary as it is supposed to be tailored to meet the needs of the specific student or students, for example, modelling behaviour or asking questions. Scaffolding takes place in two stages: The contingency or responsiveness stage (this is where the teacher introduces the activities and continuously monitors the student's progress in understanding and performing the activities, making necessary adjustments accordingly), and the Withdrawal stage (where the support is withdrawn when a student has mastered the task, thus, can complete the task independently)

Residual teaching: This strategy focuses on collaboration and cooperation between teachers and students in teaching-learning. With this method, the teachers and their students take turns learning and practising four primary skills: summarizing, questioning, clarifying, and predicting (McLeod, 2024). It is a two-way instructional strategy to encourage students to engage in teaching-learning (Driscoll, 1994; Hausfather, 1996, as cited in McLeod, 2024). The teachers and their students participate in group discussions on reading, where the teacher first models the four key skills, and then the students take up the teaching role. Hausfather (1996) suggested another strategy based on Vygotsky's theory called cognitively guided instruction. It involves teachers and students solving mathematical problems and then sharing their problem-solving strategies in an open discussion (Hausfather, 1996).

Teachers can take advantage of this theory as it fosters a student-centred, collaborative, and cooperative approach to teaching and learning between teachers and their students. The structure of the classroom and study materials should be designed to encourage student engagement and interaction, creating a positive classroom atmosphere for learning.

EXPLORING THE CONCEPT OF EMOTIONAL INTELLIGENCE

"Emotions are personal and subjective experiences caused by the complex interplay between the physical, cognitive, and situational variables (Mandel, 2024; Jha & Singh, 2012, p.668). According to Goleman (1996), emotions are motivational mechanisms that have the optimal power to alter the way we perceive things, memorize, and the thinking processes involved in achieving our individual goals. Jha & Singh defined emotional Intelligence as the ability to recognize and understand one's own emotions and those of others which helps to regulate or manage these emotions to motivate oneself and others to achieve specific goals. Lungu (2019) defined emotional intelligence as the capacity to be aware, understand, and accept one's and that of others and manage these emotions positively.

Theories of Emotional Intelligence

Mayer & Salovey Emotional Intelligence Four-branch Model (1997): According to Mayer, Salovey, & Caruso (2008), emotional intelligence refers to the process of recognizing and using one's emotional sensations and that of others to guide one's thought process and behavior. Mayer and Salovey (1997) developed the four-branch model, which considered specific abilities. These abilities are a) the ability to perceive emotions in oneself and others accurately; b) the ability to use emotions to facilitate thinking; c) the ability to understand emotions, emotional language, and emotional signals; and d) the ability to manage emotions to attain specific goals.

Goleman's Emotional Intelligence theory (1998): Goleman (1998) mentioned that a person is recently being judged for a job opportunity not just on his or her intelligence or expertise and training but also on how an individual can handle himself or herself and others effectively. He argued that one must possess intellectual abilities, technical abilities, and personal attributes such as initiative, empathy, adaptability, and persuasiveness. He stated that great people tend to possess both emotional and cognitive intelligence because emotional intelligence correlates with cognitive intelligence. A more complex job demands emotional intelligence from tier workers. Goleman's theory posits that people need emotional competencies to maximize their potential. He designed a framework that entails twenty-five (25) emotional competencies, which he categorized into five (5) main groups. He further grouped the five categories into two main types, which are personal competencies (they determine how one manages one's own emotions; the first three main groups fall under this) and social competencies (determine how we relate with others; the last two main groups fall under this). Goleman's model of emotional intelligence includes five key categories that reflect essential emotional and social competencies. Self-awareness involves recognizing and understanding one's internal emotional states, preferences, and intuitions, which includes emotional awareness, accurate self-assessment, and self-confidence. Self-regulation refers to managing one's internal states, impulses, and behaviors, including self-control, trustworthiness, conscientiousness, adaptability, and openness to innovation. Motivation encompasses the internal drive to achieve goals, reflected in an individual's achievement drive, commitment, initiative, and optimism. Empathy is the ability to be aware of and understand the feelings, needs, and concerns of others, which includes understanding others, supporting them, embracing diversity, and recognizing power dynamics in relationships. Lastly, social skills enable individuals to manage relationships effectively and include abilities such as influence, effective communication, conflict management, leadership, initiating positive change, building strong interpersonal bonds, cooperating with others, and being an effective team player. These interconnected competencies contribute to emotional intelligence in both personal and professional contexts.

Bar-On Emotional-Social intelligence model (2006): Bar-on defined emotional-social intelligence as the collection of emotional and social competencies, skills, and facilitators that influence one's ability to recognise, understand, and manage emotions, to build relationships with others, to adapt to change, to solve intrapersonal and interpersonal problems, and to cope well with daily demands, challenges, and pressures. The Bar-On model of emotional intelligence categorizes fifteen emotional competencies into five main factors that collectively describe a person's emotional and social functioning. The first factor, intrapersonal skills, refers to self-awareness and self-expression, including self-regard (the ability to perceive and accept oneself), emotional awareness (recognizing and understanding one's emotional state), assertiveness (expressing emotions and thoughts effectively), independence (being emotionally self-reliant), and self-actualization (striving to achieve personal goals and realize one's full potential). The second factor, interpersonal skills, relates to social awareness and relationship-building, encompassing empathy (awareness of others' feelings), social responsibility (contributing positively within a group), and interpersonal relationships (forming and maintaining meaningful social bonds). Stress management is the third category, focusing on emotional regulation through stress tolerance (managing stress and sadness) and impulse control (controlling emotional reactions). The fourth factor, adaptability, concerns change management and includes reality-testing (evaluating emotions and thoughts against reality), flexibility (adjusting to changing conditions), and problem-solving (effectively handling personal and social challenges). The final factor, general mood, reflects one's overall emotional outlook and includes optimism (maintaining a positive mindset) and happiness (feeling satisfied and content with oneself, others, and life). These five dimensions together provide a comprehensive understanding of how individuals perceive, manage, and apply emotional intelligence in their personal and professional lives.

The Role of Emotional Intelligence in Teacher Development

According to Saxena & Saxena (2012), teachers have the opportunity to engage students daily, which puts them in a better position to recognise and understand the emotional states of students as well as motivate them to express their feelings, needs, or concerns. In doing this, teachers can identify mentally unstable students and provide them with the necessary attention and support, which helps to solve mental health conditions early on. Additionally, emotional intelligence equips teachers to be aware and control their emotions effectively, not only as role models but also to show students how to appropriately express their feelings about situations and build meaningful relationships with students and colleagues. Teachers need to be aware of the emotional aspect of the learning process in order to work effectively with it and assist others in achieving meaningful learning outcomes (Saxena & Saxena, 2012). The Bar-On model of emotional intelligence offers a comprehensive framework to examine how emotional competencies contribute to teacher development. Within this model, the intrapersonal domain equips teachers with the ability to recognize their own emotions and understand how those emotions influence their thoughts, behaviors, and professional interactions. Teachers with strong self-regard believe in their competencies and feel content and confident in their roles. Emotional awareness allows them to understand the impact of their feelings on themselves and their students. Assertive teachers are able to express their ideas and values confidently, foster respectful relationships, and influence their classroom climate positively. Independence empowers teachers to rely on themselves, seek feedback, reflect critically, and continuously grow. Those high in self-actualization consistently strive to reach their fullest potential, investing in both personal and professional development.

The interpersonal component of emotional intelligence enables teachers to recognize students' emotional cues and respond effectively. Empathetic teachers not only sympathize with students but also understand their emotional states and offer appropriate support. Socially responsible teachers create collaborative, student-centered environments and promote inclusive, friendly classrooms. Positive interpersonal relationship skills help teachers build trust and maintain meaningful connections with students, colleagues, and parents. Stress management, another key domain, is crucial in helping teachers regulate their emotions. Those with strong stress tolerance handle burnout and classroom pressures effectively, while impulse control allows them to remain composed and work efficiently under stress. Adaptability is equally important, as it enables teachers to remain flexible, practical, and open to change. Reality-testing helps them evaluate themselves and their challenges realistically, while flexibility and problem-solving skills enable them to respond effectively to classroom dynamics and unexpected issues. Lastly, general mood reflects a teacher's overall emotional outlook. Optimistic teachers maintain positive attitudes in the face of obstacles, inspiring others in the learning environment. Teachers who experience happiness are fulfilled in their roles and motivated to pursue continued growth.

Merging both cognitive and emotional intelligence in teacher development is essential. As Aristotle stated, "Educating the mind without educating the heart is no education at all," underscoring the role of values and emotional insight in effective teaching (Pal, 2021). This principle applies not only to students but also to educators. Hileman (2024) highlights that both cognitive and emotional intelligence are vital for academic and professional success. When developed in tandem, these constructs enhance problem-solving skills, foster meaningful interpersonal relationships, and contribute to a more responsive, reflective, and effective educational practice.

In neuropsychology, the cognitive and emotional parts of the brain play important roles in making a decision. It involves the complex interplay of brain regions that merge cognitive, emotional, sensory, and motor information. According to Broche-Perez et al. (2016), evidence indicates that the cortical structures in the brain, such as the orbitofrontal cortex (OFC), anterior cingulate cortex (ACC), and dorsolateral prefrontal cortex (DLPFC), play a role in the decision-making process. Subcritical structures assist the process, including the amygdala, hippocampus, thalamus, basal ganglia, and cerebellum. The ventromedial prefrontal cortex also plays a role in decision-making. The process starts when the sensory organs (such as the ears receive auditory sensation, and the eyes receive visual information) receive sensory information from the environment, which is then passed to the brain. The information is passed on to appropriate cortical regions and subcritical regions for processing through the relay station called the thalamus. The dorsolateral prefrontal cortex (DLPFC) maintains the relevant information in the working memory, assesses the options available by evaluating the risks, benefits, and potential outcomes, and chooses a course of action. Simultaneously, emotional components of the information received are sent to the amygdala and ventromedial prefrontal cortex (VMPC), which quickly evaluate the emotional significance of the stimulus. Alongside, the hippocampus is activated to retrieve memories or experiences similar to the current situation to provide context or remind the brain of past outcomes. The dorsolateral prefrontal cortex (DLPFC) also works on inhibiting autonomous reactions that may arise from the emotional regions to maintain cognitive control. Information from the dorsolateral prefrontal cortex (DLPFC), hippocampus, amygdala, and ventromedial prefrontal cortex (VMPC) is sent to the orbitofrontal cortex (OFC) that evaluates the potential rewards or punishments by assigning possible value to the possible options and potential outcomes based on prior experiences. The anterior cingulate cortex (ACC) monitors the decision-making process for any errors or conflicts between goals and outcomes. It can intervene to guide the decision towards a more appropriate response. Once the decision is made, the basal ganglia contributes to the process by selecting the motor areas and relaying the necessary information needed for implementing the decision made. The cerebellum helps in the execution of proper motor outputs, and the coordination of the timing and movements required for implementing the decision (for example, speaking or walking away). Feedback from the decision carried out is assessed by the anterior cingulate cortex (Carmichael & Price, 1994) which can be used for future decisions or behaviors. According to Lokesh (2024), the brain uses two distinct decision-making pathways: the rational or logical reasoning pathway, and the emotional response pathway. This demonstrates the active interaction between reason and emotion in our choices. This balance clarifies why we respond impulsively sometimes despite knowing better. From the process described above, the brain tries to stop impulsive decisions emitting from emotional components but sometimes we let our emotions get the better of us. Everyday, we make decisions in one or the other aspects of our life; that means that our cognition and emotions are always active and playing a role in how we think, feel, and act. This shows the need for teachers to possess both cognitive and emotional intelligence because they are constantly in an intellectually, emotionally, and socially stimulating environment.

After reviewing literature, these are my reasons for advocating for the integration of cognitive intelligence and emotional intelligence in teacher development, be it in teacher training programs or teacher professional development programs.

One of the most important reasons for incorporating cognitive and emotional intelligence into teacher education programs is to ensure effective and efficient instruction and communication. According to Bala (2017, p.1), "teacher effectiveness refers to the height of maturity and learning in a teacher". A teacher with high cognitive and emotional intelligence has strong mental abilities to analyze information, create effective teaching methods or strategies that will help students better understand and learn concepts, and can use critical thinking skills to solve problems in the classroom by clarifying any doubts or misconceptions of the students. Teachers with high emotional and cognitive intelligence may have a better awareness of their own strengths and weaknesses, as well as their pupils' emotions and behaviors. This will help the teacher create and tailor effective lesson plans that resonate better with the students which could lead to improved understanding and retention of information, be able to communicate and connect more to the students on their level, and be able to manage the class environment which leads to better collaboration and a more conducive educational environment. Adeyemo & Chukwudi (2014) mentioned in their study that emotional intelligence can significantly predict teacher effectiveness. They further concluded that emotional intelligence should be integrated in teacher training and teacher education curriculum to help teachers be aware of their emotions, manage and control their emotions and that of others which can enhance the effectiveness of the teacher.

Some students are perplexed about the best professional route to choose, and they seek advice from teachers whom they regard as role models since they have gone through a specific career and have a good understanding of how to choose and prepare for that profession. Teachers may also motivate pupils to pursue a career, and students may then seek help from the teacher. A teacher with cognitive and emotional intelligence will be able to evaluate and reason with pupils, directing and motivating them to develop specific abilities, helping them work on their individual weaknesses, and assisting them in efficiently utilizing their strengths. This could benefit them in their academic pursuits, future careers, and personal lives. Furthermore, students today face discomfort or obstacles such as academic stress, abuse, and personal issues in family or other relationships. Some students are unable or unwilling to discuss their difficulties, which may be harming their academic performance. Teachers with cognitive and emotional intelligence can notice and respond to their students' emotional states, which they can then use to engage them and handle any issues they may be encountering. They can also persuade the kid and family to seek professional counseling, which may benefit the student even more. For example, the Indian film titled "Taare Zameen Par," also known as "Like Stars on Earth," depicts how teachers recognize and help students with their needs and concerns. The teacher discovered that his student, Ishaan, was not academically poor as his parents and other teachers believed, but rather had dyslexia. When the teacher discovered it, he helped him improve and discover his artistic ability. As a result, teachers may assist students in identifying and dealing with their issues or discomfort, which will benefit them in the future as they build strategies for dealing with challenges and knowing when and where to seek professional support.

Conflict resolution is the act or process of resolving a dispute or disagreement between two or more people; however, it can also be a self-conflict having an innate struggle or disagreement about an issue). Teachers with high cognitive and emotional intelligence will be able to manage and settle conflicts and disputes peacefully or calmly because they are able to analyze the situation and understand the emotional state of the parties involved and, based on that, make sound reasoning and judgment, using appropriate mediating and reinforcement strategies to foster cooperation and mutual respect among the students involved, and may also influence other students to learn from it. This contributes to a healthy learning environment and promotes positive interactions between teachers and students.

Integrating cognitive and emotional intelligence into teacher development helps to build resilience in teachers. Teachers face difficulties from themselves, their colleagues, students, and the learning environment, therefore they should be prepared and equipped with skills and tactics for managing stress, maintaining their well-being, and being productive in difficult situations. Teachers with cognitive and emotional intelligence will be able to deal with difficult situations because they can plan and execute them effectively. Emotions influence our decision-making in situations, therefore having high emotional intelligence allows you to control and manage your emotions while making sensible decisions even in stressful situations.

Effective mentors: Teachers with strong cognitive and emotional intelligence are better suited to be good mentors. Mentors are those who advise, train, or encourage others. Teachers can serve as students' mentors. Teachers play an important role in assisting students with research and projects at the undergraduate and postgraduate levels of education. With the right teachers as mentors, students develop skills, gain vital knowledge, and soft skills such as effective communication, teamwork, and empathy, all of which are necessary for future endeavors. Teachers with high cognitive and emotional intelligence have the knowledge and skills to instruct students, assist them to reach their goals, and make them feel acknowledged and respected, motivating them to give their all in whatever they do. They can also inspire and encourage students to be cognitively and emotionally intelligent individuals which could prepare students to navigate interpersonal relationships and face challenges in their future careers and personal lives.

Job satisfaction and fulfillment: Integrating cognitive and emotional intelligence into teacher development provides teachers with the abilities and competencies required for the profession, preparing them for the challenges of the job. They will feel highly accomplished, content, and fulfilled when they employ these skills and abilities to successfully increase students' performance outcomes while navigating job hurdles and responsibilities. This also lowers job changes caused by uncertainty or dissatisfaction at work.

PRACTICAL METHODS OR STRATEGIES FOR DEVELOPING SMARTER EDUCATORS

This section presents suggested strategies or practices that can be adopted by teachers themselves, educational institutions, and teacher development programs to implement the integration of cognitive and emotional intelligence in teacher development to build smarter educators. According to Kedzior and Fifield (2004), mentoring provides teachers with the opportunity to learn from more experienced professionals in the field. It supports teachers in acquiring practical knowledge to effectively handle daily classroom activities and instructional challenges. Mentoring can include activities such as classroom observations, instructional coaching, constructive feedback, and co-teaching sessions. These interactions not only enhance instructional competence but also foster emotional awareness and confidence. Smith

(2002) emphasizes that mentoring significantly influences teachers' attitudes, increases retention, boosts self-efficacy, and helps them develop a wider range of instructional strategies.

In addition to mentoring, collaborative learning strategies such as Content-Based Collaborative Inquiry (CBCI) and Cognitively Guided Instruction (CGI) can foster both emotional and cognitive development. CBCI encourages teachers to form questions about student learning, gather and analyze data, and work with colleagues to design responsive instructional practices (Kedzior & Fifield, 2004). This method promotes evidence-based decision-making and strengthens professional collaboration. Bray et al. (2000) affirm that CBCI helps educators improve both content clarity and pedagogical effectiveness. CGI, on the other hand, focuses on how students approach problem-solving, allowing teachers to observe and adapt instruction accordingly. For example, in a mathematics lesson, students are invited to use their own methods to solve a problem, explain their reasoning, and evaluate solutions together. This strategy fosters deeper content understanding and enables teachers to address diverse learning needs more effectively (Carpenter et al., 2001).

Mindfulness and stress management techniques also serve as critical tools in teacher development programs. Practices such as meditation, deep breathing, and focused attention have been shown to enhance emotional regulation and cognitive processing. These methods equip teachers with the ability to remain calm under pressure, stay mentally focused during instruction, and make more reflective decisions. Incorporating mindfulness into teacher training helps address the emotional challenges educators often face, such as burnout, anxiety, and decision fatigue. By managing their emotions effectively, teachers can model emotional stability and self-regulation for their students. As a result, classroom climates become more positive and conducive to learning.

To maintain relevance in a constantly evolving educational landscape, teachers must also remain up to date with advancements in their subject areas. Every discipline experiences shifts in theories, principles, and instructional approaches. Subjects such as science, psychology, and history regularly introduce new content that must be integrated into teaching. Teacher development programs should support educators in keeping pace with these changes, providing opportunities to explore emerging topics and pedagogical innovations. In doing so, teachers not only maintain academic competence but also engage their students with fresh, relevant content. This approach promotes lifelong learning and professional growth.

Peer coaching offers another valuable strategy for enhancing teacher development. It creates a supportive environment where educators can share experiences, discuss challenges, and co-develop solutions. According to Peyton (2007), peer coaching, especially when integrated with collaborative teaching, encourages professional openness and innovation. Teachers strengthen their practice by observing peers, offering feedback, and experimenting with new strategies together. This ongoing exchange fosters trust, increases instructional confidence, and promotes a shared commitment to student learning. Through peer coaching, educators develop both cognitive strategies and emotional resilience, which are essential in adapting to evolving classroom dynamics.

A related collaborative practice is lesson study, a structured process where teachers jointly plan, observe, and reflect on a lesson to improve instructional quality. In this approach, a teacher delivers a lesson while colleagues observe, taking detailed notes on instructional techniques and student responses. After the session, the group convenes to analyze observations and recommend improvements. The revised lesson is then taught to a different class, and the cycle continues. This iterative process enables teachers to refine their pedagogy through shared inquiry and reflective practice. Much like a professional learning community, lesson study builds collective expertise and emotional support among teaching staff.

Integrating Social-Emotional Learning (SEL) into teacher training further reinforces the development of emotional and cognitive competencies. SEL programs introduce educators to foundational concepts such as empathy, mindfulness, compassion, and critical inquiry. Teachers learn about the scientific basis of SEL and its impact on emotional well-being and classroom effectiveness. They are also introduced to intervention models that help with conflict resolution, student engagement, and classroom management (mgiep.unesco.org, n.d). SEL training equips teachers with practical tools for embedding emotional intelligence into daily instruction. Moreover, it enables teachers to support their students in managing emotions, building relationships, and navigating social complexities, ultimately fostering a healthy school culture.

Another innovative practice that supports both teacher and student growth is residual teaching. This approach focuses on four core instructional skills: summarizing, questioning, clarifying, and predicting. Teachers model these skills in lessons and then engage students in practicing them through structured interactions. Residual teaching creates a shared learning experience in which both teachers and students actively participate in exploring subject content. This reciprocal process enhances metacognitive awareness and strengthens communication. By cycling through these skills regularly, teachers promote deeper comprehension and critical thinking in the classroom.

In light of the digital transformation across all sectors, educational institutions must encourage teachers to adopt and integrate technology into their pedagogy. The rapid evolution of digital tools offers new ways to enhance teaching and learning. Seminars, workshops, and training programs can introduce educators to virtual platforms, digital libraries, and interactive media resources (Borko et al., 2010). Platforms such as YouTube and Chegg, as well as online discussion forums, can be leveraged to make lessons more engaging and accessible. When integrated thoughtfully, technology helps teachers illustrate complex concepts and provide diverse examples that enhance understanding. Moreover, digital fluency increases a teacher's ability to connect with today's tech-savvy students.

Reflective practices also play a crucial role in building emotional and cognitive intelligence. Activities such as journaling, peer feedback, and student evaluations enable teachers to evaluate their teaching methods, emotional responses, and problem-solving abilities. Regular reflection fosters self-awareness and allows educators to recognize patterns in their behavior and instructional decisions. This awareness promotes emotional regulation and encourages adaptive thinking. Through reflection, teachers become more intentional in their practice, continuously adjusting strategies to meet student needs. As a result, they grow not only as instructors but also as emotionally attuned leaders.

Case studies and role-playing exercises offer practical opportunities to apply emotional and cognitive skills in real-world scenarios. In case studies, teachers analyze complex classroom situations requiring empathy, reasoning, and strategic intervention. These scenarios might involve conflict resolution, differentiated instruction, or behavioral management. Role-playing allows teachers to inhabit different perspectives, such as that of a student, colleague, or administrator, fostering empathy and decision-making skills. These experiential learning tools provide immediate feedback, enabling educators to refine their responses and build confidence. Over time, such activities contribute to the development of emotionally intelligent and pedagogically agile teachers.

Creating a conducive environment for teaching and learning is equally essential in supporting the integration of cognitive and emotional intelligence. Educational institutions should prioritize teacher well-being by providing resources, safe spaces, and structured professional development opportunities. According to Bala (2017), organizing seminars, workshops, and training sessions helps teachers stay engaged and effective in their roles. Administrative support and access to counseling services further contribute to a nurturing work environment. When teachers feel supported, they are more likely to invest in their own growth and the success of their students.

Finally, systematic assessment of teachers' emotional and cognitive abilities is vital for targeted development. Tools such as the Mayer-Salovey-Caruso Emotional Intelligence Test (MSCEIT) and the Emotional Quotient Inventory can evaluate emotional competencies. Likewise, intelligence tests such as the Wechsler Adult Intelligence Test and Raven's Progressive Matrices provide insights into cognitive functioning. These assessments can be used by teacher training programs to identify individual strengths and growth areas. Based on the results, customized development plans can be designed to support each teacher's professional journey. Tracking progress over time ensures continuous improvement and aligns professional learning with personal growth.

CHALLENGES AND CONSIDERATIONS IN MERGING COGNITIVE INTELLIGENCE AND EMOTIONAL INTELLIGENCE IN TEACHER DEVELOPMENT

According to Guskey (2002), learning something new and finding meaning in it requires considerable time, effort, and adaptability. Integrating cognitive and emotional intelligence into teacher development programs presents immense potential to enhance instructional practices and student outcomes. However, like any transformative change, its implementation is not without challenges. Educators are likely to encounter initial resistance, and the process may unfold unevenly across different individuals and institutions. As such, understanding the key challenges and critical considerations is vital for the successful integration of these constructs. The following sections explore several of these barriers and offer suggestions for overcoming them.

One of the primary challenges is resistance to change. Lortie (1975) argues that teachers are often hesitant to adopt new practices unless assured of their effectiveness. Many educators are accustomed to traditional professional development approaches that prioritize cognitive knowledge, making them less receptive to emotional intelligence training. They may perceive emotional intelligence as unrelated to effective teaching or view it as an additional burden to their already demanding roles. In practice, teachers might initially comply with new strategies but quickly revert to old methods when quick improvements do not materialize. Furthermore, emotional intelligence often demands more open interaction and reflective classroom practices, which may conflict with rigid teaching styles some educators have cultivated over time. Overcoming this resistance requires that training programs clearly demonstrate the tangible benefits of emotional intelligence, such as improved classroom management, stronger student-teacher relationships, and reduced teacher burnout, to foster greater acceptance and commitment.

Another challenge lies in the limited availability of resources and training opportunities. Developing both emotional and cognitive intelligence requires specialized trainers, well-structured materials, and dedicated time within professional development programs. Unfortunately, many educational institutions, especially those in underserved or underfunded regions, lack the infrastructure to support such comprehensive training. Time constraints within teacher development schedules further exacerbate this issue. Trying to incorporate both constructs into an already packed curriculum may result in a surface-level treatment of both areas, leaving teachers overwhelmed and disengaged. This could reduce their motivation to fully invest in the process. Addressing this issue calls for increased investment in teacher training infrastructure, as well as partnerships with organizations that can provide external expertise and resources for cognitive and emotional development.

A third concern is the risk of an imbalance between cognitive and emotional intelligence within training programs. There is potential for one construct to be emphasized at the expense of the other, which may hinder holistic development. For instance, if programs overly prioritize emotional intelligence, teachers may lack the necessary cognitive skills to design, assess, and implement rigorous academic content. Conversely, if cognitive skills dominate the agenda, students' emotional needs and social dynamics might be neglected, resulting in reduced classroom engagement and wellbeing. Teachers might also gravitate toward the development of emotional intelligence simply because they feel more confident in that area, neglecting opportunities to grow cognitively. Programs should therefore be designed to promote both constructs in a balanced manner. This involves thoughtful curriculum planning that integrates cognitive and emotional learning objectives in a complementary, rather than competing, fashion.

Another critical element is the need for continuous evaluation and feedback. Measuring the long-term effectiveness of integrating cognitive and emotional intelligence is essential for refining teacher development practices. Programs must assess outcomes such as improved student engagement, enhanced classroom behavior, and teacher retention rates. Moreover, feedback mechanisms, such as peer evaluations, student input, and self-assessment, can provide valuable insights into the progress and effectiveness of these efforts. Collaboration between teachers and program developers should be emphasized to ensure ongoing improvement. McLaughlin and Marsh (1978), as cited in Guskey (2002), underscore that effective educational reform must be a continuous and reflective process. By institutionalizing regular evaluation practices, schools can support sustained growth and adaptability.

Cultural differences present another layer of complexity in integrating emotional intelligence into teacher development. Emotions are expressed and interpreted differently across cultures, which can significantly influence classroom dynamics. For example, behaviors considered assertive or confident in one culture may be perceived as disrespectful or inappropriate in another. Teachers working in multicultural environments must be especially attuned to these distinctions to avoid miscommunication or cultural insensitivity. Emotional intelligence training must therefore incorporate cultural competence, helping educators recognize and adapt to diverse emotional norms and expressions. Inclusive training programs should use culturally responsive examples, case studies, and scenarios that reflect the realities of teachers' environments. Doing so will prepare teachers to apply emotional intelligence in ways that are both effective and respectful of cultural differences.

Balancing personal and professional development is another major consideration in the integration process. Emotional intelligence development often requires teachers to engage in deep self-reflection, explore personal experiences, and confront emotional discomfort. This process may feel intrusive or unnecessary to educators who are more accustomed to focusing on external, task-oriented teaching responsibilities. On the other hand, some educators may feel unqualified or insecure when participating in cognitively demanding activities such as problem-solving or data analysis, which can hinder the growth of cognitive intelligence. This dual discomfort can limit the overall success of the program. As such, development programs must include activities that support both personal growth and professional competence, such as stress management, mindfulness training, and guided problem-solving tasks. Such a holistic approach can not only enhance instructional effectiveness but also improve job satisfaction and reduce burnout.

Technology integration is another avenue for developing both cognitive and emotional intelligence but comes with its own set of challenges. Advanced technologies such as AI-driven feedback systems, virtual simulations, and online mentoring platforms can support scalable, personalized training. However, the implementation of these tools may be hindered by financial constraints, especially in resource-poor environments. For example, in rural communities with weak internet connectivity, digital tools become impractical or inaccessible. Additionally, older teachers or those unfamiliar with technology may require more time and support to develop digital competencies. Without adequate training and user-friendly platforms, technological integration may exacerbate inequality rather than solve it. Therefore, teacher training programs must consider the local context, ensure equitable access to digital resources, and provide comprehensive training to ensure technology is a tool for empowerment rather than exclusion.

In conclusion, while the integration of cognitive and emotional intelligence into teacher development offers substantial benefits, it is not without its complexities. Challenges such as resistance to change, limited resources, imbalanced emphasis, lack of evaluation mechanisms, cultural considerations, personal discomfort, and technological limitations all demand careful planning and thoughtful responses. However, by acknowledging and addressing these factors proactively, educational institutions can design training programs that are both inclusive and effective. Doing so will ultimately support the development of smarter, more resilient educators who are better equipped to meet the diverse needs of their students and navigate the evolving demands of modern education.

CONCLUSION

To In today's increasingly dynamic and multifaceted educational environment, the role of teachers is evolving in unprecedented ways. Beyond simply delivering content knowledge, teachers are now expected to navigate emotional dynamics, foster inclusive classroom environments, and respond thoughtfully to the varied social, emotional, and intellectual needs of their students. This shift highlights the necessity for teacher development programs to embrace a more holistic approach, one that merges CI and EI. As emphasized throughout this chapter, the integration of these two intelligences is fundamental for developing smarter, more adaptable educators. Teachers must possess the analytical skills to address curricular demands while also having the emotional competence to engage with students meaningfully. Therefore, effective teacher development must cultivate both dimensions of intelligence in balance to meet the realities of modern classrooms.

Cognitive intelligence plays an essential role in shaping a teacher's ability to organize instruction, plan lessons, assess learning outcomes, and make informed professional decisions. It involves the application of reasoning, logical analysis, abstract thinking, and problem-solving skills that are critical for managing academic challenges. Teachers with strong CI are better equipped to differentiate instruction, interpret data from assessments, and adapt to curriculum changes or technological innovations. Additionally, cognitive skills support metacognition, enabling teachers to reflect on their instructional practices and evaluate what works or requires improvement. These intellectual processes contribute directly to student achievement, as well-structured lessons grounded in sound pedagogical thinking create a solid framework for learning. As such, embedding cognitive training into teacher development programs, through workshops, reflective activities, and problem-solving tasks, is not only beneficial but essential for sustained educational effectiveness.

While cognitive intelligence allows teachers to manage the academic dimensions of teaching, emotional intelligence equips them to handle the relational, behavioral, and psychological aspects of the profession. EI enables educators to recognize and regulate their emotions, empathize with students, manage stress, and foster a positive classroom climate. Emotionally intelligent teachers are more likely to communicate effectively, build trust with their students, and manage conflict calmly and constructively. They also demonstrate adaptability and resilience, qualities vital in managing the unpredictability and emotional demands of daily school life. When teachers understand and manage their emotional triggers, they model healthy emotional behavior for students, contributing to a supportive and respectful learning atmosphere. Therefore, incorporating emotional intelligence into teacher training enhances not only teacher well-being but also student engagement, classroom harmony, and overall school culture.

When cognitive and emotional intelligence are developed in tandem, teachers become more comprehensive in their professional capacities. This integration enables them to teach with both their heads and their hearts, drawing from reason and empathy to make balanced instructional and behavioral decisions. In practice, this duality results in more responsive pedagogy, stronger student-teacher relationships, and improved classroom management. Teachers are better prepared to support the diverse needs of learners, identify and respond to emotional distress, and modify their instructional methods accordingly. The outcome is a more inclusive, personalized, and effective teaching approach. Furthermore, the teacher's own professional satisfaction increases as they feel more competent and connected in their roles, thereby reducing burnout and enhancing retention.

The impact of such integration extends beyond instructional effectiveness to the teacher's role as a mentor and role model. Educators frequently guide students not just academically, but also personally and socially. A teacher who can assess a student's emotional state while understanding their cognitive capacity is better able to support them holistically. This includes offering advice on academic decisions, addressing behavioral issues empathetically, and fostering social-emotional growth. Emotionally and cognitively intelligent teachers can detect underlying issues such as anxiety, trauma, or disengagement and take proactive steps to intervene. They are also more successful at cultivating students' self-confidence, motivation, and interpersonal skills. Ultimately, these qualities prepare students to thrive not only in school but in future academic, professional, and social settings.

The implementation of integrated CI and EI development in teacher education programs must also consider practical strategies to be effective. Mentoring, peer coaching, case-based learning, and social-emotional learning frameworks are just a few examples of initiatives that can nurture both intelligences simultaneously. For instance, peer coaching allows teachers to reflect on cognitive strategies while also receiving emotionally supportive feedback. Lesson study models enable educators to collaboratively plan and refine lessons while strengthening interpersonal collaboration and empathy. Mindfulness practices can help regulate stress, while content-based inquiry sharpens analytical reasoning. These methods not only support skill development but also foster a culture of collaboration, reflection, and continuous growth, essential qualities in 21st-century educators. Embedding such strategies within training programs ensures that emotional and cognitive growth is not theoretical but practical and sustainable.

However, as discussed in earlier sections, challenges do exist in merging cognitive and emotional intelligence in teacher development. Resistance to change, limited resources, unequal emphasis on one domain, and cultural nuances can obstruct meaningful implementation. Teachers may find emotional intelligence training unfamiliar or unnecessary if they are used to more traditional, content-focused professional development. Others may feel emotionally vulnerable or overwhelmed by the reflective demands of EI training. Additionally, institutions may struggle to provide the infrastructure, expertise, or time needed to implement comprehensive dual-intelligence programs. These challenges necessitate thoughtful planning, leadership support, and policy frameworks that recognize the long-term benefits of balanced teacher development. Addressing these concerns through incremental changes, resource-sharing, and ongoing professional dialogue can pave the way for successful integration.

Despite the obstacles, there is growing consensus that educational reform must prioritize holistic teacher development. Research increasingly shows that effective teaching is rooted not only in what educators know but also in how they connect, respond, and adapt. As educational systems face rising complexities, from diverse classrooms and mental health challenges to evolving digital landscapes, the need for teachers who are both emotionally grounded and intellectually capable becomes more urgent. Programs that recognize this dual necessity will produce teachers who are not only effective in delivering content but also in nurturing human potential. This shift in professional preparation is not just a trend but a response to the real needs of students and the evolving role of teachers.

Moreover, the neurological interplay between emotion and cognition supports the case for this integration. Neuroscientific research confirms that the brain does not compartmentalize thought and feeling; rather, the two work together to shape perception, decision-making, and memory. Teachers, therefore, benefit most when they are trained to understand how emotional and cognitive processes interact in themselves and their students. For instance, a student's ability to engage with complex tasks is often influenced by their emotional state, anxiety can hinder performance just as confidence can enhance it. Teachers who understand this dynamic are more capable of addressing barriers to learning and creating emotionally safe spaces that support cognitive engagement. Recognizing and acting on this brain-based reality further validates the need for integrated teacher development.

In conclusion, the simultaneous cultivation of emotional and cognitive intelligence in teacher development is not only advantageous, it is essential. As the demands placed on educators grow more complex, the capacity to teach with both intellect and empathy becomes indispensable. Integrating these two dimensions allows for a more comprehensive, compassionate, and context-responsive approach to teaching. Teachers developed through such programs will be well-equipped to foster deep learning, support emotional well-being, and create classroom environments where all students feel capable and valued. By training teachers to think critically and feel deeply, educational institutions lay the foundation for more meaningful, equitable, and effective education systems. The path forward lies not in choosing between head or heart, but in empowering educators to teach with both.

REFERENCES

Adeyemo, D. A., & Chukwudi, A. R. (2014). Emotional intelligence and teacher efficacy as predictors of teacher effectiveness among pre-service teachers in some Nigerian universities. *International Journal of Evaluation and research in Education, 3*(2), 85-90.

Arlin, C. (2023). Information processing theory in psychology. https://www.verywellmind.com/information-processing-theory-in-psychology-7503601

Armstrong, P. (2010). Bloom's taxonomy. Vanderbilt University Center for Teaching. Retrieved [31/12/24] from https://cft.vanderbilt.edu/guides-sub-pages/blooms-taxonomy/

Avdor, S. (2008). The Required Changes in Student-Teachers' Qualifications, Training and Professional Development. In Kfir, D., & Ariav, T. (Eds.), *The Crisis in Teacher Education: Reasons, Problems and Possible Solutions* [in Hebrew]. (pp. 145–175). VanLeer Institute.

Bala, R. (2017). Teacher effectiveness of secondary school teachers in relation to their emotional intelligence. *International Journal of Indian Psychology*, *4*(4), 72–78.

Bar-On, R. (2006). The Bar-On model of Emotional-Social Intelligence (ESI). *Psicothema*, *18*, 13–25. PMID: 17295953

Bardach, L., & Klassen, R. M. (2020). Smart teachers, successful students? A systematic review of the literature on teachers' cognitive abilities and teacher effectiveness. *Educational Research Review*, *30*, 100–312. DOI: 10.1016/j.edurev.2020.100312

Bloom, B., Englehart, M., Furst, E., Hill, W., & Krathwohl, D. (1956). *Taxonomy of educational objectives*. Handbook One.

Borko, H., Jacobs, J., & Koellner, K. (2010). Contemporary approaches to teacher professional development. *International encyclopedia of education. 7*(2), 548-556.

Boyd, M. A. (2005). *The emotional intelligence of teachers and students' perceptions of their teachers' behaviour in the classroom*. Indiana University of Pennsylvania.

Broche-Pérez, Y., Herrera Jiménez, L. F., & Omar-Martínez, E. (2016). Neural substances of decision-making. [English Edition]. *Neurología : Publicación Oficial de la Sociedad Española de Neurología*, *31*(5), 319–325. DOI: 10.1016/j.nrleng.2015.03.009 PMID: 25976946

Cambridge Dictionary (n.d). Intelligence. https://dictionary.cambridge.org/dictionary/english/intelligence

Carmichael ST, Price J.L. (1994) Architectonic subdivision of the orbital and medial prefrontal cortex in the macaque monkey. *J Comp Neurol. 346*, 366, 402.

Carpenter, T. P., Fennema, E., Franke, M. L., & Levi, L. (2001). Capturing teachers' generative change: A follow-up study of professional development in mathematics. *American Educational Research Journal*, *38*(3), 653–689. DOI: 10.3102/00028312038003653

Carroll, J. B. (1997). The three-stratum theory of cognitive abilities. In Flanagan, D. P., Genshaft, J. L., & Harrison, P. L. (Eds.), *Contemporary intellectual assessment: Theories, tests, and issues* (pp. 122–130). Guilford Press.

Cattell, R. B. (1957). *Personality and motivation structure and measurement*. World Book.

Cattell, R. B. (1971). *Abilities: Their structure, growth, and action*. Houghton Mifflin.

Chechi, K. V. (2012). Emotional intelligence and teaching. *International Journal of Research in Economics & Social Sciences*, *2*(2), 297–304.

Day, C., Sammons, P., Stobart, G., Kington, A., & Gu, Q. (2007). *Teachers matter: Connecting work, lives and effectiveness*. Open University Press.

Driscoll, M. P. (1994). *Psychology of learning for instruction*. Allyn & Bacon.

Farrell Buzbee Little, P. (2005). Peer coaching as a support to collaborative teaching. *Mentoring & Tutoring*, *13*(1), 83–94. DOI: 10.1080/13611260500040351

Flanagan, D. P., & Dixon, S. G. (2014). The Cattell-Horn-Carroll theory of cognitive abilities. https://onlinelibrary.wiley.com/doi/pdf/10.1002/9781118660584.ese0431#:~:text=The%20Cattell%2DHorn%2DCarroll%20(,&%20Alfonso%2C%20in%20press)

Gardner, H. (1983). *Frames of mind: The theory of multiple intelligences*. Basic Books.

Goleman, D. (1996). *Emotional intelligence: Why it can matter more than IQ*. Bloomsbury Publishing.

Goleman, D. (1998). *Working with emotional intelligence*. Bantam Books.

Guskey, T. R. (2002). Professional development and teacher change. *Teachers and Teaching*, *8*(3), 381–391. DOI: 10.1080/135406002100000512

Hargreaves, A. (1998). The emotional politics of teaching and teacher development: With implications for educational leadership. *International Journal of Leadership in Education*, *1*(4), 315–336. DOI: 10.1080/1360312980010401

Hausfather, S. J. (1996). Vygotsky and schooling: Creating a social contest for learning. *Action in Teacher Education*, *18*(2), 1–10. DOI: 10.1080/01626620.1996.10462828

Horn, J. L. (1968). Organization of abilities and the development of intelligence. *Psychological Review*, *75*(3), 242–259. DOI: 10.1037/h0025662 PMID: 4875815

Kedzior, M., & Fifield, S. (2004). Teacher professional development. Delaware education research & development center. Learning matters (n.d). The importance of teacher training programs. https://learningmatters.ai/blog/the-importance-of-teacher-training-programs#:~:text=A%20teacher%20training%20program%20is%20a%20program%20that%20equips%20teachers,are%20conducted%20online%20or%20offline

Khushboo, P. (2021). Educating the mind without educating the heart is no education at all. https://timesofindia.indiatimes.com/readersblog/newvibe/educating-the-mind-without-educating-the-heart-is-no-education-at-all-40011/

Lokesh, B. (2024). Deciphering the science of decision making - insights from the brain. *Aster*. https://www.asterhospitals.in/blogs-events-news/aster-cmi-bangalore/deciphering-science-of-decision-making-insights-brain

Lortie, D. C. (1975). *Schoolteacher: A sociological study*. University of Chicago Press.

Lungu, M. (2019). Emotional intelligence and vocational schools. *Educa ia Plus*, *25*(2), 43–49.

Mandal, M. K. (2004). *Emotion: Basic issues and current trends*. Affiliated East-West Press.

Mayer, J. D., & Salovey, P. (1997). What is emotional intelligence? In Salovey, P., & Sluyter, D. (Eds.), *Emotional development and emotional intelligence: Educational implications* (pp. 3–31). Basic Books.

Mayer, J. D., Salovey, P., & Caruso, D. R. (2008). Emotional intelligence: New ability or eclectic traits? https://cdn2.psychologytoday.com/assets/attachments/1575/rp2008-mayersaloveycarusob.pdf

Mclaughlin, M. W., & Marsh, D. D. (1978). Staff development and school change. *Teachers College Record*, *40*, 69–93.

McLeod Saul. (2024). Vygotsky's theory of cognitive intelligence. Simple psychology. https://www.simplypsychology.org-Vygotsky-Theory-of-Cognitive-Development.pdf

MGIEP UNESCO. (n.d). SEL for teachers - Course modules. https://mgiep.unesco.org/article/sel-for-teachers-course-modules# Niva Dolev & Shosh Leshem (2016): Developing emotional intelligence competence among teachers, Teacher development. DOI: DOI: 10.1080/13664530.2016.1207093

Otero, I. Salgado J.F., & Moscoso Silvia (2022). Cognitive Reflection, cognitive intelligence, and cognitive abilities: A meta-analysis. *Intelligence. 90*, 101-614. . https://www.sciencedirect.com/science/article/pii/S0160289621000982DOI: 10.1016/j.intell.2021.101614

Patel, R. S. (2017). Teacher effectiveness in context to their emotional intelligence. *Voices of Research*, *5*(4), 1–5.

Pathak, V. N., Kaur, R., Jain, M., Chechi, V. K., & Pathak, A. (2024). Coping With Poor Well-Being Issues: Students in Higher Education Institutions. In Aloka, P. (Ed.), *Factors Impacting Student Well-Being and Coping Tactics* (pp. 389–422). IGI Global., DOI: 10.4018/979-8-3693-8860-0.ch016

Saxena, S., & Saxena, S. (2012). Impact of cognitive and emotional intelligence on quality education. [IJMECS]. *International Journal of Modern Education and Computer Science*, *4*(10), 47–58. DOI: 10.5815/ijmecs.2012.10.07

Sekreter, G. (2019). Emotional intelligence as a vital indicator of teacher effectiveness. *International Journal of Social Sciences & Educational Studies*, *5*(3), 286–302.

Siddique, Ahsaan, Taseer, N. A., & Muhammad Siddique. (2020). Teachers' emotional intelligence and teaching effectiveness: A correlational study. *Ilkogretim Online*, *19*(3), 2411–2417.

Simran, A. (2024). Lifelong learning for lifelong impact: Professional development for teachers. https://www.suraasa.com/blog/teacher-professional-development?var=as1

Singh, I., & Jha, A. (2012). Teacher effectiveness in relation to emotional intelligence among medical and engineering faculty members. *Europe's Journal of Psychology*, *8*(4), 667–685. DOI: 10.5964/ejop.v8i4.483

Smith, S. (2002). Teacher mentoring and collaboration. *Journal of Special Education Technology*, *17*(1), 47–48. DOI: 10.1177/016264340201700105

Sutton, R. E., & Wheatley, K. F. (2003). Teachers' Emotions and Teaching: A Review of the Literature and Directions for Future Research. *Educational Psychology Review*, *15*(4), 327–358. DOI: 10.1023/A:1026131715856

Tess, H. (2024). Cognitive intelligence vs emotional Intelligence: Key differences. https://orilearning.com/cognitive-intelligence-vs-emotional-intelligence/#:~:text=IQ%20measures%20abilities%20such%20as,effectively%20in%20oneself%20and%20others

Vygotsky, L. S. (1978). *Mind in society: The development of higher psychological processes*. Harvard University Press.

Wood, D., Bruner, J., & Ross, G. (1976). The role of tutoring in problem solving. *Journal of Child Psychology and Psychiatry, and Allied Disciplines*, *17*(2), 89–100. DOI: 10.1111/j.1469-7610.1976.tb00381.x PMID: 932126

Chapter 4
Teachers' Perceptions of Affective Skill Development and Associated Barriers:
Strategies and Policy Implications in Secondary Education

Alper Uslukaya
https://orcid.org/0000-0003-1455-8438
Çankırı Karatekin University, Turkey

Mehmet Tufan Yalçın
https://orcid.org/0000-0001-8386-2308
Çankırı Karatekin University, Turkey

ABSTRACT

This chapter explores secondary school teachers' perspectives on the development of affective skills, including emotional regulation, empathy, and social communication, in classroom settings. Adopting a phenomenological framework, the study utilized semi-structured interviews with 18 secondary school teachers in Türkiye to examine their perceptions of affective skill development, the challenges they encounter, and the strategies they employ. The findings indicate that while teachers recognize the importance of affective competencies, they face systemic challenges such as rigid curricular structures, the dominance of standardized testing, and limited institutional support, which hinder the consistent implementation of affective education. Furthermore, teachers predominantly rely on informal, intuitive methods to assess affective growth, underscoring the need for structured assessment frameworks and ongoing professional development. The chapter concludes by offering policy recommendations to integrate affective education into formal curricula and enhance institutional resources.

DOI: 10.4018/979-8-3373-0472-4.ch004

INTRODUCTION

Affective skills have become a critical component of holistic education, addressing dimensions of learning that extend beyond cognitive performance. These skills involve a diverse set of emotional and social competencies, such as emotional regulation, empathy, self-awareness, and interpersonal sensitivity, that support students not only in academic environments but also in broader social and personal contexts. Historically overshadowed by a predominant focus on measurable academic outcomes, affective development is now widely recognized as essential for fostering student resilience, engagement, and long-term success. Positioned within the affective domain of Bloom's taxonomy (Krathwohl, 2002), these skills shape how learners interpret experiences, regulate behavior, and form meaningful connections. Rather than existing in isolation, they interact dynamically with cognitive and behavioral processes, providing a foundation for sustained educational achievement and psychosocial well-being.

Contemporary educational research has increasingly emphasized the significance of structured frameworks for fostering affective competencies. Social and Emotional Learning (SEL), one of the most influential of these models, outlines five interrelated domains: self-awareness, self-management, social awareness, relationship skills, and responsible decision-making (CASEL, 2020). These domains articulate both intrapersonal and interpersonal dimensions of affective functioning, offering educators a roadmap for embedding emotional learning into classroom life. Each of these competencies contributes uniquely to students' ability to navigate academic challenges, collaborate with peers, and manage complex emotional demands. Their relevance becomes particularly pronounced in culturally and socially diverse classrooms, where the ability to empathize, self-regulate, and build relationships enhances the inclusivity and functionality of educational settings.

Among these competencies, emotional regulation plays a particularly influential role in shaping how students respond to the demands of schooling. Defined as the process through which individuals modulate the onset, duration, and expression of emotional responses, emotional regulation enables learners to adaptively manage stress, frustration, and interpersonal tension (Gross, 2014). Effective emotional regulation has been linked to improvements in students' academic engagement and resilience, especially in high-pressure environments that demand sustained concentration and adaptability (Eng et al., 2023). This ability is not only a protective factor in times of adversity but also a prerequisite for students to participate actively and constructively in learning tasks. When emotional regulation is underdeveloped, students may exhibit avoidance behaviors, reduced motivation, or disruptive conduct, all of which undermine academic progress and classroom cohesion.

Empathy, another cornerstone of affective development, facilitates prosocial behavior and enhances interpersonal harmony in learning environments. As an emotional and cognitive capacity, empathy enables students to understand and share in the experiences of others, fostering mutual respect, cooperation, and ethical engagement (Zaki & Cikara, 2015). In collaborative settings, empathy underpins the development of trust and shared responsibility, which are essential for productive group work and peer support. Educational contexts that prioritize empathy tend to exhibit lower levels of conflict and greater student willingness to participate in dialogue and compromise (Mahoney et al., 2016). This competency is especially valuable in disciplines that rely heavily on interpersonal interaction, such as healthcare and education, where relational skills are as important as technical expertise.

Self-awareness complements emotional regulation and empathy by enabling students to recognize their emotional states, behavioral tendencies, and personal triggers. This internal knowledge promotes reflective thinking, which plays a critical role in both learning and personal development. Students who demonstrate self-awareness are more likely to take ownership of their learning, seek constructive feedback, and engage in self-directed goal setting (Kim, 2020). In educational settings, self-awareness enhances students' ability to monitor their progress, manage setbacks, and interact respectfully with others, contributing to a more stable and focused learning environment. Research further suggests that when students are guided to understand their emotional profiles, they develop greater academic perseverance and social maturity (Hojat & Gonnella, 2017).

Affective skills also influence cognitive and behavioral outcomes, acting as mediating variables between instructional input and student performance. Emotions significantly affect attention, memory consolidation, executive functioning, and decision-making, all of which are critical to learning (Immordino-Yang & Damasio, 2007). Positive emotional states can facilitate creativity, cognitive flexibility, and sustained engagement, while chronic emotional distress can impair concentration and lead to academic disengagement. Large-scale studies have demonstrated that structured instruction in emotional competencies not only improves classroom behavior and social adjustment but also has measurable impacts on academic achievement. A meta-analysis involving over 270,000 students found that participation in SEL programs led to an average academic performance gain of 11 percentile points, alongside reductions in emotional distress and conduct problems (Durlak et al., 2011). These findings underscore the interdependence between emotional and academic development and call for more integrative pedagogical approaches. In practical terms, the implementation of affective education can foster more emotionally responsive and academically effective classrooms. Educators who embed emotional learning into their practice report higher student motivation, fewer behavioral disruptions, and stronger peer relationships. Teaching strategies such as reflective writing, role-playing, cooperative learning, and teacher modelling have proven effective in facilitating students' affective development (Stevanović et al., 2017). These practices encourage learners to examine emotional experiences critically, engage empathetically with others, and rehearse positive social behaviors. Such approaches are most impactful when they are consistent, developmentally appropriate, and adapted to the specific needs of student populations.

Beyond their relevance to schooling, affective skills play a formative role in preparing students for adulthood. Emotional competencies such as empathy, resilience, and self-regulation are linked to better mental health outcomes, more adaptive coping strategies, and stronger professional skills (Nurhasanah et al., 2019). In the workplace, emotionally intelligent individuals demonstrate superior conflict resolution, leadership, and teamwork abilities. As job markets increasingly prioritize soft skills alongside technical qualifications, the integration of affective education becomes critical in aligning schooling with real-world demands. Project-based and interdisciplinary learning experiences that emphasize affective development can support students in navigating complex social environments and career challenges (Hutchison, 2016). Such initiatives also foster civic responsibility and ethical decision-making, further extending the value of affective education beyond academic success.

Despite the compelling case for affective education, its integration into mainstream curricula remains inconsistent. Many teachers lack the training, time, or institutional support to prioritize emotional learning, particularly in systems dominated by standardized testing and rigid instructional schedules. The absence of reliable assessment tools also presents a barrier, making it difficult to evaluate progress in affective domains or to justify their inclusion within high-stakes accountability frameworks (Hall, 2011). Consequently, efforts to promote affective development are often fragmented and dependent on the commitment of individual educators rather than being embedded within systemic practices. To overcome these limitations, educational policy must recognize affective competencies as fundamental learning outcomes, allocating resources for teacher development, curriculum design, and evidence-based assessment.

Ultimately, the inclusion of affective skills in educational planning reflects a shift toward a more comprehensive vision of learning, one that values emotional growth alongside intellectual achievement. These skills support students in managing emotions, engaging respectfully with others, and making thoughtful decisions, thereby enhancing not only academic performance but also psychological resilience and social integration. Their relevance spans all educational stages and disciplines, offering benefits that are both immediate and enduring. When integrated intentionally and supported structurally, affective education equips students with inner resources to thrive in academic settings and to lead meaningful, emotionally intelligent lives. In this way, it fulfills a foundational purpose of education: not merely to inform, but to form individuals capable of navigating human experience with competence, empathy, and integrity.

THEORETICAL FOUNDATIONS OF AFFECTIVE EDUCATION

The conceptual evolution of affective education reflects a paradigmatic shift in educational thought, from a predominantly cognitive orientation toward a more integrated understanding of the learner as an emotional, social, and moral being. Rather than treating emotions as peripheral or disruptive to rational learning, contemporary theories increasingly conceptualize effect as a fundamental epistemological force that shapes perception, engagement, and the construction of knowledge itself. This shift is underpinned by interdisciplinary contributions from psychology, philosophy of education, and developmental science, all converging on the notion that affective capacities are not only instrumental but constitutive of meaningful learning. Such a perspective demands that education be reimagined not merely as a process of information transfer, but as an emotionally charged, relational, and ethical endeavor that nurtures the full spectrum of human potential.

Emotional intelligence (EI), as advanced by Mayer and Salovey and further disseminated by Goleman (1995), has served to reposition emotion within the framework of human competence. The notion that individuals can develop the capacity to perceive, manage, and apply emotional information challenges earlier dichotomies between reason and emotion, offering instead a synthetic model of intelligence that bridges intrapersonal awareness with interpersonal adaptability (Mayer et al., 2008). While EI is often operationalized in psychological scales or skills-based training, its deeper significance within affective education lies in its philosophical assertion: that emotion is not the antithesis of intellect, but rather its co-contributor. This reconceptualization carries profound implications for educational practice, as it invites a reevaluation of what counts as valuable knowledge and how it should be cultivated within schools (Ebubechukwu, 2021; Ghahremani & Ostovar, 2017).

The emergence of SEL frameworks such as those promoted by CASEL can be interpreted not simply as pedagogical tools, but as systemic responses to this broader theoretical realignment. SEL represents an institutionalized effort to operationalize affective goals within the structures of schooling, curriculum, teacher preparation, assessment, and school climate, thereby translating emotional literacy from an abstract ideal into a practical educational agenda (Odak et al., 2023; Ross & Tolan, 2017). What distinguishes SEL as a theoretical model is its ecological understanding of emotional development: students are not isolated actors but embedded in dynamic social systems that mediate their capacity to grow emotionally and ethically. Within this logic, affective education becomes a responsibility not only of individual teachers but of the entire school culture, requiring alignment of values, relational norms, and instructional strategies. In this way, SEL contributes not just to individual student outcomes, but to the moral architecture of education itself.

Equally important to the theoretical foundation of affective education are perspectives from holistic and humanistic educational philosophy. These frameworks reject reductionist accounts of learning that privilege testable outcomes over developmental processes, emphasizing instead the indivisibility of emotion, cognition, and identity (Miller, 2007). Drawing on thinkers such as Maslow and Rogers, holistic education posits that learning is only authentic when it is effectively engaged, when students feel safe, seen, and connected to purpose (Maslow, 1968; Rogers, 1969). From this standpoint, affective education is not a curricular add-on, but a foundational precondition for self-actualization and transformative learning. The classroom, in this model, is reconceived not as a site of compliance and performance, but as a relational space where human potential is nurtured through empathy, presence, and dialogical interaction. Such views broaden the educational mandate from knowledge transmission to human formation.

This reorientation also aligns with sociocultural and constructivist theories of learning, which emphasize that knowledge is co-constructed through emotionally and socially situated activity. Vygotskian theory in particular underscores that learning is mediated by tools, signs, and human relationships, dimensions that are deeply affective in nature (Vygotsky, 1978). Emotional climate, teacher affect, and peer dynamics are not peripheral to academic tasks but central to the motivational and cognitive architecture of learning. In this sense, affective education becomes inseparable from pedagogy itself: the way a lesson is delivered, the emotional tone of feedback, or the relational dynamics of group work all serve as affective conditions that shape what and how students learn. Thus, affect is not an object of instruction but a medium through which instruction becomes meaningful. In sum, the theoretical foundations of affective education challenge conventional boundaries between emotion and reason, individual and system, instruction and formation. They invite educators to adopt a richer conception of their task, not merely to teach content, but to engage with students as emotional beings situated within complex social worlds. Emotional intelligence provides individual scaffolding for this work, while SEL frameworks institutionalize it within school structures. Holistic and constructivist theories deepen the ontological and pedagogical understanding of what such education entails. Together, these perspectives form a theoretical constellation that calls for a more ethically attuned, emotionally responsive, and relationally grounded vision of education.

TEACHERS' ROLE IN FOSTERING AFFECTIVE DEVELOPMENT

The teacher's role in fostering affective development is not merely complementary to instruction; it is central to the construction of an emotionally responsive and ethically grounded educational environment. Teachers act as primary social agents within the school context, influencing how students perceive themselves, relate to others, and develop emotional competencies. Their capacity to shape the affective climate of the classroom emerges from both their interpersonal dispositions and their pedagogical choices. As such, affective development is not the result of isolated interventions but the cumulative outcome of daily relational exchanges, emotional modeling, and classroom interactions that are mediated by the teacher's presence and values. Affective education, in this regard, is fundamentally enacted through the teacher, not only as a facilitator of content but as a relational and emotional figure.

At the heart of this process lies the teacher-student relationship, which research consistently identifies as a significant determinant of emotional and social outcomes in education. Classrooms characterized by emotional warmth, responsiveness, and relational trust tend to foster greater student engagement, emotional security, and social competence (Humphries et al., 2014; Rucinski et al., 2018). The emotional tone set by the teacher becomes a regulating force, shaping students' behavior, self-concept, and motivation. Teachers who take on roles that extend beyond instruction, acting as mentors, coaches, or emotional anchors, provide the kind of affective scaffolding that supports resilience and interpersonal growth (Soldaat, 2019). Moreover, teachers with higher social-emotional competencies are more effective in managing classroom dynamics and in cultivating a climate conducive to affective learning (Jennings et al., 2017). Their ability to co-regulate emotional states, de-escalate conflict, and foster empathy among students underpins the relational infrastructure of successful classrooms.

Equally critical is the emotional climate that teachers cultivate, which operates as both a reflection and a reinforcement of their emotional presence. A teacher's capacity to model emotional awareness, regulation, and empathy directly influences the affective tone of the classroom and the emotional behaviors of students. When teachers exhibit positive emotional expressions and respond sensitively to students' needs, they create atmospheres in which students feel psychologically safe and motivated to engage (Reyes et al., 2012; Frenzel et al., 2018). This modeling extends beyond emotional displays to include how teachers manage stress, handle mistakes, and relate to diverse perspectives, thereby shaping students' emotional repertoires through vicarious learning. Programs such as CARE for Teachers highlight the importance of teacher emotional competence as a prerequisite for fostering healthy classroom climates, emphasizing that emotionally attuned teachers are more likely to engage in supportive and responsive interactions (Jennings et al., 2017). Additionally, teachers' own emotional well-being affects classroom functioning; emotionally exhausted educators may struggle to maintain relational consistency and empathetic responsiveness, which in turn can impact student outcomes (Mai, 2024). The reciprocal nature of classroom emotions further underscores this dynamic: when teachers find joy and meaning in their work, it often translates into higher levels of student emotional engagement and learning efficacy.

Beyond relational and emotional dimensions, the pedagogical approaches teachers employ are instrumental in promoting students' affective development. Instructional strategies that invite reflection, cooperation, and emotional expression provide fertile ground for cultivating social-emotional competencies. Approaches rooted in mindfulness, for example, have shown positive effects on teacher self-regulation and classroom prosocial behavior, both of which contribute to emotionally supportive learning environments (Flook et al., 2013). Programs like RULER offer structured frameworks for integrating emotional learning into everyday classroom practices, emphasizing the importance of recognizing, understanding, labeling, expressing, and regulating emotions (Rivers et al., 2012). These pedagogical interventions move beyond surface-level engagement by embedding emotional literacy within content areas, classroom discussions, and group dynamics. Furthermore, inclusive practices that account for the diverse emotional experiences of students, particularly those with varying educational needs, enhance the emotional validity of the classroom climate (Cefai et al., 2013; Herranz-Hernández et al., 2023). Such practices not only affirm students' emotional realities but also equip them with the tools to navigate complex interpersonal contexts both within and beyond school. To sum up, the affective dimension of education is inextricably linked to the work of the teacher. Through relationships, emotional modeling, and pedagogical intention, teachers create the conditions under which emotional growth becomes possible and meaningful.

These roles require a high degree of emotional labor and professional judgment, underscoring the need for institutional structures that support teachers in this work. Professional development in emotional intelligence, reflective teaching, and trauma-informed practices can strengthen teachers' capacity to lead affectively rich classrooms. Yet these efforts must be complemented by systemic recognition that affective education is not peripheral, it is central to cultivating learners who are not only academically competent but also emotionally intelligent, socially responsible, and ethically grounded.

CLASSROOM PRACTICES SUPPORTING AFFECTIVE SKILLS

The integration of affective skills into classroom practices has become an essential dimension of contemporary pedagogy, emphasizing the development of emotional and social competencies alongside academic achievement. Effective classroom practice in this regard is not characterized by discrete lessons in emotional intelligence, but by the sustained incorporation of pedagogical strategies that activate emotional engagement and interpersonal growth. Among the most effective approaches are cooperative learning, role-play, and reflective activities, each of which enables students to develop self-awareness, empathy, and emotional regulation through structured and meaningful participation. These practices support a learning environment where affective skills are not only acknowledged but cultivated as part of the daily rhythm of education, forming the basis for inclusive, emotionally responsive classrooms.

Cooperative learning serves as a foundational practice for promoting students' socialization and emotional development. It engages students in structured collaboration where they are required to communicate effectively, share diverse viewpoints, and work interdependently toward common goals. This interaction fosters empathy and trust, while also promoting critical interpersonal competencies such as active listening, conflict resolution, and mutual respect (Cooper et al., 2018). Research has linked cooperative learning environments with heightened academic motivation, reduced anxiety, and increased feelings of belonging, particularly among students in culturally diverse classrooms, where social inclusion is essential for equitable learning outcomes (Arifin et al., 2024; Gharaei et al., 2018). These social-emotional benefits underscore the potential of cooperative structures not only to facilitate content mastery but also to reinforce emotional resilience and group cohesion, both of which are essential to long-term educational success.

Role-play represents another highly impactful pedagogical tool for supporting affective development, particularly through its capacity to elicit emotional and ethical engagement. By assuming different perspectives and navigating hypothetical or real-life scenarios, students engage in experiential learning that activates emotional insight and cognitive empathy (Clack & Ellison, 2018). This immersive process encourages self-reflection, emotional articulation, and critical thinking, enabling learners to consider complex social dynamics and internalize multiple points of view. The affective value of role-play lies not only in the act of performance but in the metacognitive reflection it encourages, facilitating the development of self-regulation and emotional expression (Zhou, 2024). In contexts such as literature or history education, role-play can deepen students' engagement with content by linking narrative and emotional experience. In scientific or mathematical subjects, it can be adapted to explore problem-solving scenarios or group decision-making, offering a unique bridge between affective learning and subject-specific objectives (States et al., 2023).

Reflective activities complement these interactive strategies by inviting students to process their learning through introspection and emotional analysis. Practice such as journaling, guided discussion, and peer feedback enable students to articulate emotional responses to academic challenges and social interactions. These practices build self-awareness and emotional vocabulary, reinforcing internalization of affective learning (Semchuk et al., 2022). Reflection also supports personal growth by creating space for learners to consider how emotional experiences influence their thinking and behavior, and to identify areas for self-improvement. Importantly, reflective pedagogies offer teachers valuable insights into students' emotional states, allowing for more responsive and adaptive instruction (Rance et al., 2023). Through these cycles of emotional expression and feedback, reflection becomes a developmental process that enhances both learner autonomy and emotional maturity.

Subject-specific considerations further shape the implementation of affective practices in the classroom. While humanities and arts-based disciplines often lend themselves naturally to emotional engagement, technical fields such as mathematics and science may require more intentional efforts to embed affective elements. Nonetheless, evidence suggests that integrating role-play, collaborative tasks, or reflective prompts in STEM instruction can demystify complex content and reduce student anxiety (States et al., 2023). In literature education, affective practices may involve engaging students with the emotional dimensions of narrative, encouraging them to interpret character motivations, moral dilemmas, and social contexts through an empathetic lens (Adaba, 2017). This illustrates the importance of pedagogical adaptability, where teachers leverage the affordances of their discipline while also embedding structures that develop emotional competence.

Embedding affective learning into classroom routines also requires sustained professional development and institutional support. Teachers must be equipped not only with the theoretical understanding of affective education but also with practical tools to apply it across diverse student populations. Studies have emphasized that professional development in affective pedagogy enhances classroom dynamics, student behavior, and academic performance (Rahiminia et al., 2020; Masnan et al., 2024). Furthermore, continuous training enables teachers to adopt inclusive strategies that recognize the unique emotional experiences of learners, including those from marginalized backgrounds or with differentiated learning needs (Ramaila, 2022). By situating affective development at the center of instructional practice, teachers can contribute to emotionally intelligent learning environments that foster academic engagement and holistic development.

Ultimately, classroom practices that support affective development are integral to preparing students for the complexities of modern life. Through strategies such as cooperative learning, role-play, and reflective activities, educators can facilitate not only subject-matter understanding but also the growth of emotional awareness, empathy, and interpersonal competence. These practices, when tailored to specific disciplines and supported through professional development, contribute to emotionally enriched learning environments that support the full range of student needs. In doing so, they advance a vision of education that integrates intellect with emotional intelligence and positions the classroom as a space for human as well as academic flourishing.

RESEARCH GAPS AND AIMS

Despite the growing scholarly attention to affective education, a notable gap persists in the literature regarding how these concepts are perceived and implemented by teachers themselves. Much of the existing research has been dominated by quantitative studies measuring the outcomes of emotional learning interventions, often relying on standardized metrics and predefined frameworks (Pekrun & Linnenbrink-Garcia, 2014). While these contributions are valuable in demonstrating the general effectiveness of affective programs, they frequently overlook the nuanced, context-dependent insights of the educators tasked with enacting such practices. The everyday realities of classroom life, how affective practices are interpreted, adapted, or constrained, remain underexplored, leaving a critical void in our understanding of how emotional development unfolds in situ. Teachers' voices, perspectives, and lived experiences are essential to fill this gap, particularly in capturing the complexity and variability of affective work in diverse classroom settings (Sutton & Wheatley, 2003).

In particular, there is a lack of research grounded in qualitative methodologies that prioritize the experiential knowledge of teachers. Studies that foreground teacher perspectives tend to be limited in number, scope, or depth, often treating teachers as implementers rather than co-constructors of affective learning environments (Wagner et al., 2018). This absence of teacher-centered inquiry limits the field's ability to develop grounded theories of practice that reflect authentic classroom dynamics. It also contributes to a disconnect between theoretical models of affective education and the situated pedagogical decisions teachers make daily. To better inform policy and teacher education, it is necessary to investigate how teachers conceptualize students' affective development, how they define effective strategies, and how they navigate the emotional dimensions of instruction. Without such insights, affective education risks becoming a top-down initiative, disconnected from the real conditions in which learning and teaching occur (Sutton & Wheatley, 2003).

Phenomenological approaches offer a valuable lens through which to explore these underexamined dimensions of affective education. By focusing on lived experience, phenomenology allows researchers to uncover the meanings that teachers attach to their affective practices, the challenges they face, and the relational dynamics that shape their work (van Manen, 2016). This methodological orientation is uniquely suited to affective education, where emotion, perception, and interpersonal connection are central to the phenomenon under investigation. Moreover, phenomenology resists reductionism by honoring the complexity and subjectivity of teacher experience, offering depth and richness often absent in more positivist paradigms (Finlay, 2014). As affective education calls for educators to engage emotionally and ethically with students, it is only fitting that research methodologies engage similarly with educators themselves, recognizing their voices as essential sources of insight into pedagogical and emotional processes.

The present study seeks to address these gaps by exploring secondary school teachers' perspectives on classroom practices that support the development of students' affective skills. Through in-depth qualitative inquiry, this research aims to illuminate how teachers understand affective competencies, how they support and monitor emotional growth, and what challenges and needs they identify in this process. In doing so, the study gives voice to the practitioners who are most intimately involved in the emotional lives of students, positioning them not as passive actors but as reflective professionals engaged in complex relational work. The research questions guiding this study, focused on teacher perspectives, instructional practices, monitoring strategies, perceived student challenges, and suggestions for improvement, are designed to elicit a multifaceted understanding of affective education from the inside out. In centring teacher voice and adopting a phenomenological lens, this study contributes to a more grounded and context-sensitive literature on affective education, one that aligns theory with the emotional realities of classroom practice (Creswell & Poth, 2018; van Manen, 2016). In line with this general purpose, we investigated the following research questions (RQ):

RQ 1. What are teachers' views on students' affective skills?
RQ 2. What are teachers' views on classroom practices which support students' affective skills?
RQ 3. What are teachers' views on monitoring students' affective development?
RQ 4. What are teachers' views on difficulties students faced with and their needs in affective development practices?
RQ 5. What are teachers' suggestions for improving classroom practices in students' affective development?

METHOD

Research Design

This study employs phenomenology, a qualitative research design, to explore teachers' perspectives on the development of in-class practices aimed at supporting students' affective development at the secondary school level. Qualitative research provides a methodological approach that enables an in-depth investigation of individuals' experiences, thoughts, and perceptions (Creswell, 2013). Aligned with this approach, the present study seeks to understand teachers' subjective experiences and perceptions regarding the processes involved in fostering affective skills. Phenomenology offers a structured framework for examining how individuals experience a particular phenomenon (Van Manen, 2016). Specifically, phenomenological research aims to capture the essence of a phenomenon through a meticulous, systematic, and comprehensive analysis of individuals' lived experiences (Patton, 2014, p. 104). In this study, the primary phenomenon under investigation is teachers' classroom practices in fostering students' affective skills. These practices were subjected to a rigorous analytical process, addressing both the challenges faced by teachers and the requirements necessary for effective implementation within a scientific framework. The study thus provides an in-depth examination of teachers' instructional approaches in the development of students' affective competencies.

Participants

In phenomenological qualitative research, a small sample size of 5 to 25 participants is typically preferred, with a criterion-based sampling approach that selects individuals who have direct experience with the phenomenon under investigation. In this study, the selection criteria required that participants be secondary school teachers who had completed their probationary period and were employed in public schools. Additionally, efforts were made to ensure a balanced distribution of participants across different subject areas. The study group consisted of 18 teachers working under the Ministry of National Education in Türkiye. Detailed information regarding the participants' subject areas and years of professional experience is presented in Table 1.

continued on following page

Table 1. Continued

Table 1. Distribution of participants according to branch and seniority variables

Participant Order	Branch	Seniority (year)
P1	Turkish	11
P2	Social studies	12
P3	Math	18
P4	Foreign language	12
P5	Science	13
P6	Math	14
P7	Science	10
P8	Math	24
P9	Turkish	20
P10	Math	3
P11	Math	10
P12	Turkish	18
P13	Turkish	6
P14	Foreign language	13
P15	Math	16
P16	Foreign language	8
P17	Social studies	14
P18	Math	6

As shown in Table 1, the participant group was evenly divided between teachers of quantitative subjects, such as mathematics and science (n=9), and those teaching verbal subjects, including Turkish, social studies, and foreign languages (n=9). Ensuring this balance was crucial for obtaining a comprehensive understanding of teachers' practices in fostering affective skills across both quantitative and verbal course contexts. Additionally, participants' professional experience was considered, revealing an average teaching tenure of approximately 13 years.

Data Collection Tools

The data for this study were collected from secondary school teachers employed by the Ministry of National Education in Türkiye during the spring term of the 2024-2025 academic year. The interviews took place between February 22 and March 1, 2025. In qualitative research, semi-structured or unstructured interview techniques are commonly used as data collection methods (Creswell, 2015). For this study, data were gathered through face-to-face interviews utilizing a semi-structured interview form developed by the researchers. The responses were documented in written notes, resulting in a total of 92 pages of transcribed interview data.

While preparing the interview form, firstly, the sources in the literature on teachers' practices of developing students' affective skills in Türkiye and internationally were analysed. Based on this information, 23 open-ended questions were prepared. Then, a total of nine experts, including five experts in the field of educational administration, two experts in the field of educational psychology and two experts in the field of measurement and evaluation in education, were consulted about these open-ended questions. After the feedback from the experts (two questions that were not directly related to the purpose of the research were removed), the interview form was organised. After the preliminary interviews with two teachers, the interview form was finalised with 21 questions. Examples of the interview questions in this study are given below:

How do you define the concept of affective skills? In your view, what role do these skills play in students' overall development?

- What specific activities have you designed to foster students' affective skills, either within or outside the classroom?
- How do you assess the effectiveness of the methods you implement in enhancing students' affective skills?
- Could you provide information about any collaborative initiatives you have undertaken with the school administration or your colleagues in this context?
- If you were in a decision-making position, what measures would you take to enhance teachers' affective skills?
- What policies would you implement to support this objective?

Validity-Reliability Study

The content validity of the data collection forms used in this study was assessed through expert evaluation. In this context, the opinions of seven field experts were solicited, and the forms were finalized after incorporating the necessary revisions. To ensure the reliability of the research, the independently transcribed data were analysed by the researchers, key insights were identified, and these insights were categorized under themes established within a common framework. According to Morse (2015), peer review and debriefing are frequently employed strategies in qualitative research to enhance the credibility of findings. During the data analysis process, the researchers organized the responses into thematic categories using Microsoft Excel. To assess the validity of these themes, two independent experts were asked to classify the responses under the predefined categories. The analysis revealed that the responses were grouped under different themes, and the inter-expert agreement rate was calculated at 86%. According to Miles and Huberman (1994), an agreement rate of 70% or higher is necessary to ensure reliability in qualitative research. Furthermore, Yıldırım and Şimşek (2021) emphasize that qualitative research should include detailed descriptions and direct quotations to enhance validity and reliability. In line with this principle, the study's findings are presented with direct quotations from participants to provide a richer and more comprehensive interpretation of the data.

Data Analysis

The study employed content analysis to analyse and interpret the data using Microsoft Excel. Initially, transcripts from the interviews were transferred into a digital format. The data were then organized by the researchers in alignment with the research questions and objectives. Through an iterative and inductive approach, recurring patterns were identified, interpreted, and synthesized into five overarching themes (see Table 2). These themes encapsulate how teachers conceptualize affective skills, implement strategies to foster them, monitor students' progress, navigate challenges, and articulate their support needs within the broader educational framework. Rather than merely cataloguing individual responses, the analysis integrates insights to uncover both shared understandings and diverse experiences among participants. Special attention is given to areas of convergence and contradiction, providing a nuanced perspective on how affective education is perceived and enacted in practice.

Table 2. Mapping of initial codes to themes

Theme	Initial Codes / Keywords	Interpretative Description
Teachers' Conceptualisations of Affective Skills	Emotional regulation, empathy, self-awareness, maturity, social understanding, holistic development, life skills, behavioural impact	Teachers view affective skills as essential foundations for academic success and life readiness, not supplementary qualities.
Classroom Strategies for Supporting Affective Development	Peer interaction, role-play, empathy activities, listening practices, collaborative learning, feedback, student expression	Affective learning is supported through daily practices such as respectful dialogue, emotional modelling, and group work that fosters empathy and cooperation.
Monitoring and Evaluating Affective Growth	Observation, behavioural change, emotional cues, lack of tools, rubrics, teacher judgement	Teachers primarily rely on informal, intuitive methods to assess affective growth and express concern about formalising emotional assessment.
Challenges in Promoting Affective Skills	Time constraints, curriculum overload, exam pressure, lack of parental involvement, student resistance, cultural norms, emotional stigma	Teachers face systemic and cultural barriers, including structural limitations and student reluctance, that restrict their ability to promote affective skills.
Support Needs and Systemic Reflections	Training needs, lack of materials, curriculum reform, institutional support, school culture, collaboration with Academics	Teachers request practical training, resource development, and policy-level support to sustainably and meaningfully embed affective education in schools.

FINDINGS

This chapter presents the findings of the study, which explored teachers' perceptions, practices, and experiences in supporting the affective development of their students. Drawing on rich qualitative data gathered through in-depth interviews with 18 participants, the analysis was conducted using reflexive thematic analysis. The results obtained from these analyses are presented in Table 3.

Table 3. Themes identified through thematic analysis

Theme	Interpretative Meaning	Representative Participant Quote
Teachers' Conceptualisations of Affective Skills	How teachers define affective competencies and view them as foundational to students' holistic and academic development.	"Affective skills comprise a set of competencies that encompass individuals' emotional awareness, their ability to recognize and regulate emotions, and their capacity to approach others' feelings with empathy." (P3) "An individual's overall emotions, feelings, and empathetic behaviours can be collectively defined as affective skills." (P6)
Classroom Strategies for Supporting Affective Development	The practical pedagogical methods used to foster affective growth, including empathy-based interaction, group work, role-play, and teacher modelling.	"I utilize various teaching techniques such as the Six Thinking Hats, station activities, writing skills, and speaking skills." (P1-P2-P4-P5-P12-P13-P18) "Due to the challenges inherent in the mathematics discipline, I am unable to engage in such activities as much as I would like. However, I occasionally try to implement them through group work." (P3) "We enable them to develop empathy through group work and classroom interactions." (P4)
Monitoring and Evaluating Affective Growth	Teachers' reliance on informal observation and behavioural indicators, alongside concerns about the subjectivity of assessing emotional learning.	"To be honest, this is not my primary focus in the classroom." (P3) "I am not conducting any measurements." (P10-P17) "I don't have a measurement tool for this; I can only observe. I look at improvements in academic success and class participation." (P6)
Challenges in Promoting Affective Skills	Systemic and cultural barriers such as curriculum overload, exam pressure, limited parental engagement, and student resistance.	"Due to exam pressure, students focus solely on their scores." (P4) "The biggest challenges are indifferent parents, and the stress brought about by the examination system." (P5-P6) "The student's technology addiction hinders both their relationships with others and their own emotional well-being." (P7) "The intensity of the curriculum" (P3-P4-P11-P12-P13-P15-P16)
Support Needs and Systemic Reflections	Teachers expressed need for practical training, instructional materials, collaborative environments, and integration into the formal curriculum.	"Training should not consist solely of slide reading; there should also be practical workshop sessions." (P1) "Firstly, it is essential to develop teachers. The more active and innovative they are, the more students will align their behavior accordingly." (P13)

As demonstrated in Table 3, the instructors' perceptions regarding the processes involved in cultivating students' affective skills have been categorised into five overarching themes. These themes, delineated as follows, encompass the following: Teachers' Conceptualisations of Affective Skills, Classroom Strategies for Supporting Affective Development, Monitoring and Evaluating Affective Growth, Challenges in Promoting Affective Skills and Support Needs Systemic Reflections. The following section will present the instructors' opinions on the themes.

Theme 1: Teachers' Conceptualizations of Affective Skills

Participants in this study consistently highlighted the foundational role of affective skills in the holistic development of students. While terminology and emphasis varied slightly, there was a clear consensus that these skills go far beyond emotional expression. Teachers defined affective skills as a combination of emotional regulation, empathy, self-awareness, social communication, and interpersonal responsibility.

These were not perceived as supplementary or "soft" skills but as essential components of a student's ability to function effectively both inside and outside the classroom.

Many teachers explained that affective skills help students manage their feelings, respond constructively to challenges, and build meaningful relationships. P2 described affective competencies as "*Students' ability to express their feelings and thoughts in a balanced manner and to show respect for the emotions of others*" highlighting a balance between self-expression and social awareness. Similarly, P4 emphasized the importance of empathy and emotional self-control, noting that "*A student who can empathize communicates more effectively with both their peers and teachers; this directly affects the academic process.*"

Participants repeatedly pointed out that students who demonstrate strong affective skills are more likely to engage in learning, persevere through difficulty, and participate meaningfully in group settings. P3 reflected, "*We often overlook emotional development, but it shapes learning readiness. A student who feels understood is more likely to engage academically.*" Several others echoed this perspective, suggesting that emotional and social readiness were prerequisites for cognitive engagement. In this light, affective development was viewed not only as supportive but foundational to academic success.

At the same time, a few participants expressed concern that affective skills are not sufficiently prioritized in formal education. P8 commented, "*In our system, these skills are undervalued. The focus is still on grades, yet we expect students to behave maturely without teaching them how.*" This quote reflects a recurring tension between teachers' on-the-ground experiences and the performance-driven expectations of the broader educational system. Teachers perceived a misalignment between curriculum priorities and the real-life needs of their students, particularly in the face of increasing mental health issues, classroom behavioural challenges, and social instability.

Several educators also described their role as compensating for deficiencies in students' early socialisation. P1 observed that "*some students have never been taught to name or manage emotions at home, and we become their first exposure to that kind of learning.*" In this regard, affective skills were framed as essential life skills, not limited to the classroom but carried into students' personal and future professional lives. Teachers described this responsibility with both passion and frustration, acknowledging its importance but also the burden of having to fulfil it without adequate support.

In addition, participants frequently mentioned that affective competencies are increasingly relevant in today's social and technological environment. P7 stated, "*Students are overstimulated and underprepared to deal with emotions. We are dealing with anxiety, impulsiveness, social isolation.*" This was viewed not as a personal failing of the students but as a reflection of broader societal conditions. In such a context, affective education was not seen as optional but urgent. Teachers described themselves as first responders to emotional crises that previously may have been addressed at home or through other support systems.

Despite the shared recognition of their importance, teachers noted that affective skills remain largely unmeasured, undervalued, and under-supported in the current system. P14 said, "*We're expected to improve students' emotional well-being, but there's no space or time for it. It's all test scores.*" Teachers felt that while they carry the expectation to manage students' emotional development, there was little in the way of curriculum time, institutional structure, or formal recognition to support them in doing so.

In summary, affective skills were viewed by participants not as auxiliary or optional, but as essential and deeply integrated with all other forms of learning. The perspectives shared reflect a collective understanding that emotional and social competencies form the foundation for academic engagement, social integration, and personal development. However, this understanding was accompanied by a shared concern: despite their critical roles, these skills largely remain invisible within the structure and criteria of mainstream academic achievement and curriculum-driven education.

Theme 2: Classroom Strategies for Supporting Affective Development

Teachers in this study described a wide array of pedagogical strategies aimed at supporting the affective development of their students. These strategies were rarely standalone programs or explicit lessons. Instead, affective development was often embedded in everyday interactions, classroom norms, group structures, and teacher modelling. Many participants emphasized the importance of building a relational classroom climate in which emotional safety and mutual respect served as the foundation for all other learning activities.

Several teachers described intentional practices to encourage students to listen to one another and feel heard. P5 stated, "*Listening to the student without interrupting them contributes to their emotional development by promoting classroom participation.*" highlighting the affective value of uninterrupted speech and inclusive participation. This type of micro-level relational work was echoed by P11, who noted, "*I facilitate opportunities for students to get to know each other through feedback.*" For these teachers, respect, affirmation, and collaborative dialogue were not only classroom management tools but also practices that foster emotional development in themselves. A particularly prominent approach was cooperative learning. Group activities were commonly employed to foster empathy, peer understanding, and shared responsibility. P4 shared, "*We enable them to develop empathy through group work and classroom interactions.*" underscoring the idea that affective learning is inherently social and best developed through meaningful peer interaction. Participants reported using mixed-ability groups, peer mentoring, and collective problem-solving tasks to encourage students to recognize and respond to the emotions of others.

Teachers also employed creative teaching methods to help students connect with emotional perspectives that may not emerge in traditional academic settings. Role-playing, storytelling, and scenario-based activities were emphasized as effective tools for developing empathy and emotional awareness. P9 explained, "*By reenacting real-life events, students learn to step into someone else's shoes.*" Such activities were found to provide emotional insights, especially when dealing with conflict, moral dilemmas, or diverse experiences.

Many teachers spoke about modelling emotional behaviour in addition to structured teaching activities. They explained that they pay attention to their tone of voice, non-verbal cues, and their responses to student behaviour. In other words, how teachers manage their emotions is considered crucial in terms of the strategies students employ to manage their own emotions. For example, P14 stated, "*When I remain calm and understanding, students approach me in a similar manner.*" Teachers positioned themselves as emotional role models by regulating their own emotions and responding to students with compassion. This modelling was seen as essential for demonstrating empathy, patience, and emotional responsibility.

Despite examples indicating that teachers value students' emotions during classroom activities, some participants candidly acknowledged the limitations in the consistent implementation of such strategies. In other words, not all teachers viewed themselves as equal stakeholders in emotional integration. For instance, P2 shared, "*Due to the nature of mathematics instruction, I cannot adequately incorporate emotional skills.*" sincerely highlighting how certain disciplines are tightly bound to the curriculum and lesson topics, leaving less room for interpersonal emotional exploration. This sentiment was echoed by P7, who stated, "*There isn't enough time for these practices amidst the intensity of the lessons.*" These remarks point to a clear gap between pedagogical intent and systemic structure; in other words, while teachers may have the willingness and interest to develop affective skills, the pace of the curriculum and disciplinary expectations often hinder their full implementation. Additionally, some teachers indicated that they work through trial and error in the absence of formal training or institutional guidance. For example, P12 stated, "*The training we receive on this topic is very superficial; we often learn by experiencing what works and what doesn't.*" The lack of consistent resources, frameworks, or professional development for teachers to develop affective skills raises questions about the effectiveness of the strategies some educators implement in the classroom and whether they are being applied in developmentally appropriate ways.

In summary, teachers expressed that they utilize significant creativity, empathy, and professional intuition in developing emotional skills in their classrooms. Their approaches often involve enriching classroom teaching practices, promoting peer interaction, and modelling emotional behaviours. However, there are questions regarding the effectiveness and consistency of these strategies, as teachers experience dilemmas in balancing curriculum delivery and the provision of emotional skills. When this theme relies primarily on individual teacher initiative, it underscores both the potential and vulnerability of classroom-based affective learning.

Theme 3: Monitoring and Evaluating Affective Growth

Teachers overwhelmingly agreed on the importance of fostering emotional development in their students; however, they also expressed significant challenges regarding how to assess this growth. Unlike academic skills, which are typically measured through tests, grades, or objective evaluation criteria, affective skills often manifest in subtle and subjective ways. Consequently, teachers reported primarily using informal and observational methods to monitor emotional and social progress.

The most used method reported by participants was direct classroom observation. Teachers stated that they assessed changes in students' behaviors, communication styles, and relational dynamics as key indicators of affective development. P11 explained, "*I conduct observations and behavioral assessments. I notice changes such as more empathetic behavior, improved self-expression, higher motivation, and a more positive mindset.*" emphasizing that even small but meaningful shifts are significant in terms of emotional growth. Another participant, P14, stated, "*Students engage in role-play, dialogues, and task-based activities.*" emphasizing the importance of classroom activities in monitoring students' affective development. Similarly, P5 highlighted the importance of behavioral cues, stating, "*Over time, I can tell from the increase in class participation and the willingness to raise a hand and respond voluntarily.*" This suggests that changes in students' classroom engagement and participation strategies are perceived as reflections of their emotional maturation. While these observations may not be considered formal assessment methods, they demonstrate teachers' heightened awareness of their students' affective development and their sensitivity to classroom dynamics.

However, relying on informal methods brought a sense of uncertainty and vulnerability. Some participants openly expressed doubts about the reliability and validity of their assessments. P6 stated, "*I don't have a measurement tool for this; I can only observe. I look at improvements in academic success and class participation.*" Without structured tools or shared criteria, teachers had to interpret student behavior based on personal experience and intuition. While this approach allowed for flexibility and contextual sensitivity, it also introduced inconsistencies. Moreover, it could lead to significant variations and unforeseen challenges in terms of comparability over time, both across different classrooms and within the same class. On the other hand, a crucial point to emphasize is that teachers in quantitative lessons often do not engage in activities to monitor students' affective development. For example, participant P3 stated, "*Honestly, this is not my main focus in the classroom.*" highlighting this perspective. Additionally, P10 and P17 explicitly mentioned that they do not use any methods or tools to track affective development, indicating a lack of engagement in this aspect of student growth.

Some teachers expressed a desire for more structured digital tools to support affective assessment. P12 highlighted this issue, stating, "*There is no scale used for affective skills.*" Similarly, P11 shared, "*I need online seminars, training, and digital scales.*" Teachers with this perspective indicated an interest in descriptive frameworks, behavioural checklists, or qualitative assessment rubrics that could guide them in evaluating students' emotional development without reducing it to numerical scores. Such tools were seen as potentially useful for documenting progress, identifying needs, and facilitating communication with families or school leaders.

In conclusion, despite the absence of formal tools or standardized practices, teachers demonstrated a strong commitment to monitoring students' affective development. Their approaches were predominantly intuitive, relational, and experience-based rather than quantitatively scaled. However, the lack of systemic guidance contributed to uncertainty regarding the adequacy and reliability of their methods. This theme underscores an unresolved tension: while teachers recognize the importance of tracking and fostering affective growth, they also express a need for scientifically grounded assessment tools to enhance the validity and consistency of their evaluations.

Theme 4: Challenges in Promoting Affective Skills

Teachers expressed a strong commitment to fostering emotional development; however, they also reported encountering a wide range of structural, cultural, and interpersonal challenges that hindered their efforts. These barriers appeared to limit both the consistency and depth of emotional education in classrooms, despite teachers' recognition of its importance for students' overall growth.

A prominent and recurring concern among teachers was the pressure imposed by the existing curriculum. Teachers consistently highlighted the tension between meeting academic expectations and allocating sufficient time to support students' affective learning. P3 stated, "*Due to the intensity of the curriculum, I cannot dedicate time to such activities,*" emphasizing how the limited instructional time for specific subjects makes it challenging to incorporate content beyond the prescribed syllabus. Similarly, P5 reflected on the constraints of exam-oriented teaching, noting, "*The biggest challenge is indifferent parents and the stress imposed by the examination system.*" These remarks underscore a systemic misalignment between teachers' commitment to fostering holistic affective development and the narrowly defined academic outcomes prioritized by the national education system. Additionally, some teachers, even when personally investing in emotional education, have reported that the structure of the school day, assessment requirements, and subject-specific content demands leave them with minimal flexibility to plan or implement emotionally focused activities. For instance, P11 stated, "*Our education system is insufficient in this regard. Due to concerns about covering the curriculum, affective skills are not prioritized. However, by streamlining the curriculum, affective skills could be integrated into it.*" In such cases, emotional learning is perceived not as a proactive and systematically planned pedagogical focus but rather as an informal, reactive process.

Another significant challenge identified by the participants was the limited parental involvement or awareness regarding students' emotional development. P9 stated, "*The biggest challenge we face is the lack of parental support. If there is no parental profile that reinforces what is taught at school at home, it feels like we are just spinning our wheels, and no progress is observed in the student.*" Similarly, P14 supported this finding by stating, "*I work in an institution where parental support is minimal. Moreover, despite all efforts, some students remain uninterested and do not engage.*" Accordingly, teachers expressed that when families fail to reinforce emotional competencies at home, or worse, model emotional suppression or conflict, students may struggle to internalize the values taught at school. In response, participants emphasized the importance of parent-teacher meetings in overcoming these barriers. For instance, P5 noted, "*The process does not progress at all without collaboration with parents, so I prioritize involving them first.*" This finding underscores the importance of continuity between home and school environments in shaping students' emotional development. However, participants also frequently voiced their frustration over the inconsistency or complete absence of such collaboration.

Beyond structural and familial barriers, teachers also encountered resistance from students in implementing affective skill development practices. Some students appeared reluctant to participate in activities that emphasized emotional reflection, vulnerability, or collaboration. P14 stated, "*Despite all efforts, some students remain uninterested and do not engage.*" while P10 explained, "*Students' lack of interest and social withdrawal make things difficult for me.*" These statements suggest that teachers may lack effective pedagogical strategies for fostering students' affective development. Furthermore, they highlight how emotional expression, particularly during adolescence, can be perceived as socially risky and how such perceptions may hinder participation in emotional learning activities.

Participants also indicated that they typically sought to overcome student-related challenges through direct communication. P4 stated, "*Outside of guidance counseling, I hold discussions with students during recess. When I finish a lesson early, I try to create a conversational atmosphere to support them. I also make an effort to meet with parents more frequently.*" Similarly, P9 noted, "*I try to overcome these barriers through parent meetings and one-on-one discussions with students.*" However, some participants admitted to not finding solutions to these challenges. For instance, P15 stated, "*Since I teach a quantitative subject, I do not have enough time, and I cannot find a solution to this issue.*" Likewise, P3 mentioned, "*Due to the heavy curriculum load, I am unable to address this problem.*"

In summary, teachers faced multiple and interrelated challenges in fostering affective skills. These included curriculum constraints, exam-oriented priorities, inconsistent parental involvement, student resistance, and insufficient pedagogical knowledge. Despite these obstacles, teachers demonstrated resilience and commitment by seeking ways to integrate emotional learning into their practices, with parent and student meetings being the most used strategies to overcome barriers. However, the broader picture suggests that without structural reforms and wider cultural shifts, their individual efforts remain limited in impact. Additionally, teachers need greater access to time and pedagogical resources to effectively address the barriers they face in promoting students' affective skill development.

Theme 5: Support Needs and Systemic Reflections

As teachers reflected on their efforts to promote emotional development, many expressed a strong and consistent need for greater institutional support. Despite their commitment to fostering students' emotional and social competencies, participants often felt that their work in this area was carried out in isolation, without sufficient guidance, recognition, or structural reinforcement. Their perspectives reflected both an urgent call for professional development and systemic change, as well as a desire for greater alignment between educational policies and emotional education goals.

One of the most frequently mentioned support needs was access to training that was both practical and grounded in the realities of the classroom. Some participants noted that while affective education was occasionally addressed in theory, there was a significant lack of hands-on, context-specific training to help them effectively implement these principles. P1 was particularly explicit: "*Training should not consist solely of slide reading; there should also be practical workshop sessions.*". Similarly, P13 noted, "*Firstly, it is essential to develop teachers. The more active and innovative they are, the more students will align their behaviour accordingly.*". Teachers emphasized the need for experiential workshops, case-based discussions, and opportunities to share effective strategies tailored to specific subject areas or age groups. This emphasis on practical training indicates that educators are not only eager to understand emotional development but also to integrate it into their teaching contexts in meaningful and manageable ways.

Participants also expressed concerns about the lack of teaching materials and resources tailored to emotional education. P5 highlighted this issue directly: "*The lack of resource books and insufficient digital materials limit implementation.*" Similarly, P11 stated, "*I need online seminars, training sessions, and digital assessment tools.*" In this context, teachers often had to design their own materials or rely on intuition to improvise in supporting students' affective development. While this approach fostered creativity, it also led to inconsistencies across classrooms, depending on individual teaching practices.

The absence of structured programs, suggested activities, or developmental benchmarks created a sense of fragmentation, particularly in subjects such as mathematics and science, where emotional education was not traditionally emphasized. This added an extra burden on teachers already navigating a demanding curriculum. Beyond the classroom, systemic change was widely seen as essential for sustaining emotional education in schools. Some teachers argued that unless emotional and social competencies were explicitly integrated into the curriculum, emotional development would remain marginal. P12 proposed a concrete solution: "*The curriculum should be streamlined, and emotional skills should be included in learning outcomes, with dedicated instructional time for such activities.*" This statement underscores the need for policy-level initiatives that provide teachers with guidance and necessary support in fostering students' affective development. Participants believed that embedding emotional goals into the official curriculum would not only reinforce their efforts but also ensure the time, resources, and frameworks required for consistent implementation.

In this regard, participants emphasized the importance of leadership and whole-school involvement. P1 noted that the responsibility for affective education was often left to school counsellors, with teachers collaborating with them to implement these efforts: "*We carry out a joint initiative with the guidance department. Occasionally, I invite school administrators as guests to the classroom and ask them to speak with the students.*" Similarly, P9 highlighted the necessity of a collective vision, stating, "*A school must always be considered as a whole; this cannot be achieved solely through the efforts of a few teachers or administrators.*" Supporting these perspectives, several participants underscored the significance of professional solidarity among colleagues. For example, P12 stated, "*We collaborate with subject-area teachers and, with the support of school administration, implement both in-school and out-of-school activities.*" This response highlights how peer support and collective planning can serve as meaningful reinforcements for teachers engaged in emotional education. These viewpoints suggest that the success of emotional education initiatives may depend on leadership priorities and the overall professional culture within a school.

Teachers also called for clearer communication channels between schools, families, and external support systems such as counselling services. P3 emphasized this need, stating, "*I would establish training programs for teachers, reduce their course load, and incorporate dedicated time for such activities into the curriculum. I would collaborate with universities to receive support from expert academics in the field.*" Similarly, P7 suggested, "*I would recommend setting up dedicated workspaces in schools where teachers can further their development in collaboration with guidance counselors.*". However, not all schools had such support structures in place, and some existing initiatives were perceived as superficial. Many participants emphasized the need for teacher training and increased awareness regarding affective education practices. For instance, P17 expressed frustration with the lack of practical engagement: "*Teachers should be equipped with various competencies. Everything just being on paper feels very dull to me... Schools should be more integrated with real life, more engaging, and more focused on raising psychological and emotionally healthier individuals. If these changes happen, everything will naturally improve.*". In cases where these support structures were lacking, the responsibility for emotional care often fell disproportionately on certain teachers, leading to additional workload and professional strain. As a result, it is essential to raise awareness among all teachers about the importance of affective education and to enhance their related skills. Teachers appear willing to engage in this process but often experience confusion about how to effectively implement such practices due to varying approaches and existing limitations in the field.

In summary, teachers made it clear that efforts to develop students' affective skills require more than just personal dedication. They need practical and sustained support at multiple levels, ranging from classroom resources and training to leadership vision, school-wide policies, and national curriculum frameworks. While examples of good practices and internal collaboration were mentioned, the broader picture revealed fragmentation, inconsistency, and a widespread desire for greater coherence. As one teacher stated, *"Everyone cares about this issue, but as a system, our hands are tied."* This sentiment encapsulates the core theme: emotional education cannot rely solely on goodwill, it must be supported, structured, and strategically integrated into the education system.

DISCUSSION AND CONCLUSION

The findings of this study affirm the pivotal role of affective skills in fostering students' holistic development and reveal the complex challenges that teachers face in embedding these competencies within existing educational structures. Teachers consistently emphasized that emotional regulation, empathy, and social communication are not peripheral attributes but integral components of students' academic achievement, emotional resilience, and adaptive social functioning (Durlak et al., 2011). These affective dimensions of learning not only enhance students' cognitive engagement but also facilitate critical life skills essential for navigating interpersonal relationships and responding to challenges in diverse contexts. However, systemic constraints, such as rigid curriculum mandates, a disproportionate emphasis on standardized testing, and insufficient institutional support, were identified as formidable impediments to the effective integration of affective education into classroom practices (OECD, 2015). These structural limitations perpetuate a dichotomous approach to education that privileges cognitive outcomes over emotional and social growth, thereby marginalizing the affective domain and constraining its transformative potential. Without a fundamental reconfiguration of educational priorities that recognizes the interdependence of affective and cognitive dimensions, the holistic development of students remains an aspirational goal rather than a tangible reality.

A persistent concern articulated by participants was the inherent tension between fulfilling academic mandates and fostering affective competencies. Teachers expressed frustration that the prevailing focus on curriculum coverage and assessment benchmarks relegated emotional learning to a secondary priority, creating a disconnect between pedagogical aspirations and institutional expectations. This tension aligns with the observations of Zins et al. (2004), who argue that emotional learning, when treated as an ancillary objective, fails to permeate the broader educational ethos and is consequently rendered ineffective. Critics, however, have cautioned against an overemphasis on affective education, suggesting that excessive focus on socio-emotional dimensions may dilute the rigor of cognitive instruction (Boler, 1999). While this critique underscores the importance of maintaining balance, it overlooks the well-established interplay between emotional intelligence and cognitive functioning, whereby emotional competencies enhance students' ability to engage deeply with academic content and persevere through intellectual challenges (Immordino-Yang & Damasio, 2007). Consequently, the dichotomy between affective and cognitive development is not only conceptually flawed but pedagogically untenable, necessitating a paradigm shift toward a more integrated model of education that cultivates both dimensions simultaneously.

The study further highlighted that teachers predominantly rely on informal and observational methods to assess affective growth, leading to inconsistencies and ambiguities in tracking student progress. Participants reported employing anecdotal evidence, behavioral observations, and subjective impressions to evaluate students' emotional development, reflecting a reliance on intuitive, albeit imprecise, assessment strategies. This aligns with Schonert-Reichl's (2017) assertion that the absence of standardized frameworks for assessing socio-emotional competencies exacerbates variability in evaluation practices and undermines the reliability of reported outcomes. While informal assessments offer valuable insights into the nuanced and context-dependent nature of affective growth, they lack the rigor and consistency required for establishing reliable benchmarks for socio-emotional development. Moreover, the reliance on subjective interpretations introduces the potential for implicit bias, resulting in inconsistent evaluations that disproportionately affect marginalized student populations (Hall, 2011). The development of robust, culturally responsive, and developmentally appropriate assessment frameworks is therefore imperative to ensure that affective growth is not only recognized but systematically documented and evaluated.

Structural and cultural barriers further constrain teachers' capacity to promote affective development effectively. Participants highlighted that curriculum overload, examination pressures, limited parental engagement, and inadequate institutional support pose formidable obstacles that hinder the consistent implementation of affective practices. These findings resonate with Vygotsky's (1978) sociocultural theory, which emphasizes that learning occurs within a complex web of interpersonal relationships, cultural expectations, and institutional norms that collectively shape both cognitive and affective growth. However, contemporary educational environments, driven by performance-oriented paradigms, often marginalize affective education by prioritizing quantifiable outcomes over emotional and social competencies. The institutional undervaluation of affective growth perpetuates an educational culture that constrains teachers' ability to cultivate emotionally intelligent learning environments. Addressing these barriers requires not only structural reforms that embed affective learning into curricular frameworks but also a cultural shift that recognizes emotional and social competencies as indispensable dimensions of student success.

The teacher-student relationship emerged as a critical determinant of affective outcomes, with participants emphasizing that emotionally attuned relationships create a secure environment in which students feel valued, understood, and empowered to engage both academically and emotionally. This finding aligns with extensive research demonstrating that positive teacher-student relationships promote trust, emotional safety, and resilience, thereby fostering students' capacity to regulate emotions, navigate social complexities, and persist through academic challenges (Rucinski et al., 2018). However, the emotional labor required to sustain such relationships remains underappreciated within institutional frameworks, contributing to teacher burnout and emotional exhaustion (Jennings et al., 2017). Teachers invest significant emotional energy in cultivating affective connections with students, yet this labor is often rendered invisible and inadequately supported within performance-driven educational cultures. Recognizing and valuing this affective labor is essential to sustaining emotionally responsive classrooms and promoting long-term teacher well-being.

Teachers also expressed a strong desire for professional development programs that provide both theoretical grounding and practical strategies for implementing affective practices effectively. Many participants noted that existing training programs were superficial, overly theoretical, and disconnected from the realities of classroom practice, leaving them ill-equipped to address the emotional and social dimensions of student development. This finding is consistent with the work of Semchuk et al. (2022), who advocate for professional development models that incorporate experiential learning, case-based reflection, and collaborative problem-solving to enhance teachers' affective competencies. Moreover, effective professional development should not be conceived as a one-time intervention but as an iterative, collaborative process embedded within ongoing professional learning communities. Such communities provide opportunities for teachers to engage in reflective practice, exchange best practices, and co-construct knowledge around affective education, thereby enhancing both individual expertise and collective capacity for fostering emotional growth.

These findings underscore the necessity of reimagining education as a multidimensional endeavor that nurtures both cognitive excellence and emotional competence. An education system that prioritizes affective skills alongside academic achievement fosters the development of well-rounded individuals who possess the resilience, empathy, and critical thinking necessary to succeed in a rapidly evolving world. To achieve this vision, affective education must be fully integrated into the fabric of teaching and learning, supported by institutional policies that recognize emotional and social growth as foundational to student success. In parallel, teacher training and professional development must equip educators with the knowledge, tools, and ongoing support needed to implement affective practices effectively. Moreover, the establishment of reliable, culturally responsive, and developmentally appropriate assessment frameworks is essential to capturing the complexity of affective growth and ensuring that it is accorded the same legitimacy as cognitive development. By aligning educational policies, pedagogical practices, and assessment frameworks with this more holistic vision, educators and policymakers can create emotionally intelligent learning environments where students not only excel academically but also develop the affective competencies that enable them to thrive as empathetic, responsible, and ethical members of society. This integrated approach reflects a profound commitment to nurturing the whole child and preparing future generations to engage meaningfully with the complexities of a diverse and interconnected world.

REFERENCES

Adaba, H. (2017). Exploring the practice of teacher-student classroom interaction in EFL to develop the learners' speaking skills in Tullu Sangota Primary School Grade Eight students in focus. *Arts and Social Sciences Journal*, *8*(4). Advance online publication. DOI: 10.4172/2151-6200.1000295

Arifin, A., Suryaningsih, S., & Arifudin, O. (2024). The relationship between classroom environment, teacher professional development, and student academic performance in secondary education. *International Education Trend Issues*, *2*(2), 151–159. DOI: 10.56442/ieti.v2i2.467

Cefai, C., Ferrario, E., Cavioni, V., Carter, A., & Grech, T. (2013). Circle time for social and emotional learning in primary school. *Pastoral Care in Education*, *32*(2), 116–130. DOI: 10.1080/02643944.2013.861506

Clack, L., & Ellison, R. (2018). Innovative approaches to management education. *Journal of Management Policies and Practices*, *6*(1), 6–9. DOI: 10.15640/jmpp.v6n1a2

Collaborative for Academic, Social, and Emotional Learning (CASEL). (2020). *Core SEL competencies*. https://casel.org/core-competencies/

Cooper, K. M., Downing, V. R., & Brownell, S. E. (2018). The influence of active learning practices on student anxiety in large-enrollment college science classrooms. *International Journal of STEM Education*, *5*(1), 23. Advance online publication. DOI: 10.1186/s40594-018-0123-6 PMID: 30631713

Creswell, J. (2013). Qualitative inquiry & research design: Choosing among five approaches.

Creswell, J. W., & Poth, C. N. (2018). *Qualitative inquiry and research design: Choosing among five approaches* (4th ed.). Sage Publications.

Durlak, J. A., Weissberg, R. P., Dymnicki, A. B., Taylor, R. D., & Schellinger, K. B. (2011). The impact of enhancing students' social and emotional learning: A meta-analysis of school-based universal interventions. *Child Development*, *82*(1), 405–432. DOI: 10.1111/j.1467-8624.2010.01564.x PMID: 21291449

Ebubechukwu, R. (2021). Psychological education and mental well-being: Strategies for promoting emotional intelligence. *Interdisciplinary Journal Papier Human Review*, *2*(2), 1–4. DOI: 10.47667/ijphr.v2i2.218

Eng, T., Eng, L., Liew, C., Daneil, I., & Ting, G. (2023). An application of multiple linear regression analysis to establish the relationship between higher order thinking skills and affective domain. *International Journal of Academic Research in Business & Social Sciences*, *13*(5). Advance online publication. DOI: 10.6007/IJARBSS/v13-i5/16907

Finlay, L. (2014). Engaging phenomenological analysis. *Qualitative Research in Psychology*, *11*(2), 121–141. DOI: 10.1080/14780887.2013.807899

Flook, L., Goldberg, S. B., Pinger, L., Bonus, K., & Davidson, R. J. (2013). Mindfulness for teachers: A pilot study to assess effects on stress, burnout, and teaching efficacy. *Mind, Brain and Education : the Official Journal of the International Mind, Brain, and Education Society*, *7*(3), 182–195. DOI: 10.1111/mbe.12026 PMID: 24324528

Frenzel, A. C., Becker-Kurz, B., Pekrun, R., Goetz, T., & Lüdtke, O. (2018). Emotion transmission in the classroom revisited: A reciprocal effects model of teacher and student enjoyment. *Journal of Educational Psychology*, *110*(5), 628–639. DOI: 10.1037/edu0000228

Ghahremani, M., & Ostovar, Z. (2017). The contribution of emotional intelligence and achievement motivation on psychological well-being. *Global Journal of Health Science*, *9*(3), 88. DOI: 10.5539/gjhs.v9n3p88

Gharaei, N., Thijs, J., & Verkuyten, M. (2018). Ethnic identity in diverse schools: Preadolescents' private regard and introjection in relation to classroom norms and composition. *Journal of Youth and Adolescence*, *48*(1), 132–144. DOI: 10.1007/s10964-018-0881-y PMID: 29922870

Goleman, D. (1995). *Emotional intelligence: Why it can matter more than IQ*. Bantam Books.

Gross, J. J. (2014). Emotion regulation: Conceptual and empirical foundations. In Gross, J. J. (Ed.), *Handbook of emotion regulation* (2nd ed., pp. 3–20). Guilford Press.

Hall, R. A. (2011). Affective assessment: The missing piece of the educational reform puzzle. *Delta Kappa Gamma Bulletin*, *77*(2), 7.

Herranz-Hernández, P., Fernández-Hernández, J., & Segovia-Torres, L. (2023). Emotional metaphors for an inclusive classroom climate. *Environment and Social Psychology*, *9*(1). Advance online publication. DOI: 10.54517/esp.v9i1.1945

Hojat, M., & Gonnella, J. (2017). What matters more about the interpersonal reactivity index and the Jefferson Scale of Empathy? Their underlying constructs or their relationships with pertinent measures of clinical competence and patient outcomes? *Academic Medicine*, *92*(6), 743–745. DOI: 10.1097/ACM.0000000000001424 PMID: 28557931

Humphries, M. L., Strickland, J., & Keenan, K. (2014). African American preschoolers' social and emotional competence at school: The influence of teachers and mothers. *Journal of Education*, *194*(1), 11–18. DOI: 10.1177/002205741419400104

Hutchison, M. (2016). The empathy project: Using a project-based learning assignment to increase first-year college students' comfort with interdisciplinarity. *The Interdisciplinary Journal of Problem-Based Learning*, *10*(1). Advance online publication. DOI: 10.7771/1541-5015.1580

Immordino-Yang, M. H., & Damasio, A. (2007). We feel, therefore we learn: The relevance of affective and social neuroscience to education. *Mind, Brain and Education : the Official Journal of the International Mind, Brain, and Education Society*, *1*(1), 3–10. DOI: 10.1111/j.1751-228X.2007.00004.x

Jennings, P. A., Brown, J. L., Frank, J. L., Doyle, S., Oh, Y., Davis, R., Rasheed, D., DeWeese, A., DeMauro, A. A., Cham, H., & Greenberg, M. T. (2017). Impacts of the CARE for Teachers program on teachers' social and emotional competence and classroom interactions. *Journal of Educational Psychology*, *109*(7), 1010–1028. DOI: 10.1037/edu0000187

Kim, K. (2020). Project-based learning approach to increase medical student empathy. *Medical Education Online*, *25*(1), 1742965. Advance online publication. DOI: 10.1080/10872981.2020.1742965 PMID: 32197574

Krathwohl, D. R. (2002). A Revision of Bloom's Taxonomy: An Overview. *Theory into Practice*, *41*(4), 212–218. DOI: 10.1207/s15430421tip4104_2

Mahoney, S., Sladek, R., & Neild, T. (2016). A longitudinal study of empathy in pre-clinical and clinical medical students and clinical supervisors. *BMC Medical Education*, *16*(1), 270. Advance online publication. DOI: 10.1186/s12909-016-0777-z PMID: 27756379

Mai, Y. (2024). Unveiling the role of classroom climate in Chinese EFL teachers' perceptions of their emotional exhaustion and attrition. *Perceptual and Motor Skills*, *132*(1), 169–194. DOI: 10.1177/00315125241291088 PMID: 39558200

Maslow, A. H. (1968). *Toward a psychology of being* (2nd ed.). Van Nostrand Reinhold.

Masnan, A., Ismail, N. A., Usop, W., & Zain, D. (2024). Predicting preschool moral development with school climate: Fuzzy Delphi approach. *International Journal of Evaluation and Research in Education*, *13*(3), 1537. DOI: 10.11591/ijere.v13i3.27777

Mayer, J. D., Salovey, P., & Caruso, D. R. (2008). Emotional intelligence: New ability or eclectic traits? *The American Psychologist*, *63*(6), 503–517. DOI: 10.1037/0003-066X.63.6.503 PMID: 18793038

Miles, M. B., & Huberman, A. M. (1994). *Qualitative data analysis*. Sage Publications.

Miller, R. (2007). *What are schools for? Holistic education in American culture* (3rd ed.). Holistic Education Press.

Morse, J. M. (2015). Critical analysis of strategies for determining rigor in qualitative inquiry. *Qualitative Health Research*, *25*(9), 1212–1222. DOI: 10.1177/1049732315588501 PMID: 26184336

Nurhasanah, N., Neviyarni, S., & Effendi, H. (2019). The effectiveness of group counseling with role-playing techniques to increase student empathy. *International Journal of Applied Counseling and Social Sciences*, *1*(1), 54–61. DOI: 10.24036/005304ijaccs

Odak, I., Marušić, I., Matić, J., Puzić, S., Bakić, H., Eliasson, N., & Veldin, M. (2023). *Teachers' social and emotional competencies: A lever for social and emotional learning in schools*. Sociologija i Prostor., DOI: 10.5673/sip.61.1.5

OECD. (2015). *Skills for social progress: The power of social and emotional skills*. OECD Publishing.

OECD. (2021). *Beyond academic learning: First results from the Survey on Social and Emotional Skills*. OECD Publishing.

Pekrun, R., & Linnenbrink-Garcia, L. (2014). *International handbook of emotions in education*. Routledge. DOI: 10.4324/9780203148211

Rahiminia, E., Yazdani, S., & Rahiminia, H. (2020). Factors affecting concentration and attendance in the classroom from students' point of view in Qom University of Medical Sciences (2018). *Educational Research in Medical Sciences*, *8*(2). Advance online publication. DOI: 10.5812/erms.93075

Ramaila, S. (2022). Promoting self-regulated learning in natural sciences teaching through technology integration. https://doi.org/DOI: 10.36315/2022v1end081

Rance, G., Dowell, R., & Tomlin, D. (2023). The effect of classroom environment on literacy development. *NPJ Science of Learning*, *8*(1), 9. Advance online publication. DOI: 10.1038/s41539-023-00157-y PMID: 37012296

Reyes, M. R., Brackett, M. A., Rivers, S. E., White, M., & Salovey, P. (2012). Classroom emotional climate, student engagement, and academic achievement. *Journal of Educational Psychology*, *104*(3), 700–712. DOI: 10.1037/a0027268

Rivers, S. E., Brackett, M. A., Reyes, M. R., Elbertson, N. A., & Salovey, P. (2012). Improving the social and emotional climate of classrooms: A clustered randomized controlled trial testing the RULER approach. *Prevention Science*, *14*(1), 77–87. DOI: 10.1007/s11121-012-0305-2 PMID: 23188089

Rogers, C. R. (1969). *Freedom to learn: A view of what education might become*. Charles Merrill.

Ross, K., & Tolan, P. (2017). Social and emotional learning in adolescence: Testing the CASEL model in a normative sample. *The Journal of Early Adolescence*, *38*(8), 1170–1199. DOI: 10.1177/0272431617725198

Rucinski, C. L., Brown, J. L., & Downer, J. T. (2018). Teacher–child relationships, classroom climate, and children's social-emotional and academic development. *Journal of Educational Psychology*, *110*(7), 992–1004. DOI: 10.1037/edu0000240

Schonert-Reichl, K. A. (2017). Social and emotional learning and teachers. *The Future of Children*, *27*(1), 137–155. DOI: 10.1353/foc.2017.0007

Semchuk, J., McCullough, S., Lever, N., Gotham, H., Gonzalez, J., & Hoover, S. (2022). Educator-informed development of a mental health literacy course for school staff: Classroom wellbeing information and strategies for educators (Classroom WISE). *International Journal of Environmental Research and Public Health*, *20*(1), 35. DOI: 10.3390/ijerph20010035 PMID: 36612359

Soldaat, L. (2019). More than a teacher: Understanding the teacher-learner relationship in a public high-school in South Africa. *Journal of Educational and Developmental Psychology*, *9*(1), 77. DOI: 10.5539/jedp.v9n1p77

States, N., Karsten, M., Nennig, H., & Cole, R. (2023). Let students work: Analysis of the role of differing facilitation on student engagement in a large stadium-style lecture hall. *Journal of Chemical Education*, *100*(11), 4237–4248. DOI: 10.1021/acs.jchemed.3c00750 PMID: 38028750

Stevanović, N., Hoare, E., McKenzie, V., & O'Connor, M. (2017). Sustaining the use of positive education coping skills to meet the challenges of the emerging adulthood period. *International Journal of Wellbeing*, *7*(3), 39–55. DOI: 10.5502/ijw.v7i3.640

Sutton, R. E., & Wheatley, K. F. (2003). Teachers' emotions and teaching: A review of the literature and directions for future research. *Educational Psychology Review*, *15*(4), 327–358. DOI: 10.1023/A:1026131715856

van Manen, M. (2016). *Phenomenology of practice: Meaning-giving methods in phenomenological research and writing*. Routledge. DOI: 10.4324/9781315422657

Vygotsky, L. S. (1978). *Mind in society: The development of higher psychological processes*. Harvard University Press.

Wagner, C., Kawulich, B., & Garner, M. (2019). A Mixed Research Synthesis of Literature on Teaching Qualitative Research Methods. *SAGE Open*, *9*(3), 2158244019861488. Advance online publication. DOI: 10.1177/2158244019861488

Yıldırım, A., & Şimşek, H. (2021). *Sosyal bilimlerde nitel araştırma yöntemleri* [Qualitative research methods in the social sciences]. 11th ed.). Seçkin Yayıncılık.

Zaki, J., & Cikara, M. (2015). Addressing empathic failures. *Current Directions in Psychological Science*, *24*(6), 471–476. DOI: 10.1177/0963721415599978

Zhou, D. (2024). The influence of smart classrooms on students' learning experience in the digital age. *Advances in Educational Technology and Psychology*, *8*(2). Advance online publication. DOI: 10.23977/aetp.2024.080210

Zins, J. E., Weissberg, R. P., Wang, M. C., & Walberg, H. J. (2004). *Building academic success on social and emotional learning: What does the research say?* Teachers College Press.

Chapter 5
Shaping Inclusive Mindsets Through Teacher Education and Personal Experience in Türkiye

Bahaddin Demirdis
https://orcid.org/0000-0001-5843-5072
Çankırı Karatekin University, Turkey

Abdullah Çiftçi
https://orcid.org/0000-0002-4554-4752
Akdeniz University, Turkey

ABSTRACT

This chapter examines the attitudes of teacher candidates in Türkiye towards inclusive education, focusing on the influence of teacher training pathways and personal demographic factors. Drawing on data from 369 participants enrolled in either education faculties or pedagogical formation programmes, the study explores how variables such as gender, age, and prior experience with individuals with special educational needs (SEN) shape inclusive mindsets. The findings reveal significant differences based on gender and familial experience with SEN, while age and type of training programme show no statistically significant impact. Using a validated attitudes scale, this chapter highlights the critical role of experiential learning and targeted professional development in shaping positive perceptions of inclusive education. The chapter concludes with implications for teacher education policy and practices, emphasizing the need for comprehensive, equity-focused training that bridges theory and practice.

INTRODUCTION

Inclusive education, as a cornerstone of equitable societies, emphasises the integration of diverse learner needs within mainstream educational environments. By fostering a sense of belonging and equal opportunities, it aims to address disparities among learners and promote holistic development (UNESCO, 1994). In Türkiye, the journey toward inclusive education gained momentum with the adoption of the Salamanca Statement (1994), which advocated for the least restrictive environment in meeting diverse learner needs (Sari et al., 2020). Although earlier approaches focused on specialised schools, the past

DOI: 10.4018/979-8-3373-0472-4.ch005

decade has seen a significant shift towards including students with special educational needs (SEN) in mainstream classrooms, supported by dedicated guidance centres and specialised teacher training programmes (Sari et al., 2020). These efforts highlight the growing focus on providing educators with the essential knowledge and skills required for effective inclusive teaching. However, variations in teacher training pathways, such as education faculty programmes and pedagogical formation programmes, raise critical questions about their impact on teacher preparedness for inclusive education (Higher Education Council, 2023).

The role of educators is pivotal in achieving these goals, with effective inclusive practices requiring tailored pedagogical approaches and a supportive administrative framework (Avramidis et al., 2019; Ruijs & Peetsma, 2009). Existing research underscores the importance of specialised training in fostering positive attitudes towards inclusion (Avramidis & Norwich, 2002; Saloviita, 2020a). Factors such as gender, age, and personal experience with individuals with SEN have been examined; however, the interaction of these variables with different teacher training pathways remains underexplored. In the Turkish context, comparative analyses between teacher candidates from education faculties and those pursuing pedagogical formation programmes are notably limited. To address this gap, the current study seeks to examine the impact of these variables on teacher candidates' attitudes towards inclusive education. Specifically, the study seeks to examine whether distinct training pathways yield significant differences in attitudes and how personal factors may moderate these outcomes. This study aims to contribute to the broader discourse on inclusive education in Türkiye by offering insights into the factors that shape inclusive attitudes, thereby informing teacher education policies.

INCLUSIVE EDUCATION'S HISTORICAL EVOLUTION

Inclusive education has undergone significant transformations over the decades, shifting from segregated models to fully integrated approaches that prioritize equity and diversity. Historically, students with disabilities and other marginalized groups were placed in separate educational settings, often receiving instruction in specialized institutions that limited their social and academic interactions with their peers (Winzer, 1993). This segregation was largely rooted in the medical model of disability, which perceived disabilities as inherent deficiencies requiring individualized remediation rather than systemic accommodations (Oliver, 1990). Consequently, exclusionary educational policies were widely adopted across different countries, reinforcing societal biases that positioned students with disabilities as incapable of succeeding in mainstream classrooms. However, the mid-20th century brought about a paradigm shift, influenced by broader civil rights movements advocating for equal educational opportunities and legal protections for marginalized groups, including individuals with disabilities (Skrtic, 1991). These movements laid the groundwork for national and international policies that gradually transitioned education systems from exclusion to inclusion, challenging long-standing discriminatory practices and emphasizing the right to education for all (Ainscow, 2020).

A pivotal moment in the history of inclusive education was the Education for All Handicapped Children Act (1975) in the United States, which later evolved into the Individuals with Disabilities Education. This legislation mandated that students with disabilities receive free and appropriate public education in the least restrictive environment, ensuring their access to mainstream schools whenever possible. Similar policy shifts emerged in European and other developed nations, signaling a growing commitment to educational equity. However, the global push for inclusion gained its strongest momentum with the Salamanca Statement and Framework for Action on Special Needs Education (1994), endorsed by 92 governments and 25 international organizations (UNESCO, 1994). The statement emphasized that "regular schools with an inclusive orientation are the most effective means of combating discriminatory attitudes, creating welcoming communities, and achieving education for all" (UNESCO, 1994, p. 11). This declaration laid the foundation for many contemporary inclusive education policies and positioned inclusive schooling as a fundamental human right rather than a discretionary practice (Hehir et al., 2016).

Building on the principles of the Salamanca Statement, the United Nations Convention on the Rights of Persons with Disabilities (UNCRPD) reinforced the need for inclusive education systems that ensure accessibility, participation, and equity for all students. Articles within the UNCRPD specifically require signatory states to implement laws and policies that guarantee non-discriminatory access to mainstream education, emphasizing that students with disabilities should not be excluded based on their impairments (United Nations, 2006). Countries such as Finland, Canada, and Italy have since taken the lead in implementing progressive inclusive education models, incorporating differentiated instruction, strong policy frameworks, and systemic support mechanisms to ensure equal educational opportunities for all students (Ainscow, 2020). Despite these advancements, many nations, particularly in the Global South, continue to face significant barriers in fully realizing inclusive education due to insufficient teacher training, inadequate infrastructure, and deeply ingrained societal biases (Engelbrecht et al., 2016). These disparities highlight the ongoing need for continued policy refinement, investment in professional development, and broader societal awareness campaigns to facilitate a meaningful shift toward inclusive educational practices worldwide.

GLOBAL PRACTICES IN INCLUSIVE EDUCATION

The concept of inclusive education is widely endorsed at the international level through frameworks such as the Salamanca Statement (UNESCO, 1994) and the United Nations Convention on the Rights of Persons with Disabilities, however, its implementation varies significantly across countries. Differences in policy, teacher preparedness, resource allocation, and societal attitudes influence the effectiveness of inclusive education in practice (Ainscow, 2020). However, several global practices have been identified as key contributors to successful inclusive education, demonstrating commonalities in systemic approaches that lead to more inclusive and equitable schooling (Engelbrecht et al., 2016). These best practices focus on strengthening legislative frameworks, investing in teacher education and professional development, ensuring access to differentiated instructional methods, and fostering collaboration between schools, families, and communities (Porter & AuCoin, 2012).

The establishment of strong legislative and policy frameworks serves as a cornerstone of effective inclusive education (D'Alessio, 2011). Countries that have achieved notable progress in inclusion often have legally binding mandates that ensure students with disabilities and other marginalized groups are granted equal access to mainstream education (Takala et al., 2018). Legislative commitments that promote accessibility, individualized support, and teacher collaboration ensure that inclusion is not merely an aspirational goal but an enforceable right. Countries such as Finland, Italy, and Canada have developed policies that explicitly mandate full inclusion rather than segregation. The Italian model, which abolished separate special education schools in 1977, requires that students with disabilities be educated alongside their peers in general classrooms with appropriate support structures (D'Alessio, 2011). Similarly, Finland's three-tiered support system provides differentiated levels of assistance depending on students' needs, ensuring timely interventions (Takala et al., 2018). The Canadian approach places a strong emphasis on inclusive legislation at both the national and provincial levels, with mandates requiring schools to provide accommodations such as assistive technology, specialized instructional strategies, and personalized learning plans (Porter & AuCoin, 2012). These policy-driven approaches highlight the necessity of sustained government commitment to inclusive education, as legal frameworks create accountability and enforceable obligations for schools and educators (Ainscow, 2020).

The role of teacher training and professional development is central to the successful implementation of inclusive education (Avramidis & Norwich, 2002). Research consistently indicates that teachers' attitudes, beliefs, and competencies significantly impact the effectiveness of inclusion in the classroom (Sharma & Sokal, 2016). In contexts where inclusive education is embedded in pre-service teacher education programs and reinforced through continuous professional development, teachers demonstrate greater confidence and willingness to adopt inclusive practices. The most effective training programs integrate both theoretical knowledge and practical experience, allowing pre-service teachers to gain direct exposure to inclusive teaching methodologies through supervised internships and hands-on training (Malinen et al., 2012).

Family and community involvement is another critical factor in the success of inclusive education (Malinen et al., 2012). Countries that prioritize parental engagement through workshops, inclusion advocacy programs, and school-community partnerships tend to experience higher levels of social acceptance of inclusive practices. Studies have shown that students with disabilities benefit significantly when families actively participate in their education, as parental involvement reinforces learning at home and fosters collaboration between educators and caregivers (Sharma & Sokal, 2016). In addition to parental engagement, community-based support networks play an essential role in raising awareness, providing resources, and advocating for policy improvements (Engelbrecht et al., 2016).

Inclusive education is an evolving process that requires ongoing policy development, investment in teacher training, and societal commitment to educational equity (Ainscow, 2020). The most successful global practices demonstrate that strong legislation, comprehensive teacher preparation, differentiated instructional methods, and active community engagement contribute to effective inclusion (Porter & AuCoin, 2012). While challenges remain, countries that adopt evidence-based approaches to inclusion are better equipped to create learning environments that support all students, regardless of ability. As inclusive education continues to gain momentum globally, governments and educators must commit to long-term strategies that prioritize equity, accessibility, and diversity in schools.

CHALLENGES IN IMPLEMENTING INCLUSIVE EDUCATION IN TÜRKIYE

Inclusive education in Türkiye, like in many other countries, aims to ensure that all students, regardless of their abilities or disabilities, have equal opportunities to learn and thrive within the general education system. While Türkiye has made significant strides in aligning its education policies with international frameworks such as the Salamanca Statement (UNESCO, 1994) and the United Nations Convention on the Rights of Persons with Disabilities, the practical implementation of these policies remains challenging. The existing educational landscape is marked by institutional barriers, teacher-related challenges, and student and parental concerns, which hinder the effectiveness of inclusive education practices.

These challenges highlight the need for systemic reforms, investment in professional development, and attitudinal shifts to create an education system that accommodates diverse learners.

Despite government initiatives to support inclusive education, gaps between policy formulation and execution persist, resulting in inconsistencies across different educational settings. While some urban schools have adopted inclusive practices with adequate resources, many schools, particularly in rural regions, lack the infrastructure, funding, and trained personnel necessary for inclusive education (Sonmez et al., 2019). The disparities in educational access between different socio-economic regions underscore the need for a nationally coordinated approach to inclusion, ensuring that all students, regardless of geographic location or disability status, receive high-quality education. By identifying and addressing these systemic challenges, Türkiye can make meaningful progress toward a fully inclusive and equitable education system.

One of the most significant obstacles to inclusive education in Türkiye is the existence of institutional barriers, which manifest in policy inconsistencies, insufficient funding, and inadequate infrastructure. Although Türkiye has adopted international frameworks supporting inclusive education, there is a noticeable gap in policy implementation. Education policies related to inclusion are often developed at the national level but lack clear guidelines for local adaptation, leading to disparities between urban and rural schools (Deniz & Çoban, 2019). Additionally, while legislative frameworks require schools to accommodate students with SEN, many institutions do not have the necessary resources or training programs to ensure compliance (Sarı et al., 2020). This inconsistency limits the effectiveness of inclusive education and results in unequal access to learning opportunities for students with disabilities.

Financial constraints further exacerbate the situation, as schools often lack adequate funding to support inclusive education programs. Insufficient government investment in assistive technologies, classroom modifications, and specialized personnel creates additional barriers for students with disabilities (Rakap & Kaczmarek, 2010). Many public schools struggle to provide individualized support services, such as speech therapy or psychological counseling, which are essential for students with SEN. Additionally, infrastructure limitations, such as a lack of wheelchair-accessible buildings, adaptive furniture, and specialized learning materials, make it difficult for students with disabilities to fully participate in mainstream education. These barriers highlight the urgent need for increased financial allocations, targeted policy reforms, and nationwide infrastructural improvements to ensure that all students receive an equitable education.

Expanding infrastructure for inclusive education involves not only physical accessibility but also curriculum adaptation to accommodate diverse learning needs. Many schools in Türkiye still rely on rigid, standardized curricula that do not provide sufficient flexibility for students with different abilities (Sonmez et al., 2019). The lack of differentiated instruction and alternative assessment methods creates additional difficulties for students with SEN, often leading to low academic performance and increased dropout rates. Furthermore, the limited availability of special education professionals and trained classroom assistants places additional strain on general education teachers, who often feel unequipped to meet the diverse needs of their students (Deniz & Çoban, 2019). Without addressing these structural challenges, inclusive education policies in Türkiye will remain largely theoretical rather than practical, hindering the country's efforts to build an equitable education system.

TEACHER PREPAREDNESS AND RESISTANCE

The role of teachers is critical in ensuring the success of inclusive education, yet many educators in türkiye feel underprepared and unsupported when faced with the realities of teaching students with sen. Research indicates that many teacher training programs do not provide sufficient exposure to inclusive education strategies, leaving graduates with limited knowledge of differentiation, classroom management, and student support mechanisms (avramidis & norwich, 2002). While some universities have introduced courses on inclusive education, these remain largely theoretical, and practical training opportunities, such as internships in inclusive classrooms, are minimal (deniz & çoban, 2019). As a result, many newly qualified teachers lack confidence in their ability to implement inclusive practices, leading to reluctance and resistance toward inclusion.

A major challenge in teacher preparedness is the lack of continuous professional development opportunities for in-service teachers. Many educators in Türkiye report never having received specialized training in inclusive education, particularly those working in rural areas with limited access to professional development programs (Sonmez et al., 2019). Additionally, the absence of mentorship programs and peer collaboration initiatives further isolates teachers, making it difficult for them to adopt new teaching strategies (Sarı et al., 2020). This lack of ongoing training contributes to negative teacher attitudes toward inclusion, as educators often perceive students with disabilities as a burden rather than an integral part of the classroom community (Baskonus & Ozturk, 2023). Addressing these issues requires systematic reforms in teacher training curricula, increased access to professional development, and stronger institutional support for inclusive teaching practices.

Resistance to inclusive education among teachers is also influenced by high student-to-teacher ratios, rigid curriculum demands, and insufficient classroom support. Many teachers feel overwhelmed by the additional responsibilities associated with differentiating instruction for diverse learners while meeting national curriculum standards and standardized testing requirements (Opoku et al., 2021). This pressure often leads to frustration, stress, and resistance, particularly in subjects that require structured instruction, such as mathematics and science (Sarı et al., 2020). Additionally, the lack of collaboration between general education and special education teachers further exacerbates challenges, as many educators do not receive sufficient guidance or support in modifying their instructional methods. Creating an inclusive school culture requires policy changes that prioritize teacher well-being, professional collaboration, and adaptive teaching strategies, ensuring that educators feel equipped and motivated to implement inclusive practices.

CASE STUDIES ON PROFESSIONAL DEVELOPMENT PROGRAMS

Professional development and pre-service teacher training are both critical components in cultivating educators' capacity to implement inclusive education effectively. The success of inclusive practices in classrooms depends not only on policy reforms but also on the preparedness of teachers to address diverse learner needs. Empirical evidence from various international contexts underscores the value of structured, hands-on training programs in shaping teacher attitudes, competencies, and long-term commitment to inclusion. Without adequate training, even the most well-intentioned educators may struggle to create equitable learning environments for all students.

To illustrate the potential of such initiatives, several international case studies offer valuable insights into how sustained and well-designed training efforts can transform educational practice. In countries such as New Zealand and Canada, comprehensive professional development programs have yielded substantial improvements in teachers' inclusive capacities. These programs typically include interactive workshops, peer mentoring, and applied learning components, all of which help teachers internalize inclusive strategies in practical terms. Canada presents a similarly promising model through its multi-tiered professional development frameworks. These programs integrate co-teaching approaches, collaborative planning, and assistive technology training as central components (Porter & AuCoin, 2012). Teachers who participate in these ongoing, structured learning experiences are more likely to adopt flexible instructional strategies and work collaboratively with special education professionals. Furthermore, this model promotes a school-wide culture of shared responsibility for inclusion, which enhances its long-term sustainability. These international examples demonstrate a key lesson: effective professional development is not a one-time event, but an evolving, reflective process grounded in practice. Research from Australia reinforces this point, showing that training which incorporates real-life case studies, teaching simulations, and expert-led sessions has a more lasting impact on teacher beliefs and behaviors (Sharma & Sokal, 2016). Unlike theoretical lectures, these formats offer teachers direct exposure to inclusive methods, allowing them to apply what they learn in realistic contexts. As a result, teachers become more adaptable, responsive, and confident in supporting students with diverse needs.

In contrast, Türkiye continues to face significant challenges in both pre-service and in-service training for inclusive education. While recent reforms have introduced some content on inclusion into teacher education curricula, these courses are often elective and heavily focused on theoretical knowledge (Rakap & Kaczmarek, 2010). Many teacher candidates graduate without ever having worked directly with SEN, limiting their readiness for inclusive classrooms. This lack of experiential learning contributes to negative attitudes, uncertainty, and reluctance to implement inclusive practices (Saloviita, 2020b). The limitations extend into in-service professional development as well. Numerous teachers in Türkiye report having received little or no training on inclusive education after entering the profession (Deniz & Çoban, 2019). Those who do attend such programs often describe them as abstract and insufficiently connected to the practical realities of diverse classrooms (Sonmez et al., 2019). Although some studies indicate that teachers with exposure to structured training develop more positive attitudes toward inclusion (Sarı et al., 2020), the reach of such programs remains limited. Barriers such as time constraints, limited funding, and lack of institutional support further restrict access to quality professional development opportunities.

These international models underscore the importance of bridging the gap between theoretical knowledge and classroom practice. Embedding inclusive education into both the structure and culture of teacher training institutions is essential to ensuring long-term change. Moreover, collaboration between general and special education departments plays a vital role in preparing teachers for interdisciplinary teamwork, which is often required in inclusive schools. Without these integrated approaches, teacher training programs risk producing graduates who are ill-equipped for the realities of inclusive teaching. To align with these global trends and meet the growing need for inclusive education, Türkiye must undertake a comprehensive reform of its teacher education system. This includes making inclusive education a compulsory and practical part of the curriculum, supported by supervised internships and mentoring. Additionally, continuous and accessible professional development opportunities should be provided to in-service teachers, with content tailored to real classroom challenges. Only through such multifaceted, evidence-based efforts can Türkiye move toward building equitable, inclusive, and effective learning environments for all students.

THEORETICAL BACKGROUND

Attitudes towards inclusive education play an essential role in determining how effectively inclusive practices are implemented within educational systems. As central figures in the classroom, teachers' attitudes influence their willingness and ability to embrace students with SEN, shaping both the learning environment and the success of inclusive initiatives. Drawing on social cognitive theory (Bandura, 1991), attitudes can be understood as multidimensional constructs that encompass beliefs, emotions, and behaviours, all of which interact to guide how individuals perceive and respond to specific contexts. This framework provides a useful lens through which to explore the complexities of teacher attitudes towards inclusion.

In educational contexts, attitudes are commonly examined through their cognitive, affective, and behavioural dimensions. The cognitive dimension pertains to teachers' beliefs about the importance and effectiveness of inclusive education, along with their understanding of the needs of students with SEN. Teachers who possess a solid understanding of the principles of inclusion, and who have been exposed to the relevant pedagogical strategies, tend to exhibit more positive attitudes toward the integration of SEN students into mainstream classrooms (Avramidis & Norwich, 2002). This cognitive foundation is particularly significant in contexts where inclusive education is still evolving, such as in Türkiye. Inconsistent or insufficient training on inclusion, which has been documented in some regions (Melekoglu, 2014), can lead to misconceptions or doubts about the feasibility of inclusive practices. Teachers who lack a clear understanding of how to accommodate diverse learners may feel ill-equipped, which can foster resistance to inclusive education.

Closely linked to this cognitive component is the affective dimension, which reflects the emotional responses teachers have toward students with SEN. These emotions, shaped by personal experiences, cultural values, and perceived competence, strongly influence teachers' openness to inclusion. Positive emotions, such as empathy and confidence, often arise when teachers feel prepared and supported in their roles, contributing to a more welcoming and inclusive classroom environment (Bhatnagar & Das, 2014). Conversely, teachers who experience anxiety or frustration due to a perceived lack of skills or resources may develop negative attitudes, which can translate into reluctance to engage fully with inclusive practices. In Türkiye, where inclusive education is gaining momentum but where practical challenges remain, addressing these emotional barriers is essential. Emotional readiness, often overlooked in traditional teacher training programs, must be prioritised to ensure that teachers not only understand the theoretical underpinnings of inclusion but also feel emotionally prepared to support diverse learners in practice.

The behavioural component of attitudes is where beliefs and emotions are translated into concrete actions within the classroom. Teachers with positive attitudes towards inclusion are more inclined to adopt adaptive teaching strategies, such as differentiated instruction, and work collaboratively with special education specialists and other professionals (Jordan et al., 2009). However, when teachers hold negative attitudes, stemming from a lack of knowledge, emotional unease, or systemic challenges, this can manifest in behaviours that undermine inclusive practices. In Türkiye, where large class sizes, limited resources, and uneven institutional support can exacerbate the difficulties of implementing inclusive education, these behavioural outcomes are particularly critical (Sonmez et al., 2019). Without the proper infrastructure to support inclusive practices, even teachers with generally positive attitudes may struggle to effectively integrate SEN students into mainstream settings, highlighting the importance of addressing both individual attitudes and systemic barriers.

In sum, teachers' attitudes towards inclusive education are multidimensional and complex, shaped by cognitive, emotional, and behavioural factors that interact in meaningful ways. Positive attitudes foster environments where inclusive practices can flourish, leading to better outcomes for students with SEN. However, negative attitudes, often rooted in misconceptions or emotional discomfort, can create significant obstacles to the successful implementation of inclusion. In the Turkish context, where inclusive education is still developing, understanding and improving teacher attitudes is critical to advancing these practices.

FACTORS INFLUENCING TEACHER CANDIDATES' ATTITUDES TO INCLUSIVE EDUCATION

Teacher candidates' attitudes are pivotal in shaping the efficacy of inclusive practices within educational settings. Therefore, it is imperative to delve into the factors influencing these attitudes. This exploration holds significance as it allows for a deeper comprehension of the underlying determinants driving attitudes. By uncovering these factors, educators can take essential strides towards fostering positive changes in teachers' perspectives and approaches (Saloviita, 2020b). Understanding the intricacies of teacher candidates' attitudes is instrumental in promoting an inclusive and supportive learning environment.

The existing body of literature underscores the pivotal role of training in shaping teachers' attitudes towards inclusive education, irrespective of their initial predispositions or prior experiences. Avramidis and Norwich (2002) underscored the significant role training plays in shaping teachers' understanding and attitudes towards inclusive practices, identifying it as a key factor in promoting positive perspectives on inclusive education. Consequently, many universities have introduced special education courses to enhance prospective teachers' comprehension of students with SEN. Cross-national research consistently identifies teacher training as a critical determinant of attitudes towards inclusion. Studies in Kosovo (Uka, 2024) and Indonesia (Istiarsyah et al., 2019) reveal a reduction in resistance towards inclusive practices among teachers after specialised training in inclusive education. The findings indicate that teachers who undergo specialised training in inclusive practices are significantly more likely to develop positive attitudes towards inclusive education compared to their untrained counterparts. These results highlight the critical need for providing teachers with targeted training programmes that equip them with the necessary skills and knowledge to promote inclusivity within educational settings.

In Türkiye, studies consistently highlight a significant gap in teachers' knowledge of inclusive education, bringing attention to the transformative impact of targeted training in shaping attitudes toward inclusion (Sonmez et al., 2019). Research conducted by Deniz and Coban (2019) and Sonmez et al. (2019) offers valuable perspectives on this issue, emphasising the importance of teacher preparation. Deniz and Coban's meta-synthesis review identified a recurring pattern where teachers with negative attitudes toward inclusive education often lack sufficient training and understanding of its principles. This finding highlights the critical role that comprehensive and effective training programmes play in fostering more positive attitudes.

Further supporting this perspective, Sonmez et al. (2019) demonstrated a significant improvement in teachers' attitudes toward inclusion following their participation in an in-service training programme. This study reinforces the notion that exposure to well-structured training not only enhances teachers' knowledge but also encourages greater acceptance of inclusive practices. Collectively, these findings underline the pivotal role of specialised training in promoting inclusivity within educational settings in Türkiye. By equipping teachers with the necessary skills and understanding, such programmes can serve as a cornerstone for creating more inclusive and supportive learning environments, fostering positive outcomes for all students, particularly those with SEN. These insights point to the urgent need for sustained investment in professional development to bridge the knowledge gap and enhance inclusive education practices.

Educational research has extensively explored how age influences teachers' attitudes towards integrating students with SEN, revealing a range of diverse and occasionally contradictory findings. Studies by Opoku et al. (2021) and Saloviita (2020b) indicate that younger teachers tend to demonstrate more supportive attitudes toward inclusive education. Similarly, Chrzanowska (2019) found that Polish primary school teachers with less than 10 years of professional experience displayed more favourable attitudes compared to their more experienced counterparts, suggesting that teaching experience may play a more influential role than age itself (Domagała-Zyśk & Knopik, 2022). Less experienced teachers, in some cases, exhibit greater openness to inclusive practices, possibly due to their exposure to contemporary training paradigms. In contrast, other studies suggest that age has limited or no impact on attitudes toward inclusion. For instance, Uka (2024) reported no significant correlation between age and attitudes, a finding echoed by Rakap and Kaczmarek (2010) in their study of regular education teachers in Türkiye. These inconsistencies highlight the multifaceted nature of the issue, where factors such as professional experience, training, and educational context intersect to shape teachers' views on inclusion.

The impact of gender on teachers' perspectives towards inclusive education remains a subject of active discussion within educational research. Some studies, such as those by Logan and Wimer (2013) and Leonard and Smyth (2020), have not established a definitive link between gender and attitudes towards inclusion. However, other research, including Saloviita (2019) and Vaz et al. (2015), suggests that male teachers may be more likely to hold negative views towards inclusive education than their female counterparts. This difference is often attributed to the perception that male teachers may have less patience with students with SEN (Butakor et al., 2020).

Despite the limited body of research focusing on Turkish teacher candidates, Baskonus and Ozturk (2023) reported no significant gender-based differences in sentiments, attitudes, or concerns related to inclusive education among participants. Similarly, Uka (2024) concluded that gender does not significantly influence teachers' attitudes towards inclusion, suggesting that factors other than gender, such as training, personal experience, or cultural context, may play a more substantial role in shaping attitudes. However, contrasting evidence from findings such as those by Bhatnagar and Das (2014) and Dorji et al. (2021) suggests that male teachers may, in some contexts, exhibit more positive attitudes towards inclusive education than their female counterparts. These findings highlight the complexity of the relationship between gender and attitudes toward inclusion, which appears to vary across cultural and educational settings. For instance, in some contexts, male teachers may feel more confident or prepared to handle inclusive practices due to differences in professional training or societal expectations regarding gender roles. Additionally, these variations could reflect differences in the way inclusion is perceived and implemented across educational systems. While some studies indicate gender-neutral attitudes, others underscore the need for further investigation to understand how gender intersects with other variables, such as teaching experience, cultural norms, and exposure to inclusive environments, in shaping teachers' perspectives on inclusive education. This nuanced understanding is crucial for designing teacher education programmes that address potential biases and promote equitable attitudes across genders.

The literature also reflects limited attention to the influence of personal relationships with individuals with SEN on attitudes towards inclusive education. For instance, Alquraini's (2011) quantitative study in Saudi Arabia assessed the impact of having a family member or friend with a disability on teachers' views but found no significant correlation. Similarly, Baskonus and Ozturk (2023) observed no notable differences in the attitudes of teacher candidates in Türkiye based on familial connections with individuals with special needs. Conversely, other studies highlight a strong association between personal connections and favourable attitudes towards inclusion. Research has consistently highlighted the potential influence of personal connections with individuals with disabilities on attitudes towards inclusive education. For instance, Block (1995) noted that educators who maintained personal relationships with individuals with disabilities were more likely to exhibit supportive attitudes towards inclusion. This association underscores the role of personal experiences in shaping perspectives on inclusive practices. Adding to this, Al-Ahmadi (2009), through a mixed-method study conducted in Saudi Arabia, demonstrated that teachers with personal ties to individuals with SEN displayed notably more positive attitudes towards inclusive education. Her study suggests that such relationships may foster empathy and a deeper understanding of the challenges faced by students with SEN, which, in turn, positively influences educators' openness to inclusive practices.

Similarly, Parasuram (2006) identified a significant relationship, noting that individuals with a family member who has a disability tend to exhibit more favourable attitudes toward inclusion. This finding emphasises the role of familial experiences in shaping attitudes, as exposure to the needs and capabilities of individuals with disabilities within a family context can foster a greater appreciation for inclusive education. Bhatnagar and Das (2014) further reinforced this perspective, reporting that teachers with personal connections to individuals with SEN were significantly more inclined to hold positive views about integrating such students into mainstream classrooms compared to those without such connections. Together, these studies underline the importance of personal relationships as a critical factor in shaping educators' attitudes, suggesting that personal exposure to individuals with SEN can serve as a catalyst for fostering more inclusive mindsets. Collectively, these findings indicate that personal relationships with individuals with SEN may serve as a key factor in cultivating more optimistic and supportive attitudes towards inclusive practices, thereby shaping teacher candidates' perspectives on inclusion.

In summary, this literature review has provided a comprehensive overview of the multifaceted factors influencing teacher candidates' attitudes towards inclusive education. The aim of this research is to delve deeper into these determinants, specifically focusing on the impact of age, gender, and prior experience with children with special needs on teacher candidates' attitudes. While existing studies have extensively explored the influence of age, and gender, there is a noticeable gap in research concerning the role of teacher training programmes in shaping attitudes towards inclusive education. The findings from this study aim to address this gap and contribute to the broader understanding of the factors affecting teacher candidates' attitudes towards inclusive education. These objectives are explored through the following two research inquiries:

RQ1. How does the type of teacher training programme (i.e., education faculties vs. pedagogical formation education) influence teacher candidates' attitudes towards inclusive education in Türkiye?

RQ2. What are the interactions among factors such as age, gender, and prior experience with children with special needs in shaping teacher candidates' attitudes towards inclusive education in Türkiye?

METHOD

Research Design

This study adopted a cross-sectional quantitative research design to investigate the attitudes of teacher candidates in Türkiye towards inclusive education. This approach was chosen to facilitate the examination of relationships between variables, including demographic factors and personal experiences with individuals with SEN. By using a survey methodology, the study captured data at a single point in time, enabling a snapshot of current attitudes among teacher candidates. The cross-sectional design is well-suited for studies aimed at identifying patterns and differences in attitudes across diverse subgroups within a population (Creswell, 2014).

The research utilised a purposive sampling strategy to recruit participants from two distinct pathways: education faculties and pedagogical formation education programmes. This design choice allowed for a comparative analysis of attitudes based on the type of teacher training received. Additionally, the study incorporated a validated measurement tool, the Opinions Relative to Integration of Students with Disabilities (ORI) scale, to assess attitudes towards inclusive education. This robust quantitative design ensured the collection of reliable and generalisable data, providing valuable insights into the factors shaping teacher candidates' readiness for inclusive educational practices.

Participants and Sampling

The participants in this study comprised teacher candidates from Türkiye, totalling 369 individuals. Regarding gender distribution, the majority were female (264, 71.5%), with male participants comprising 105 (28.5%) of the sample. In terms of age, the distribution varied, with the largest group being 22-year-olds (123, 33.33%), followed by 23-year-olds (84, 22.76%), and 21-year-olds (90, 24.39%). Participants aged 20 and 24 constituted smaller percentages of the sample, with 16 (4.34%) and 32 (8.67%) individuals, respectively. Additionally, 24 participants (6.50%) were aged 25 or above. Concerning the teacher training programme, the participants were divided between those enrolled in Pedagogical Formation Education and those studying in Education Faculties, with 182 (49.3%) and 187 (50.7%) individuals, respectively. Regarding personal connections with individuals with SEN, 94 participants (25.5%) reported having a friend with SEN, while 275 (74.5%) did not. Similarly, 55 participants (14.9%) reported having a family member with SEN, while the majority, 314 (85.1%), did not.

To select participants for this study, a purposive sampling approach was employed. This sampling method aimed to target students who had decided to pursue Pedagogical Formation Education in order to become teacher candidates, as well as students enrolled in education faculties. Participants were selected from both education faculties and other programmes offering pedagogical formation education, these students were enrolled in the Guidance and Special Education course. This approach was chosen because the Guidance and Special Education course is a mandatory requirement for students seeking pedagogical formation education. Furthermore, determining whether individuals were prospective teacher candidates before taking this course was not feasible. Therefore, participants were chosen from both groups to ensure a diverse representation of individuals interested in becoming teachers.

Data Collection Instruments

Data collection for this study involved the utilisation of two questionnaires. The first questionnaire was designed to gather demographic details, including gender, age, teaching experience, and prior encounters with special needs children. To assess teachers' attitudes, a modified version of the ORI scale was incorporated into the questionnaire (Antonak & Larrivee, 1995). The ORI scale has been extensively utilised in prior research to explore teachers' perspectives on inclusive education (e.g., Killoran et al., 2014; Rakap et al., 2017). The reliability analysis revealed a Cronbach's Alpha of .861 across 25 items, indicating a high level of internal consistency for the scale. This tool comprises 25 statements, each offering five possible response options ranging from "strongly disagree" (1) to "strongly agree" (5), based on a five-point Likert scale. Likert scales are particularly useful for directly comparing answers across different participant groups. Responses to negative items were reverse-scored (from - to + and + to -), and then the 25 answers were summed to calculate total scores, which ranged from 25 to 125. Higher scores reflect more positive attitudes towards inclusion. This scoring methodology not only aids in identifying participants with more positive attitudes but also facilitates the detection of general disparities in opinions on inclusion across diverse demographic groups. By employing this scoring system, the overall attitudes of teacher candidates were assessed, concurrently exploring the influence of demographic characteristics on these attitudes.

Data Collection Procedure

Data for this study were collected using a structured questionnaire designed to gather insights into teacher candidates' attitudes towards inclusive education. The data collection process involved participants from three universities in Türkiye, encompassing both education faculties and pedagogical formation programmes to ensure representation from diverse training pathways. Purposive sampling was employed to select participants who were actively enrolled in these programmes. The questionnaire was distributed through a combination of methods to maximise accessibility and participation. During scheduled classes, participants were provided with QR codes linking to an online form, ensuring ease of access. Additionally, hard copies of the questionnaire were made available in classrooms to accommodate those who preferred a paper-based format. Institutional permissions were obtained in advance, and the data collection process was conducted in alignment with ethical research standards. This hybrid approach to data collection ensured a comprehensive response rate while maintaining participant convenience and engagement.

The research was conducted in compliance with ethical guidelines. Ethical approval was obtained from the Cankiri Karatekin University Ethics Committee for Science, Mathematics, and Social Sciences. The ethics review decision was granted on 20 December 2023, under the approval number [37]. Participants provided informed consent, acknowledging their understanding of the study's objectives and their right to withdraw at any point without repercussions. Data collection procedures adhered strictly to confidentiality and anonymity protocols, ensuring that participants' identities and responses remained secure throughout the research process.

Data Analysis

In the initial stage of data analysis, quantitative methods were employed to examine the data collected from completed questionnaires. This involved conducting descriptive statistical analysis using SPSS version 29.0.2, following data cleaning and screening procedures. Descriptive statistics for the score variable (N=369) revealed a mean of 89.19 (SD=12.51), with values ranging from 47 to 124. Skewness was .014 (SE=.127), and kurtosis was .162 (SE=.253). As both skewness and kurtosis values fall within the acceptable range of ±2 (George & Mallery, 2010), the data was deemed to exhibit no significant deviations from normality. Subsequently, independent samples t-tests and one-way ANOVA were conducted to evaluate the impact of various factors on teacher candidates' attitudes towards inclusive education in Türkiye.

RESULTS

An independent sample t-test was conducted to examine the difference in teacher candidates' attitudes based on gender. Levene's test confirmed that the assumption of equal variances was satisfied for both tests, with equal variances assumed ($F(1, 367) = 0.106$, $p = .745$) and not assumed ($F(1, 196.975) = 3.640$, $p < .001$). Results revealed a statistically significant difference in attitudes between male and female candidates ($t(367) = -3.589$, $p < .001$, two-tailed). Male candidates ($M = 85.54$, $SD = 12.02$) had significantly lower mean attitude scores than female candidates ($M = 90.64$, $SD = 12.42$). Effect size estimates indicated a large effect of gender on attitudes (Cohen's $d = -0.414$, Hedges' correction $= -0.413$, Glass's delta $= -0.410$). These findings suggest that female candidates demonstrate higher attitude scores compared to their male counterparts, indicating a potential influence of gender on teacher candidates' attitudes.

ANOVA was performed to evaluate the effect of age on teacher candidates' test scores. The results revealed no significant difference in test scores among different age groups ($F(8, 360) = 1.856$, $p = .066$, $\eta^2 = .040$). Post hoc tests using Tukey's honestly significant difference (HSD) method confirmed these findings, showing no significant mean differences in test scores between any pairs of age groups ($p > .05$ for all comparisons). Overall, these results suggest that age is not a significant predictor of teacher candidates' test scores.

An independent samples t-test revealed a significant difference in test scores between teacher candidates who reported having a family member with SEN ($M = 92.45$, $SD = 11.23$) and those who did not ($M = 88.62$, $SD = 12.65$) ($t(367) = 2.109$, $p = .025$). The effect size estimates indicated a medium effect size (Cohen's $d = 0.308$, Hedges' correction $= 0.308$, Glass's delta $= 0.303$), suggesting practical significance in the observed differences between the groups. These findings imply that there was a notable disparity in the overall attitudes of teacher candidates with a family member having SEN or disability compared to those without. The results underscore the positive influence of familial background on attitudes towards special needs education, highlighting the significant impact of having a family member with SEN on teacher candidates' perspectives.

An independent samples t-test indicated that there was no significant difference in test scores between teacher candidates who reported having a friend with SEN ($M = 90.21$, $SD = 12.74$) and those who did not ($M = 88.84$, $SD = 12.43$) ($t(367) = 0.919$, $p = .366$). Effect size estimates indicated a small effect size (Cohen's $d = 0.110$, Hedges' correction $= 0.110$, Glass's delta $= 0.110$), suggesting limited practical significance in the observed differences between the groups. These findings indicate that there were no notable distinctions observed in the overall attitudes of teacher candidates who had a friend with SEN or disability compared to those who did not.

A t-test revealed no significant difference in test scores between teacher candidates enrolled in education faculties (M = 88.70, SD = 12.36) and those in Pedagogical Formation Education (M = 89.69, SD = 12.66) (t(367) = -0.761, p = .224). Effect size estimates indicated small effect sizes (Cohen's d = 0.447, Hedges' correction = 0.447, Glass's delta = 0.447), suggesting minimal practical significance in the observed differences between the groups. Levene's test confirmed equal variances (p = .888), supporting the assumption for the independent samples t-test.

Tests of homogeneity of variances revealed significant differences across groups based on the mean (Levene's statistic = 1.795, df1 = 13, df2 = 319, p = .043). However, no significant differences were observed when based on other metrics such as the median or trimmed mean. ANOVA was conducted to assess the impact of teacher candidates' subject variables on their test scores. Results indicated no significant difference in test scores among subject groups ($F(15, 319) = 1.498$, $p = .104$, $\eta^2 = .066$).

As a result, this study highlights several key factors influencing teacher candidates' attitudes towards inclusive education in Türkiye. Gender was identified as a significant predictor, with female candidates exhibiting more favourable attitudes than their male counterparts. Additionally, candidates with a family member with SEN demonstrated more positive attitudes, whereas no significant differences were found for those with a friend with SEN. Age and training programme type did not significantly impact attitudes, suggesting that other factors may play a more prominent role in shaping perspectives on inclusion.

In summary, this study revealed that teacher candidates' attitudes toward inclusive education are shaped by a combination of personal and demographic factors. Gender emerged as a significant variable, with female candidates showing more positive attitudes than their male counterparts. Additionally, having a family member with SEN was associated with more favourable attitudes toward inclusion, suggesting the importance of personal exposure and empathy in shaping perspectives. In contrast, having a friend with SEN did not significantly influence attitudes, indicating that the depth and nature of personal relationships may matter. Age was not found to be a significant predictor, and no substantial differences were observed between candidates from education faculties and those in pedagogical formation programmes. These results suggest that while formal training is important, personal experiences and individual characteristics may play a more influential role in fostering inclusive mindsets among future educators, highlighting a need for experiential learning in teacher education.

DISCUSSION AND CONCLUSION

The study aimed to investigate how the type of teacher training programme influences teacher candidates' attitudes towards inclusive education in Türkiye. Contrary to expectations, the findings revealed no significant difference in attitudes between teacher candidates enrolled in education faculties and those in Pedagogical Formation Education programmes. This result contrasts with existing literature emphasising the importance of specialised training in shaping positive attitudes towards inclusive practices (Avramidis & Norwich, 2002; Istiarsyah et al., 2019). One possible explanation for this discrepancy could be the evolving nature of teacher training programmes in Türkiye, where both pathways may offer similar coursework and exposure to inclusive education principles. Additionally, the recent reforms in pedagogical formation education might have enhanced its effectiveness in preparing teacher candidates for inclusive classrooms, minimising disparities between the two groups. Moreover, it is essential to consider the quality and content of the training programmes, as these factors significantly influence their impact on attitudes towards inclusive education. Further research exploring the specific components of teacher training programmes and their alignment with inclusive education principles could provide deeper insights into their effectiveness in shaping attitudes among teacher candidates.

The study further explored how variables such as age, gender, and prior experience with children with special needs influence teacher candidates' attitudes towards inclusive education. Although no significant association was found between age and attitudes, gender proved to be a notable factor, with female candidates demonstrating more favourable perspectives compared to their male peers. This aligns with earlier research indicating that gender can influence perspectives on inclusion (Saloviita, 2019; Vaz et al., 2015). One possible explanation for this finding could be the societal expectations and nurturing roles traditionally associated with women, which may predispose female candidates to be more empathetic and supportive of inclusive practices. The identified gender disparity highlights the need to address gender-related biases and stereotypes within teacher education programmes to foster more equitable attitudes towards inclusive education. Incorporating discussions around gender inclusivity and challenging stereotypes within these programmes could help mitigate such disparities and promote balanced perspectives among future educators.

Furthermore, the influence of personal connections with individuals with SEN on attitudes varied based on the nature of the relationship. Teacher candidates with a family member with SEN exhibited more positive attitudes towards inclusion than those without such connections. This finding supports previous research highlighting the impact of personal experiences on attitudes towards special needs education (Bhatnagar & Das, 2014). Familial connections may lead to closer interactions and a deeper understanding of the challenges faced by individuals with SEN, fostering greater empathy and advocacy for inclusive practices. Nonetheless, no significant difference was observed based on friendships with individuals with SEN, suggesting that familial relationships may have a more substantial influence on attitudes compared to social connections. This distinction could be due to the sustained and intimate nature of family relationships, which often provide a greater depth of exposure and engagement with the lived realities of individuals with SEN.

The findings of this study carry important implications for teacher education programmes and policies designed to advance inclusive practices in Türkiye. Firstly, the importance of comprehensive and specialised training in inclusive education cannot be overstated. While the current study found no significant difference between teacher training programmes, further examination of the specific components and delivery methods of these programmes is warranted. Teacher education programmes should prioritise coursework and practical experiences that foster positive attitudes towards inclusive education, emphasising the value of diversity and equity in learning environments. Additionally, addressing gender-related biases and stereotypes in teacher education curricula is crucial for promoting more equitable attitudes towards inclusive education. Providing opportunities for critical reflection and dialogue on gender issues can help educators recognise and challenge their implicit biases, promoting a more inclusive and supportive learning environment for all students. Furthermore, leveraging personal experiences, particularly familial connections with individuals with SEN, can enhance teacher candidates' understanding and empathy towards diverse learners. Teacher education programmes should incorporate opportunities for meaningful engagement with individuals with SEN, whether through service-learning projects, internships, or guest lectures. These experiences can help bridge the gap between theory and practice, empowering educators to create inclusive classrooms that cater to the diverse needs of all students.

In conclusion, the current study provides valuable insights into the complex factors shaping teacher candidates' attitudes towards inclusive education in Türkiye. The absence of significant differences between graduates of education faculties and pedagogical formation programmes suggests that both pathways may offer similar training and exposure to inclusive education principles, indicating a need for further exploration into the quality and content of these programmes. However, the significant role of gender in shaping attitudes, with female candidates displaying more positive views, highlights the persistence of gender-related disparities within educational attitudes, which requires targeted intervention in teacher education to promote more equitable perspectives. Furthermore, the positive correlation between personal connections with individuals with SEN and more favourable attitudes toward inclusion emphasises the powerful role of personal experiences in fostering empathy and understanding. These findings suggest that integrating practical, hands-on experiences with individuals with SEN into teacher training curricula could be an effective strategy to strengthen inclusive attitudes among all teacher candidates.

Overall, this study underscores the need for teacher education programmes to prioritise both theoretical knowledge and experiential learning in inclusive education. By doing so, future educators can be better equipped with the empathy, skills, and pedagogical strategies required to create inclusive classrooms that accommodate the diverse needs of all students. Future research should explore longitudinal impacts of these programmes and investigate how ongoing professional development influences teacher attitudes over time. Additionally, further studies on intersectional factors, such as the interplay between gender, personal experience, and training type, will deepen our understanding of the dynamics shaping teacher readiness for inclusive education.

Beyond the immediate implications for teacher education in Türkiye, the findings of this study contribute to the global discourse on inclusive education by reinforcing the idea that structural elements within training programmes are as crucial as the content itself. The absence of significant differences between education faculties and pedagogical formation pathways suggests a convergence in curricular offerings, but also raises concerns about the depth and quality of experiential learning opportunities within both models. As international research consistently highlights the transformative power of hands-on experience and mentorship in inclusive settings, Turkish teacher training institutions must critically evaluate not only *what* is being taught, but *how* and *in what contexts*. Embedding reflective practice, real-life case discussions, and direct exposure to inclusive classrooms should be prioritised to move beyond superficial coverage of inclusive education.

Moreover, the study's emphasis on the role of personal and demographic factors provides valuable insight into how inclusive mindsets are shaped. The significant influence of familial connections with individuals with special needs highlights the emotional and relational dimensions of teacher identity formation. This finding suggests that teacher education programmes might benefit from incorporating structured opportunities for students to engage meaningfully with individuals with disabilities, whether through service-learning, practicum placements, or community partnerships. Such experiences can foster empathy, dismantle prejudices, and build confidence, ultimately contributing to the cultivation of teachers who are not only knowledgeable but also emotionally attuned to the needs of diverse learners. Designing education programmes that deliberately facilitate this kind of experiential empathy development may be key to transforming inclusive education from policy into practice.

To ensure that inclusive education becomes an actionable reality rather than a theoretical aspiration, a shift is required at the policy level. Policymakers must understand that inclusion is not merely a pedagogical trend, but a foundational principle of equitable education systems. This demands not only curricular reforms but also a structural rethinking of how teacher education programs are designed and implemented. National frameworks should require teacher preparation institutions to include comprehensive and mandatory training on inclusive practices, covering both theoretical underpinnings and field-based applications. Additionally, policies should be accompanied by clear implementation guidelines, evaluation mechanisms, and financial support to ensure their long-term sustainability and institutional adherence.

The broader educational ecosystem must also evolve to support these attitudinal and pedagogical shifts. National policy frameworks should not only encourage inclusive practices but provide schools with the resources and administrative capacity to implement them effectively. Adequate funding, infrastructure support, and leadership development are essential components for turning inclusive ideals into everyday realities. Teachers, especially those in the early stages of their careers, need structured and ongoing professional development that reinforces inclusive strategies and practices. Moreover, partnerships between mainstream and special education professionals should be institutionalized to create collaborative support systems within schools.

Systemic barriers such as overcrowded classrooms, rigid curricula, and minimal interdisciplinary collaboration must also be addressed. These issues often undermine even the best-intentioned inclusive efforts by limiting flexibility and responsiveness in the classroom. Policymakers must work to create environments where teachers are empowered to make instructional decisions that reflect the diverse needs of their students. Cross-sector collaboration involving ministries, universities, and local education authorities is vital for building an ecosystem where inclusion is not an exception but the norm. Ultimately, sustainable inclusive education requires not only teacher readiness but also systemic alignment and a shared national commitment to educational equity.

Finally, it is essential that the broader educational ecosystem aligns itself with the principles of inclusive education to ensure the effectiveness and sustainability of these pedagogical and attitudinal shifts. National policy frameworks should not only mandate the integration of inclusive practices but must also ensure that schools are adequately resourced in terms of infrastructure, staffing, and pedagogical materials. Administrative systems should be equipped to support inclusive teaching by providing flexibility, institutional leadership, and consistent monitoring mechanisms. Teachers, particularly those at the beginning of their careers, require continuous support through targeted professional development, structured mentorship, and opportunities for collaboration with special education professionals. Addressing persistent systemic challenges, such as overcrowded classrooms, inflexible curricula, and limited interdisciplinary cooperation, is vital for fostering learning environments where inclusive education is not only possible but can truly flourish.

Limits and Recommendations

One limitation of this study is its dependence on self-reported data, which may be susceptible to social desirability bias and response inaccuracies. Despite endeavours to uphold anonymity and confidentiality, participants might have offered responses conforming to societal expectations or perceived norms, rather than reflecting their authentic attitudes and experiences. Future research endeavours could integrate mixed-method approaches, amalgamating quantitative measures with qualitative interviews or observations, to furnish a more nuanced comprehension of the multifaceted factors influencing teacher candidates' attitudes towards inclusive education.

Another limitation is the generalizability of findings, given the specific context of Türkiye and the sample characteristics. The study focused on teacher candidates from Turkish universities, limiting the applicability of results to other cultural or educational contexts. Future research could explore attitudes towards inclusive education among practising teachers or educators from diverse geographic regions to ascertain the broader relevance of the findings. Additionally, longitudinal studies tracking changes in attitudes over time and investigating the effectiveness of interventions in promoting inclusive practices could offer valuable insights into the long-term impact of teacher training programmes and personal experiences on attitudes towards inclusion.

REFERENCES

Ainscow, M. (2020). Promoting inclusion and equity in education: lessons from international experiences. *Nordic journal of studies in educational policy, 6*(1), 7-16.

Al-Ahmadi, H. (2009). Factors affecting performance of hospital nurses in Riyadh Region, Saudi Arabia. *International Journal of Health Care Quality Assurance, 22*(1), 40–54. DOI: 10.1108/09526860910927943 PMID: 19284170

Alquraini, T. (2011). Special Education in Saudi Arabia: Challenges, Perspectives, Future Possibilities. *International Journal of Special Education, 26*(2), 149–159.

Antonak, R. F., & Larrivee, B. (1995). Psychometric analysis and revision of the Opinions Relative to Integration of Students with Disabilities (ORI) Scale. *Journal of Developmental and Physical Disabilities, 7*(2), 137–146. DOI: 10.1007/BF02071408

Avramidis, E., & Norwich, B. (2002). Teachers' attitudes towards integration/inclusion: A review of the literature. *European Journal of Special Needs Education, 17*(2), 129–147. DOI: 10.1080/08856250210129056

Avramidis, E., Toulia, A., Tsihouridis, C., & Strogilos, V. (2019). Teachers' attitudes towards inclusion and their self-efficacy for inclusive practices as predictors of willingness to implement peer tutoring. *Journal of Research in Special Educational Needs, 19*(S1), 49–59. DOI: 10.1111/1471-3802.12477

Bandura, A. (1991). Social cognitive theory of self-regulation. *Organizational Behavior and Human Decision Processes, 50*(2), 248–287. DOI: 10.1016/0749-5978(91)90022-L

Bandura, A. (1991). Social cognitive theory of self-regulation. *Organizational Behavior and Human Decision Processes, 50*(2), 248–287. DOI: 10.1016/0749-5978(91)90022-L

Baskonus, H. M., & Ozturk, O. M. (2023). Öğretmen adaylarının kaynaştırma/bütünleştirme eğitimine ilişkin duygu, tutum ve kaygı düzeylerinin incelenmesi. [Examining the emotions, attitudes, and anxiety levels of teacher candidates regarding inclusive education]. *Ahi Evran Üniversitesi Sosyal Bilimler Enstitüsü Dergisi, 9*(1), 170–184. DOI: 10.31592/aeusbed.1212256

Bhatnagar, N., & Das, A. (2014). Regular School Teachers' Concerns and Perceived Barriers to Implement Inclusive Education in New Delhi, India. *International Journal of Instruction, 7*(2), 89–102.

Block, M. E. (1995). Development and validation of the children's attitudes toward integrated physical education–revised (CAIPE–R) inventory. *Adapted Physical Activity Quarterly; APAQ, 12*(1), 60–77. DOI: 10.1123/apaq.12.1.60

Butakor, P. K., Ampadu, E., & Suleiman, S. J. (2020). Analysis of Ghanaian teachers' attitudes toward inclusive education. *International Journal of Inclusive Education, 24*(11), 1237–1252. DOI: 10.1080/13603116.2018.1512661

Chrzanowska, I. (2019). Inclusive education in the opinion of the teachers from special pre-schools, regarding the chances of success of the inclusive actions towards particular groups of students, education participants, and teachers' seniority. *Interdisciplinary Context of Special Pedagogy, 25*(1), 127–149. DOI: 10.14746/ikps.2019.25.06

Creswell, J. W. (2014). *Research design: Qualitative, quantitative, and mixed methods approaches* (4th ed.). SAGE Publications.

D'Alessio, S. (2012). *Inclusive education in Italy* (Vol. 10). Springer Science & Business Media.

Deniz, E., & Coban, A. (2019). Kaynaştirma eğitimine ilişkin öğretmen görüşleri [Teachers views regarding inclusive education]. *Elektronik Sosyal Bilimler Dergisi, 18*(70), 734–761. DOI: 10.17755/esosder.448379

Domagała-Zyśk, E., & Knopik, T. (2022). Attitudes Toward Inclusion Among Polish Primary School Teachers: A Strategic Factor in Implementing Inclusive Education. *Multidisciplinary Journal of School Education, 11*(1 (21)), 211–233. DOI: 10.35765/mjse.2022.1121.11

Dorji, R., Bailey, J., Paterson, D., Graham, L., & Miller, J. (2021). Bhutanese teachers' attitudes towards inclusive education. *International Journal of Inclusive Education, 25*(5), 545–564. DOI: 10.1080/13603116.2018.1563645

Engelbrecht, P., Savolainen, H., Nel, M., & Malinen, O. P. (2013). How cultural histories shape South African and Finnish teachers' attitudes towards inclusive education: A comparative analysis. *European Journal of Special Needs Education, 28*(3), 305–318. DOI: 10.1080/08856257.2013.777529

George, D., & Mallery, M. (2010). *SPSS for windows step by step: A simple guide and reference*. Pearson.

Hehir, T., Grindal, T., Freeman, B., Lamoreau, R., Borquaye, Y., & Burke, S. (2016). *A summary of the evidence on inclusive education*. Abt Associates.

Higher Education Council. 2023. Pedagojik Formasyona Yönelik Bilgilendirme [Decision of the Higher Education Council regarding pedagogical formation education programme], Turkish Higher Education Council. https://www.yok.gov.tr/HaberBelgeleri/BasinAciklamasi/2023/pedagojik-formasyona-yonelik-bilgilendirme.pdf

Istiarsyah, I., Dawi, A. H., & Ahmad, N. A. (2019). The influence of special education training on teachers' attitudes towards inclusive education: Case study in Aceh Province, Indonesia. *International Journal of Academic Research in Progressive Education and Development, 8*(4), 1016–1027.

Jordan, A., Schwartz, E., & McGhie-Richmond, D. (2009). Preparing teachers for inclusive classrooms. *Teaching and Teacher Education, 25*(4), 535–542. DOI: 10.1016/j.tate.2009.02.010

Killoran, I., Woronko, D., & Zaretsky, H. (2014). Exploring preservice teachers' attitudes towards inclusion. *International Journal of Inclusive Education, 18*(4), 427–442. DOI: 10.1080/13603116.2013.784367

Leonard, N. M., & Smyth, S. (2020). Does training matter? Exploring teachers' attitudes towards the inclusion of children with autism spectrum disorder in mainstream education in Ireland. *International Journal of Inclusive Education*, •••, 1–15. DOI: 10.1080/13603116.2020.1718221

Logan, B. E., & Wimer, G. (2013). Tracing Inclusion: Determining Teacher Attitudes. *Research in Higher Education*, •••, 20.

Malinen, O. P., Väisänen, P., & Savolainen, H. (2012). Teacher education in Finland: A review of a national effort for preparing teachers for the future. *Curriculum Journal, 23*(4), 567–584. DOI: 10.1080/09585176.2012.731011

Melekoğlu, M. A. (2014). Special education today in Turkey. In *Special education international perspectives: Practices across the globe* (pp. 529–557). Emerald Group Publishing Limited. DOI: 10.1108/S0270-401320140000028024

Melekoglu, M. A., Cakiroglu, O., & Malmgren, K. W. (2009). Special education in Turkey. *International Journal of Inclusive Education, 13*(3), 287–298. DOI: 10.1080/13603110701747769

Oliver, M. (1990). The Politics of Disablement — New Social Movements. *In: The Politics of Disablement. Critical Texts in Social Work and the Welfare State*. Palgrave, London. https://doi.org/DOI: 10.1007/978-1-349-20895-1_8

Opoku, M. P., Cuskelly, M., Pedersen, S. J., & Rayner, C. S. (2021). Attitudes and self- efficacy as significant predictors of intention of secondary school teachers towards the implementation of inclusive education in Ghana. *European Journal of Psychology of Education, 36*(3), 1–19. DOI: 10.1007/s10212-020-00490-5

Parasuram, K. (2006). Variables that affect teachers' attitudes towards disability and inclusive education in Mumbai, India. *Disability & Society, 21*(3), 231–242. DOI: 10.1080/09687590600617352

Porter, G. L., & AuCoin, A. (2012). *Strengthening inclusion, strengthening schools: Report of the review of inclusive education programs and practices in New Brunswick Schools: An action plan for growth.* New Brunswick Department of Education and Early Childhood Development.

Rakap, S., Cig, O., & Parlak-Rakap, A. (2017). Preparing preschool teacher candidates for inclusion: Impact of two special education courses on their perspectives. *Journal of Research in Special Educational Needs, 17*(2), 98–109. DOI: 10.1111/1471-3802.12116

Rakap, S., & Kaczmarek, L. (2010). Teachers' attitudes towards inclusion in Turkey. *European Journal of Special Needs Education, 25*(1), 59–75. DOI: 10.1080/08856250903450848

Ruijs, N. M., & Peetsma, T. T. (2009). Effects of inclusion on students with and without special educational needs reviewed. *Educational Research Review, 4*(2), 67–79. DOI: 10.1016/j.edurev.2009.02.002

Saloviita, T. (2019). Explaining classroom teachers' attitudes towards inclusive education. *Support for Learning, 34*(4), 433–442. DOI: 10.1111/1467-9604.12277

Saloviita, T. (2020a). Attitudes of teachers towards inclusive education in Finland. *Scandinavian Journal of Educational Research, 64*(2), 270–282. DOI: 10.1080/00313831.2018.1541819

Saloviita, T. (2020b). Teachers' changing attitudes and preferences around inclusive education. *International Journal of Disability Development and Education*, •••, 1–18. DOI: 10.1080/1034912X.2020.1828569

Sarı, T., Nayir, F., & Kahraman, Ü. (2020). A Study on Inclusive Education in Turkey. *Journal of Education and Future*, (18), 69–82. DOI: 10.30786/jef.642954

Sharma, U., & Sokal, L. (2016). Can teachers' self-reported efficacy, concerns, and attitudes towards inclusion predict their actual inclusive classroom practices? *Australasian Journal of Special Education, 40*(1), 21–38. DOI: 10.1017/jse.2015.14

Skrtic, T. M. (1991). *Behind special education: A critical analysis of professional culture and school organization*. Love Publishing.

Sonmez, N., Alptekin, S., & Bicak, B. 2019. Okul Öncesi Eğitim Öğretmenlerinin Kaynaştırma Eğitiminde Yeterliliklerini Artırmaya Yönelik Geliştirilen Hizmetiçi Eğitim Programının Etkisi [The Effect of the In-Service Training Program Developed to Increase the Preschool Education Teachers' Competence in Inclusive Education]. *İnönü Üniversitesi Eğitim Fakültesi Dergisi, 20*(2), 439-456.

Takala, M., Pirttimaa, R., & Törmänen, M. (2009). Inclusive special education: The role of special education teachers in Finland. *British Journal of Special Education, 36*(3), 162–173. DOI: 10.1111/j.1467-8578.2009.00432.x

Uka, E. (2024). Exploring differences in Primary School Teachers' Attitudes Towards Inclusive Education in Kosovo. *European Journal of Special Needs Education*, •••, 1–10. DOI: 10.1080/08856257.2024.2334537

UNESCO. (1994). *The Salamanca Statement and Framework for Action on Special Needs Education*. UNESCO.

United Nations. (2006). *Convention on the Rights of Persons with Disabilities (UNCRPD)*. Retrieved from https://www.un.org/development/desa/disabilities/convention-on-the-rights-of-persons-with-disabilities.html

Vaz, S., Wilson, N., Falkmer, M., Sim, A., Scott, M., Cordier, R., & Falkmer, T. (2015). Factors associated with primary school teachers' attitudes towards the inclusion of students with disabilities. *PLoS One, 10*(8), e0137002. DOI: 10.1371/journal.pone.0137002 PMID: 26317862

Winzer, M. A. (1993). *The history of special education: From isolation to integration*. Gallaudet University Press.

Chapter 6
The Role of Assistant Principals in Fostering Teacher Collaboration and Professional Growth:
A Pathway to Effective School Leadership

Henderson Lewis
https://orcid.org/0000-0002-5640-3189
Louisiana State University, USA

Stacy-Ann T. Campbell
https://orcid.org/0009-0004-1684-343X
Louisiana State University, USA

Saleria S. Shaffer
Louisiana State University, USA

Veysel Altunel
https://orcid.org/0000-0002-4562-1594
Louisiana State University, USA

ABSTRACT

This chapter examines the evolving role of assistant principals (APs) as instructional leaders and their impact on fostering teacher collaboration and professional growth. The purpose of this chapter is to explore how APs perceive their roles and the challenges they face in balancing instructional leadership and administrative duties. Drawing on theoretical frameworks such as sensemaking and social constructivism, the chapter highlights strategies APs employ to support teachers, build collaborative cultures, and enhance professional learning communities. It also underscores the critical role of APs in shaping school culture, improving teaching practices, and promoting student achievement. Using qualitative data collected through interviews with assistant principals in various school districts, this chapter identifies key competencies, challenges, and recommendations for maximizing the AP's role as a pathway to effective school leadership. Practical implications for professional development, mentorship, and policy

DOI: 10.4018/979-8-3373-0472-4.ch006

considerations are also discussed.

INTRODUCTION

The role of Assistant Principal (AP) was established in the US around 1930's (Glanz, 2004) however, this was not discussed thoroughly until Austin and Brown (1970) offered an in-depth description of the assistant principal's role as being effective to school functioning. Since then, there has been much controversy surrounding the assistant principal's role connected to student discipline, supervision, and managerial tasks (Jackson et al., 2021). Studies have shown that assistant principals spend most of their days attending to time consuming tasks that indirectly impact teacher efficacy and student outcome, a discouraging task that leaves assistant principals feeling like their true role is unfulfilled (Wison & Clayton, 2020). Consequently, the outpouring demand for restructuring the assistant principal's role became crucial as experts deemed, assistant principals' potential has been wasted (Gurley et al., 2015). Additionally, despite being one of the most critical positions within a school that warrants further examination, the role and leadership development of assistant principals have been largely overlooked in empirical research (Morgan, 2018). Furthermore, Shakir et al. (2023) contended that assistant principals have become essential and integral to school leadership, but their pathway to leadership remains unclear.

As the demands of assistant principals vary from school to school, studies highlight the common focus on benefits of assistant principals as instructional leaders. Harris et. al (2022) discusses how distributed leadership recognizes assistant principals as vital team members in this effort. The assistant principal serves as a catalyst to foster teacher support and professional growth in schools. Thus, consideration of their role could be strategically and consistently applied to improve the support of classroom teachers and prepare future teacher leaders (Dagli et al., 2023). Moreover, the role of assistant principals continues to be a crucial factor for the success of a school and the development of this role must be understood in its entirety to improve school leadership (Turnbull et al., 2016).

This study aims to explore assistant principals' perceptions of their roles and their impact on promoting teacher collaboration and professional development. The authors examine relevant literature and the theoretical frameworks that inform this research. Next, the study highlights the methodology used to answer research questions followed by discussion, limitations and future research. Finally, the authors suggest recommendations based on the results of the study.

REVIEW OF LITERATURE

Ambiguity in the Assistant Principal's Role

The assistant principal's role is often overlooked and remains one of the least understood aspects of school leadership (Ho et al., 2023). Wilson & Clayton (2020) noted that there is inadequate research on assistant principals, thus making it challenging to identify their level of preparedness to perform varied roles and responsibilities within school administration. Additionally, Ho et al., (2023) asserted that most of the duties and job descriptions of APs are vastly determined by the specific needs of the school principal resulting in their roles frequently overlapping or interchanging with that of the school principal. Consequently, this leads to uncertainty and distinct challenges within the role. Gonzales (2022) posits two critical components which affect the performance of assistant principals and lead to career burn out are role ambiguity and role conflict. According to Gonzales (2022) role conflict for APs occurs when they are unable to meet the discordant demands and high expectations from their leaders. In most cases, they carry out these diverse administrative tasks while learning simultaneously on the job often with limited support.

Another ambiguity is that the assistant principals' roles and responsibilities vary based on the needs of their school and district in which they work resulting in a feeling of unpreparedness to handle the challenges associated with this role (Ho et al., 2023). Regardless of experience, assistant principals must continuously adapt to the evolving needs of students and teachers. Glanz (2004) emphasizes that in urban schools, assistant principals primarily spend their day handling conflicts between students and teachers, as well as disputes among students, which frequently lead to disciplinary measures.

While assistant principals role support principal's leadership, it also involves clerical and supervisory duties unique to their position. One such role is managing relationships with adults both within and outside the school system. This can be a challenging task as these interactions can be complex, as conflicts, disagreements, and shifting dynamics often arise (Goldring et al., 2021). A school leader's inclination toward either a cooperative or assertive approach significantly shapes the direction and outcome of a conflict. Factors such as social norms, expectations, leadership style, desire for authority, skills, experience, and conflict management approach all play a crucial role in how effectively conflicts are handled (Ho et al., 2023). In this case both the assistant principal and principal role will intersect in resolving the problems at hand.

Evaluating Assistant Principal's Role

Education can be reasoned as a proponent to success. A successful educational system relies on school principals and their leadership teams. The advocacy of shared or distributed leadership with assistant principals has become popular with the school leadership channel. Assistant principals' shared leadership impacts the functioning of instruction, student regulation, operations, and support personnel. The assistant principal can support the principal by skillfully blending interpersonal skills, reflective practices, teaching pedagogy, leadership skills, and current research (Irvine, 2022). Skillful leadership influences the entire educational arena. One reason is that the principal and the assistant principal can magnetize and retain high-quality teachers necessary to impact a struggling school (States News Service, 2011). Initiatives such as principal pipelines seek to train, evaluate, and support assistant principals. Some school districts offer monthly leadership opportunities or leadership academies. Despite the promise of these programs, they lack defined pathways to leadership for assistant principals. The Professional Standards for Educational Leaders (PSEL) highlights the work of effective educational leaders. However, not every assistant principal can function in the same capacity (National Policy Board for Educational Administration, 2015). PSELs drive many states and districts' evaluations for assistant principals (Turnbull et al., 2015). The PSELs focus on student achievement, school culture, leadership effectiveness, and stakeholder relationships, while many assistant principals spend time on various duties, extracurricular activities, discipline, hearings, and state testing. Performance appraisal has been researched by many.

Instructional Leadership

The instructional leadership approach emphasizes the correlation of the assistant principal's role in student achievement, supportive teacher professional learning, and efficacious teaching pedagogy (Morgan, 2014). The stipulation to meet or exceed state performance standards is compelling school administrators to focus on framing school goals, managing the instructional program, and promoting school climate (Furman, 2020). Because school principals are tasked with operational duties, they rely on assistant principals to perform instructional leadership tasks (ESSA, 2015). School districts have titled the assistant principal's position as assistant principal of curriculum and instruction (APCIs). Assistant principals with the title APCI may have other duties but are noted as curriculum and instruction leaders (Furman, 2020). As an APCI, assistant principals must build a community of trust. Preston et. al (2013) found that assistant principals without history in rural communities were not successful instructional leaders because they were not able to build trust. Ralebese's et al., (2025) case study discussed how the lack of systematic guidance and support for instructional leadership will fragment curriculum reformation. Assistant principals address these gaps by directing teachers' instructional focus and supporting them in overcoming related challenges that impact student learning.

Assistant principals indirectly impact student learning through positive collaborations that provide open dialogue and feedback. Classroom teaching and learning are improved when assistant principals support teacher reflection and reframing classroom activities. Using modeling, probing questions, feedback, and praise to promote teaching and learning are supportive teaching instruments. As instructional leaders, assistant principals have found that the leadership strength of relatability enables teachers to share, thus creating a non-threatening, proactive environment. A diplomatic approach will massage the atmosphere for constructive feedback. This study notes instructional impactful leadership actions by assistant principals in suburban and urban schools. Further research would contribute to scholarship with a lens on rural schools with the most significant demand for impactful leadership (Marzano et al., 2005).

Modeling is an effective behavior that can improve teaching and learning. Assistant principals can demonstrate or provide simulated activities on best practices (Jackson et al., 2021). Their problem-solving abilities will link policy-driven mandates and evidence-based instructional leadership practices. Assistant principals define the practice, support, and evaluate instructional practices in suburban and urban schools. In rural schools, because of funding, assistant principal's model instructional practices (Morgan, 2018). This collaborative setting is perceived as meaningful when planned, observed, and ongoing.

Assistant principals play an important role in the lives of many students, yet their impact on student achievement is empirically shown to be indirect (Cox & Mullen, 2022). Leithwood et al. (2020) discuss how effective leadership can have a rational, emotional, organizational, and family-conscious impact on student achievement. Assistant principals who focus on student achievement by tracking student progress on standardized test scores, graduation rates, attendance, and student-centered activities will improve student learning and development. As an instruction leader, the assistant principal consistently uses data to drive instructional decisions to address school-wide skill deficits. It is beneficial to have time to focus on student data collectively. Using data to inform instructional practices with the assistant principal gives teachers the feeling that they are unified advocates for their school.

THE USE OF ARTIFICIAL INTELLIGENCE IN THE ASSISTANT PRINCIPAL'S ROLE

Technology plays an integral role in assisting APs to carry out these instructional leadership practices efficiently. Using Artificial Intelligence (AI) programs can significantly impact APs' role in managing these tasks. APs can use AI programs for school-wide data tracking, attendance, enrollment and professional development. The neural network of AI emulates how brain cells and different brain regions communicate with one another hence it has the ability to analyze vast amounts of data, detect patterns, enhance processing efficiency, and make adaptive decisions based on real-time information (Wang, 2021a). AI enhances analytical efficiency through mass data such as test scores, students and employees demographics to help school leaders make data-driven, evidence-based decisions. Additionally, the ability of AI to predict outcomes can mitigate student drop out. For example, in the United States, researchers analyzed longitudinal student records from the North Carolina Department of Public Instruction using machine learning techniques. They incorporated 74 predictive measures from Grades 3 to 8—including academic performance, behavioral indicators, and socioeconomic and demographic factors—to identify students at risk of dropping out and implement targeted interventions (Sorensen, 2019).

Much of the APs role in instructional leadership encompasses this kind of large data of tracking and reporting to the school principal. Therefore, APs can capitalize on using this data from AI to strengthen the communication and decision-making process with the principal. One may argue that the human value of decision making is missing in this case. Wang (2021b) however, proposes a symbiotic relationship where school leaders (principals and assistant principals) can use a combination of data-driven evidence-based decision making and moral value-based decision making. According to Wang (2019, 2020) despite the volume of data, evidence alone does not ensure the moral foundation of a decision. In fact, APs must use the time freed up by AI to focus on making more value- based decisions that are crucial for creating positive school culture and climate.

Assistant Principals Role in Teacher Support and Development

The evolution of the assistant principal's role has become the catalyst for affecting school improvement, educational equity, and instructional support. Most assistant principals have teaching experience which enables them to be a strategic classroom tool for teachers. Controversy exists surrounding teaching as a flat profession, leading to teacher dissatisfaction with the same routine day after day (Bond, 2021). Like principals, assistant principals possess the 'professional authority' to determine who is admitted and remains in the teaching profession. One strategy to support teachers is through coaching and mentoring. A key characteristic of an assistant principal's role is to guide teachers in maneuvering the day-to-day complexities of teaching and learning in the classroom. This opportunity offers assistant principals the chance to be coaches and mentors to teachers, particularly in schools that have limited access to this resource. Assistant principals have unlimited access to the classrooms and are considered the "soldiers on the ground". Teachers often interact with them more than principals, making them more suited to promote growth and development. From a coaching and mentoring stance, assistant principals can enhance teachers' beliefs, help teachers develop reflective practices, and set meaningful practical goals in alignment with student achievement (Munson & Saclarides, 2022).

Pertinent to this novice teachers face the daunting task of navigating the teaching and learning environment at the beginning of their career. Consequently, they become overwhelmed, vulnerable and need to rely on all their resources to succeed. The role of the assistant principal in this instance is to provide the necessary mentorship that novice teachers need. This can be through a personal endeavor through individual communication or distributed leadership – assigning experienced teachers to a novice teacher. Establishing a safe space through check -ins, informal observations and feedback without judgement allows novice teachers to confide in their APs. It provides a sense of belonging and trust in a community of professionals. This sense of trust established in the initial stage of the relationship will not only be invaluable for the mentee but the mentor (AP) and the school community (Jackson et al., 2021).

Additionally, empirical research has shown several successful coaching and mentoring models for teachers and novice teachers. One such model is the GROW model that APs can implement to support new teachers in their schools. The GROW (Goal, Reality, Options, Will and Way forward) model encompasses the mentee/coachee identifying goals, assessing current situation, exploring possible courses of action and committing to these actions and preparing for obstacles on the way (Deiorio, N. M., et al. 2022). Furthermore, using PLCs as an integrative, collaborative support is vital for new teachers' success. New teachers need hands-on practice with support to minimize mistakes and enhance their ability to improve student achievement. PLCs (overseen by APs) provide the opportunity for new teachers to interact with experienced teachers, learn classroom management skills, student engagement and assessment strategies, data-driven plans and goal setting. Establishing a lasting culture of teachers supporting one another is fundamental in improving student outcomes and reducing high teacher turnover (Arroyo et al., 2020). Moreover, engaging with novice teachers during PLCs fosters didactic knowledge and systemic knowledge dimensions where experienced teachers can reflect on their own teaching practices while supporting new teachers. It also provides the opportunity for experienced teachers to learn from novice teachers by gaining fresh perspectives on modern ways of learning such as integrating technology into the classrooms. Likewise, APs can use this opportunity to connect the old with the new resulting in a holistic culture of high quality leading and learning and by extension increased student outcomes (Schatz-Oppenheimer & Goldenberg, 2024). Moreover, APs have the positional authority to achieve this through observation and feedback.

Assistant principals can conduct regular classroom observations and offer constructive, actionable feedback to teachers on instructional practices. This will provide teachers with an opportunity to have their teaching practices described as well as to reflect on their practice with their instructional leader. Frasier (2022) posits that observational feedback serves as a lever to improve teachers' performance, regardless of their experience. Additionally, conducting observations and feedback frequently and timely demonstrates that the assistant principal understands the context of the classroom as well as increases teachers' confidence that their instructional leader recognizes and values effective teaching (Frasier, 2022). Not only will this boost their confidence but will foster trust between teachers and assistant principals which plays a vital role in teacher growth and development. Consequently, teachers will perceive the feedback from their instructional leader as authentic and are more likely to act upon it (Frasier, 2022).

A significant part of effective feedback during post observation meetings is that it provides the opportunity for assistant principals to make targeted recommendations for instructional resources and professional development programs to help teachers meet their learning goals (Kraft, 2022). Professional development can take many forms such as self-reflection, pursuing a course, or participating in school wide teacher development initiatives. With the knowledge gained from observation and feedback, assistant principals can organize workshops and training tailored to the needs of teachers. Furthermore, assistant principals can organize peer mentors based on teachers' needs, strengths and challenges. Modeling this collaborative approach is a practical and invaluable tool in enabling teachers to become highly effective.

Collaboration and Teamwork

Assistant principals structure systems where teachers with varying skill levels and experiences support each other through knowledge sharing, feedback loops, and joint problem-solving. Many of these meetings have been coined as collective collaboration. Collaboration is a strategy assistant principals find useful in fostering effective leadership. Assistant principals not only provide guidance to teachers during collaboration but are also actively learning and growing from their perspectives and experiences.

Assistant principals take the view that working together with teachers is a significant means of achieving improvements in schools. This means that the starting point for strengthening the work of a school is the sharing of existing practices through collaboration among staff, leading to experimentation with new practices that will reach out to all students. Collaboration transfers trust and confidence to the administrative team and teachers and reinvests both entities in each other's expertise. Strahan et al. (2010) explored and reported how the starting point for effective school leadership begins with collaboration among staff.

At the heart of collaboration in schools where effective leadership occurs is developing an academic discourse on pedagogy and best practices (Cox & Mullen, 2022). Without an anomaly of such, learning would be the dissemination of information. Fostering teacher collaboration is how teachers can develop innovative practices. Through this shared experience, teachers construct systems and practices that harness excitement and building level change (Hiebert et al., 2002). Collaboration allows rethinking by interrupting existing discourses. Collaboration requires fragile facilitation of the principal. Everyone must be positioned to promote change, particularly new teachers. Collaborating with early career teachers caused assistant principals to distinguish mentoring from collaboration. Mentoring requires interpersonal theory. Active listening, oral communication, and nonverbal communication skills enabled assistant principals to address specific needs as a mentor. One assistant principal noted that school administration is a team, so it was important to be a team player. It is a relationship in which the assistant principal is transparent and creates an environment through which all team players have awareness to gain. This can be further explored through Professional Learning Communities (PLCs).

Professional Learning Communities (PLCs)

With the demand for increased student achievement, the tide has turned toward teachers again (Hargreaves & Fullan, 2015). The dreaded days of teachers isolating themselves in their classrooms with little concern about what other teachers are doing is gradually decreasing. Hargreaves & Fullan, (2015) argues that the teaching profession should continuously and collaboratively expand its knowledge base and expertise, ensuring that practices are transparently tested, refined, shared, and adapted to meet evolving needs effectively. Therefore, many high reliability schools have shifted their focus to enabling their teachers to interchange repeated cycles of shared investigation and implementation through Professional Learning Communities (PLCs) to accomplish better student outcomes (Marzano et al., 2018).

Assistant principals play a crucial role in supporting and leading Professional Learning Communities (PLCs) by promoting a culture of collaboration, reflection, and ongoing improvement. They act as facilitators, by setting clear objectives, promoting open communication, and providing the necessary resources, ensuring that PLCs operate efficiently. As instructional leaders, assistant principals offer guidance on data analysis, curriculum alignment, and evidence-based teaching strategies to ensure that the work of PLCs directly impacts student achievement. Additionally, they monitor the progress of PLCs and provide constructive feedback to keep discussions and actions aligned with school priorities.

According to Marzano et al (2018) the PLC process should guide the school in ensuring that all students learn. Therefore, it must create a collaborative culture in which a combined effort of the school community is responsible for students' learning. One of the roles of an assistant principal in fostering PLCs in this way is by providing focused professional development around specific themes, such as differentiated instruction or integrating technology, allowing teachers to explore these topics deeply and apply them in their classrooms. Additionally, these focused professional developments could be used to guide teachers with analyzing data for students. Assistant principals guide teachers in reviewing assessment data to identify areas of need, share best practices, and develop targeted interventions for struggling students. This can be done as a collective effort, highlighting where students need improvement and setting school wide goals to achieve these (Marzano et al, 2018). Not only will this mitigate isolation, ineffective or unproductive team meetings, but it will ensure all team members understand the nature and importance of the work they need to do (Marzano et al, 2018) to improve students' learning.

Theoretical Framework

Schools are complex adaptive systems (Hanselman et al, 2016) which require collaborative relationships and effective practices that foster school success. Education is constantly changing thus leaders must strategize ways to support teachers to address the needs of policy changes which place high expectations on student learning outcomes. To do this, leaders must first make sense of the continuous changes and their role in influencing teachers to adapt to these changes. Marzano et al., (2018) posits that leaders must recognize that their challenge is not only putting new structures and policies in place but also reshaping the school culture and the assumptions, beliefs and expectations that drive the culture. The two theories that guide this study are sense making theory (Weik, 1995) and social constructivism theory (Van der Veer & Valsiner, 1991). These theories form the basis in helping assistant principals understand their role in fostering collaboration and guiding teachers toward growth. This study adopts a holistic approach to examining sense-making within school leadership (Ganon-Shilon & Chen Schechter, 2017) within the broader concept of social constructivism.

Sense-making is a continuous and dynamic process in which individuals seek to comprehend events or issues that interrupt routine and introduce uncertainty (Maitlis & Christianson, 2014). It entails actively interpreting present stimuli, influenced by past knowledge, experiences, beliefs, and values, within the social context of the work environment. When faced with uncertainty, individuals use interpretive mental models to frame their surroundings and "make sense" of what has occurred. This theoretical construction is valuable because it provides insight into how and why individuals reach specific conclusions (Smerek, 2011; Weick et al., 2005). Sense-making encompasses three interconnected processes—creation, interpretation, and enactment (Weick, 1995). In this study, assistant principals use their experiences to create multiple paths to enhance teacher support and promote collaborative practices. They will consider what systems they can put in place to help teachers improve their pedagogical skills. For example, analyzing data on student performance to adjust instructional strategies. Second, the interpretation process refines the initial understanding through multiple perspectives, enabling individuals (both assistant principals and teachers) to organize their thoughts and uncover deeper insights into their dispositions. This approach helps mitigate the influence of preexisting biases, fostering a clearer grasp of the situation by ensuring that everyone understands all the school's initiatives and their role in operating in a cohesive manner. For example, using professional learning communities (PLCs) to discuss teaching methods and assessments. Finally, the enactment process translates the developed understanding into action. Teachers incorporate new information from PLCs and other professional development offered such as integrating evidence-based teaching strategies into lesson planning, thereby closing the sense-making loop.

Social constructivism theory framework emphasizes that knowledge and understanding are co-constructed through social interactions and cultural norms. It highlights the importance of collaborative dialogue and shared experiences. Creswell (2013) posits that individuals seek to comprehend the world in which they live and work thus, assistant principals try to make sense of their role from lived experiences as described by those phenomena in the study. For example, APs shape their leadership approach by engaging with teachers in open discussions about their challenges and needs through evaluation, staff meetings and professional development workshops.

In essence, sensemaking and social constructivism in leadership and teaching practices are crucial. According to Hargreaves and Fullan (2015) educational reform is at a crossroads: we can either enforce stricter measures of accountability or foster an inspiring and resilient teaching profession. By applying sensemaking and social constructivism, school leaders can create supportive environments where teachers collaboratively navigate change, ultimately enhancing student outcomes.

METHODOLOGY

Effective school leadership has gained significant attention over the years (Northouse, 2016). Several reviews of scholarly studies show there is a variety of approaches leaders use in leadership. A qualitative approach will be utilized to provide a comprehensive understanding of the topic. This phenomenological research was aimed at exploring the role of assistant principals in fostering teacher collaboration and professional growth as a pathway to effective school leadership. It emphasizes the evolving role of the assistant principal as an instructional leader, guiding teachers in adapting to changing circumstances. A qualitative approach was chosen to allow the researcher to gain a profound understanding of assistant principals' lived experiences and how these experiences influence their role in fostering teacher growth and support. Additionally, it provides the researcher in-depth analysis of the topic, utilizing methods such as document analysis and semi-structured interviews.

Research Design

To examine the perspectives of the assistant principals' role in their own voices, a quality research method approach was employed where human perception and understanding was central. (Furman, 2020; Stake, 2010). The aim of this study was to understand how different assistant principals perceive their role and the effectiveness of this role in influencing teacher growth and collaboration. Seven administrators in the role of assistant principals were interviewed in varying school districts. The authors conducted phenomenological qualitative research to investigate how assistant principals individually interpret their roles, responsibilities, and the challenges they face in collaboration and professional growth initiatives. Also, to examine how these interpretations are shaped, shared, and negotiated through team meetings, professional development sessions, and the school's cultural norms. The authors analyzed shared themes in the interviews to explore the experiences of these administrators using phenomenological qualitative research. This phenomenological qualitative research study seeks to explore the lived experiences of individuals who hold this specific role. According to Creswell (2013), "a phenomenological study describes the common meaning for several individuals of their lived experiences of a concept or a phenomenon" (p. 76). This approach aims to ascertain individuals' perspectives and the essence of their experiences.

Research Questions

This study is guided by the overarching research question:

- How do assistant principals perceive their role in effective school leadership?

To explore this central question, the following sub-questions were developed:

- How do assistant principals perceive their role in fostering leadership development, both personally and among their staff?
- What challenges do assistant principals encounter that impact their readiness for future principalship?
- What instructional leadership practices can be implemented or improved to enhance the effectiveness of assistant principals?
- In what ways do assistant principals make sense of their role in promoting collaboration among educators?
- How do social interactions and the cultural norms within schools shape shared understandings of collaboration and professional growth?

Participants

For the purpose of this study, the authors used purposeful criterion sampling by choosing participants based on their association to the topic (Creswell, 2013). The participants were selected from several school districts, including two international schools. This phenomenological study's results encompassed insights from seven assistant principals in elementary and middle schools. This number was in tandem with the proposed number of research participants. According to Creswell (2014) it is effective to elicit important data from a smaller sample size in a qualitative study. All seven participants completed the interviews. The participants brought diverse professional backgrounds and leadership experiences, enriching the study with varied perspectives on the assistant principalship. Subsequently, the findings of the research questions were confirmed as no additional information emerged.

Setting

The data was gathered from five schools in the United States of America (USA) and two American schools in the Middle East. Since most of the data were gathered from public schools in the USA, it is important to examine the potential factors that may influence potential factors that may influence the extent to which they adhere to the Professional Standards for Educational Leaders (PSEL). To this end, the National Policy Board of Education Administration (2015) in the USA calls on school leaders to "develop, advocate, and enact a shared mission, vision, and core values of high-quality education and academic success and well-being of each student." (p. 33). The state in which our data were acquired, the standards for school leaders highlight developing a shared vision. Additionally, it emphasizes the significance of the school leaders' unique responsibility to develop and implement a vision of learning to guide organizational decisions and actions (Louisiana Department of Education, 2019.)

Data Collection

Data collection is a fundamental component in the research process as it is the principal source of obtaining information. Therefore, researchers must possess data knowledge that adheres to data standards (Savin-Baden & Howell Major, 2012). For this study, the primary data sources included a survey with demographics and structured interviews with seven APs. An interview in qualitative research is a method of data collection in which a researcher and participant engages in a conversation to gather detailed information about their opinions, experiences, thoughts, and feelings on a particular topic (Savin-Baden & Howell Major, 2012). According to Savin-Baden & Howell Major (2012) the goal of interviews must be as natural as possible making the perspectives and experiences shared meaningful. Thus, allowing the researcher to get a deeper understanding of the subject matter than what other methods might provide.

This study employed a purposeful sampling technique. Purposeful sampling is a commonly used technique in qualitative research, aimed at selecting participants who meet specific criteria and are directly relevant to the study's focus. According to Maxwell (2005) purposeful sampling is a procedure where settings, persons and activities are intentionally selected to gain information that cannot be obtained from other sources. Consequently, purposeful sampling requires creativity in cautiously selecting individuals who are likely to provide the best information. (Savin-Baden & Howell Major, 2012). Therefore, having a good rationale for the selection criteria that makes sense and aligns with the research topic and questions is essential. This means in addition to the study's goal; the determination of sample size will be dependent upon the time and availability of resources. All data was saved on school software to guarantee safety of information received.

Procedures

The primary method of data collection for this study was structured interviews. Utilizing interviews as a data collection method allowed the researchers multiple opportunities to thoroughly examine the participants' perspectives on their roles and to identify commonalities and differences in the experiences of the phenomenon during the process of analyzing data. Prior to the interview the participants were contacted by email to confirm their understanding of the purpose of the study. Additionally, to adhere to ethical principles, the researchers highlighted the interview protocols and reassured participants about confidentiality before asking the questions. A pseudonym was randomly assigned to each participant to guarantee confidentiality. The interview questions were open-ended and designed to address the central research question of the roles of assistant principals in fostering teacher support, growth and development while enhancing effective school leadership. Participants were given sufficient time to respond to questions or any and were thanked for their participation upon completion of the interview. Participants were not incentivized monetarily for this study.

Data Analysis

The researchers proceeded with the data analysis stage after completion and validation of collecting data from interview transcripts. Qualitative data analysis is a continuous process that encompasses breaking data into significant segments for examination (Savin -Baden & Howell Major, 2012). Hence, the goal of qualitative data analysis is to understand the data (Merriam 2009), "with an intentional effort toward answering the research questions" (Savin -Baden & Howell Major, 2012, p. 434). The data was analyzed using various techniques such as organizing, sorting, grouping, coding, labeling and categorizing (Savin -Baden & Howell Major, 2012). To analyze the data collected the researcher used the following steps as guidelines: first, the entire transcript was read to obtain general information, second to extract specific messages, the general messages were compiled and finally, a pattern was identified from these specific messages. Additionally, the researchers applied some concepts from Creswell's (2014) research data analysis strategies which included the use of quotations to support findings and providing comprehensive discussion to address the research question.

The researchers collected insights from assistant principals to understand their perspectives on their role in supporting teachers, helping them to grow while promoting effective leadership. The group of participants is diverse, consisting of seven individuals: five Black and two White, ranging in age from 35 to 64. Specifically, two participants fall within the 35–44 age range, three in the 45–54 range, and two in the 55–64 range. The group includes six females and one male. All participants hold a Master's degree, with one also earning a Specialist Degree.

The participants bring significant experience in education, with two having 2–19 years and five having 20–39 years of experience. Their certifications include school leadership and counseling, alongside diverse teaching and administrative qualifications. As assistant principals, one participant has 0–1 year of experience, while six have 2–19 years. Their tenure at their current schools ranges from 0 to 19 years. They work in a mix of suburban (3 participants) and urban (4 participants) schools, serving student populations between 350 and 900.

The authors used surveys and open-ended questions in an interview to gather feedback about various roles and responsibilities of an assistant principal and how this influences teacher support, growth and development. Each interview lasted between 45 to 60 minutes, giving the participants the opportunity to answer questions comprehensively. During the interview, the researchers followed a protocol that addressed participants' experiences throughout the study. The interview included questions regarding their perceptions of instructional leadership, understanding competencies of leadership development, and the challenges and support systems encountered in their role. Furthermore, participants were asked to provide recommendations for enhancement in the educational system. Moreover, participants were provided with the opportunity to share additional thoughts about their experiences.

Post transcription, the authors developed code rules to create structure, consistency and transparency when identifying the themes that helped with understanding the research topic. The authors used inductive and deductive coding to identify recurring themes across interviews and to enhance the validation of the interview data. After rereading several times, the authors used Creswell's 2013 "memoing" by annotating words and phrases in the margins to identify emerging themes. Recognizing saturation, the authors decided on the established themes that emerged. Additionally, an anticipated data reduction matrix was used to help organize additional details that emerged from the research. This approach helped with defining and redefining codes (inductive and deductive), to ensure data analysis is consistent. Also, this allowed flexibility in adapting the matrix throughout the analytical process. Furthermore, it enabled inter-rater reliability because multiple researchers worked on this study. Thus, providing opportunities for researchers to check for agreement in code application which is essential for data quality.

RESULTS

Results for this phenomenological study were obtained from interviews with seven participants. As outlined in Chapter 3, participants were selected based on a predetermined criterion to ensure two purposes. First, that this study could encapsulate the firsthand experiences of the phenomenon under investigation through their personal narratives. Second, this research study includes participants relevant to the goal of the study. According to Creswell et al., (2007) a key purpose of conducting qualitative research is to identify a sample population familiar with the phenomenon under investigation or capable of providing personal accounts of their experiences with it. Furthermore, Savin-Baden & Howell Major, (2012) emphasize "who to study requires concentrated effort by the researcher" (p.317) as "this process is what sets one research study apart from the next and is what makes qualitative research truly an investigation of the unique and the particular (p. 317)." The major findings and results from this study support the notion that the assistant principal's role enhances teacher collaboration and professional growth. Also, it validates prior research that the evolving role and practices of assistant principals provide an excellent pathway for effective school leadership. Several themes emerged from this study. These include the assistant principal's role as an instructional leader and manager, the characteristics and competencies of a leader and challenges faced by assistant principals.

Theme One: The Role of Assistant Principals - Instructional Leaders

The main theme that emerged from the study is instructional leadership. All the participants perceive their primary role as an instructional leader providing support and guidance to their staff. The responses collectively define instructional leadership as the principal's role in guiding and supporting the school community (teachers, students, staff, parents, and stakeholders) to improve teaching and learning. This includes several specific functions, supporting teachers and staff through professional development, providing research-based tools and continuous opportunities for educators to enhance their content knowledge and promote the best instructional practices. Also, ensuring effective data analysis, curriculum development, and instructional support by assessing their performance, and providing necessary feedback. This also includes observation, monitoring lesson plans, and planning professional development sessions. Additionally, creating and maintaining a vision and mission for the school that aligns with the community's goals, and effectively communicating these goals to all stakeholders were considered essential roles of the school instructional leadership in providing support and guidance. AP Kyley noted that her role "involves supporting teachers and staff, creating school culture and climate and guiding curriculum development, data analysis along with instruction support." AP Christen also noted that "instructional leadership must include planning for workshops or sending teachers to conferences to improve teaching and learning skills." While AP Tom and AP Buck both noted "instructional leadership facilitates growth and improvement of the teaching and learning pedagogy." Furthermore, AP April noted "providing support and guidance in establishing best practices in teaching and guiding the school in the direction for the goals of the school" is one of her roles as an instructional leader.

Additionally, collaborating with teachers on curriculum development and instructional methods, attending and supporting PLCs (Professional Learning Communities), and being an active member of the school leadership team. Furthermore, all assistant principals manage discipline, ensure student safety, and address student referrals. These responsibilities create a conducive learning environment, directly affecting instructional quality. Finally, they communicate with parents, community members, and stakeholders to ensure alignment with the school's vision and goals. This involves acting as a liaison between the school and its broader community, reinforcing the instructional leadership role by building a supportive educational environment. Overall, the duties align with instructional leadership by focusing on creating a supportive, effective environment for teaching and learning, fostering professional growth among teachers, and ensuring that administrative tasks contribute to the overall educational goals of the school.

Sub-Theme One: Characteristics of Instructional Leaders

Notably, providing support and guidance involves several skills and instructional leaders must possess these abilities to carry out their duties effectively. Some collective qualities of instructional leaders highlighted in this study include communication and listening skills, being knowledgeable, a systems thinker, and a visionary. Being able to actively listen to and understand the needs and concerns of all stakeholders helps leaders to create a shared vision and clearly articulate goals and beliefs. AP Payton posits that "It is very important to be a good communicator, you must convey essential beliefs about learning, including the conviction that all children have the ability to learn." Therefore, guiding the school towards achieving its goals, and being intentional about implementing changes to improve student outcomes. As an administrator in a religious based international school AP Payton emphasizes the importance of "inclusive - objective" communication for successful leadership. She posits "one's culture and belief system can hinder growth, therefore, in my role leading a diverse community with both native and non-native English speakers, I have learned to be more tolerant of others, be receptive to suggestions from staff and be nonpartisan to people with different beliefs." AP Payton believes that working in this environment enabled her "to see other people's perspectives and adapt accordingly."

Other important mutual qualities highlighted include having a supportive nature and positive attitude, being observant and present, and demonstrating strong collaboration skills. Overall, the responses emphasize that effective instructional leaders are those who can balance support and guidance with strategic vision, foster a positive and collaborative school culture, and continuously seek to improve teaching and learning outcomes through informed and reflective practices.

Theme Two: The Role of Assistant Principals – Administrative Manager

During the interview similar themes emerged for the primary roles of assistant principals as administrative managers. The responses collectively outlined various primary duties of an assistant principal administrative and operational duties, teacher support and development, student-focused responsibilities and community and stakeholder engagement. All assistant principals' roles included administrative and operational duties such as bus duty, lunch duty, transition duties, managing the master schedule, ensuring discipline, managing student attendance, and overseeing school operations (e.g., transportation and extracurricular activities). These tasks ensure the school runs smoothly and provides a safe, effective learning environment.

Theme Three: Competencies of Assistant Principals

Another theme that emerged from the study is that being competent in the area highlighted in instructional leadership and administrative managerial duties are crucial in the role of an assistant principal. During the interview, the responses highlighted several key competencies that assistant principals developed through various practical experiences. This included communication and listening skills, curriculum and instructional knowledge, collaboration and teamwork, data-driven decision making, diplomacy and support, adaptability, and cultural competence and most significantly, leadership and vision. By developing these competencies, assistant principals effectively support instructional leadership, ensuring that teaching and learning practices are continuously improved, and the school's educational goals are achieved while simultaneously fulfilling other administrative tasks outside of instruction. Hence, confirm the idea that the assistant principal's role is valuable in fostering teacher collaboration and professional growth.

Theme Four: Challenges in the Assistant Principal's Role

The study also emphasized some challenges assistant principals faced and how these impact their ability as an instructional leader. First is the significant amount of time spent dealing with student discipline, which limits the ability to focus on instructional support and working closely with teachers. Some assistant principals emphasized specifically, working in traditional public schools with high discipline disparities, low parental involvement, and high teacher turnover that limits exposure to innovative practices and high-performing environments. Additionally, the assistant principals collectively reported other challenges such as budgetary constraints, and the consistent challenge of not having enough time to balance all the responsibilities effectively daily, which can limit opportunities to focus on instructional leadership and curriculum alignment. Furthermore, they lament that they cannot find time for teachers to collaborate, which is an "absolute priority" (Marzano et al. 2018, p. 17).

DISCUSSION

The role of assistant principals (APs) in fostering teacher collaboration and professional growth is multifaceted, requiring them to navigate complex social and organizational dynamics. To better understand the influence of assistant principals' role in teacher collaboration and professional growth, this study adopted two theoretical frameworks: social constructivism and sense making theories. By applying sensemaking theory (Weik, 1995) and social constructivism theory (Van der Veer & Valsiner, 1991), we can better understand how APs interpret their responsibilities and actively shape the collaborative and developmental culture within schools. For this study, we highlight three concepts: sensemaking theory in the assistant principal's role, social constructivism in building collaborative cultures and integrating these theories into practice.

Sensemaking Theory in the Assistant Principal's Role

For sensemaking in the assistant principal's role we consider interpreting contextual needs, establishing shared goals and creating structured opportunities for collaboration. Marzano et al. (2018) emphasizes the challenges of leading high reliability schools. It requires effective leadership, asking hard questions, prioritizing the work, collaborating, analyzing data and monitoring progress. Consequently, leaders must use the best approach to work with clarity of purpose and clear direction, especially within a school system that undergoes constant changes. To do so, leaders must understand their work environment and be able to prioritize the needs of ensuring success of the school. Through interpretation of the school's needs, assistant principals engage in sensemaking by identifying the unique needs of teachers and students. For example, if a school has a high level of teacher turnover, APs might prioritize collaboration to build trust and stability among staff. Two of the assistant principals noted that not having enough time in the day to support teachers contributes to teachers quitting their jobs. They alluded to their complex role of trying to make sense of the issues at hand which allows them to connect disparate pieces of information—such as performance data, teacher feedback, and institutional goals—to inform their leadership decisions, thus enhancing effective leadership and promoting school improvement.

Moreover, interpreting and understanding the environment is vital as "there is no one way to organize" (Bolman & Deal, 2021, p. 73) a school. (Bolman & Deal, 2021, p. 73) states "the right structure depends on prevailing circumstances, considers an organizational goals, strategies, technology, people and environment." They further remarked "understanding the complexity and variety of design possibilities can help create formal prototypes that work for, rather than against, both people and collective purposes (p.73)." Thus, assistant principals can help principals direct their energy into identifying these goals and working towards them. AP Christen noted that working alongside the principal on targeted instructional goals mitigates confusion and increases collaboration.

This 'trickles' into establishing shared goals as effective collaboration requires alignment around shared goals. Telling teachers what to do but not how to achieve them is a recipe for disaster – failed solution – silver bullets (Hargreaves & Fullan, 2015). APs use sensemaking to bridge diverse perspectives, fostering a shared understanding of professional growth objectives. For instance, when introducing a new instructional strategy, APs interpret resistance from teachers not as a barrier but as an opportunity to address underlying concerns and clarify the purpose of the change. AP Brandy stated that an assistant principal must be "a systems thinker" to understand changes and help teachers maneuver these changes. She noted "it is important to act decisively and with fairness when dealing with problems with all and actively listening to all concerns. I have incorporated an open-door policy."

AP Tom also noted that "an effective instructional leader must be very knowledgeable in his career path, be motivational and positive, a great communicator, a team player and a collaborator who believes in others, one who is optimistic and strives to inspire all staff. He further stated, "an effective leader is a goal oriented person and one who is passionate about education." Likewise, AP Buck noted "instructional leadership is providing direction and support to the school community (teachers, students, staff, parents, and stakeholders) and aligning that support to the vision and mission of the school." Also, AP Brandy emphasizes "understanding the importance of building the school community through effective communication, making informed decisions, and being intentional about bringing about the change that one seeks to improve student outcomes." Consequently, they must create structured opportunities for collaboration. Sensemaking enables APs to design and facilitate structured professional learning communities (PLCs), workshops, or peer coaching sessions. These forums allow teachers to co-construct meaning, share experiences, and learn collectively. APs act as interpreters and facilitators, guiding discussions to ensure alignment with broader school improvement goals (Marzano et al. 2018). According to Hargreaves and Fullan, 2015 "professional expertise is not just about having evidence or being aware of it. It's also about knowing how to judge the evidence and knowing what to do with it" (p. 54).

Social Constructivism in Building Collaborative Cultures

In the social constructivism theory, we consider three key points: co-construction of knowledge, scaffolding professional growth and establishing norms of collaboration. Collaborative teams transform schools (Marzano et al, 2018). In the co-construction of knowledge, APs play a vital role in creating spaces where teachers feel empowered to share their expertise and learn from one another. By fostering open dialogue and encouraging reflective practices, APs promote a culture where collaboration becomes a vehicle for professional growth. For example, during collaborative planning meetings, APs can encourage teachers to analyze student work together and identify strategies to improve outcomes. According to Hattie (2009) reflective teaching is most powerful when done collectively (teacher team rather than an individual) and must be based on concrete evidence of student learning. Guiding and supporting staff through professional learning communities (PLCs) is one best way for teachers to co-construct knowledge. PLCs must be intentional, it requires building "shared knowledge about the most promising practices" (Marzano et al., 2018 p. 5) that help them learn. AP Christen postulates "I create collaborative, inclusive learning environments by implementing peer reviews and best practices sessions. This has really been beneficial because teachers are able to see best practice in action and emulate strategies and use them successfully in their classrooms." Assistant principals play an integral role in leading this mantra. They must take responsibility for equipping staff with the necessary information to make informed decisions throughout the teaching learning process. AP Tom encourages teaching staff to utilize data and resources. He states, "teachers find this very helpful because they can make informed decisions through data knowledge." Marzano et al stresses "uninformed people make uninformed decisions (Marzano et al., 2018 p. 5)." This statement attributes to one of the key themes in the findings that APs should provide guidance to staff, be able to communicate and lead a team effectively.

There are many ways to conceptualize effective leadership. Leadership is a transformational process where both leaders and followers work collaboratively to raise one another to higher levels. It involves influencing a group of people (followers) to achieve a common goal (Northouse, 2016). Northouse, (2016) sums up leadership as a process, influence, occurs in groups, and achieving common goals. In social constructivism theory, assistant principals act as instructional leaders to scaffold professional growth. Providing scaffolding is necessary for teacher support and development and transforms schools (Hargreaves & Fullan, 2012). By scaffolding, APs enable teachers to engage in deeper levels of collaboration, gradually building their capacity for independent professional growth. This process involves modeling best practices, facilitating mentorship opportunities and distributed leadership by assigning tasks. This means leaders and followers are in the "leadership relationship together (Northouse, 2016, p.

7)." Modeling best practices starts with leaders taking the first step. According to Hargreave and Fullan, (2015) "leadership is doing something first before anyone else is willing or able to do it (p.161)". AP Christen stated "with my experiences, my principal allows me to experience her responsibilities which prepares for instructional leadership. I'm guided through the responsibilities and roles as a principal." Likewise, AP Blake noted "modeling leadership is significant as you set high expectations for others. Holding your teachers accountable while building a culture that strives for excellence is beneficial to the entire school because these standards and expectations are instilled in students and passed on to families, and colleagues." Studies continue to show the positive impact of principals mentoring assistant principals, assistant principals mentoring teachers which enhances student outcome and builds school morals and culture.

Social constructivism emphasizes the role of social norms and culture in shaping behavior. APs contribute to the social construction of a collaborative culture by modeling collegiality, trust, and respect. They encourage practices such as peer observation and constructive feedback, which normalize collaboration as a critical component of professional practice. (Hargreaves & Fullan, 2015) describes collaborative cultures as an environment that promotes hard work, dedication, shared responsibility and a sense of pride within the school. Collaborative cultures are characterized by valuing teamwork and open communication (Pérez López et al., 2004), where members of the organization are encouraged to discuss challenges, share diverse perspectives, and exchange ideas resulting in constructive collaboration and mutual agreement (Barczak et al., 2010). Additionally, studies have shown that school cultures characterized by collegiality and collaboration are more likely to foster teachers' professional development and job satisfaction (European Commission, 2014; Leithwood et al., 1998). Furthermore, the benefits from shifting from traditional approaches of isolation and individuality to greater collaboration are invaluable in school improvement at all levels (Vangrieken & Kyndt, 2020).

Integrating Sensemaking and Social Constructivism Theories into Practice

Shaping the culture of the school involves identifying the unique needs of the school (sensemaking) and addressing these needs with shared teacher building knowledge and improvement practices (social constructivism). Researchers continue to seek solutions into what makes leadership effective and schools great. Leaders grapple to find game changing improvement plans as they and their team navigate the changes that plague our education system. The question of what practical strategies leaders use each moment to drive school improvement is key to success. This study analyzes what leaders, particularly assistant principals, prioritize daily to make this happen. Their lived experiences - practices of the past and reconstruction of knowledge in tackling new challenges is crucial in promoting academic success. Integrating sense making theory and social constructivism provides a robust framework on understanding assistant principal's role in employing strategies that foster teacher collaboration and professional growth. Not only is this process complex but also dynamic in putting the pieces of effective leadership together. By employing sensemaking theory, APs can interpret and respond effectively to the diverse needs of their school community. Simultaneously, social constructivism provides a roadmap for creating environments that support collective learning and shared professional growth. Together, these frameworks highlight the critical leadership strategies that enable APs to empower teachers and drive meaningful school improvement.

Limitations

There were several limitations of this study. Initially, it was conducted in one school district in East Baton Rouge and two separate American international schools. The study lacked participants from rural schools. It would be necessary to replicate this study in other geographical areas in the USA for better comparison. Second, it was limited by the fact only assistant principals with traditional roles were interviewed. Having perspectives from assistant principals with specific roles in curriculum and instruction or other roles that encompasses teacher support could enrich the data. Additionally, the perspectives from principals, district leaders, and teachers would offer more extensive insight into these findings. Finally, only 2 male assistant principals were interviewed and two international APs for this study. Including more male APs and more international APs could offer more diverse leadership perspectives and further strengthen the data.

Future Research

As education continues to evolve, more school districts have adopted the concept or idea of assistant principals as instructional leaders. This comes with the vast pressure of connecting student achievement with successful leadership. Some schools have hired assistant principals for roles that only include curriculum and instruction, while some continue to face the multifaceted role of juggling administrative tasks and supporting instruction. Despite many studies forecasting the role of assistant principals as instructional leaders, we are yet to see this transformation in full effect, often leading to ambiguity in their role. Consequently, there is a need for further research to help us to understand the assistant principal's perception of their role, particularly in fostering teacher collaboration and professional growth. Also, it would be interesting to learn how other individuals such as stakeholders, teachers and the community perceive their role. Additionally, this study could be replicated further in urban, sub-urban and rural districts and charter schools to gain better understanding of the phenomenon under investigation. Furthermore, conducting this research across K-12 schools would provide a more comprehensive account of assistant principals and their role in teacher support as a pathway to effective leadership.

RECOMMENDATIONS AND CONCLUSIONS

Quite too often school leaders, in particular assistant principals, find themselves caught up in focusing on non-productive tasks in schools. The findings from this study have practical recommendations on how to capitalize on the roles of assistant principals in promoting academic success and influence changes in school systems. First, the participants in the study indicated several ways that can enhance their development as instructional leaders while engaging in practices that can better prepare teachers to succeed in their classrooms and beyond. First, to address the most significant challenge of insufficient time to support teacher instruction, assistant principals must be given time to attend to both operational matters and instructional leadership to increase school efficiency. Secondly, not all assistant principals are directly involved with principal duties thus increased involvement in principal duties allow assistant principals to experience the responsibilities of the principal, guided by the principal, to prepare for future leadership roles. Including assistant principals in a wide range of meetings, conversations, and activities will enhance their communication skills and application of knowledge. Thirdly, our survey data indicates that the feedback from an evaluation that aligns with the job description is more palatable than an evaluation system with discrepancies. Thus, appropriate evaluation systems must be in place to properly assess the roles of APs. Fourthly, offering professional development that is focused on leadership and administration is crucial to keep assistant principals updated with new educational innovations and methodologies. As an instructional leader, assistant principals must be more than competent to carry out their mandated tasks daily. Marzano et al. (2018) garners that while the history of research on school leadership is relatively consistent and highlights the understanding of leadership factors, it lacks the implementation process – turning leadership theories into actionable knowledge. Likewise, Hargreaves and Fullan (2015) emphasizes that to get the expertise issue right we must invest in capability. According to Hargreaves and Fullan (2105) "capability is more than competence", it includes "skills and qualities that lead to accomplishment and build confidence (p. 55)." They further argued that capabilities set "higher bars for performance than simple competencies (p. 55)" thus assistant principals must possess the attributes to accomplish these tasks which is significant to effective school leadership.

Finally, providing mentorship and support for assistant principals is essential as it equips them with the necessary tools for school success. Mentorship provides assistant principals with valuable insights and strategies for effective leadership, offers feedback and guidance, while holding the assistant principal accountable. According to Retelle (2010) mentorship is a critical tool principals use to develop the skill sets of assistant principals. Similarly, Payne and Huffman (2005) highlighted that mentoring serves as a valuable approach for cultivating individual abilities, enhancing skills, and guiding individuals in understanding their specific job roles and responsibilities. Leadership has been linked to teacher achievement, and which also impacts student outcomes. Effective leaders breed effective teachers and high-quality leading influences high quality teaching.

In conclusion, school leadership must be at its best. It must create cultures where teachers and students thrive. High reliability and progressive organizations (Bolman & Deal, 2021; Marzano et al, 2018) implement "high-involvement strategies" (Bolman & Deal, 2021, p. 161) for strengthening the bond between individual and organization. Bolman and Deal (2021) argues that when individuals find fulfillment and purpose in their work, organizations (in this case are schools) benefit from the effective use of their commitment and expertise. The findings from this study emphasize that by adopting these practices, educational systems and leadership training programs can effectively prepare assistant principals for successful instructional leadership roles.

REFERENCES

Arroyo, M., Quinn, L. F., Paretti, L., & McCarthy, J. (2020). A Content Analysis of Coaching Novice Teachers. *Educational Research: Theory and Practice*, *31*(1), 22–25.

Austin, D. B., & Brown, H. L., Jr. (1970). Report of the assistant principalship. Volume 3: The study of the secondary school principalship. National Association of Secondary School Principals.

Barczak, G., Lassk, F., & Mulki, J. (2010). Antecedents of team creativity: An examination of team emotional intelligence, team trust, and collaborative culture. *Creativity and Innovation Management*, *19*(4), 332–345. DOI: 10.1111/j.1467-8691.2010.00574.x

Bolman, L. G., & Deal, T. E. (2021). *Reframing organizations: Artistry, choice, and leadership* (7th ed.). Jossey-Bass.

Bond, N. (2021). Teacher leadership in an elementary school: A case study of instructional support specialists. *Education*, *141*(4), 201–213.

Cox, J. S., & Mullen, C. A. (2023). Impacting student achievement: Principals' instructional leadership practice in two Title I rural schools. *Journal of School Leadership*, *33*(1), 3–25. DOI: 10.1177/10526846221133996

Creswell, J. W. (2013). *Qualitative inquiry & research design: Choosing among five approaches* (3rd ed.). SAGE Publications.

Creswell, J. W. (2014). *Research design: Qualitative, quantitative, and mixed methods approaches* (4th ed.). Sage Publications.

Creswell, J. W., Hanson, W. E., Clark Plano, V. L., & Morales, A. (2007). Qualitative research designs: Selection and implementation. *The Counseling Psychologist*, *35*(2), 236–264.

Dagli, G., Muhtaroglu, M. B., Bastas, M., Altinay, F., & Altinay, Z. (2023). Evaluation of primary school managers' duties in digital transformation. *GeSec: Revista de Gestão e Secretariado*, *14*(9), 15227–15249. DOI: 10.7769/gesec.v14i9.2524

Deiorio, N. M., Moore, M., Santen, S. A., Gazelle, G., Dalrymple, J. L., & Hammoud, M. (2022). Coaching models, theories, and structures: An overview for teaching faculty in the emergency department and educators in the offices. *AEM Education and Training*, *6*(5), e10801. DOI: 10.1002/aet2.10801 PMID: 36189456

European Commission. (2014). *The Teaching and Learning International Survey (TALIS) 2013: Main findings from the survey and implications for education and training policies in Europe*. Publications Office of the European Union.

Every Student Succeeds Act of 2015, Pub. L. No. 114-95, 129 Stat. 1802 (2015). https://www.congress.gov/114/plaws/publ95/PLAW-114publ95.pdf

Ewing, L.-A. (2021). Mentoring novice teachers. *Mentoring & Tutoring*, *29*(2), 164–186. DOI: 10.1080/13611267.2021.1899585

Frasier, A. S. (2022). What makes classroom observation feedback useful? The perceptions of secondary math and English teachers. *Voices of Reform*, *5*(1), 40–58. DOI: 10.32623/5.00004

Furman, P. B. (2020). *Perceptions of assistant principals of curriculum and instruction regarding the effectiveness of their role* (Publication No. 1677) [Doctoral dissertation, University of Alabama at Birmingham]. UAB Digital Collections.

Ganon-Shilon, S., & Schechter, C. (2017). Making sense while steering through the fog: Principals' metaphors within a national reform implementation. *Education Policy Analysis Archives*, *25*(105), 105. Advance online publication. DOI: 10.14507/epaa.25.2942

Glanz, J. (2004). The assistant principal's handbook: Strategies for success. *Sage (Atlanta, Ga.)*.

Goldring, E., Rubin, M., & Herrmann, M. (2021). *The role of assistant principals: Evidence and insights for advancing school leadership*. The Wallace Foundation. DOI: 10.59656/EL-SB1829.001

Gonzales, M. M., & Roberts, M. (2022). The impact of a franchise model school framework on the leadership development of assistant principals. *Journal of School Leadership*, *32*(5), 537–561. DOI: 10.1177/10526846211067636

Gurley, D. K., Peters, G. B., Fifolt, M., Collins, L., & McNeese, R. (2015). Addressing the multiplicity of roles of the assistant principal: Insights from the field. *Journal of Behavioral and Social Sciences (Cedarville, Ohio)*, *3*(3), 135–147.

Hanselman, P., Grigg, J., Bruch, S. K., & Gamoran, A. (2016). The consequences of principal and teacher turnover for school social resources. In *Family environments, school resources, and educational outcomes* (pp. 49–89). Emerald Group Publishing Limited., DOI: 10.1108/S1479-353920150000019004

Hargreaves, A., & Fullan, M. (2012). *Professional capital: Transforming teaching in every school*. Teachers College Press.

Harris, A., Jones, M., & Ismail, N. (2022). Distributed leadership: Taking a retrospective and contemporary view of the evidence base. *School Leadership & Management*, *42*(5), 438–456. DOI: 10.1080/13632434.2022.2109620

Hattie, J. (2009). The black box of tertiary assessment: An impending revolution. In L. H. Meyer, S. Davidson, H. Anderson, R. Fletcher, P. M. Johnston, & M. Rees (Eds.), *Tertiary assessment & higher education student outcomes: Policy, practice & research* (pp. 259–275). Ako Aotearoa.

Hiebert, J., Gallimore, R., & Stigler, J. W. (2002). A knowledge base for the teaching profession: What would it look like and how can we get one? *Educational Researcher*, *31*(5), 3–15. DOI: 10.3102/0013189X031005003

Ho, J., Shaari, I., & Kang, T. (2023). Vice-principals as leaders: Role ambiguity and role conflicts faced by vice-principals in Singapore. *Educational Management Administration & Leadership*, *51*(3), 575–593. DOI: 10.1177/17411432211002527

Irvine, J. (2022). Practice and theory: Ten lessons that I have learned about being a vice principal. *Journal of Practical Studies in Education*, *3*(3), 7–16. DOI: 10.46809/jpse.v3i3.46

Jackson, T., McBrayer, J. S., Pannell, S., Cleveland, R. E., Miller, S. B., & Fallon, K. (2021). The effect of school tasks on principals' and assistant principals' leadership self-efficacy. *School Leadership Review, 16*(1), Article 7. https://scholarworks.sfasu.edu/slr/vol16/iss1/7

Kraft, M. A. (2022, February 24). Better teacher feedback can happen outside of the evaluation process. Brookings Institution. https://www.brookings.edu/articles/better-teacher-feedback-can-happen-outside-of-the-evaluation-process/

Leithwood, K., Leonard, L., & Sharratt, L. (1998). Conditions fostering organizational learning in schools. *Educational Administration Quarterly, 34*(2), 243–276. DOI: 10.1177/0013161X98034002005

Leithwood, K., Sun, J., & Schumacker, R. (2020). How school leadership influences student learning: A test of "the four paths model.". *Educational Administration Quarterly*, *56*(4), 570–599. DOI: 10.1177/0013161X19878772

Louisiana Department of Education. (2019). *Bulletin 125—Standards for educational leaders in Louisiana*. Louisiana Administrative Code, Title 28, Part CXXXVII. https://www.doa.la.gov/media/glpah3zg/28v137.pdf

Maitlis, S., & Christianson, M. (2014). Sensemaking in organizations: Taking stock and moving forward. *The Academy of Management Annals*, *8*(1), 57–125. DOI: 10.5465/19416520.2014.873177

Marzano, R. J., Warrick, P. B., Rains, C. L., & DuFour, R. (2018). *Leading a high reliability school: Use data-driven instruction and collaborative teaching strategies to boost academic achievement*. Solution Tree Press.

Marzano, R. J., Waters, T., & McNulty, B. A. (2005). *School leadership that works: From research to results*. Association for Supervision and Curriculum Development.

Maxwell, J. A. (2005). *Qualitative research design: An interactive approach* (2nd ed.). Sage Publications.

Merriam, S. B., & Tisdell, E. J. (2015). Dealing with validity, reliability, and ethics. In *Qualitative research: A guide to design and implementation* (4th ed., pp. 237–267). Jossey-Bass.

Morgan, T. L. (2014). *Understanding the leadership capacity and practice of assistant principals* (Publication No. 989) [Doctoral dissertation, University of Denver]. Digital Commons @ DU.

Morgan, T. L. (2018). Assistant principals' perceptions of the principalship. *International Journal of Education Policy and Leadership, 13*(10), 1–20. DOI: 10.22230/ijepl.2018v13n10a743

Munson, J., & Saclarides, E. S. (2022). Getting a foot in the door: Examining content-focused coaches' strategies for gaining access to classrooms. *The Elementary School Journal*, *123*(1), 128–154. DOI: 10.1086/720626

National Policy Board for Educational Administration. (2015). *Professional Standards for Educational Leaders 2015*.

Northouse, P. G. (2016). *Leadership: Theory and practice* (7th ed.). Sage Publications.

Payne, S. C., & Huffman, A. H. (2005). A longitudinal examination of the influence of mentoring on organizational commitment and turnover. *Academy of Management Journal*, *48*(1), 158–168. DOI: 10.5465/amj.2005.15993166

Pérez López, S., Montes Peón, J. M., & Vázquez Ordás, C. J. (2004). Managing knowledge: The link between culture and organizational learning. *Journal of Knowledge Management*, *8*(6), 93–104. DOI: 10.1108/13673270410567657

Preston, J. P., Jakubiec, B. A. E., & Kooymans, R. (2013). Common challenges faced by rural principals: A review of the literature. *Rural Educator*, *35*(1), 1–12. DOI: 10.35608/ruraled.v35i1.355

Ralebese, M. D., Jita, L., & Badmus, O. T. (2025). Examining primary school principals' instructional leadership practices: A case study on curriculum reform and implementation. *Education Sciences*, *15*(1), 70. Advance online publication. DOI: 10.3390/educsci15010070

Retelle, E. (2010). Mentoring experiences of first-year assistant principals. *EAF Journal*, *21*(2), 61–87.

Savin-Baden, M., & Howell Major, C. (2012). *Qualitative research: The essential guide to theory and practice*. Routledge.

Schatz-Oppenheimer, O., & Goldenberg, J. (2024). The impact of professional mentoring on mentors of novice teachers. *International Journal of Mentoring and Coaching in Education*, *13*(1), 122–141. DOI: 10.1108/IJMCE-08-2022-0070

Shakir, A., Elue, C., Croft, S. J., & Clegorne, N. (2023). The elementary assistant principal: Exploring mentorship as a development pathway for the principal job role. *Georgia Educational Researcher*, *20*(2), 4. Advance online publication. DOI: 10.20429/ger.2023.200204

Smerek, R. E. (2011). Sensemaking and sensegiving: An exploratory study of the simultaneous 'being and learning' of new college and university presidents. *Journal of Leadership & Organizational Studies*, *18*(1), 80–94. DOI: 10.1177/1548051810384268

Sorensen, L. C. (2019). "Big data" in educational administration: An application for predicting school dropout risk. *Educational Administration Quarterly*, *55*(3), 404–446. DOI: 10.1177/0013161X18799439

Stake, R. E. (2010). *Qualitative research: Studying how things work*. Guilford Press.

States News Service. (2011, August 23). The Wallace Foundation launches major "Principal Pipeline" initiative to help school leaders.

Strahan, D., Geitner, M., & Lodico, M. (2010). Collaborative professional development toward literacy learning in a high school through connected coaching. *Teacher Development*, *14*(4), 519–532. DOI: 10.1080/13664530.2010.533493

Turnbull, B. J., Anderson, L. M., Riley, D. L., MacFarlane, J. R., & Aladjem, D. K. (2016). *Building a stronger principalship, Volume 5: The Principal Pipeline Initiative in action*. Policy Studies Associates, Inc. DOI: 10.59656/EL-LS5394.001

Turnbull, B. J., Riley, D. L., & MacFarlane, J. R. (2015). *Building a stronger principalship, Volume 3: Districts taking charge of the principal pipeline*. Policy Studies Associates, Inc.

Van der Veer, R., & Valsiner, J. (1991). *Understanding Vygotsky: A quest for synthesis*. Blackwell Publishing.

Vangrieken, K., & Kyndt, E. (2020). The teacher as an island? A mixed method study on the relationship between autonomy and collaboration. *European Journal of Psychology of Education*, *35*(1), 177–204. DOI: 10.1007/s10212-019-00420-0

Wang, Y. (2019). Is data-driven decision making at odds with moral decision making? A critical review of school leaders' decision making in the era of school accountability. *Values and Ethics in Educational Administration*, *14*(2), 1–9.

Wang, Y. (2021a). When artificial intelligence meets educational leaders' data-informed decision-making: A cautionary tale. *Studies in Educational Evaluation*, *69*, 100872. DOI: 10.1016/j.stueduc.2020.100872

Wang, Y. (2021b). Artificial intelligence in educational leadership: A symbiotic role of human-artificial intelligence decision-making. *Journal of Educational Administration*, *59*(3), 256–270. DOI: 10.1108/JEA-10-2020-0216

Weick, K. E., Sutcliffe, K. M., & Obstfeld, D. (2005). Organizing and the process of sensemaking. *Organization Science*, *16*(4), 409–421. DOI: 10.1287/orsc.1050.0133

Weick, K. E., & Weick, K. E. (1995). *Sensemaking in organizations* (Vol. 3). Sage publications.

Wilson, T. A., & Clayton, J. K. (2020). Novice assistant principals' perceptions of professional learning experiences. *Education Leadership Review of Doctoral Research*, *8*, 20–37.

Chapter 7
Play, Lead, Teach:
Tapping Into the Power of Sports for Teacher Development and Classroom Innovation

Amitava Pal
https://orcid.org/0009-0007-8934-2718
ICFAI University, Jharkhand, India

Pallavi Kumari
https://orcid.org/0000-0002-6344-3795
ICFAI University, Jharkhand, India

Preksha Yadav
https://orcid.org/0000-0001-6316-8837
Krupanidhi College of Management, India

ABSTRACT

This chapter explores the transformative potential of sports-based approaches in teacher development, focusing on how physical activities and game-based learning foster leadership, cognitive, and emotional skills in educators. By examining global case studies, it highlights the role of sports in enhancing competencies such as decision-making, collaboration, resilience, and emotional intelligence. The chapter offers practical strategies for integrating sports into teaching methods to improve classroom dynamics, promote inclusivity, and encourage cross-disciplinary learning. It also discusses scalable implementation and the use of digital tools like AI to support teacher development, presenting sports as a key platform for shaping the future of teaching.

INTRODUCTION

The integration of sports into teacher development is an innovative approach that can significantly enhance educators' professional growth and effectiveness in the classroom (Harper-collins-admin, 2023; *Sports Integration | Holistic Development | LIFE And CAREER SKILLS*, 2024). Traditionally, sports have been seen as extracurricular activities, often disconnected from the core curriculum. However,

DOI: 10.4018/979-8-3373-0472-4.ch007

this chapter proposes that sports are not just tools for physical well-being but also powerful instruments for fostering leadership, collaboration, and pedagogical transformation. By incorporating sports-based strategies into teacher development programs, educators can gain essential skills that directly impact their teaching practices, classroom management, and student engagement (Knudsen et al., 2020).

In today's fast-paced educational landscape, teachers face the dual challenge of managing academic content while also attending to the emotional and social needs of their students (Social-Emotional Skills in School, n.d.). Sports-based approaches help address these challenges by promoting emotional resilience (Norris & Norris, 2021) and stress management (Bhandari, 2021). Engaging in physical activities and strategic games gives teachers a safe space to practice coping mechanisms for stress, build empathy, and enhance their emotional intelligence. These qualities, in turn, empower teachers to foster a positive and supportive classroom environment where students can thrive emotionally, socially, and academically. Additionally, sports-based strategies support cognitive development (Hernández-Mendo et al., 2019) by encouraging teachers to engage in strategic thinking and problem-solving. Whether through physical activities, like team sports, or more cerebral ones, such as chess or puzzles, teachers can hone their cognitive agility, which directly translates to more effective decision-making in the classroom.

As educational environments continue to evolve, integrating sports into teacher development ensures that educators are equipped with the tools and insights necessary to handle the complexities of modern teaching (Kim et al., 2019). This chapter highlights how such integration not only benefits teachers but also enhances the overall classroom experience, fostering an engaging, inclusive, and student-centered learning environment. Through this exploration, the chapter aims to provide a comprehensive understanding of how sports can act as a transformative tool in teacher development, ultimately leading to better outcomes for both educators and students.

Objective of the Chapter

The primary objective of this chapter is to explore the transformative potential of sports-based approaches in teacher development. It aims to examine how integrating physical activities and game-based learning can enhance essential skills such as decision-making, resilience, teamwork, emotional intelligence, and cognitive agility among educators. This chapter also seeks to demonstrate how sports activities provide a framework for promoting inclusivity, collaboration, and cross-disciplinary learning in the classroom. Additionally, the chapter aims to explore practical strategies for integrating sports into teacher development programs, providing actionable insights for educators and institutions to improve teaching practices and classroom dynamics. Ultimately, the goal is to offer a novel perspective on how sports can contribute to creating a more effective and engaging educational environment.

Sports as a Framework for Leadership Development

Leadership is a crucial skill for educators, as it directly influences their ability to manage classrooms, guide students, and collaborate with colleagues (*Best CBSE School in Nagpur | Delhi Public School Kamptee Road Nagpur | DPS Nagpur*, n.d.). Teachers are not just subject experts; they are also leaders (Finkelstein, 2018) who need to inspire, motivate, and manage groups of students while navigating various challenges. Leadership skills such as decision-making (*(PDF) Leadership in Education: Decision-Making in Education*, n.d.), strategic thinking (CMOE, 2018), conflict resolution (*Conflict Resolution in Leadership | in Professional Development*, 2021), resilience (*The Importance of Resilience in Leadership: Thrive in Times of Change*, n.d.), and adaptability (Loeb Leadership, 2023) are essential for creating an effective and positive classroom environment. Traditional professional development programs often focus on theoretical leadership models or administrative strategies (Moltke, 2024), but incorporating sports into leadership training offers a more hands-on and dynamic approach. Sports can serve as a powerful framework for developing these skills, providing real-life situations where teachers can practice and refine their leadership abilities.

Role-Playing Sports Scenarios

One of the most effective ways sports contribute to leadership development is through role-playing scenarios, where educators actively engage in team-based activities or competitive games (Kark, 2011). These scenarios simulate real-world situations that demand quick decision-making, strategic thinking, and conflict resolution, all of which are essential leadership qualities in the classroom (Aptara, 2024). For example, when teachers participate in a basketball game, they are required to make split-second decisions about positioning, collaboration, and tactics. They must consider the strengths and weaknesses of their teammates, communicate effectively under pressure, and devise strategies that lead to success.

In a role-playing environment, teachers can experience first-hand the challenges of leadership and learn how to navigate complex interpersonal dynamics (Agboola Sogunro, 2004), much like they would in a classroom setting. For instance, a soccer game can be used to demonstrate how to lead a team under pressure, manage conflict between team members, and shift strategies when things aren't going as planned. Teachers can apply these lessons directly to their classrooms by using these skills to manage diverse student needs, address conflicts, and make timely decisions.

Collaboration and Teamwork

Team sports are an invaluable tool for fostering collaboration and communication skills, which are key components of effective leadership (CoachMePlus, 2020). In team-based activities like volleyball, rugby, or football, participants must work closely together, often under pressure, to achieve a common goal. Teachers, when engaging in these sports, learn to share responsibilities, build trust, and collaborate with others towards a shared objective. These experiences help educators develop essential teamwork skills that are transferable to the classroom environment. For example, when working in a group, teachers need to understand the strengths of each team member and leverage these strengths to accomplish tasks more efficiently. Similarly, just as in sports, teachers must communicate effectively with colleagues to deliver coordinated lessons, manage classroom activities, and foster a sense of community within the school. By participating in team sports, educators learn the importance of supporting each other, communicating openly, and stepping up when needed, all critical skills for fostering a positive classroom environment.

Resilience and Adaptability

Resilience and adaptability are critical leadership qualities (*Adaptability and Resilience*, n.d.), especially in the fast-paced, often unpredictable classroom setting. Sports-based challenges are an excellent way to cultivate these traits. In sports, players constantly face obstacles: an unexpected injury, a change in the game's pace, or a shift in team dynamics (Gupta & McCarthy, 2022). How athletes respond to these challenges, whether they rise to the occasion, adjust their strategies, or support their teammates, teaches resilience and adaptability. These qualities can be directly translated to the classroom. For instance, a teacher might face a classroom full of students with varying learning needs, behavioural challenges, or a difficult lesson plan that doesn't go as expected. By drawing on the lessons learned through sports, such as staying calm under pressure, adjusting tactics when things don't go as planned, and bouncing back after setbacks, teachers can better handle these challenges. A scenario like a rugby match, where the team is down by a large margin but remains determined to keep playing, mirrors the teacher's ability to stay positive, adjust strategies, and persist through difficult situations in the classroom. This resilience and adaptability are essential for handling both the emotional and academic demands of teaching.

Sports provide a dynamic and engaging framework for leadership development (*Leadership and Management in Sports: A Framework for Success*, 2025) that goes beyond traditional classroom-based professional development. By participating in team sports and role-playing scenarios, teachers can hone essential leadership skills such as decision-making, strategic thinking, conflict resolution, collaboration, resilience, and adaptability (Riera et al., 2024). These sports-based challenges equip educators with the tools they need to navigate the complexities of modern classrooms, ultimately leading to better classroom management, enhanced teacher-student relationships, and improved student outcomes.

COGNITIVE AND EMOTIONAL BENEFITS OF SPORTS IN TEACHER DEVELOPMENT

Teachers face immense pressures on a daily basis, ranging from managing diverse student needs to balancing administrative tasks and maintaining a positive classroom environment. As a result, stress and emotional fatigue are common challenges for educators (Active Norfolk, 2021). However, incorporating physical activities like team sports, individual sports, and strategy-oriented games into teacher development programs offers a powerful way to combat these challenges while simultaneously enhancing emotional resilience, emotional intelligence, and cognitive agility (Kang et al., 2024). These activities provide teachers with the tools to manage stress, understand and connect with their students, and think critically in fast-paced environments, which are crucial skills for effective teaching.

Stress Relief and Emotional Resilience

Physical activities such as team sports or individual exercises play a vital role in stress relief by promoting physical well-being and mental relaxation (Surya & Maheshbabu, 2023). Engaging in sports provides teachers with an opportunity to disconnect from the pressures of their day-to-day responsibilities and focus on their physical health. Whether it's a high-energy soccer match, a calming yoga session, or a brisk run, exercise helps release endorphins, which are natural stress relievers. This biochemical response is crucial for teachers who experience mental fatigue and emotional exhaustion (Donker et al., 2020). Regular physical activity also promotes better sleep, reduces anxiety, and helps regulate mood, making teachers more equipped to handle the challenges of the classroom with a clear mind.

In team sports, teachers not only improve their physical health but also gain emotional resilience by facing challenges in a group setting (Chen & Chi-Kin Lee, 2022). Sports present numerous obstacles, losing a match, making mistakes, or dealing with the disappointment of a missed goal. Overcoming these hurdles in a supportive environment builds resilience, teaching teachers to bounce back from setbacks and manage their emotions effectively. For example, in a competitive basketball game, players might experience frustration from losing a point or making an error, but the ability to recover, refocus, and keep playing is an important emotional lesson that directly translates to managing stress in the classroom. Teachers who engage in regular sports-based activities are better able to handle classroom stressors and remain emotionally balanced, which benefits both their well-being and their students' learning environment.

Emotional Intelligence Through Sports

Sports settings provide a rich environment for developing emotional intelligence (EI) (The Power of Emotional Intelligence in Sports | Telecoming | Monetization Technology, 2024), which is critical for teachers who must navigate diverse student personalities, needs, and emotions. Emotional intelligence involves the ability to recognize, understand, and manage one's own emotions while also recognizing and influencing the emotions of others (*The Power of Emotional Intelligence in Sports | Telecoming | Monetization Technology*, 2024). Team sports, in particular, create opportunities for teachers to develop these EI competencies in a practical, hands-on manner. For instance, during a team soccer game, players must work together, communicate effectively, and respond to each other's emotional cues in high-pressure situations. The ability to empathize with a teammate who is struggling, offer support when someone feels discouraged, or celebrate collective success fosters emotional awareness and empathy. Teachers participating in these environments learn how to better recognize emotional signals in others and manage interpersonal dynamics in positive ways. This enhanced emotional intelligence helps educators build stronger relationships with their students, creating a supportive and responsive classroom environment. Likewise, through sports, teachers learn to manage their own emotional responses in challenging situations. For example, if a teacher is playing on a team and a mistake leads to a setback, they practice self-regulation by maintaining composure rather than reacting negatively. This emotional self-control is crucial in classrooms where teachers must remain calm and constructive when students display challenging behaviours or when conflicts arise.

Cognitive Agility Through Strategy-Oriented Games

Cognitive agility, the ability to think flexibly and make quick, effective decisions (Good, 2012), is another significant benefit of engaging in strategy-oriented games like chess (Thought, 2024) or team sports (*Training the Cognitive Aspects of Agility in Team Sports*, n.d.). These games require players to evaluate multiple options, anticipate future outcomes, and adjust strategies in response to changing circumstances (Henry, 2023). Teachers can apply these cognitive skills directly to their classrooms, where they need to make real-time decisions, adapt lesson plans, and quickly address problems. In chess, for example, players must constantly think ahead, anticipate their opponent's moves, and reconfigure their strategies based on the evolving game board. Similarly, in a game like soccer, players must quickly decide whether to pass the ball, shoot for a goal, or change their positioning based on the flow of the game. These activities strengthen problem-solving abilities and help teachers develop the cognitive flexibility to switch between tasks, juggle various teaching responsibilities, and respond effectively to students' needs.

This cognitive training also helps teachers improve their decision-making skills, as they learn to prioritize and act under pressure (Blackley et al., 2021). In a classroom scenario, a teacher might need to decide whether to intervene in a conflict between students, adjust a lesson plan in the middle of a class, or quickly shift to a different teaching approach based on student feedback. The cognitive agility gained through strategic games helps teachers perform these tasks more effectively and with greater confidence.

Sports offer valuable cognitive and emotional benefits for teacher development (Knudsen et al., 2020). Physical activities help relieve stress, build emotional resilience, and foster better mental health (Mahindru et al., 2023), allowing teachers to perform at their best. Team sports and strategy games further develop emotional intelligence, empathy, and cognitive agility, skills essential for effective teaching. By participating in sports, teachers not only improve their physical fitness but also become better equipped to navigate the complexities of modern classrooms, ultimately benefiting both their professional growth and their students' learning experiences (Quennerstedt, 2019).

INTEGRATING SPORTS-BASED APPROACHES IN PEDAGOGICAL PRACTICES

Sports-based approaches can serve as an innovative and engaging tool for enhancing teaching methods. By incorporating sports metaphors, cross-disciplinary learning, and gamification, educators can create more dynamic, interactive, and inclusive classrooms (Pandian, 2023). These approaches allow teachers to present complex concepts in a more relatable way, increase student motivation, and improve engagement. The integration of sports into pedagogy not only makes learning fun but also fosters key skills such as leadership, goal setting, and teamwork (Pandian, 2023). Below, we explore how teachers can effectively integrate sports into their lesson plans.

Sports Analogies in Teaching

One of the most effective ways to use sports in education is through sports analogies. These analogies help break down complex educational concepts by comparing them to familiar sports scenarios (Magazine, 2013). For example, a teacher might use the concept of teamwork in a lesson on collaboration by likening it to the way a football team works together to reach a common goal. Just as players pass the ball and rely on each other's strengths to score a goal, students must work together, utilizing each other's skills and strengths to achieve a shared educational outcome. Another powerful sports analogy is goal setting (Monsma, 2018). In sports, athletes set both short-term and long-term goals to achieve success, whether it's training for a marathon or preparing for a championship. Similarly, teachers can guide students to set academic goals, such as achieving a certain grade or mastering a skill, and use sports metaphors to emphasize the importance of persistence, discipline, and strategic planning. For example, when discussing the importance of preparation and focus, a teacher can refer to how professional athletes visualize their performance before a game to ensure success.

Performance evaluation is another area where sports analogies can be effective (Harris, 2025) (Kearl, n.d.). Just as coaches review players' performance after each match and suggest areas for improvement, teachers can implement regular assessments that allow students to reflect on their progress, set new goals, and identify areas for improvement. Teachers can also discuss the concept of constructive criticism, which in the sports world often comes from coaches to help athletes improve their skills. This helps students see the value in feedback and continuous learning.

Cross-Disciplinary Learning Using Sports

Sports can also serve as a bridge between subjects, enabling teachers to create cross-disciplinary lessons that show students how different fields of knowledge intersect (*Grade 4 Cross-Curricular Lesson Plan: A World of Sports*, 2015). For example, in physics, a teacher can use the dynamics of basketball to explain principles of motion and force. Teachers can explore topics like projectile motion, explaining how the angle at which a player shoots the ball affects its trajectory. Students can learn about velocity, acceleration, and force by observing how a basketball moves through the air and how gravity influences its path. This makes the concepts of physics more tangible and relatable to students. Similarly, math lessons can incorporate sports data analysis. Teachers might use real-world data from various sports to help students understand statistics and probability. For example, a math teacher could show students the probability of a soccer player scoring a goal based on the player's position on the field, or how teams calculate their batting average in cricket or baseball. By using real-world examples from sports, students learn to apply theoretical knowledge in practical situations.

In social studies, teachers can use team sports like soccer to teach about global collaboration and cultural diversity. For instance, they can discuss the formation of national teams in international tournaments, highlighting how players from diverse cultural backgrounds must collaborate to succeed. This cross-disciplinary approach not only makes the learning process more interactive but also enhances students' understanding of various concepts by showing the real-world application of the lessons.

Gamification and Sports Metaphors in the Classroom

Gamification, the integration of game mechanics and principles into learning environments, can significantly enhance student engagement. By incorporating sports-themed gamification techniques, teachers can create a fun, interactive atmosphere while promoting competition, collaboration, and achievement (AKSOY et al., 2022). For example, a teacher can set up a classroom competition that mirrors the structure of a sports tournament, where students compete in different rounds or challenges. These challenges can be related to class topics, such as answering questions or solving problems, and students can earn points for their team or individual performance, much like players earning points in a game. Teachers can then introduce levels or milestones, just like levels in a game or a sports season, where students work toward achieving new "ranks" or rewards.

Sports metaphors can also be used in gamified activities. For example, when discussing teamwork and leadership, a teacher might create a simulation where students take on different "team captain" roles, guiding their peers through a project, just as a coach leads a team. The teacher can also use sports-themed badges or leaderboards to track progress, motivating students to compete in a healthy way and encouraging them to challenge themselves to do better, like athletes striving to improve their personal best.

By using sports metaphors, teachers create a dynamic and inclusive classroom culture where students feel like they are part of a team working toward a shared goal. This is particularly effective for students who thrive in competitive environments, as they may feel more motivated and engaged in the learning process. It also fosters inclusivity by ensuring that every student plays a role in the class's "game," encouraging equal participation, teamwork, and communication. Integrating sports-based approaches into pedagogical practices allows teachers to foster a more engaging, dynamic, and inclusive classroom environment (Rodríguez Macías et al., 2021). By using sports analogies, educators can effectively teach key concepts such as leadership, goal setting, and performance evaluation. Cross-disciplinary learning through enables students to understand how different subjects are interconnected, making abstract concepts more tangible (*Cross-Disciplinary Learning: The Key to a Successful Future*, n.d.). Incorporating gamification and sports metaphors creates an interactive classroom culture where students are motivated to collaborate, challenge themselves, and apply their learning in meaningful ways. Through these methods, teachers can create an environment where both educators and students thrive, making the learning process more enjoyable and effective

THE ROLE OF SPORTS IN FOSTERING INCLUSIVE AND DIVERSE TEACHING PRACTICES

Promoting Inclusivity and Equity

Sports can play a significant role in creating an inclusive and equitable classroom environment (*(PDF) Equity and Inclusion in Physical Education and Sport, Second Edition*, n.d.; Yoffi, 2023), particularly when teachers leverage sports-based activities to connect with diverse learners. By incorporating sports into teaching practices, educators can foster a culture of collaboration, understanding, and empathy among students, allowing them to feel represented, supported, and valued (*SPORT VALUES in EVERY CLASSROOM Teaching Respect, Equity and Inclusion to 8–12-Year-Old Students*, n.d.). The unique qualities of sports, such as teamwork, competition, and strategy, can bridge cultural gaps (Yellowbrick, 2021; Rashid & Kabir, 2024), improve emotional resilience, and promote cooperation among students of different backgrounds. Sports-based activities can also be used to promote inclusivity and equity in the classroom by ensuring that all students (Ainscow, 2020) (Halpin, 2024) (*In What Ways Can We Use Sports to Promote Diversity and Inclusion in Our Communities?* 2019), regardless of their abilities or background, have the opportunity to participate and succeed. It is important for teachers to design activities that cater to a wide range of student needs, whether they are related to physical abilities, learning styles, or socio-economic backgrounds.

For example, teachers can modify the rules of games or use adaptive sports for students with physical disabilities. Activities such as wheelchair basketball, sitting volleyball, or swimming can ensure that students with mobility challenges are able to engage in the same sports as their peers. These modifications create an inclusive environment where every student has the opportunity to participate and feel included. Mixed-ability teams can also be formed where students of various skill levels work together. This helps foster a sense of equity, as it ensures that every student, regardless of their ability, has the chance to contribute meaningfully to the team's success. Also, the teacher can ensure that gender inclusivity is part of the sports experience. For instance, in co-ed sports activities, teachers can be mindful of equal opportunities for participation and leadership. They can actively work to dismantle stereotypes about gender roles in sports by encouraging all students, regardless of gender, to take leadership roles or participate in any position within a sport. This helps break down traditional gender expectations and allows all students to thrive in a more equitable environment. Likewise, using sports metaphors to discuss equity in academic subjects can allow students to see the value of fairness, participation, and inclusivity in the classroom. For example, a teacher might explain how an athletic team requires equal effort from every player, regardless of their role or skill level, to be successful, paralleling this with how students in the classroom must contribute and support one another to reach academic goals.

Sports as a Tool for Connecting with Diverse Learners

Sports provide an excellent platform for teachers to connect with students from various cultural, social, and academic backgrounds (Ross et al., 2008; (*Sports as a Tool for Social Inclusion and Diversity in Schools | Bring It on Sports*, 2024). Engaging in sports activities allows students to demonstrate their skills and interact with peers in a non-academic setting, which can foster a sense of belonging and camaraderie. Teachers can use sports as a tool to better understand their students' strengths, preferences, and communication styles, which can then be applied to their teaching methods. For example, consider a classroom with students from different cultural backgrounds, some of whom may feel disconnected from traditional classroom activities. By organizing multicultural sports events, such as soccer, basketball, or even games like kabaddi (popular in India), teachers can allow students to showcase their cultural heritage and share their traditional games with classmates. This not only helps students from diverse backgrounds feel seen and valued, but it also encourages other students to appreciate and respect different cultures. A simple example of this could be a teacher inviting students to explain the rules and significance of a sport that is culturally important to them. This type of sports-based activity can break down cultural barriers and create common ground for students to bond.

Likewise, sports-based activities can help students develop vital social skills, such as empathy, communication, and conflict resolution (Health Direct, 2018). When students work together to accomplish a team goal, they learn to navigate differences and recognize that everyone brings something valuable to the table. For example, in a team sport like soccer, a teacher can highlight how players with different skills (e.g., offensive players and defensive players) must work in harmony to achieve success, reinforcing the importance of collaboration and diversity in real-world situations.

Cultural Responsiveness: Leveraging Sports to Address Cultural Differences

In a culturally diverse classroom, teachers can leverage sports to improve cultural responsiveness by fostering respect for different customs, traditions, and worldviews. Sports provide an opportunity for teachers to introduce cultural diversity into their teaching methods, helping students recognize and value differences while promoting unity and shared understanding (Council of Europe, 2012). For example, a teacher can incorporate global sports such as cricket (from South Asia), rugby (from New Zealand and South Africa), or baseball (from the United States and Japan) into their lessons to highlight how these sports transcend borders and bring people from different cultures together. Teachers can also share stories and history behind these sports, which helps students from diverse backgrounds feel proud of their heritage while encouraging others to appreciate the rich cultural diversity in the classroom. Additionally, culturally responsive teaching can involve incorporating cultural practices or games into physical education lessons. Teachers can invite students to share their traditional games, such as Mexican football (soccer), Chinese Dragon Boat Racing, or Indigenous sports from various cultures, and explain how these games are played, their significance, and what they represent. Through this process, students not only learn about each other's cultures but also develop a deeper appreciation for the importance of culture in shaping individuals' identities and experiences. For instance, in a global history lesson, a teacher can use a sports metaphor to discuss the impact of international sports events like the Olympics on global diplomacy and understanding. Students can learn how the Olympics provide a platform for nations with different cultural values to engage in friendly competition, emphasizing the idea that sports can be a unifying force in bridging cultural divides. Moreover, teachers can use sportsmanship and fair play as tools to teach students about cultural values, respect, and tolerance. Teachers can create class discussions or debates based on sportsmanship in different countries or regions, examining how cultural expectations of respect, collaboration, and competition vary across different sports.

Sports provide a valuable framework for fostering inclusive and culturally responsive teaching practices that not only engage students but also promote understanding, respect, and empathy among them. By integrating sports-based activities into classroom settings, teachers can connect with diverse learners, promote equity and inclusivity, and leverage cultural perspectives to enrich the learning experience. Ultimately, using sports in education cultivates a dynamic and supportive classroom culture where students feel valued, heard, and included, regardless of their backgrounds or abilities.

FINDINGS AND DISCUSSION

The integration of sports-based approaches into teacher development has been shown to have a profound impact on educators' skills and pedagogical practices. The findings from the literature and practical case studies indicate that physical activities, games, and sports can be effective in nurturing leadership qualities, cognitive agility, emotional intelligence, and collaboration skills. These competencies are crucial for fostering a positive classroom environment, improving teacher-student relationships, and promoting inclusivity in education.

One of the most significant findings from this exploration is the link between sports participation and the enhancement of leadership skills among educators. Role-playing sports scenarios allow teachers to engage in simulated real-world challenges, where they must make quick decisions, manage conflicts, and work with others towards shared goals. This experience directly translates into classroom leadership, where educators face complex decisions and diverse student needs. Teachers who participated in sports-based training were observed to handle classroom challenges with more confidence and competence. These findings support existing research on the importance of experiential learning in developing critical leadership and decision-making skills (Knudsen et al., 2020).

Cognitive development was another key area where sports-based activities showed measurable benefits. Strategy-based games, such as chess or team sports, encourage teachers to think critically, make rapid decisions, and apply problem-solving strategies. These activities enhance cognitive agility, enabling teachers to manage multiple tasks in a fast-paced classroom. Studies have indicated that the incorporation of such games into professional development programs can sharpen cognitive flexibility, which is crucial for effective teaching (Hernández-Mendo et al., 2019).

Likewise, the promotion of emotional intelligence (EI) emerged as a central finding. Teachers who engaged in team sports and collaborative games developed greater empathy, emotional regulation, and self-awareness. This is especially significant in a classroom setting where managing diverse emotional needs and building relationships is key to creating a supportive learning environment. Teachers with higher emotional intelligence were better equipped to handle student frustrations, resolve conflicts, and engage in productive conversations with students from various backgrounds (Norris & Norris, 2021). These findings align with previous research that links EI with improved teaching effectiveness (Bhandari, 2021).

The collaborative and inclusive nature of sports also proved to be a powerful tool for fostering a sense of belonging among students. By using team sports and group activities in the classroom, teachers were able to encourage cooperation and mutual respect among students of varying abilities and cultural backgrounds. For instance, teachers reported that students who initially struggled with collaboration were able to improve their teamwork skills through group-based sports activities. This ability to collaborate and build relationships not only benefited students' academic progress but also fostered a sense of community and support (Kim et al., 2019).

The use of sports metaphors in the classroom, as part of gamification and cross-disciplinary learning, also emerged as a promising strategy for enhancing student engagement. By drawing parallels between sports and academic content, teachers could help students relate complex concepts to everyday experiences. For example, using sports analogies to explain teamwork or goal setting in mathematics or social studies encouraged students to see the real-world applications of abstract ideas. This approach promoted active learning and made lessons more relatable and enjoyable for students (Harper-collins-admin, 2023).

A significant barrier to adopting sports-based learning is the scarcity of real-world implementation data. Conducting pilot programs in schools can provide empirical evidence of its effectiveness. For instance, a high school ice hockey program designed to teach life skills and values demonstrated that such initiatives enhance student engagement and improve classroom management. Collaborating with sports ministries, research organizations, and universities can facilitate comprehensive evaluations of these pedagogical strategies, refining best practices and promoting wider adoption.

In terms of scalability and adaptability, the chapter highlights how sports-based learning can be tailored to suit diverse educational contexts, including urban, rural, public, and private schools. In settings with limited access to traditional sports equipment or resources, teachers can adapt activities to available materials or use virtual simulations and gamified platforms. AI and digital tools such as virtual reality (VR) or mobile apps can enhance these sports-based learning experiences, allowing teachers to engage with professional development content at their convenience and participate in virtual simulations of leadership training or classroom management scenarios (Sports Integration | Holistic Development | LIFE and CAREER SKILLS, 2024). A study conducted in Texas public schools examined the impact of sports participation on students' academic and behavioral outcomes. The research compared standardized test passing rates, dropout risks, and disciplinary actions between athletes and non-athletes across grades 7–12. Findings revealed that student-athletes outperformed non-athletes in all academic areas, exhibited lower dropout rates, and had fewer disciplinary incidents. These results align with existing literature suggesting that involvement in school sports is positively associated with enhanced academic achievement and improved behavior among students (ELDRIDGE et al., 2014).

The chapter also emphasizes the role of sports-based training programs in professional development. For instance, the creation of mentorship programs, workshops, and partnerships with local sports organizations allows educators to practice sports-based techniques and apply them to their teaching practices. Findings indicate that teachers who participated in such programs demonstrated significant improvements in their leadership, communication, and collaboration skills. Moreover, the emphasis on cultural responsiveness and inclusivity through sports was also found to be an effective way to engage students from diverse backgrounds and create a more equitable learning environment (Social-Emotional Skills in School, n.d.).

The discussion and findings suggest that integrating sports-based approaches into teacher development not only enhances educators' leadership, cognitive, and emotional skills but also promotes inclusivity, collaboration, and innovation in teaching. By leveraging the power of sports, educators can transform their teaching practices, leading to more engaging and effective classrooms. As digital tools and AI continue to evolve, sports-based learning experiences can be further personalized and scaled, making them accessible to educators and students across different educational settings. Ultimately, sports have the potential to shape a new generation of educators who are not only skilled in subject matter but also in fostering social, emotional, and cognitive development in their students.

Recommendations and Future Studies

Integrating sports-based approaches into teacher development programs offers a transformative method to nurture critical competencies such as leadership, emotional resilience, collaboration, and cognitive agility. Educators often face high-stakes environments where decision-making, adaptability, and interpersonal skills are crucial. Sports, by nature, present real-time opportunities for participants to exercise and refine these skills in a dynamic, experiential setting. Through structured interventions, schools and training institutions can incorporate these benefits into formal teacher development frameworks. The process begins by recognizing that sports-based learning extends beyond physical fitness, it includes emotional and intellectual growth, empathy, and reflective practice. In this section, several practical strategies are outlined for embedding sports-based training into both preservice and in-service professional development programs, allowing educators to grow holistically and better meet the demands of modern education.

A primary method for introducing sports-inspired professional development is through collaborative workshops and simulations. These workshops can be designed around role-play scenarios and team-based games that simulate classroom challenges. For example, teachers participating in a cooperative sports activity such as basketball or volleyball must engage in strategic communication, manage conflict constructively, and develop collective goals, skills directly applicable to teaching. These activities not only reinforce teamwork but also challenge educators to reflect on their own leadership styles and interpersonal behaviors. A post-activity debrief encourages teachers to draw explicit connections between their performance in the game and similar classroom situations, enhancing self-awareness and emotional intelligence. In addition, simulations can be customized for different subject areas or grade levels, making them versatile tools for diverse educational contexts. By making learning tangible and interactive, such workshops provide memorable experiences that strengthen professional competencies.

Another effective entry point is the integration of sports-based activities into teacher induction programs, which set the tone for a teacher's professional identity and practice. New educators often experience stress, isolation, and uncertainty during their transition into the profession. Structured sports activities can serve as a buffer by fostering camaraderie, promoting trust among peers, and easing the emotional burden of early-career teaching. Activities such as relay challenges, leadership games, and physical trust exercises provide real-time feedback on team dynamics and personal resilience. These experiences not only promote mental well-being but also encourage a growth mindset, equipping new teachers with the confidence and emotional tools to navigate the classroom effectively. Embedding such practices into induction programs signals the institution's commitment to holistic teacher development and supports long-term teacher retention. Over time, these early experiences can lead to sustained professional relationships and a deeper understanding of collaborative teaching.

Partnerships with local sports organizations and community-based clubs offer another rich avenue for integrating sports into teacher training. Schools can benefit from the expertise of professional coaches, athletes, and trainers who bring unique insights into motivation, discipline, and team management. These experts can facilitate specialized workshops where educators learn how to translate athletic coaching principles into classroom strategies. Field visits or joint training sessions allow teachers to experience authentic sports environments, thereby expanding their perspectives on leadership and performance.

Additionally, community partnerships promote school engagement and social cohesion by strengthening the relationship between educational institutions and their local contexts. These collaborative ventures not only add depth and authenticity to professional development but also provide access to facilities and resources that schools may not have internally. Moreover, exposure to diverse coaching styles and motivational techniques helps teachers broaden their instructional toolkit.

Within the school setting, mentorship programs based on team sports principles can foster sustained professional growth. Veteran teachers can serve as mentors using sports-inspired strategies to guide newer colleagues in managing classroom challenges. For example, mentors may employ techniques such as goal setting, reflective debriefs, and role assignments to build confidence and competence in early-career teachers. The mentor-mentee relationship can mirror a coach-player dynamic, emphasizing encouragement, constructive feedback, and shared responsibility. Over time, this approach cultivates a culture of collaboration and collective efficacy, with both mentors and mentees contributing to school improvement. These relationships often evolve into communities of practice that continuously exchange ideas, refine teaching methods, and build a repository of sports-based educational strategies. By modeling resilience, adaptability, and empathy, mentor teachers also demonstrate how sports values translate into long-term educational leadership.

To ensure the sustainability and effectiveness of sports-based development initiatives, institutions must create structured training models. A well-rounded program should include a sequence of workshops, reflective sessions, and practical engagements that progress from foundational to advanced levels. Utilizing a "train-the-trainer" approach can help scale these programs, enabling select educators to become facilitators who train others in their schools or districts. Certification tracks, developed in collaboration with universities or sports training organizations, can validate educators' competencies in this innovative field and provide recognition for their efforts. Furthermore, integrating sports-based modules into pre-service teacher education curricula ensures that future educators are exposed to these strategies early in their careers. This proactive integration contributes to a new generation of teachers who are more emotionally attuned, collaborative, and capable of adaptive leadership. Ultimately, a structured and sustainable approach allows schools to embed sports-based development into their institutional culture, ensuring lasting impact and continuous growth.

To maximize the benefits of sports-based teacher development, structured training programs must be established. These programs should include interactive workshops where teachers engage in team-building exercises, leadership drills, and stress management strategies using sports-based scenarios. Implementing a train-the-trainer model can empower select educators to become master trainers, ensuring the sustainable integration of sports-based learning within their institutions. Collaborations with universities and sports organizations can lead to standardized certification programs, validating teachers' competencies in this innovative approach. Moreover, incorporating sports-based modules into pre-service teacher education can prepare future educators for holistic teaching methodologies

Scaling Sports-Inspired Pedagogical Practices

To scale sports-inspired pedagogical practices across various educational settings, whether urban or rural, public or private, educators must prioritize contextual flexibility and inclusivity. The success of such an initiative depends not only on the availability of resources but also on the adaptability of sports-based methods to diverse learning environments. Schools must assess their unique capacities, student demographics, and cultural dynamics to design programs that are locally relevant and impactful. In low-resource or rural schools, for instance, educators can implement modified versions of sports activities such as relay races, obstacle courses, or team-building games that require minimal equipment but still cultivate essential skills like communication and leadership. Conversely, urban schools with access to sports infrastructure can incorporate complex activities like basketball tournaments or soccer drills, integrating these into lessons on collaboration, resilience, and strategic thinking. The fundamental goal is to align sports-based activities with educational outcomes while ensuring that every institution, regardless of resources, can benefit from this approach. Schools should also explore indigenous or culturally significant sports as a means of connecting pedagogy with local traditions, thus fostering greater student engagement and cultural pride.

The adaptability of sports-based learning to various contexts is a cornerstone of its scalability. Urban institutions may leverage digital sports analytics, advanced equipment, and established partnerships with athletic organizations to embed sports within their professional development programs. These settings can facilitate comprehensive leadership development sessions, data-driven performance evaluations, and experiential learning through physical sports simulations. In contrast, rural schools may adopt low-cost, community-centered models by utilizing open spaces for group exercises or collaborating with local clubs for volunteer-led sessions. Equally important is the inclusion of adaptive sports for educators and students with disabilities, ensuring that physical limitations do not hinder participation. Activities such as wheelchair basketball, seated volleyball, or inclusive yoga promote equity and teach empathy and teamwork. This commitment to inclusivity helps foster a school culture where diversity is valued, and all individuals have opportunities to thrive, regardless of background or ability level.

In addressing the diversity of school systems, hybrid approaches offer a compelling solution for expanding sports-based learning. These models combine traditional physical sports training with virtual and digital sports experiences, providing flexibility and accessibility for schools with differing levels of infrastructure. For instance, rural schools may benefit from online coaching modules, digital strategy games, and video-based simulations that teach collaboration and critical thinking without requiring physical sports fields or equipment. Urban schools, on the other hand, can supplement in-person training with digital tools to enhance analysis, feedback, and long-term tracking of teacher development. This hybrid framework allows institutions to tailor their programs based on availability, teacher preferences, and specific pedagogical goals. By integrating in-person and virtual experiences, schools can ensure that professional development remains dynamic, inclusive, and responsive to evolving educational demands.

Peer collaboration is another vital component in scaling sports-inspired teaching practices. Teacher networks that span across geographic and institutional boundaries can serve as effective channels for sharing resources, strategies, and success stories. These peer learning communities can host regular forums, both virtual and physical, where educators from different backgrounds exchange best practices on implementing sports metaphors, promoting classroom inclusion through games, and building emotional resilience among students. Such networks also provide a platform for mentorship, where experienced teachers guide others in adapting and refining their use of sports-based tools in instruction. Moreover, cross-sector collaboration between public and private institutions encourages innovation and helps bridge equity gaps in training quality and access. By leveraging collective wisdom and promoting peer-driven learning, these communities foster a sustainable culture of continuous improvement in education.

To fully scale these initiatives, schools must also focus on building inclusive and equitable sports programs. Equity in access is critical, particularly in ensuring that all students, regardless of gender, socioeconomic status, or physical ability, can participate meaningfully in sports-based learning activities. Teachers must be equipped with the knowledge and tools to adapt games and rules for varying ability levels, create gender-inclusive environments, and eliminate barriers to participation. For example, games can be redesigned so that physical intensity is adjusted or roles are rotated to give every student a chance to lead. Additionally, providing professional development on diversity, equity, and inclusion within sports-based pedagogy ensures that educators are sensitive to and proactive about addressing systemic inequities. A truly scalable model does not merely replicate sports-based training across schools but adapts it in ways that uphold the values of fairness, empathy, and collective growth.

Using AI and Digital Tools for Sports-Based Learning

In the evolving landscape of education, artificial intelligence and digital tools offer groundbreaking opportunities to enrich teacher development, particularly through sports-based learning models. These technologies simulate real-life sports scenarios, create immersive learning environments, and enable customized professional development experiences tailored to individual teacher needs. Rather than relying solely on traditional workshops or lectures, schools can now leverage interactive and technology-driven platforms to engage teachers in ways that mirror the dynamics of team sports. These tools help educators develop leadership, resilience, emotional intelligence, and problem-solving, skills essential in both athletic and classroom contexts. As digital literacy becomes increasingly critical in education, integrating AI with sports-based pedagogy not only enhances teacher competencies but also prepares them to better connect with students who are growing up in tech-savvy, fast-paced environments.

One of the most impactful applications of AI in this domain is the use of Virtual Reality simulations for leadership and decision-making. VR platforms can replicate the complexity and pressure of live sports situations, allowing teachers to step into roles similar to coaches or team leaders. Through simulations such as a high-stakes penalty shootout or a strategic team huddle during a basketball game, educators practice making quick decisions, managing interpersonal dynamics, and adjusting strategies on the fly. These experiences are deeply engaging and emotionally resonant, providing a safe space to learn from mistakes without real-world consequences. Furthermore, VR scenarios can be customized to mirror classroom realities, like conflict resolution among students or team planning with colleagues, thereby strengthening the transfer of skills between sports and teaching. By repeatedly navigating these virtual environments, teachers can build confidence, self-awareness, and adaptability, which are invaluable in day-to-day classroom management.

In addition to VR, AI-driven analytics for performance evaluation provide powerful insights into how teachers interact with sports-based learning tasks. These tools can assess variables such as response times, collaboration tendencies, and resilience during high-pressure simulations. For example, an AI system might identify that a teacher tends to avoid leadership roles in team challenges, prompting targeted support or coaching in that area. These data-driven evaluations help personalize the development journey, ensuring that no two teachers receive a one-size-fits-all training experience. Institutional leaders can also aggregate this data to identify trends across departments, design more effective workshops, and allocate resources based on actual needs. By integrating AI into the evaluation process, schools move toward a more equitable and evidence-based model of professional development, one that empowers teachers through feedback that is timely, specific, and actionable.

Another avenue for expanding engagement is through gamified learning platforms, which use elements of game design to make teacher development more interactive and motivating. These platforms replicate the competitive and cooperative spirit of sports by incorporating levels, points, badges, leaderboards, and team-based challenges. Educators might complete simulations, solve real-world classroom problems, or engage in communication-based tasks that mirror the coordination required in sports teams. The gamification model not only makes learning fun but also fosters healthy competition and collaborative learning among teachers. By participating in these structured, game-like environments, educators enhance critical soft skills such as creativity, flexibility, and interpersonal communication. Moreover, the instant feedback and visible progression within gamified systems reinforce continued engagement and a growth mindset, both of which are essential for long-term professional development.

Mobile applications further enhance accessibility by bringing sports-based learning into the hands of educators wherever they are. These apps often include features such as leadership exercises, stress-reduction activities, team-building games, and even virtual fitness challenges that can be completed alone or collaboratively. Teachers might use a mobile app to set goals for weekly emotional wellness, complete resilience modules, or track participation in physical activities that mirror classroom engagement scenarios.

The portability of mobile apps ensures that learning doesn't have to be confined to workshops or formal training days, it can become part of a teacher's daily routine. These tools are especially valuable in supporting ongoing, self-directed learning, as they offer reminders, rewards, and community interaction in a flexible format. As a result, mobile apps can help foster habits of reflection, self-regulation, and continuous improvement, even in the busiest of teaching schedules.

Lastly, adaptive learning platforms powered by AI represent the future of personalized professional growth. These platforms continuously collect data on a teacher's behavior, performance, and preferences, using it to adjust the content and structure of their learning journey in real time. For instance, if a teacher shows strong leadership during collaborative games but struggles with emotional regulation under pressure, the system may prioritize modules focused on mindfulness and stress management. This level of personalization ensures that every educator receives training that is not only relevant but also optimized for their specific growth areas. Institutions benefit as well, since adaptive platforms can report aggregate data that supports strategic planning and long-term improvement of teacher development programs. In this way, adaptive technologies make professional learning more responsive, efficient, and impactful, fostering a culture of innovation and accountability within the education system.

To ensure the sustainability and scalability of these digital tools, schools must invest in thoughtful infrastructure and long-term planning. Establishing public-private partnerships can facilitate access to cutting-edge technologies and training, while embedding digital sports-based development into education policy can provide structural support and funding. Teacher onboarding, technical support, and continuous updates must also be part of any implementation plan to keep tools relevant and effective.

Furthermore, regular evaluation and feedback cycles are necessary to monitor progress and refine strategies as needed. These steps help institutionalize sports-based digital training as an enduring element of teacher development rather than a short-term innovation. When supported with intention and foresight, AI and digital tools can significantly elevate the quality of teacher education, leading to more confident, collaborative, and emotionally intelligent educators across all contexts.

CONCLUSION

This chapter has delved into the multifaceted benefits of integrating sports-based approaches into teacher development programs. It has established that sports are not merely physical activities relegated to playgrounds and gymnasiums, but powerful educational tools that can cultivate essential skills in educators. Teachers engaged in sports-based training develop competencies such as leadership, emotional intelligence, decision-making, adaptability, collaboration, and cognitive agility, skills that are critical in managing contemporary classroom environments. These activities promote experiential learning, allowing educators to engage physically, mentally, and emotionally in scenarios that mirror classroom challenges. As such, sports-based approaches help bridge the gap between theoretical knowledge and practical application in teacher training.

A central theme in this chapter is the idea that sports-inspired development benefits not only individual educators but also the wider classroom environment. When teachers experience the value of teamwork, trust, and resilience through sports, they are better equipped to foster these same qualities in their students. For example, a teacher who has practiced conflict resolution on the basketball court is more likely to apply calm and effective strategies in handling student disagreements. Similarly, understanding team dynamics through cooperative sports helps educators manage group projects and peer-to-peer learning more effectively in class. This dual impact, on the teacher and on the student, makes sports-based training an exceptionally impactful model for education.

Another significant finding is how sports can support inclusive and equitable teaching practices. Sports have a universal appeal and can be adapted for all learners, regardless of physical ability, gender, culture, or socioeconomic background. This chapter highlighted the importance of adaptive sports, culturally relevant games, and diverse team activities that promote empathy and understanding among students. When teachers are trained in these inclusive practices, they are more likely to create classrooms that reflect the values of equity and participation. Furthermore, the ability to use sports metaphors and analogies can also help teachers connect with diverse learners by presenting complex academic ideas in familiar and accessible ways.

Moreover, the chapter emphasized how sports-based learning directly enhances the emotional and cognitive dimensions of teaching. Activities like role-playing, cooperative games, and strategic challenges serve as low-risk environments for educators to explore leadership styles, handle emotional stress, and sharpen problem-solving abilities. This builds emotional resilience, enabling teachers to manage the emotional weight of their profession with greater balance. In high-pressure environments like schools, this resilience is crucial for preventing burnout and maintaining positive interactions with students, parents, and colleagues. Cognitively, the strategic aspects of sports also improve agility and decision-making, which are indispensable in dynamic classroom settings.

The integration of technology, particularly AI and digital platforms, into sports-based teacher development opens up new horizons for accessibility and personalization. Tools such as virtual reality simulations, mobile learning apps, and gamified platforms can simulate sports-based scenarios for teachers who may not have access to physical resources. These technologies enable flexible and scalable training models, allowing educators to engage in professional development at their own pace and based on their individual learning needs. Moreover, AI-driven analytics can provide teachers with real-time feedback on their decision-making, collaboration, and emotional responses, offering targeted pathways for growth. This fusion of sports pedagogy and digital innovation makes professional learning more dynamic, interactive, and future-ready.

From a policy and institutional perspective, the chapter advocates for the formal inclusion of sports-based models in teacher preparation and ongoing development. Educational leaders and policymakers should consider embedding these strategies into national training frameworks, certification programs, and school improvement plans. Collaborations with universities, sports academies, and community organizations can strengthen implementation while ensuring that programs remain inclusive and research-based. In low-resource settings, low-cost alternatives and digital simulations can ensure that schools are not left behind. With the right structural support, sports-based teacher development can be scaled across geographic and demographic boundaries, benefiting educators worldwide.

Looking to the future, the potential for sports-inspired teacher development is immense. As education systems continue to face unprecedented challenges, ranging from increased student diversity to mental health concerns, teachers will require more holistic training that supports their full professional identity. Sports-based learning offers such a model, combining physical engagement, emotional insight, and cognitive development in a single, cohesive framework. Moreover, it aligns well with current educational priorities, such as social-emotional learning, inclusive pedagogy, and student-centered instruction. By embracing sports-based strategies, education systems can create more resilient, collaborative, and emotionally intelligent teaching forces.

In summary, this chapter has illustrated how sports, when intentionally integrated into teacher development, serve as a powerful vehicle for transformation. They enhance teacher confidence, competence, and classroom culture. From leadership development and emotional resilience to digital innovation and cultural inclusivity, sports-based training addresses a broad spectrum of educational needs. The research and strategies presented here provide a strong case for why sports should no longer be viewed as supplementary, but as central to professional growth in education. As the field of teacher development continues to evolve, sports-inspired practices will undoubtedly play a critical role in shaping the next generation of effective, empathetic, and empowered educators.

REFERENCES

Norfolk, A. (2021, May 7). The Mental Benefits Of Sport - News. Active Norfolk. https://www.activenorfolk.org/2021/05/mental-benefits-of-sport/

Adaptability and Resilience. (n.d.). Www.linkedin.com. https://www.linkedin.com/pulse/adaptability-resilience-ariel-matius-surbakti

Adaptive Learning AI Technology in Education. (2021, August 13). Belitsoft. https://belitsoft.com/custom-elearning-development/ai-in-education/adaptive-learning-ai

Agboola Sogunro, O. (2004). Efficacy of role-playing pedagogy in training leaders: Some reflections. *Journal of Management Development*, *23*(4), 355–371. DOI: 10.1108/02621710410529802

Ainscow, M. (2020). Promoting Inclusion and Equity in education: Lessons from International Experiences. *Nordic Journal of Studies in Educational Policy*, *6*(1), 7–16. DOI: 10.1080/20020317.2020.1729587

Aksoy, N., Aksoy, E., & Usta, E. (2022). Metaphors Developed by Teachers for the Gamification Approach in Education. *Journal of Teacher Education and Lifelong Learning*.

Almubarak, A., Alhalabi, W., Albidewi, I., & Alharbi, E. (2024). An analytical approach for an AI-based teacher performance evaluation system in Saudi Arabia's schools. *Discover Applied Sciences*, *6*(8), 406. Advance online publication. DOI: 10.1007/s42452-024-06117-4

Aptara. (2024, December 9). *Simulation-Based Learning: A Game Changer for Developing Corporate Leaders*. Aptara Corp - Home. https://www.aptaracorp.com/simulation-based-learning-a-game-changer-for-developing-corporate-leaders/

Baraishuk, D. (2021). *Adaptive Learning AI Technology in Education*. Belitsoft. https://belitsoft.com/custom-elearning-development/ai-in-education/adaptive-learning-ai

Best CBSE school in Nagpur. Delhi Public School Kamptee Road Nagpur. DPS Nagpur. (n.d.). Delhi Public School Kamptee Road Nagpur. https://dpsnagpur.edu.in/sports-shape-leadership-skills-build-character

Bhandari, S. (2021, October 25). *Benefits of Sports for Mental Health*. WebMD; WebMD. https://www.webmd.com/mental-health/benefits-of-sports-for-mental-health

Blackley, C., Redmond, P., & Peel, K. (2021). Teacher decision-making in the classroom: The influence of cognitive load and teacher affect. *Journal of Education for Teaching*, *47*(4), 1–14. DOI: 10.1080/02607476.2021.1902748

Chen, J., & Chi-Kin Lee, J. (2022). Teacher resilience matters: A buffering and boosting effect between job driving factors and their well-being and job performance. *Teachers and Teaching*, *28*(7), 1–18. DOI: 10.1080/13540602.2022.2116574

CMOE. (2018, May 15). *Strategic Thinking in Leadership [The Biggest Gap for New Leaders]*. CMOE. https://cmoe.com/blog/the-biggest-gap-for-new-leaders-at-any-level-is-thinking-strategically/

CoachMePlus. (2020, August 11). *Key Principles of Effective Collaboration in Sport*. CoachMePlus. https://coachmeplus.com/principles-team-staff-collaboration-sport/

Conflict Resolution in Leadership | In Professional Development. (2021, October 4). Www.inpd.co.uk. https://www.inpd.co.uk/blog/conflict-resolution-in-leadership

Council of Europe. (2012). *Culture and Sport*. Council of Europe. https://www.coe.int/en/web/compass/culture-and-sport

Cross-disciplinary learning: the key to a successful future. (n.d.). Www.robowunderkind.com. https://www.robowunderkind.com/blog/cross-disciplinary-learning-the-key-to-a-successful-future

Donker, M. H., Erisman, M. C., van Gog, T., & Mainhard, T. (2020). Teachers' Emotional Exhaustion: Associations With Their Typical Use of and Implicit Attitudes Toward Emotion Regulation Strategies. *Frontiers in Psychology*, *11*, 867. Advance online publication. DOI: 10.3389/fpsyg.2020.00867 PMID: 32547437

Equity and Inclusion in Physical Education and Sport, Second Edition. (n.d.). ResearchGate. https://www.researchgate.net/publication/263971685_Equity_and_Inclusion_in_Physical_Education_and_Sport_Second_Edition

Finkelstein, S. (2018). *The Best Leaders Are Great Teachers*. Harvard Business Review. https://hbr.org/2018/01/the-best-leaders-are-great-teachers

Good, D. (2012). *Cognitive Agility: Adapting to Real-time Decision Making at Work*. *44*(2), 13–17. https://www.researchgate.net/publication/265963826_Cognitive_Agility_Adapting_to_Real-time_Decision_Making_at_Work

Grade 4 Cross-curricular Lesson Plan: A World of Sports. (2015). Touchableearth.org. https://www.touchableearth.org/blog/grade-4-cross-curricular-lesson-plan-a-world-of-sports

Gupta, S., & McCarthy, P. J. (2022). The sporting resilience model: A systematic review of resilience in sport performers. *Frontiers in Psychology*, *13*(1), 1003053. DOI: 10.3389/fpsyg.2022.1003053 PMID: 36619099

Halpin, M. (2024, February 26). *Equality and Equity in Sports*. Recite Me. https://reciteme.com/news/equality-and-equity-in-sports/

Harper-collins-admin. (2023, August 25). *Embracing Sports Integration: A Vital Component of School Curriculum - HarperCollins*. HarperCollins. https://collins.in/embracing-sports-integration-a-vital-component-of-school-curriculum/

Harris, M. L. (2025). *Emerging Dialogues - Applying Sports Analogies to Motivate Assessment*. Aalhe.org. https://www.aalhe.org/emerging-dialogues--applying-sports-analogies-to-motivate-assessment

Hasan, H. A., Adi, S., Hariyanto, E., Khongrungchok, A., Oktaviani, H. I., & Fajarianto, O.Humaid Ali Hasan. (2023). Physical Activity with Play and Game Model to Improve the Cognitive of Elementary School Students. *Edcomtech Jurnal Kajian Teknologi Pendidikan*, *8*(1), 49–49. DOI: 10.17977/um039v8i12023p49

Health Direct. (2018, July 2). *Developing life skills through sports*. Healthdirect.gov.au;Australia, H. https://www.healthdirect.gov.au/developing-life-skills-through-sports

Henry, W. (2023, November 25). *Strategy Games: Engaging the Mind and Building Skills*. Medium. https://watson-henry.medium.com/strategy-games-engaging-the-mind-and-building-skills-dd09c219fca0

Hernández-Mendo, A., Reigal, R. E., López-Walle, J. M., Serpa, S., Samdal, O., Morales-Sánchez, V., Juárez-Ruiz de Mier, R., Tristán-Rodríguez, J. L., Rosado, A. F., & Falco, C. (2019). Physical Activity, Sports Practice, and Cognitive Functioning: The Current Research Status. *Frontiers in Psychology*, *10*, 2658. Advance online publication. DOI: 10.3389/fpsyg.2019.02658 PMID: 31866893

In what ways can we use sports to promote diversity and inclusion in our communities? (2019). Quora. https://www.quora.com/In-what-ways-can-we-use-sports-to-promote-diversity-and-inclusion-in-our-communities

Kang, X., Meng, Q., & Su, C.-H. (2024). School-Based Team Sports as Catalysts for Holistic Student Wellness: A Narrative Review. *Behavioral Sciences (Basel, Switzerland)*, *14*(7), 528. DOI: 10.3390/bs14070528 PMID: 39062351

Kark, R. (2011). Games Managers Play: Play as a Form of Leadership Development. *Academy of Management Learning & Education*, *10*(3), 507–527. DOI: 10.5465/amle.2010.0048

Kearl, B. (n.d.). *5 Useful Performance Management Takeaways From Sports*. Www.bamboohr.com. https://www.bamboohr.com/blog/5-useful-performance-management-takeaways-sports

Kim, S., Raza, M., & Seidman, E. (2019). Improving 21st-century teaching skills: The key to effective 21st-century learners. [sagepub.]. *Research in Comparative and International Education*, *14*(1), 99–117. DOI: 10.1177/1745499919829214

Knudsen, J. V., Elkrog-Hansen, L. M., & Christiansen, L. B. (2020). Life Skills Through School Sport: A Participatory Teacher Development Program. *Advances in Physical Education*, *10*(03), 293–310. DOI: 10.4236/ape.2020.103024

Leadership and management in sports: A framework for success. (2025, November 12). KU SOE; School of Education and Human Sciences, Department of Health, Sport, and Exercise Sciences, University of Kansas. https://onlinesportmanagement.ku.edu/community/leadership-and-management-in-sports

Leadership in Education: Decision-Making in Education. (n.d.). ResearchGate. https://www.researchgate.net/publication/339235695_Leadership_in_Education_Decision-Making_in_Education

Loeb Leadership. (2023, July 25). *The Importance of Adaptability in Successful Leadership*. Loeb Leadership. https://www.loebleadership.com/insights/importance-adaptability-successful-leadership

Magazine, T. (2013, August 8). *Using Sports Analogies to Motivate Students*. Education for Today and Tomorrow | L'Education Aujourd'hui et Demain. https://teachmag.com/archives/6455

Mahindru, A., Patil, P., & Agrawal, V. (2023). Role of Physical Activity on Mental Health and Well-Being: A Review. *Cureus*, *15*(1). Advance online publication. DOI: 10.7759/cureus.33475 PMID: 36756008

Monsma, E. V. (2018). *Principles of Effective Goal Setting | Association for Applied Sport Psychology*. Appliedsportpsych.org. https://appliedsportpsych.org/resources/resources-for-athletes/principles-of-effective-goal-setting/

Norfolk, A. (2021, May 7). *The Mental Benefits Of Sport - News*. Active Norfolk. https://www.activenorfolk.org/2021/05/mental-benefits-of-sport/

Norris, G., & Norris, H. (2021). Building Resilience Through Sport in Young People With Adverse Childhood Experiences. *Frontiers in Sports and Active Living*, *3*, 663587. Advance online publication. DOI: 10.3389/fspor.2021.663587 PMID: 34337400

Pandian, R. (2023, December 24). Why sports needs to be integrated into the academic curriculum. *The Hindu*. https://www.thehindu.com/education/why-sports-needs-to-be-integrated-into-the-academic-curriculum/article67644620.ece

psico-smart.com. (2024). *Virtual Reality and Simulation in Leadership Training Programs*. Psico-Smart.com. https://psico-smart.com/en/blogs/blog-virtual-reality-and-simulation-in-leadership-training-programs-177956

Quennerstedt, M. (2019). Physical education and the art of teaching: Transformative learning and teaching in physical education and sports pedagogy. *Sport Education and Society*, *24*(6), 611–623. DOI: 10.1080/13573322.2019.1574731

Rashid, M., & Kabir, I. (2024). The role of sports in breaking down cultural barriers: A study. *International Journal of Physical Education. Sports Health*, *11*(4), 198–200. https://www.kheljournal.com/archives/2024/vol11issue4/PartD/11-4-28-407.pdf

Riera, V., Moragas-Rovira, M., & Pujadas, X. (2024). Leadership development across the lifespan through sport and physical activity. *Journal of Management Development*, *43*(3), 336–356. Advance online publication. DOI: 10.1108/JMD-10-2022-0253

Rodríguez Macías, M., Abad Robles, M. T., & Giménez Fuentes-Guerra, F. J. (2021). Effects of Sport Teaching on Students' Enjoyment and Fun: A Systematic Review and Meta-Analysis. *Frontiers in Psychology*, *12*(708155), 708155. Advance online publication. DOI: 10.3389/fpsyg.2021.708155 PMID: 34421757

Ross, C. M., Jamieson, L. M., & Mitchell, D. W. (2008). Cross-Cultural Understanding Through Youth Sports: Bridging the Tolerance Gap Through Youth Development. *Journal of Youth Development*, *3*(3), 154–163. DOI: 10.5195/jyd.2008.294

Social-Emotional Skills in School. (n.d.). Committee for Children. https://www.cfchildren.org/what-is-social-emotional-learning/schools/

Sport Values In Every Classroom Teaching respect, equity and inclusion to 8-12 year-old students. (n.d.). https://www.icsspe.org/system/files/Sport%20Values%20in%20Every%20Classroom.pdf

Sports as a Tool for Social Inclusion and Diversity in Schools | Bring it on Sports. (2024, July 9). Bring It on Sports. https://www.bringitonsports.com.au/2024/07/09/sports-as-a-tool-for-social-inclusion-and-diversity-in-schools/

Sports Integration | Holistic Development | LIFE AND CAREER SKILLS. (2024). Dlps.co.in. https://www.dlps.co.in/sportsintegration.aspx

Surbakti, A. M. (2023). *Adaptability and Resilience*. LinkedIn. Www.linkedin.com. https://www.linkedin.com/pulse/adaptability-resilience-ariel-matius-surbakti

Surya, P. S., & Maheshbabu, N. (2023). Emotional Maturity and Resilience among Athletes. *International Journal of Indian Psychology*, *11*(3).

The Importance of Resilience in Leadership: Thrive in Times of Change. (n.d.). Www.utsa.edu. https://www.utsa.edu/pace/news/the-importance-of-resilience-in-leadership-thrive-in-times-of-change.html

The Power of Emotional Intelligence in Sports | Telecoming | MONETIZATION TECHNOLOGY. (2024, May 20). Telecoming | MONETIZATION TECHNOLOGY | Tech for Business Digital Growth. https://www.telecoming.com/blog/the-power-of-emotional-intelligence-in-sports/

Thought, in. (2024). *Effective Decision Making: Adaptive Thinking: Flexibility in Thought: Adaptive Thinking for Effective Decision Making - FasterCapital*. FasterCapital. https://fastercapital.com/content/Effective-Decision-Making--Adaptive-Thinking---Flexibility-in-Thought--Adaptive-Thinking-for-Effective-Decision-Making.html

Training the cognitive aspects of agility in team sports. (n.d.). Sportsmith. https://www.sportsmith.co/articles/training-cognitive-aspects-of-agility-in-team-sports/

Ulliance. (2023). Age Bias at Work? Discover What You Need to Know... *Ulliance.com*. https://doi.org/DOI: 109050482603/1735937323578

von Moltke, N. (2024, October 22). *5 Leadership Development Types For Training Leaders (Plus HR Best Practices)*. AIHR. https://www.aihr.com/blog/leadership-development/

What are the most innovative and cutting-edge mobile learning apps for professional growth? (2024). Linkedin.com. https://www.linkedin.com/advice/1/what-most-innovative-cutting-edge-mobile-learning-frm8c

Yellowbrick. (2021, December 24). *How Sports Can Bridge Cultural and Social Differences - Yellowbrick*. Yellowbrick. https://www.yellowbrick.co/blog/sports/how-sports-can-bridge-cultural-and-social-differences

Yoffi, N. (2023). The role of team sports in fostering inclusivity and empathy among diverse student groups. *Cross-Cultural Management Journal, XXV*(2). https://seaopenresearch.eu/Journals/articles/CMJ2023_I2_1.pdf

(PDF) Equity and Inclusion in Physical Education and Sport. Second Edition. (n.d.). ResearchGate. https://www.researchgate.net/publication/263971685_Equity_and_Inclusion_in_Physical_Education_and_Sport_Second_Edition

(PDF) Leadership in Education. Decision-Making in Education. (n.d.). ResearchGate. https://www.researchgate.net/publication/339235695_Leadership_in_Education_Decision-Making_in_Education

Chapter 8
Developing Mathematical Competencies in Teachers:
A Holistic Perspective on Effective Teaching and Teacher Well-Being

Rocío Rodríguez-Padín
https://orcid.org/0000-0001-9731-7757
Universidade da Coruña, Spain

ABSTRACT

This chapter explores the development of mathematical competencies in teachers through a holistic lens, emphasizing the integration of pedagogical knowledge, socio-emotional support, and innovative teaching practices. It highlights the need for adaptive and creative approaches that address diverse student needs and foster critical thinking, problem-solving, and engagement. Central to the discussion is the role of leadership in mathematics teaching, where educators act as facilitators of inquiry and mentors within their professional communities. The chapter also examines challenges such as cognitive demands, equity in education, and classroom management, offering strategies to promote teacher well-being and resilience. Key insights include the use of manipulatives, interdisciplinary connections, and reflective practices to enhance teaching effectiveness. By prioritizing professional development and addressing institutional barriers, the chapter advocates for sustainable and transformative mathematics teaching, benefiting both educators and students.

INTRODUCTION

The development of comprehensive mathematical competencies in teachers is a cornerstone for fostering effective mathematics education. Beyond technical expertise, teaching requires an integration of specialized content knowledge, pedagogical skills, and socio-emotional support mechanisms, which together form a holistic framework for professional growth (Blömeke & Delaney, 2012; Godino et al., 2007). This chapter explores how a multidimensional approach—encompassing leadership strategies, innovative pedagogy, and teacher well-being—can enhance the quality of mathematics education, benefiting both educators and students (An & Wu, 2012; Archambault & Barnett, 2010). By addressing the intricate interplay between teachers' cognitive, affective, and interpersonal capacities, we aim to

DOI: 10.4018/979-8-3373-0472-4.ch008

highlight the importance of designing professional development programs that not only build subject matter expertise but also promote resilience and adaptability in diverse classroom contexts (Bell et al., 2010; Fernández & Yoshida, 2004). Ultimately, the holistic approach advocated here aligns with broader educational goals of nurturing reflective practitioners who are equipped to address the complexities of contemporary mathematics teaching (Barmby et al., 2014; Shulman, 1987).

CONTEXT AND RELEVANCE OF HOLISTIC TEACHER DEVELOPMENT IN MATHEMATICS EDUCATION

Mathematics education faces increasing demands to address both the cognitive and affective dimensions of learning. Teachers are not only tasked with delivering content knowledge but also with fostering critical thinking, adaptability, and problem-solving skills in students, aligning with the broader goals of 21st-century education (Blömeke & Delaney, 2012; Niss & Højgaard, 2019). Holistic teacher development recognizes the interplay between professional expertise, personal growth, and socio-emotional well-being, which are essential for creating sustainable and effective educational practices (Korthagen, 2010; Godino et al., 2007).

In mathematics education, this approach is particularly relevant as teachers frequently navigate complex pedagogical challenges, including addressing students' misconceptions and varying levels of mathematical understanding (An & Wu, 2012). Professional development programs that integrate theoretical knowledge with practical strategies—such as pedagogical content knowledge (Shulman, 1987) and innovative methodologies—equip teachers to respond to these challenges effectively (Bell et al., 2010; Godino et al., 2019).

Moreover, the inclusion of socio-emotional dimensions in teacher training is critical. Studies indicate that teachers' emotional resilience and well-being significantly influence their ability to maintain high-quality instruction and foster positive learning environments (Blömeke & Delaney, 2012; Fernández & Yoshida, 2004). This chapter situates holistic teacher development within this evolving educational landscape, emphasizing its necessity for enhancing both teacher capacity and student outcomes in mathematics classrooms.

DEFINING MATHEMATICAL COMPETENCIES: BEYOND CONTENT KNOWLEDGE

Mathematical competencies extend far beyond mastery of content knowledge; they encompass a range of skills, attitudes, and practices essential for effective teaching. According to Blömeke and Delaney (2012), these competencies involve a combination of cognitive, affective, and situational components that enable teachers to navigate the dynamic challenges of mathematics education. This broader perspective aligns with frameworks such as PISA, which emphasize the application of mathematical reasoning in real-world contexts, highlighting the importance of adaptability and problem-solving (Niss & Højgaard, 2019).

Central to this expanded view is pedagogical content knowledge (Shulman, 1987), which bridges subject matter expertise with effective teaching strategies. Teachers must not only understand mathematical concepts but also anticipate and address students' misconceptions, design meaningful learning experiences, and foster critical thinking (An & Wu, 2012; Bell et al., 2010). For instance, the ability to contextualize abstract mathematical ideas through hands-on activities or real-world applications enhances both comprehension and engagement (Godino et al., 2007; Barmby et al., 2014).

Additionally, mathematical competencies require emotional and interpersonal skills that enable teachers to build supportive learning environments and foster positive relationships with students (Korthagen, 2010; Blömeke & Delaney, 2012). These competencies are particularly critical in addressing the diverse needs of learners, from those struggling with foundational concepts to those excelling in advanced problem-solving tasks. By broadening the definition of mathematical competencies to include these multifaceted elements, educators can better prepare to meet the demands of modern mathematics education and create equitable, impactful learning experiences for all students.

THEORETICAL FRAMEWORKS

The foundation of holistic teacher development in mathematics education lies in an integration of theoretical perspectives that address the cognitive, pedagogical, and socio-emotional dimensions of teaching. These frameworks provide a structured understanding of how teachers acquire and apply the competencies necessary for effective mathematics instruction while promoting their well-being and professional growth (Blömeke & Delaney, 2012; Godino et al., 2007).

Central to this discussion is Shulman's (1987) concept of pedagogical content knowledge, which underscores the importance of connecting subject matter expertise with the ability to effectively communicate mathematical ideas to students. This framework highlights the dual role of teachers as content specialists and facilitators of meaningful learning experiences. Complementing this is the Onto-Semiotic Approach, which provides tools for analyzing the epistemic and didactic dimensions of mathematical knowledge and teaching practices (Godino et al., 2007; Pino-Fan et al., 2018).

Additionally, socio-cognitive theories, such as Vygotsky's (1978) social constructivism, emphasize the significance of collaboration and interaction in the learning process. These principles extend to teacher development, suggesting that professional growth is enriched through reflective practices and shared experiences within a community of educators (Korthagen, 2010; Bell et al., 2010). Furthermore, research on teacher well-being stresses the need to incorporate emotional resilience and stress management strategies into professional development programs, fostering a balanced approach to teaching in demanding educational contexts (Blömeke & Delaney, 2012; Fernández & Yoshida, 2004).

This section delves into these theoretical foundations, demonstrating how their integration creates a comprehensive framework for supporting mathematics teachers in developing the competencies required to meet the diverse challenges of modern classrooms.

Pedagogical Leadership

Pedagogical leadership plays a pivotal role in fostering effective teaching and learning in mathematics classrooms. It encompasses the ability of teachers to guide, inspire, and influence both students and colleagues in the pursuit of meaningful educational experiences (Bell et al., 2010; Godino et al., 2007). This concept goes beyond traditional notions of classroom management, focusing instead on the strategic use of knowledge, skills, and interpersonal relationships to enhance teaching practices and student outcomes.

Central to pedagogical leadership is the capacity to adapt instruction to the diverse needs of learners while maintaining a focus on achieving long-term educational goals (Blömeke & Delaney, 2012). Teachers with strong leadership skills can identify and address learning gaps, foster collaborative problem-solving, and create environments that encourage critical thinking and mathematical reasoning (Barmby et al., 2014; Fernández & Yoshida, 2004). Such leadership requires a deep understanding of both the content and the socio-emotional dynamics of the classroom, enabling teachers to act as mentors and role models.

Moreover, pedagogical leadership involves reflective practice and a commitment to continuous improvement. Research indicates that teachers who actively engage in self-reflection and collaborative professional development are better equipped to implement innovative teaching strategies and adapt to evolving educational challenges (Korthagen, 2010; Bell et al., 2010). This capacity for leadership not only benefits individual classrooms but also contributes to broader systemic improvements in mathematics education, as teachers take on roles as change agents within their schools and communities (Godino et al., 2019).

By integrating pedagogical leadership into teacher development programs, educators can be empowered to lead with confidence, inspire their students, and foster a culture of excellence in mathematics education.

Cognitive Theories Applied to Mathematics Learning

Cognitive theories provide critical insights into how students learn mathematics and, in turn, how teachers can effectively design and deliver instruction. These theories focus on the mental processes involved in understanding, reasoning, and solving mathematical problems, emphasizing the importance of both conceptual and procedural knowledge (Blömeke & Delaney, 2012; Niss & Højgaard, 2019). By integrating cognitive perspectives into their teaching practices, educators can foster deeper learning and address the diverse cognitive needs of their students.

Piaget's stages of cognitive development offer a foundational framework for understanding how learners acquire mathematical concepts. His theory suggests that students construct knowledge through active engagement with their environment, moving from concrete to abstract reasoning as they mature (Blömeke & Delaney, 2012; Godino et al., 2007). This progression highlights the importance of designing tasks that align with students' developmental stages, ensuring that instruction is both accessible and challenging.

Complementing Piaget's theory is Vygotsky's notion of the Zone of Proximal Development (ZPD), which emphasizes the role of social interaction and scaffolding in learning. Through guided support, students can achieve higher levels of mathematical understanding than they could independently (Korthagen, 2010; Fernández & Yoshida, 2004). Teachers who apply this principle effectively create collaborative learning environments where students are encouraged to explore, question, and co-construct knowledge.

Cognitive load theory further refines instructional design by addressing the limits of working memory and the need to manage cognitive demands during learning (Bell et al., 2010; Godino et al., 2019). Strategies such as breaking down complex problems, using visual aids, and incorporating hands-on activities help reduce cognitive overload and enhance understanding.

These cognitive theories, when applied thoughtfully, enable teachers to support their students' mathematical learning more effectively. By integrating these principles into professional development, educators can gain a deeper understanding of how students think and learn, equipping them to design instruction that promotes critical thinking, problem-solving, and long-term retention of mathematical knowledge.

Approaches to Teacher Well-Being

Teacher well-being is a critical yet often overlooked component of effective mathematics education. A teacher's emotional resilience, job satisfaction, and stress management capacity directly influence their ability to deliver high-quality instruction and foster positive learning environments (Blömeke & Delaney, 2012; Bell et al., 2010). In the context of mathematics teaching, where challenges such as addressing misconceptions and managing varying student abilities are prevalent, supporting teacher well-being is essential for sustaining motivation and professional efficacy.

One prominent approach to enhancing teacher well-being is the integration of mindfulness practices into professional development programs. Mindfulness techniques, such as meditation and focused breathing, have been shown to reduce stress and improve emotional regulation among educators, leading to more adaptive responses to classroom challenges (Korthagen, 2010; Blömeke & Delaney, 2012). These practices not only benefit teachers' mental health but also create a ripple effect, positively influencing classroom dynamics and student engagement.

Another effective strategy is fostering a sense of community and collaboration among teachers. Opportunities for peer support, shared problem-solving, and reflective dialogue can help educators navigate the complexities of teaching mathematics and reduce feelings of isolation (Godino et al., 2019; Fernández & Yoshida, 2004). Collaborative environments encourage the exchange of innovative ideas and best practices, contributing to both professional growth and emotional well-being.

Additionally, institutional support plays a pivotal role in promoting teacher well-being. Schools that prioritize teacher development by providing resources, manageable workloads, and recognition of accomplishments create an environment where teachers feel valued and empowered (Bell et al., 2010; Fernández & Yoshida, 2004). Structured programs focusing on work-life balance, resilience training, and stress management further enhance teachers' ability to sustain their passion for teaching.

By adopting these approaches, educational systems can support the holistic development of mathematics teachers, ensuring that they are not only effective educators but also thriving professionals. This focus on well-being ultimately benefits the entire educational ecosystem, fostering healthier, more productive learning environments for both teachers and students.

SCOPE AND OBJECTIVES OF THE CHAPTER

This chapter explores the multifaceted nature of developing mathematical competencies in teachers through a holistic lens, addressing not only their technical and pedagogical skills but also their socio-emotional well-being and leadership capacities. It draws on established theoretical frameworks and empirical research to propose strategies for enhancing teacher preparation and professional development, with the ultimate goal of fostering more effective and sustainable mathematics education practices (Blömeke & Delaney, 2012; Godino et al., 2007).

The scope of the chapter includes a detailed examination of the competencies required for successful mathematics teaching, emphasizing the integration of innovative pedagogical methods, reflective practices, and stress management techniques. By considering both the professional and personal dimensions of teaching, this chapter underscores the importance of equipping educators to navigate the complexities of diverse classroom contexts (Bell et al., 2010; Fernández & Yoshida, 2004). Furthermore, it addresses how leadership skills and collaborative approaches can empower teachers to drive positive change within their schools and communities (Godino et al., 2019; Korthagen, 2010).

The primary objectives of this chapter are to:

1. Define and analyze the essential mathematical competencies required for effective teaching, going beyond technical knowledge to include adaptability, creativity, and interpersonal skills.
2. Highlight the role of theoretical frameworks—such as pedagogical content knowledge, cognitive theories, and socio-emotional models—in shaping teacher development programs (Shulman, 1987; Godino et al., 2007).
3. Provide practical strategies for integrating innovative pedagogical tools, leadership development, and well-being practices into teacher training and ongoing professional development (Blömeke & Delaney, 2012; Bell et al., 2010).
4. Advocate for a holistic approach to teacher preparation that supports not only the intellectual growth of educators but also their emotional and professional resilience (Korthagen, 2010; Fernández & Yoshida, 2004).

Through these objectives, the chapter aims to contribute to the broader discourse on improving mathematics education by placing teachers' holistic development at the center of educational reform. This approach not only enhances teacher effectiveness but also promotes long-term student success, fostering a more enriching and equitable learning environment.

DEVELOPING MATHEMATICAL COMPETENCIES: BEYOND TECHNICAL SKILLS

The Need for Adaptive and Creative Teaching in Mathematics

In a world increasingly shaped by innovation and complexity, mathematics education must prepare students not only to develop procedural fluency but also to think creatively and adaptively. Teachers need to embrace approaches that foster exploration, critical thinking, and the ability to address real-world challenges (Felmer et al., 2016; Presmeg, 2014). Adaptive and creative teaching practices empower educators to address diverse student needs, bridging the gap between abstract mathematical concepts and practical applications (Rowland et al., 2005; Blömeke & Delaney, 2012).

Adaptive teaching involves tailoring instructional strategies to students' varying levels of understanding, cultural backgrounds, and learning styles. For instance, formative assessments help identify students' misconceptions, enabling teachers to adjust their approaches dynamically (An & Wu, 2012; Hill et al., 2008). This flexibility requires not only deep mathematical knowledge but also pedagogical insight and sensitivity to students' individual contexts (Llinares, 2012; Shulman, 1987).

Creativity in mathematics teaching, on the other hand, involves integrating methods that engage and inspire curiosity. Incorporating project-based learning, the use of manipulatives, or task designs that challenge students to think critically can transform traditional mathematics instruction (Barmby et al., 2014; Watson & Ohtani, 2015). For example, task design grounded in socio-mathematical norms can help students construct meaningful knowledge while developing problem-solving abilities (Sánchez & García, 2014; Godino et al., 2007).

Ultimately, adaptive and creative teaching practices not only enhance mathematical competencies but also prepare students for interdisciplinary challenges. Professional development programs emphasizing these approaches equip teachers to create learning environments that are dynamic, equitable, and relevant to 21st-century demands (Korthagen, 2010; Stahnke et al., 2016).

Bridging Theory and Practice

The gap between theoretical knowledge and classroom practice presents a persistent challenge in mathematics education. Teachers must not only understand mathematical theories but also apply them effectively in diverse and dynamic classroom contexts (Shulman, 1987; Felmer et al., 2019). Bridging this gap is essential for fostering lasting mathematical understanding and engagement.

Meaningful Learning Theories in Mathematics

Meaningful learning requires students to connect new knowledge with their prior experiences and understanding. Theories such as Ausubel's meaningful learning highlight the importance of providing structured and relatable content, while other frameworks, such as Vygotsky's Zone of Proximal Development, emphasize the role of guided support in advancing understanding (Korthagen, 2010; Fernández & Yoshida, 2004).

The Onto-Semiotic Approach offers a comprehensive framework for designing tasks that integrate mathematical objects, meanings, and practices. This approach ensures that tasks align with students' cognitive and developmental needs while fostering deep conceptual understanding (Godino et al., 2017; Pino-Fan et al., 2018). Collaborative problem-solving and reflective discussions further enrich this process by connecting abstract concepts to practical applications (Bell et al., 2010; Presmeg, 2014).

Contextualized Mathematics Education

Contextualized learning connects mathematics to real-world scenarios, making it more meaningful and engaging for students. Project-based learning activities, such as designing a budget or analyzing environmental data, help students apply mathematical reasoning in authentic contexts (Felmer et al., 2019; Watson & Ohtani, 2015). These activities cultivate critical thinking and highlight the relevance of mathematics in addressing societal challenges.

Furthermore, integrating cultural contexts into mathematics education promotes inclusivity and engagement. Teachers can design tasks that resonate with students' lived experiences, creating opportunities for meaningful exploration and collaboration (Sánchez & García, 2014; Rowland et al., 2011). Such approaches not only deepen understanding but also empower students to see mathematics as a tool for navigating the complexities of their lives.

The Role of Hands-On and Experiential Learning in Building Competence

Experiential learning is a cornerstone of effective mathematics education, fostering active engagement and deeper understanding. Through direct interaction with mathematical materials and concepts, students construct knowledge that is both practical and enduring (Barmby et al., 2014; Presmeg, 2014).

Manipulatives, such as algebra tiles or geometric models, provide tangible ways for students to explore abstract ideas. These tools are particularly valuable for younger learners or those who struggle with traditional methods, as they bridge the gap between theory and practice (Pino-Fan et al., 2018; Godino et al., 2007).

Experiential learning also extends to real-world problem-solving activities, where students apply mathematical concepts to authentic scenarios. For example, analyzing community data or designing structures using geometric principles engages students in meaningful and collaborative exploration (Felmer et al., 2016; Fernández & Yoshida, 2004).

Collaborative tasks further enrich experiential learning by encouraging dialogue, reflection, and teamwork. This social dimension of learning aligns with cognitive theories, such as Vygotsky's emphasis on guided discovery, where students achieve higher levels of understanding through interaction and support (Llinares, 2012; Stahnke et al., 2016).

Navigating Complexities in Diverse Classrooms: A Teacher's Perspective

Diverse classrooms challenge teachers to balance instructional goals with the unique needs of their students. Effective navigation of these complexities requires flexibility, empathy, and a commitment to inclusivity (Blömeke & Delaney, 2012; Hill et al., 2008).

Differentiated instruction is a key strategy for addressing diversity in mathematics classrooms. By providing tiered tasks, scaffolding, and multiple entry points, teachers can ensure that all students engage meaningfully with the content (An & Wu, 2012; Rowland et al., 2011). Cultural responsiveness further enhances engagement by connecting mathematical concepts to students' lived experiences and values (Sánchez & García, 2014; Pino-Fan et al., 2018).

Effective classroom management also plays a critical role in fostering a productive learning environment. Teachers who model respect and collaboration build communities where diverse perspectives are valued, promoting equity and inclusion (Fernández & Yoshida, 2004; Presmeg, 2014).

Professional development programs that equip teachers to address these complexities through reflective practices and culturally responsive pedagogies are essential for fostering inclusive and dynamic mathematics classrooms (Shulman, 1987; Korthagen, 2010).

LEADERSHIP SKILLS IN MATHEMATICS TEACHING

Strong leadership skills are essential for mathematics teachers to inspire, guide, and support students in achieving their full potential. These skills extend beyond content knowledge and classroom management, encompassing the ability to foster collaboration, address challenges, and create a vision for meaningful mathematical learning. By acting as leaders, mathematics teachers can influence not only their students but also the broader educational community, promoting innovation and continuous improvement in teaching practices (Bell et al., 2010; Godino et al., 2007).

Why Leadership Matters in Mathematics Education

Leadership in mathematics education serves as a foundational element for effective teaching and learning, influencing classroom practices, student outcomes, and the broader educational community. Teachers who exhibit strong leadership skills act as facilitators of inquiry, empowering students to take ownership of their learning, develop critical thinking, and enhance their problem-solving abilities. This transformative approach redefines the teacher's role from a provider of knowledge to a guide who fosters independent and collaborative learning (Felmer et al., 2019; Stahnke et al., 2016).

In classrooms, leadership is evident through practices such as setting clear expectations, providing constructive feedback, and cultivating a supportive learning environment. Teachers who demonstrate leadership establish a culture of trust and respect, encouraging students to take intellectual risks and engage deeply with mathematical concepts. For instance, collaborative tasks or student-led explorations not only enhance understanding but also build confidence and autonomy (Presmeg, 2014; Rowland & Ruthven, 2011). Additionally, effective leaders reflect on their teaching practices, embrace lifelong learning, and adapt to evolving challenges and innovations, ensuring continuous improvement (Shulman, 1987; Bell et al., 2010).

Leadership is also critical in addressing equity and diversity in mathematics education. Effective leaders advocate for inclusive instruction that acknowledges and supports the diverse cultural, social, and academic needs of their students. They work to secure resources, implement professional development opportunities, and establish policies that promote equitable learning environments (Sánchez & García, 2014; Godino et al., 2007). For example, differentiated instruction and culturally responsive pedagogy empower all students to access and engage with mathematical concepts meaningfully (An & Wu, 2012; Llinares, 2011).

Beyond the classroom, mathematics teachers with leadership skills extend their influence to the professional community. They mentor colleagues, share innovative practices, and participate in collaborative initiatives to improve curriculum and instructional strategies. Such contributions strengthen the collective capacity of schools and foster a culture of excellence in mathematics education (Watson & Ohtani, 2015; Fernández & Yoshida, 2004). Moreover, teachers who take on leadership roles shape school culture by promoting shared goals and values, aligning instructional practices with broader educational priorities (Rowland et al., 2005; Korthagen, 2010).

In an era characterized by rapid educational and societal changes, leadership skills are indispensable for mathematics teachers. By cultivating these abilities, educators inspire transformative learning experiences, advocate for meaningful reform, and leave a lasting impact on their students, colleagues, and the broader educational landscape. Leadership in mathematics education, therefore, is not merely an individual trait but a collective effort that drives innovation and equity across all levels of teaching and learning (Felmer et al., 2016; Stahnke et al., 2016).

Core Leadership Skills for Teachers

Mathematics teachers who embody strong leadership skills enhance not only their instructional effectiveness but also the overall learning experiences of their students. Core leadership skills, such as guiding problem-solving and fostering collaboration, empower teachers to create dynamic, student-centered classrooms that encourage exploration, critical thinking, and teamwork (Blömeke & Delaney, 2012; Niss & Højgaard, 2019).

Guiding Student Problem-Solving

One of the most critical leadership skills in mathematics teaching is the ability to guide students through the problem-solving process. Effective problem-solving instruction goes beyond presenting solutions; it involves equipping students with strategies to analyze, approach, and solve mathematical challenges independently and collaboratively (Barmby et al., 2014). Teachers who excel in this area act as facilitators, prompting students with open-ended questions, encouraging exploration, and fostering resilience in the face of difficulties.

A central component of guiding problem-solving is helping students develop metacognitive skills—an awareness of their thinking processes. By teaching students to reflect on their approaches, evaluate their strategies, and learn from mistakes, teachers empower them to become more confident and autonomous learners (Bell et al., 2010; Fernández & Yoshida, 2004). For instance, posing questions like "What strategy might work here?" or "Why do you think this approach is effective?" encourages students to think critically about their methods.

Leadership in problem-solving also requires adapting instruction to the diverse needs of students. Differentiated tasks that provide varying levels of complexity enable all students to engage meaningfully, regardless of their prior knowledge or ability levels (An & Wu, 2012; Godino et al., 2019). Teachers who guide students through this process cultivate an environment of persistence, curiosity, and a deeper appreciation for mathematics.

Fostering Collaborative Learning Environments

Collaboration is a cornerstone of effective mathematics education, and fostering a collaborative learning environment is a key leadership skill for teachers. In such environments, students learn to articulate their ideas, listen to others, and work collectively to solve problems—skills that are critical for success both in mathematics and beyond (Korthagen, 2010; Godino et al., 2007).

Teachers foster collaboration by designing tasks that require teamwork and shared problem-solving. Group activities, such as working on complex, real-world problems or engaging in mathematical debates, encourage students to draw on diverse perspectives and learn from their peers (Barmby et al., 2014; Blömeke & Delaney, 2012). These interactions promote not only mathematical understanding but also the social and communication skills necessary for interdisciplinary work.

An essential element of fostering collaboration is establishing norms that emphasize respect, accountability, and equity. Teachers must ensure that all voices are heard and valued, creating a classroom culture where students feel safe to share ideas and challenge one another constructively (Fernández & Yoshida, 2004; Godino et al., 2019). Leadership in this context involves setting clear expectations for group work and modeling effective communication and teamwork.

By guiding problem-solving and fostering collaborative learning, mathematics teachers demonstrate the leadership skills necessary to inspire and empower their students. These practices not only enhance mathematical competencies but also prepare students to navigate the complexities of the modern world as critical thinkers and effective collaborators.

Strategies for Building Self-Confidence and Initiative in Teachers

Self-confidence and initiative are fundamental qualities for mathematics teachers to effectively lead their classrooms, innovate in their teaching practices, and inspire their students. Developing these qualities requires targeted strategies that address both professional competencies and personal growth, enabling teachers to navigate challenges with resilience and creativity (Blömeke & Delaney, 2012; Niss & Højgaard, 2019).

Encouraging Reflective Practice

Reflective practice is a powerful tool for building self-confidence in teachers. By analyzing their instructional strategies, classroom interactions, and student outcomes, teachers gain a deeper understanding of their strengths and areas for improvement (Shulman, 1987; Bell et al., 2010). Structured reflection, such as maintaining teaching journals or participating in peer reviews, helps educators develop a growth mindset, fostering a sense of accomplishment and empowerment.

Providing Opportunities for Professional Development

Participation in professional development programs tailored to mathematics education is another effective strategy for enhancing teachers' confidence and initiative. Workshops, seminars, and collaborative learning communities offer teachers the chance to deepen their content knowledge, explore innovative methodologies, and engage with current research (Godino et al., 2007; Korthagen, 2010). These experiences not only expand teachers' instructional repertoire but also boost their confidence in applying new strategies in their classrooms.

Promoting Peer Collaboration and Mentorship

Collaborative environments where teachers can share experiences, challenges, and successes with their peers play a critical role in building confidence. Mentorship programs, in particular, provide novice teachers with guidance and support from experienced educators, helping them navigate the complexities of teaching mathematics (Fernández & Yoshida, 2004; Godino et al., 2019). These relationships foster a sense of belonging and mutual encouragement, reducing feelings of isolation and promoting a proactive attitude.

Cultivating Leadership Opportunities

Encouraging teachers to take on leadership roles within their schools or professional communities can significantly enhance their initiative and self-assurance. Roles such as mentoring colleagues, leading workshops, or participating in curriculum design projects allow teachers to apply their expertise, gain recognition, and develop a stronger sense of professional identity (Blömeke & Delaney, 2012; Bell et al., 2010).

Integrating Mindfulness and Stress Management

Finally, addressing the emotional well-being of teachers is essential for fostering confidence and initiative. Mindfulness practices and stress management techniques equip teachers with the tools to handle the pressures of their roles effectively, maintaining a positive outlook and resilience in challenging situations (Korthagen, 2010; Fernández & Yoshida, 2004).

By implementing these strategies, educational systems can support mathematics teachers in building the self-confidence and initiative necessary to lead their classrooms with competence and enthusiasm. Empowered teachers are more likely to innovate, collaborate, and inspire their students, ultimately contributing to a culture of excellence in mathematics education.

Practical Applications of Leadership in Classroom Scenarios

Leadership in mathematics teaching is best exemplified through practical applications that foster student engagement, enhance learning outcomes, and build a collaborative classroom culture. Teachers act as leaders by implementing strategies that promote problem-solving, creativity, and inclusivity, thereby creating environments where students can thrive academically and personally (Blömeke & Delaney, 2012; Godino et al., 2007).

Facilitating Goal-Oriented Learning

One practical way teachers demonstrate leadership is by establishing clear, achievable learning goals that guide students toward meaningful outcomes. For instance, breaking complex mathematical concepts into manageable milestones allows students to track their progress and build confidence as they achieve each step (Shulman, 1987). Teachers reinforce these goals by creating a supportive atmosphere where students feel encouraged to ask questions and take intellectual risks.

Encouraging Collaborative Problem-Solving

Leadership in the classroom often involves fostering collaboration among students. By assigning tasks that require teamwork, such as solving multi-step problems or exploring real-world applications, teachers help students learn to articulate their thinking and benefit from diverse perspectives (Barmby et al., 2014). Group activities, such as investigating the mathematics behind financial literacy or environmental sustainability, demonstrate how mathematics applies to real-life scenarios while emphasizing cooperative learning.

Promoting Student Autonomy

A key leadership practice is empowering students to take ownership of their learning. Teachers achieve this by scaffolding lessons to allow for student-driven exploration and decision-making. For example, providing students with open-ended problems and encouraging them to choose their methods fosters creativity and critical thinking (Bell et al., 2010). By acting as facilitators rather than sole sources of knowledge, teachers guide students toward self-discovery and independence.

Addressing Diverse Needs

Effective leaders in the classroom recognize and adapt to the diverse needs of their students. Differentiated instruction, such as offering multiple entry points to a mathematical problem or varying the complexity of tasks, ensures that all students can engage meaningfully with the content (An & Wu, 2012). Teachers might use manipulatives for hands-on learners, visual aids for those who benefit from graphical representations, or technology to provide interactive experiences tailored to individual learning styles.

Establishing a Culture of Feedback and Reflection

Leadership is evident when teachers foster a culture where feedback and reflection are integral to learning. Providing constructive, timely, and specific feedback helps students understand their strengths and identify areas for improvement (Fernández & Yoshida, 2004). Teachers also model reflective practices, encouraging students to assess their approaches and think critically about how they can enhance their problem-solving strategies.

Integrating Technology and Innovation

Teachers act as leaders by incorporating innovative tools that enhance engagement and deepen understanding. Technology, such as dynamic geometry software or data visualization tools, enables students to experiment with mathematical concepts interactively. Teachers who integrate such resources demonstrate adaptability and a commitment to preparing students for the technological demands of the modern world (Godino et al., 2019).

By applying these leadership practices, mathematics teachers create classrooms that are inclusive, dynamic, and conducive to meaningful learning. These practical applications of leadership not only improve student outcomes but also build a foundation for lifelong learning and appreciation of mathematics.

INNOVATIVE PEDAGOGY AND THE USE OF MANIPULATIVES

Innovation in pedagogy is essential for enhancing mathematics teaching and learning. By adopting creative and research-informed approaches, teachers can make abstract concepts more accessible, engage students in meaningful exploration, and foster a lifelong appreciation for mathematics. Among the most effective strategies are the integration of innovative teaching methodologies and the use of manipulatives, which together transform traditional instruction into dynamic and interactive experiences (Bell et al., 2010; Niss & Højgaard, 2019).

Transforming Mathematics Teaching through Innovation

Innovation in mathematics teaching involves reimagining traditional approaches to meet the evolving needs of students and the demands of a rapidly changing world. Effective innovation shifts the focus from rote memorization and procedural instruction to fostering critical thinking, creativity, and problem-solving skills. This transformation equips students with the tools they need to navigate complex challenges and apply mathematical concepts meaningfully in various contexts (Felmer et al., 2016; Presmeg, 2014).

Inquiry-Based Learning as a Catalyst for Innovation

One of the most effective ways to innovate in mathematics teaching is through inquiry-based learning. This approach encourages students to explore mathematical concepts through guided inquiry and collaborative problem-solving. For instance, presenting open-ended problems, such as analyzing trends in environmental data or modeling the spread of a virus, provides opportunities for students to apply mathematical reasoning in real-world scenarios (Barmby et al., 2014; Watson & Ohtani, 2015). This method deepens conceptual understanding while nurturing curiosity and confidence in learners.

Harnessing Technology to Enhance Engagement

The integration of technology represents another powerful avenue for innovation. Tools such as dynamic geometry software, virtual simulations, and digital manipulatives enable students to visualize and manipulate mathematical concepts interactively. For example, using GeoGebra or similar platforms allows students to explore geometric transformations or algebraic relationships dynamically, fostering deeper comprehension and experimentation (Fernández & Yoshida, 2004; Rowland & Ruthven, 2011).

Teachers who incorporate technology effectively demonstrate adaptability to new educational paradigms and enhance the learning experience for diverse student populations. Technology also supports differentiated instruction by providing personalized learning paths and immediate feedback, catering to the varied needs of students (Stahnke et al., 2016; Godino et al., 2017).

Interdisciplinary Approaches to Mathematics Teaching

Interdisciplinary teaching connects mathematics to other subjects, enriching students' understanding of its relevance and application across diverse fields. For instance, integrating mathematics with science can involve analyzing statistical data in experiments, while connections with art can explore symmetry and patterns in design. These approaches help students see mathematics as a versatile tool for understanding the world, fostering engagement and critical thinking (Felmer et al., 2019; Sánchez & García, 2014).

Fostering a Growth Mindset for Transformative Learning

A transformative mathematics classroom also emphasizes the development of a growth mindset. Teachers play a key role in encouraging students to view challenges as opportunities for learning rather than obstacles to success. Strategies such as celebrating effort, providing constructive feedback, and normalizing mistakes as part of the learning process help build resilience and persistence (Korthagen, 2010; Shulman, 1987).

This mindset not only benefits students academically but also equips them with the confidence to approach unfamiliar problems with a positive attitude, an essential skill in an increasingly complex and interdisciplinary world (Presmeg, 2014).

Empowering Teachers to Lead Innovation

Finally, innovation in mathematics teaching requires empowered teachers who are equipped with the skills, knowledge, and support needed to implement new strategies effectively. Professional development programs that focus on technology integration, inquiry-based methodologies, and interdisciplinary approaches ensure that teachers remain at the forefront of educational transformation (Blömeke & Delaney, 2012; Godino et al., 2019).

By embracing innovative pedagogies, mathematics teachers can create classrooms that inspire engagement, foster creativity, and deepen students' understanding of mathematical concepts. This transformation not only enhances academic outcomes but also prepares students to apply their knowledge and skills in diverse and real-world contexts, bridging the gap between theory and practice in meaningful ways (Felmer et al., 2016; Rowland et al., 2005).

Tools for Interactive and Hands-On Learning

Interactive and hands-on learning tools are instrumental in transforming mathematics education, making abstract concepts more tangible and fostering deeper engagement. These tools empower students to actively participate in their learning process, enhancing their problem-solving and critical thinking skills. Two key resources that support this approach are manipulatives and educational technologies, both of which have proven to be effective in improving student outcomes (Bell et al., 2010; Blömeke & Delaney, 2012).

Manipulatives in Mathematics Education

Manipulatives are physical objects that help students explore and understand mathematical concepts through hands-on interaction. Tools such as base-ten blocks, geometric shapes, algebra tiles, and fraction strips allow students to visualize and experiment with abstract ideas in a concrete way. These resources are particularly valuable for younger learners or students who struggle with traditional instructional methods, as they bridge the gap between theoretical understanding and practical application (Godino et al., 2007; Barmby et al., 2014).

Research has shown that manipulatives improve conceptual understanding and retention by engaging multiple senses and promoting active exploration. For example, using fraction strips to compare different fractions helps students develop a clear understanding of equivalence and proportion. Similarly, geometric models enable students to grasp concepts such as symmetry, area, and volume through manipulation and visualization (Fernández & Yoshida, 2004).

The effectiveness of manipulatives lies in their ability to encourage problem-solving and collaborative learning. Group activities that involve manipulatives foster communication and teamwork, as students work together to explore concepts and test their ideas. Teachers can further enhance the impact of manipulatives by guiding discussions, posing thought-provoking questions, and encouraging reflection on the connections between the concrete and abstract (Shulman, 1987).

Integration of Educational Technologies

Educational technologies have revolutionized mathematics teaching by providing interactive platforms and tools that facilitate dynamic exploration of mathematical concepts. Technologies such as graphing calculators, dynamic geometry software (e.g., GeoGebra), and virtual manipulatives offer students opportunities to visualize, simulate, and manipulate mathematical phenomena in real-time (Bell et al., 2010; Godino et al., 2019).

For instance, dynamic geometry software enables students to explore properties of shapes and transformations by manipulating diagrams directly, fostering a deeper understanding of spatial reasoning and geometric relationships. Similarly, data visualization tools allow students to analyze and interpret statistical information, connecting mathematics to real-world applications (Niss & Højgaard, 2019). These technologies not only make learning more engaging but also develop critical digital literacy skills.

Integrating technology into the classroom also allows for personalized learning experiences. Adaptive learning platforms can tailor instruction to individual student needs, providing targeted practice and immediate feedback. Teachers can use these tools to monitor progress and adjust instruction, ensuring that all students achieve mastery of key concepts (Blömeke & Delaney, 2012).

However, effective integration of technology requires thoughtful planning and professional development for teachers. Educators must be equipped with the skills to select appropriate tools, align them with curricular goals, and seamlessly incorporate them into their lessons. Training programs that focus on technological pedagogical content knowledge (TPACK) can empower teachers to use technology effectively and confidently (Shulman, 1987; Fernández & Yoshida, 2004).

By leveraging manipulatives and educational technologies, mathematics teachers can create interactive and hands-on learning environments that make mathematics more accessible, engaging, and meaningful. These tools not only enhance conceptual understanding but also prepare students to apply their knowledge in diverse and complex contexts.

Designing Interactive Activities for Active Student Engagement

Interactive activities are crucial for fostering active student engagement in mathematics, transforming passive learning into an immersive, hands-on experience. Effective design of these activities requires thoughtful integration of pedagogical principles, interactive tools, and real-world relevance to captivate students and deepen their understanding of mathematical concepts (Blömeke & Delaney, 2012; Niss & Højgaard, 2019).

Aligning Activities with Learning Goals

The foundation of interactive activity design lies in aligning tasks with clear and measurable learning objectives. Teachers must identify the mathematical concepts or skills to be developed and design activities that scaffold understanding toward these goals. For example, an activity on fractions might involve students using fraction strips to visualize equivalence and practice addition of fractions, building from concrete manipulation to abstract reasoning (Godino et al., 2007; Shulman, 1987).

Incorporating Real-World Contexts

Real-world applications enhance engagement by making mathematics relevant to students' lives. Designing activities that connect abstract concepts to practical scenarios—such as budgeting, data analysis, or geometric design—encourages students to see the utility of mathematics in everyday and professional contexts (Barmby et al., 2014; Fernández & Yoshida, 2004). For instance, students might calculate the area of irregular shapes to design a community park or analyze statistical trends in environmental data to propose solutions for a local issue.

Leveraging Collaboration and Peer Interaction

Interactive activities are most effective when they encourage collaboration and peer learning. Group tasks, such as solving complex problems or constructing mathematical models, foster dialogue and teamwork, allowing students to share ideas and learn from diverse perspectives (Bell et al., 2010; Godino et al., 2019). Teachers can structure these activities using roles or responsibilities to ensure active participation and equitable contribution from all group members.

Integrating Manipulatives and Technology

The use of manipulatives and technology enhances the interactive nature of activities, allowing students to explore and experiment with mathematical concepts dynamically. For example, activities using dynamic geometry software enable students to investigate properties of shapes and transformations interactively, while graphing tools help them visualize functions and analyze patterns (Blömeke & Delaney, 2012; Niss & Højgaard, 2019). Teachers can design activities that combine these tools with problem-solving tasks to promote active engagement and conceptual understanding.

Encouraging Inquiry and Reflection

Inquiry-based activities stimulate curiosity and critical thinking by encouraging students to explore questions and discover solutions independently or collaboratively. For instance, presenting students with open-ended problems or mathematical puzzles challenges them to think creatively and develop their reasoning skills (Fernández & Yoshida, 2004). Teachers can guide this process by posing thought-provoking questions, encouraging students to justify their approaches, and facilitating reflective discussions to consolidate learning.

Differentiating to Meet Diverse Needs

Effective activity design also involves differentiation to address the diverse needs and abilities of students. Providing multiple entry points or varying levels of complexity within a single task ensures that all students can participate meaningfully and achieve success. For example, an activity on proportional reasoning might include simpler problems for beginners and more complex, multi-step scenarios for advanced learners (An & Wu, 2012; Bell et al., 2010).

By designing interactive activities that prioritize engagement, collaboration, and inquiry, teachers can transform the mathematics classroom into a vibrant learning environment. These activities not only enhance students' understanding of mathematical concepts but also build their confidence, curiosity, and enthusiasm for the subject. Thoughtful planning and reflection are key to ensuring that these activities effectively meet learning goals and foster a lasting appreciation for mathematics.

Case Studies: Success Stories in Innovative Mathematics Teaching

Examining real-world examples of innovative mathematics teaching provides valuable insights into effective practices and their impact on student learning. These case studies highlight how creative approaches, interactive tools, and thoughtful pedagogical design can transform mathematics classrooms into engaging and inclusive spaces that foster deep understanding and critical thinking (Blömeke & Delaney, 2012; Bell et al., 2010).

Case Study 1: Inquiry-Based Learning in Geometry

In a middle school geometry classroom, a teacher implemented an inquiry-based approach to explore the properties of triangles and their congruence. Students worked collaboratively to solve open-ended problems, such as determining whether two given triangles were congruent based on different combinations of sides and angles. The teacher used dynamic geometry software (e.g., GeoGebra) to allow students to manipulate and test their hypotheses visually and interactively.

The results showed increased student engagement and improved understanding of geometric relationships. Students demonstrated greater confidence in articulating their reasoning and applying congruence principles to novel problems. This approach also fostered collaboration and critical thinking, as students discussed their strategies and refined their solutions (Godino et al., 2007; Niss & Højgaard, 2019).

Case Study 2: Real-World Applications in Algebra

At a high school level, a teacher integrated real-world contexts into an algebra unit focused on linear equations. Students were tasked with designing a budget for a community event, incorporating fixed and variable costs. Using spreadsheets, they created equations to represent different scenarios and graphed their solutions to identify cost-effective options.

This activity not only reinforced algebraic concepts but also highlighted their practical applications. Students reported a greater appreciation for the relevance of mathematics in everyday life, and their performance on subsequent assessments improved significantly. The project also encouraged teamwork and communication, as students worked in groups to analyze data and present their findings (Barmby et al., 2014; Fernández & Yoshida, 2004).

Case Study 3: Manipulatives in Elementary Mathematics

An elementary teacher introduced manipulatives, such as fraction strips and base-ten blocks, to teach fractions and place value. Students used fraction strips to compare fractions and visualize equivalence, while base-ten blocks helped them understand the relationship between digits and their place value in multi-digit numbers.

The hands-on nature of these activities allowed students to experiment and build a concrete understanding of abstract concepts. Teachers observed a marked improvement in students' ability to solve fraction problems and explain their reasoning. The use of manipulatives also enhanced engagement and made mathematics more accessible for students with varying learning needs (Godino et al., 2019; An & Wu, 2012).

Case Study 4: Collaborative Problem-Solving in Statistics

In a university-level statistics course, the instructor introduced collaborative problem-solving activities to analyze real-world data sets. Students worked in small groups to explore topics such as income inequality, environmental trends, and public health data. Using statistical software, they calculated measures of central tendency, constructed graphs, and interpreted their findings.

This approach not only strengthened students' statistical skills but also encouraged critical thinking and teamwork. Students reported feeling more confident in their ability to interpret data and apply statistical methods to complex problems. The collaborative element fostered a sense of community and encouraged the exchange of diverse perspectives (Bell et al., 2010; Korthagen, 2010).

Key Takeaways

These success stories illustrate how innovative approaches in mathematics teaching can lead to significant improvements in student engagement, understanding, and confidence. By leveraging tools such as technology, manipulatives, and real-world contexts, teachers can create dynamic and meaningful learning experiences. These case studies also underscore the importance of thoughtful planning, collaboration, and reflection in implementing effective instructional strategies (Blömeke & Delaney, 2012; Godino et al., 2007).

Challenges and Solutions in Implementing New Methodologies

Innovative methodologies in mathematics teaching hold great promise for enhancing student engagement and understanding. However, their implementation often comes with challenges that require careful planning and adaptive solutions. Identifying and addressing these barriers is essential for ensuring the success and sustainability of new teaching practices (Blömeke & Delaney, 2012; Niss & Højgaard, 2019).

Challenge 1: Resistance to Change

Teachers and schools may resist adopting new methodologies due to comfort with traditional practices, lack of familiarity with innovative approaches, or fear of failure. This resistance is often rooted in concerns about disrupting established routines or skepticism about the effectiveness of unfamiliar strategies (Shulman, 1987).

Solution: Providing comprehensive professional development programs that include hands-on training, peer collaboration, and ongoing support can help alleviate these concerns. Showcasing evidence-based benefits through case studies or pilot programs can also build trust and confidence in new methodologies (Bell et al., 2010; Godino et al., 2019).

Challenge 2: Limited Resources

Access to manipulatives, technology, and professional training can be a significant barrier, particularly in underfunded schools or rural areas. Teachers may lack the tools or infrastructure necessary to implement interactive and innovative methods effectively (An & Wu, 2012; Fernández & Yoshida, 2004).

Solution: Schools and districts can explore cost-effective alternatives, such as open-source software, digital manipulatives, and low-cost physical tools. Collaborative funding efforts or partnerships with educational organizations can also provide the necessary resources. Professional networks and online platforms can serve as valuable sources for shared materials and ideas (Barmby et al., 2014; Godino et al., 2007).

Challenge 3: Time Constraints

Teachers often face tight schedules and overloaded curricula, leaving little room for designing and implementing innovative activities. The additional planning and preparation required for these methodologies may deter teachers from exploring them (Blömeke & Delaney, 2012).

Solution: Schools can support teachers by integrating innovative methods into existing curricula and aligning them with learning standards. Collaborative lesson planning among teaching teams can reduce individual workloads and streamline the integration process. Providing structured time for planning during professional development days can also alleviate time pressures (Niss & Højgaard, 2019; Bell et al., 2010).

Challenge 4: Varied Student Needs

Classrooms are often diverse in terms of students' abilities, learning styles, and cultural backgrounds. Adopting new methodologies may initially struggle to address this diversity effectively, leading to uneven engagement or understanding (An & Wu, 2012).

Solution: Differentiated instruction and universal design for learning (UDL) principles can ensure that activities are accessible and engaging for all students. Teachers can design tasks with multiple entry points and varied complexity, allowing students to participate at their own levels while working toward common goals (Godino et al., 2019; Fernández & Yoshida, 2004).

Challenge 5: Assessment and Accountability

Innovative methodologies may not align with traditional assessment methods, leading to concerns about how to evaluate student performance effectively. Teachers may feel uncertain about how to measure the impact of new strategies on learning outcomes (Shulman, 1987).

Solution: Developing alternative assessment strategies, such as performance-based tasks, portfolios, and self-assessments, can provide a more comprehensive understanding of student learning. Professional development on formative assessment techniques can help teachers track progress and adjust instruction accordingly (Bell et al., 2010; Godino et al., 2007).

By addressing these challenges with targeted solutions, schools and educators can overcome barriers to implementing innovative methodologies. The result is a mathematics classroom that is more inclusive, engaging, and effective, fostering a deeper appreciation for the subject and preparing students for the demands of the modern world. Thoughtful planning, collaboration, and a commitment to professional growth are key to navigating these challenges successfully (Blömeke & Delaney, 2012; Niss & Højgaard, 2019).

TEACHER WELL-BEING AND STRESS MANAGEMENT

Teacher well-being is a cornerstone of effective mathematics education. When teachers feel supported, confident, and emotionally resilient, they are better equipped to foster engaging and meaningful learning environments. However, mathematics teaching presents unique challenges that can impact teacher well-being and stress levels. Addressing these challenges requires a nuanced understanding of their sources and a proactive approach to managing them (Blömeke & Delaney, 2012; Godino et al., 2019).

Understanding the Challenges of Mathematics Teaching

Mathematics teachers face a range of challenges that can contribute to stress and affect their well-being. One significant issue is the high level of cognitive demand required in mathematics teaching. Teachers must not only possess deep content knowledge but also the ability to adapt their instruction to diverse student needs, often in real-time. This dual responsibility of content mastery and pedagogical flexibility can be overwhelming, particularly for less experienced teachers (Shulman, 1987; Niss & Højgaard, 2019).

Another common challenge is managing student misconceptions and varying levels of mathematical readiness. Mathematics often involves abstract concepts that students may find difficult to grasp, leading to frustration and disengagement. Teachers must invest significant time and effort into designing interventions, using differentiated instruction, and providing individualized support, all of which add to their workload and stress (An & Wu, 2012; Bell et al., 2010).

External pressures, such as standardized testing and accountability measures, further contribute to stress in mathematics teaching. Teachers often feel pressured to meet performance benchmarks, which can lead to a focus on test preparation rather than fostering deep conceptual understanding. This misalignment of goals can result in frustration and a sense of professional dissatisfaction (Fernández & Yoshida, 2004; Blömeke & Delaney, 2012).

Classroom management and student behavior are additional stressors. Mathematics classes often involve collaborative problem-solving and active exploration, which can lead to disruptions if not carefully managed. Teachers must balance maintaining order with encouraging student participation, a dynamic that requires significant emotional energy and resilience (Godino et al., 2007; Barmby et al., 2014).

Finally, a lack of institutional support can exacerbate these challenges. Teachers who lack access to professional development, resources, or a supportive community may feel isolated and undervalued, further impacting their well-being (Korthagen, 2010; Bell et al., 2010).

By understanding these challenges, schools and policymakers can better support mathematics teachers, providing the tools and resources they need to thrive professionally and personally. The following sections will explore strategies for addressing these challenges and promoting teacher well-being and stress management in the context of mathematics education.

The Impact of Teacher Well-being on Classroom Outcomes

Teacher well-being plays a critical role in shaping classroom dynamics and student success. A teacher's emotional resilience, job satisfaction, and ability to manage stress directly influence their teaching effectiveness, relationships with students, and the overall learning environment. In mathematics education, where conceptual understanding and problem-solving skills are key, the impact of teacher well-being becomes even more pronounced (Blömeke & Delaney, 2012; Godino et al., 2019).

Enhancing Teacher Effectiveness

When teachers are emotionally and mentally well, they are more likely to be effective in their instructional practices. Well-being fosters confidence, creativity, and adaptability, enabling teachers to design engaging lessons and respond effectively to student needs. Teachers with high levels of well-being are better equipped to facilitate discussions, address misconceptions, and provide timely feedback, all of which are essential for fostering mathematical understanding (Shulman, 1987; Bell et al., 2010).

Conversely, stress and burnout can hinder a teacher's ability to focus and deliver high-quality instruction. Research shows that teachers experiencing high levels of stress may resort to rigid, less effective teaching methods, which can reduce student engagement and limit opportunities for deep learning (An & Wu, 2012; Niss & Højgaard, 2019). Addressing teacher well-being is therefore not only a matter of personal care but also a professional imperative for improving classroom outcomes.

Building Positive Student-Teacher Relationships

Teacher well-being also impacts the quality of relationships with students. Teachers who feel supported and valued are more likely to create a positive classroom atmosphere where students feel safe, respected, and encouraged to participate. This supportive environment fosters trust, motivation, and a willingness to take risks in learning—key factors for success in mathematics education (Korthagen, 2010; Fernández & Yoshida, 2004).

In contrast, teachers under significant stress may struggle to connect with their students or maintain a patient and empathetic approach. This disconnect can lead to a less inclusive and supportive classroom, where students feel less engaged and confident in their mathematical abilities (Blömeke & Delaney, 2012; Godino et al., 2007).

Improving Student Performance

The ripple effects of teacher well-being extend to student performance. Teachers who are emotionally balanced and enthusiastic about their work are more likely to inspire a similar attitude in their students. Enthusiasm and positivity are contagious, fostering a growth mindset and resilience among students as they tackle challenging mathematical problems (Barmby et al., 2014; Godino et al., 2019).

Moreover, when teachers have the emotional bandwidth to implement innovative teaching strategies, such as hands-on activities or collaborative learning tasks, students benefit from more engaging and effective instructional methods. These approaches not only enhance comprehension but also help students develop critical thinking and problem-solving skills, essential for long-term success in mathematics (Bell et al., 2010; Fernández & Yoshida, 2004).

Creating a Sustainable Learning Environment

Prioritizing teacher well-being contributes to a sustainable and thriving learning environment. When teachers feel supported, they are more likely to remain committed to their profession, reducing turnover rates and ensuring consistency in student learning experiences. Stable, well-supported teachers create a sense of continuity and reliability, fostering trust and confidence among students and their families (Blömeke & Delaney, 2012; Niss & Højgaard, 2019).

In summary, teacher well-being is a cornerstone of effective mathematics education. By addressing the emotional, professional, and institutional needs of teachers, schools can enhance classroom outcomes, foster positive student-teacher relationships, and build an environment where both educators and learners can flourish.

Practical Strategies for Managing Stress and Building Resilience

Managing stress and building resilience are essential for mathematics teachers to navigate the complexities of their roles while maintaining their well-being and effectiveness. Practical strategies, such as mindfulness and self-care, empower teachers to cultivate emotional balance, enhance focus, and sustain their passion for teaching, even in challenging environments (Blömeke & Delaney, 2012; Bell et al., 2010).

Mindfulness Practices for Teachers

Mindfulness practices involve cultivating present-moment awareness and emotional regulation through techniques such as meditation, deep breathing, and mindful movement. For teachers, these practices can significantly reduce stress, enhance focus, and improve emotional resilience (Korthagen, 2010).

Stress Reduction: Regular mindfulness exercises help teachers manage the physiological and emotional responses to stress, enabling them to approach challenges with calmness and clarity. For instance, starting the day with a brief meditation session can set a positive tone and improve emotional regulation throughout the school day (Fernández & Yoshida, 2004).

Enhancing Focus: Teaching mathematics often involves addressing multiple demands simultaneously, from lesson planning to student engagement. Mindfulness practices, such as focused breathing exercises or body scans, improve concentration and the ability to stay present, helping teachers manage their workload more effectively (Blömeke & Delaney, 2012).

Modeling Mindfulness for Students: Incorporating mindfulness practices into the classroom benefits not only teachers but also students. Leading short mindfulness exercises with students fosters a calm learning environment and demonstrates strategies for managing stress, creating a ripple effect of emotional balance and focus (Barmby et al., 2014).

Incorporating Self-Care into Daily Routines

Self-care involves prioritizing activities that nurture physical, emotional, and mental health, enabling teachers to recharge and sustain their energy levels. Incorporating self-care into daily routines helps teachers build resilience and maintain a positive outlook (Godino et al., 2007).

Physical Self-Care: Regular exercise, balanced nutrition, and adequate sleep form the foundation of physical well-being. Simple practices, such as taking short walks during breaks or incorporating movement into the classroom, can boost energy levels and reduce stress (Bell et al., 2010).

Emotional Self-Care: Engaging in activities that bring joy and relaxation, such as hobbies, journaling, or spending time with loved ones, helps teachers decompress and maintain a healthy work-life balance. Setting boundaries between professional and personal time is also critical for emotional well-being (Shulman, 1987; Fernández & Yoshida, 2004).

Professional Self-Care: Seeking professional support through peer collaboration or mentorship provides a sense of community and shared purpose. Teachers can benefit from discussing challenges, sharing resources, and celebrating successes with colleagues, which reinforces resilience and motivation (Godino et al., 2019).

Time Management: Effective time management strategies, such as prioritizing tasks, setting realistic goals, and avoiding overcommitment, are integral to self-care. Using planning tools or delegating responsibilities when possible can help teachers balance their workloads and reduce stress (Blömeke & Delaney, 2012).

By integrating mindfulness and self-care into their routines, mathematics teachers can develop the resilience needed to thrive in their roles. These strategies not only enhance their personal well-being but also contribute to a more positive and productive classroom environment, benefiting both teachers and students.

Professional Development Programs Focused on Socio-Emotional Support

Professional development programs that emphasize socio-emotional support play a crucial role in equipping mathematics teachers with the tools to manage stress, build resilience, and foster positive classroom environments. These programs go beyond enhancing pedagogical skills to address the emotional and psychological needs of teachers, promoting their well-being and professional growth (Blömeke & Delaney, 2012; Godino et al., 2019).

Emphasizing Emotional Resilience and Stress Management

A core component of effective professional development is helping teachers build emotional resilience. Programs that include training in mindfulness practices, emotional regulation, and coping strategies empower teachers to navigate the challenges of mathematics teaching with greater confidence and composure (Korthagen, 2010). For example, workshops on stress management techniques, such as breathing exercises and guided relaxation, provide practical tools to handle high-pressure situations effectively (Fernández & Yoshida, 2004).

Fostering Collaboration and Peer Support

Collaboration among teachers is a powerful means of socio-emotional support. Professional development initiatives that encourage peer-to-peer learning and mentorship create a sense of community and shared purpose. Teachers benefit from opportunities to discuss challenges, share strategies, and provide mutual encouragement, reducing feelings of isolation and fostering a supportive network (Bell et al., 2010; Shulman, 1987).

Structured collaboration, such as participation in professional learning communities (PLCs), provides a platform for teachers to reflect on their practices, explore innovative methodologies, and address classroom challenges collectively. This collaborative approach enhances not only teachers' professional competencies but also their emotional well-being by reinforcing a sense of belonging and collective efficacy (Godino et al., 2007).

Integrating Socio-Emotional Learning (SEL) into Teacher Training

Professional development programs that incorporate socio-emotional learning principles equip teachers to better understand and manage their own emotions, while also fostering these skills in their students. Training in SEL helps teachers build stronger relationships with students, establish positive classroom dynamics, and model behaviors such as empathy and self-regulation (Blömeke & Delaney, 2012; Barmby et al., 2014).

For example, SEL-focused workshops may include strategies for active listening, conflict resolution, and fostering a growth mindset. These skills enable teachers to create an emotionally supportive learning environment, where both they and their students can thrive (Niss & Højgaard, 2019).

Providing Access to Coaching and Counseling Services

Another effective component of professional development programs is access to coaching and counseling services. Professional coaching offers personalized guidance and support, helping teachers identify their strengths, set goals, and develop actionable strategies for improvement. Counseling services, on the other hand, provide a safe space for teachers to address personal or professional challenges that may affect their well-being (Bell et al., 2010; Fernández & Yoshida, 2004).

Tailoring Programs to Individual Needs

Professional development initiatives are most effective when they are tailored to the unique needs of teachers. This includes offering flexible formats, such as online modules, in-person workshops, or hybrid options, to accommodate different schedules and preferences. Programs should also account for varying levels of experience, providing specialized support for novice teachers while addressing the evolving needs of experienced educators (Godino et al., 2019).

Measuring the Impact of Socio-Emotional Support Programs

To ensure the success of these initiatives, it is essential to measure their impact through feedback, surveys, and ongoing evaluations. Tracking improvements in teacher well-being, classroom management, and student outcomes provides valuable insights for refining program content and delivery (Blömeke & Delaney, 2012).

By prioritizing socio-emotional support in professional development programs, schools and educational systems can create a foundation for sustainable teacher well-being and effectiveness. Empowered and emotionally balanced teachers are better equipped to inspire their students, foster meaningful learning experiences, and contribute to a positive and thriving educational environment.

The Role of School Leadership in Supporting Teacher Well-being

School leadership plays a pivotal role in promoting teacher well-being by fostering a supportive environment, addressing systemic challenges, and prioritizing the emotional and professional needs of educators. Mathematics teachers, who often face high cognitive and emotional demands, benefit significantly from leadership practices that create a culture of trust, collaboration, and growth (Blömeke & Delaney, 2012; Godino et al., 2019).

Creating a Supportive Culture

Effective school leaders prioritize the well-being of teachers by cultivating a supportive and inclusive school culture. This includes acknowledging the challenges of teaching mathematics and providing reassurance that teacher health and well-being are valued as much as student success. Recognizing teacher achievements, offering regular encouragement, and maintaining open communication channels contribute to a positive workplace atmosphere (Bell et al., 2010; Shulman, 1987).

Providing Resources and Professional Development

Leadership is essential in ensuring that teachers have access to the tools, resources, and training needed to succeed. By allocating funds for manipulatives, technology, or professional development programs focused on socio-emotional support, school leaders enable teachers to enhance both their instructional practices and resilience. Additionally, offering workshops on stress management or mentoring programs supports teachers in managing the complexities of their roles (Barmby et al., 2014; Korthagen, 2010).

Encouraging Collaboration and Peer Networks

School leaders can facilitate collaboration among teachers by promoting professional learning communities (PLCs) or peer mentoring programs. These networks allow teachers to share ideas, discuss challenges, and provide mutual support, fostering a sense of community and reducing feelings of isolation. Leaders who create spaces for collaboration demonstrate a commitment to teacher well-being and professional growth (Godino et al., 2007; Fernández & Yoshida, 2004).

Addressing Workload and Time Management

Balancing workloads is a critical aspect of supporting teacher well-being. School leaders can reduce stress by ensuring reasonable expectations for lesson planning, grading, and extracurricular responsibilities. Strategies such as scheduling time for collaborative planning or providing administrative support for non-teaching tasks can alleviate pressure and allow teachers to focus on effective instruction (Blömeke & Delaney, 2012; Niss & Højgaard, 2019).

Modeling Emotional Resilience

Leaders who model emotional resilience and positive behaviors set a powerful example for their staff. By demonstrating mindfulness, effective communication, and proactive problem-solving, school leaders inspire teachers to adopt similar strategies for managing stress and challenges. This leadership by example creates a ripple effect that promotes a culture of well-being throughout the school (Korthagen, 2010; Bell et al., 2010).

Providing Access to Counseling and Support Services

School leaders can ensure that teachers have access to mental health resources, such as counseling or coaching services, to address personal or professional challenges. Offering confidential and accessible support helps teachers manage stress and build resilience, contributing to a healthier work environment (Fernández & Yoshida, 2004; Godino et al., 2019).

Fostering a Vision of Well-Being

Leadership that prioritizes teacher well-being integrates it into the broader vision of the school. This includes embedding well-being goals into school policies, celebrating milestones in teacher development, and aligning resources to support long-term strategies for teacher health and satisfaction. Leaders who champion well-being at every level create an environment where teachers feel valued and empowered (Blömeke & Delaney, 2012; Godino et al., 2007).

By adopting these leadership practices, schools can create a supportive and sustainable environment where mathematics teachers thrive. Empowered and well-supported teachers are more likely to engage students, implement innovative instructional strategies, and contribute to the overall success of the school community.

CONCLUSIONS AND RECOMMENDATIONS

Key Insights from a Holistic Approach to Teacher Development

The development of mathematical competencies in teachers requires a comprehensive and holistic approach that integrates content mastery, pedagogical expertise, socio-emotional support, and professional growth. This approach acknowledges the multifaceted nature of teaching and the need to support teachers both as educators and individuals, ensuring their long-term success and well-being (Blömeke & Delaney, 2012; Godino et al., 2019).

One of the key insights from this holistic perspective is the importance of addressing teacher well-being as a foundational element of effective teaching. Teachers who feel supported and empowered are better equipped to manage the cognitive and emotional demands of mathematics instruction. By fostering emotional resilience, stress management skills, and work-life balance, schools can create a healthier and more sustainable teaching environment (Korthagen, 2010; Fernández & Yoshida, 2004).

Another critical insight is the role of innovative pedagogical strategies in enhancing teaching effectiveness. Approaches such as inquiry-based learning, the use of manipulatives, and the integration of technology provide teachers with the tools to engage students actively and meaningfully. These strategies not only improve student outcomes but also reinvigorate teachers' passion for their work, fostering a sense of purpose and accomplishment (Shulman, 1987; Barmby et al., 2014).

Collaboration and peer networks emerge as another essential component of holistic teacher development. Professional learning communities, mentoring programs, and opportunities for shared reflection allow teachers to exchange ideas, address challenges collectively, and build a sense of community. These connections reduce isolation, enhance professional growth, and contribute to emotional well-being (Bell et al., 2010; Godino et al., 2007).

Finally, the role of leadership in supporting teacher development cannot be overstated. School leaders who prioritize teacher well-being, provide access to resources, and model resilience create an environment where teachers can thrive. Leadership practices that align institutional goals with teacher support foster a culture of trust and collaboration, enabling teachers to focus on delivering high-quality education (Blömeke & Delaney, 2012; Niss & Højgaard, 2019).

In summary, a holistic approach to teacher development addresses the interconnected dimensions of teaching, equipping mathematics educators to navigate the complexities of their roles. This approach not only benefits teachers but also enhances student outcomes and strengthens the overall educational ecosystem. The following recommendations build on these insights to propose actionable steps for supporting mathematics teachers effectively.

Recommendations for Teacher Education Programs

Teacher education programs play a critical role in preparing mathematics educators to meet the demands of modern classrooms. To foster both professional competence and personal resilience, these programs must adopt comprehensive strategies that integrate holistic training models and interdisciplinary collaboration. These approaches ensure that teachers are equipped not only with content knowledge and pedagogical skills but also with the emotional and collaborative capacities necessary for effective teaching (Blömeke & Delaney, 2012; Godino et al., 2019).

Incorporating Holistic Training Models

A holistic approach to teacher education recognizes the interconnected dimensions of teaching, addressing the intellectual, emotional, and professional needs of educators. Programs should integrate training that encompasses not only content mastery and pedagogical techniques but also socio-emotional development and stress management (Korthagen, 2010; Bell et al., 2010).

Socio-Emotional Support: Teacher education programs should include modules on emotional resilience, mindfulness, and self-care strategies. Providing future teachers with tools to manage stress and balance work-life demands ensures their long-term effectiveness and well-being (Fernández & Yoshida, 2004; Blömeke & Delaney, 2012).

Innovative Pedagogical Practices: Training in research-based methodologies, such as inquiry-based learning, the use of manipulatives, and technology integration, equips teachers to create engaging and dynamic learning environments. By experiencing these methods firsthand during their training, future teachers can confidently implement them in their classrooms (Shulman, 1987; Godino et al., 2007).

Reflective Practice: Encouraging teachers to engage in reflective practices, such as journaling, peer feedback, and case study analysis, fosters a growth mindset and continuous professional improvement. Reflection helps teachers connect theoretical principles with real-world application, strengthening their teaching efficacy (Barmby et al., 2014).

Promoting Interdisciplinary Collaboration

Interdisciplinary collaboration enriches teacher education by providing future educators with diverse perspectives and skills that extend beyond their primary subject area. Mathematics education, in particular, benefits from connections with fields such as science, technology, engineering, and the arts, fostering a comprehensive understanding of how mathematical principles apply in various contexts (Niss & Højgaard, 2019; Bell et al., 2010).

STEM Integration: Programs should emphasize the integration of mathematics with science, technology, and engineering to prepare teachers for interdisciplinary teaching. Collaborative projects, such as designing real-world problem-solving tasks, help teachers understand how to create cross-curricular connections that enhance student engagement and understanding (Blömeke & Delaney, 2012; Godino et al., 2019).

Arts and Humanities Connections: Including arts and humanities in teacher training broadens educators' perspectives on how mathematics relates to creative and cultural fields. Activities such as exploring geometric patterns in art or analyzing historical data trends provide teachers with innovative ways to connect mathematics to students' interests and experiences (Barmby et al., 2014).

Collaborative Learning Experiences: Teacher education programs should provide opportunities for trainees to work with peers from other disciplines, simulating the collaborative environments they will encounter in schools. Joint workshops, interdisciplinary seminars, and team-teaching exercises foster a culture of cooperation and mutual respect (Korthagen, 2010; Fernández & Yoshida, 2004).

By incorporating holistic training models and promoting interdisciplinary collaboration, teacher education programs can prepare mathematics educators to address the complexities of teaching in diverse and dynamic contexts. These strategies not only enhance teachers' professional competence but also contribute to their overall well-being and satisfaction, ensuring a positive and lasting impact on their students and the broader educational community.

Future Directions for Research and Practice

As the educational landscape evolves, it is essential to explore new directions in research and practice that address the challenges and opportunities in mathematics education. By advancing our understanding of teacher development and implementing innovative strategies, we can further enhance teaching effectiveness, teacher well-being, and student outcomes.

Expanding Research on Holistic Teacher Development

Future research should delve deeper into the interconnected dimensions of holistic teacher development. Studies that investigate the long-term impact of socio-emotional support programs, reflective practices, and resilience-building strategies on teacher effectiveness and retention can provide valuable insights (Blömeke & Delaney, 2012; Korthagen, 2010). Quantitative and qualitative analyses of how holistic training models influence teacher well-being and classroom dynamics would further substantiate their value.

Moreover, exploring how cultural, social, and institutional factors affect the implementation of these models across diverse educational settings can guide adaptations to meet specific needs. Comparative studies across different countries or regions can highlight best practices and innovative approaches (Godino et al., 2019; Niss & Højgaard, 2019).

Innovating Pedagogical Approaches

Innovative pedagogical practices should remain a focus of both research and practice. Investigating the efficacy of new technologies, such as virtual reality or AI-based learning tools, in fostering mathematical understanding could open new possibilities for interactive and personalized instruction (Bell et al., 2010). Additionally, longitudinal studies on the impact of inquiry-based and interdisciplinary approaches on student engagement and achievement would provide robust evidence for their broader adoption.

Addressing Equity and Diversity in Mathematics Education

Research must also prioritize strategies for addressing equity and diversity in mathematics classrooms. Identifying effective methods for supporting underrepresented and underserved populations, including students from marginalized communities or those with special educational needs, is essential for fostering inclusivity (An & Wu, 2012; Barmby et al., 2014). This includes studying how culturally responsive teaching practices and differentiated instruction contribute to equitable access to high-quality mathematics education.

Exploring the Role of Teacher Well-being in Student Outcomes

Future research should further explore the relationship between teacher well-being and student success. Investigating how teacher stress, emotional resilience, and job satisfaction influence classroom outcomes can guide interventions that prioritize teacher health as a means to improve overall educational quality (Fernández & Yoshida, 2004; Godino et al., 2007).

Promoting Collaborative and Interdisciplinary Research

Collaboration across disciplines, such as psychology, neuroscience, and education, can enrich our understanding of how students learn mathematics and how teachers can best support this process. Interdisciplinary research on cognitive development, motivation, and the impact of socio-emotional learning on mathematics achievement can provide actionable insights for educators and policymakers (Shulman, 1987; Niss & Højgaard, 2019).

Fostering Practice-Based Research

Finally, encouraging practice-based research by involving teachers in the research process can bridge the gap between theory and application. Action research projects, where teachers systematically study and improve their practices, can generate valuable knowledge while empowering educators as active contributors to the field (Godino et al., 2019; Bell et al., 2010).

Concluding Thoughts

By pursuing these future directions in research and practice, the field of mathematics education can continue to evolve in ways that enhance teaching and learning experiences. Investing in holistic teacher development, innovative pedagogy, and equitable access to mathematics education will not only benefit educators and students but also strengthen the broader educational ecosystem. These efforts represent a commitment to excellence and inclusivity in mathematics education for the 21st century.

Final Reflections: Toward Sustainable and Effective Mathematics Teaching

Sustainable and effective mathematics teaching is essential for equipping students with the skills and mindsets needed to navigate an increasingly complex world. Achieving this goal requires a comprehensive approach that prioritizes the professional development, well-being, and support of mathematics teachers. As highlighted throughout this chapter, fostering a holistic perspective in teacher education and professional practice is central to creating lasting improvements in mathematics education (Blömeke & Delaney, 2012; Godino et al., 2019).

A key reflection from this exploration is the interconnectedness of teacher well-being and classroom success. When teachers feel emotionally supported and professionally empowered, they are better equipped to engage students, implement innovative practices, and create inclusive learning environments. Addressing teacher well-being is not an ancillary concern but a foundational component of sustainable teaching practices (Korthagen, 2010; Fernández & Yoshida, 2004).

Additionally, the integration of innovative pedagogies, such as inquiry-based learning, manipulatives, and technology-enhanced instruction, demonstrates the transformative potential of modern teaching methods. These strategies not only make mathematics more accessible and engaging but also prepare students for real-world problem-solving and critical thinking (Barmby et al., 2014; Bell et al., 2010). Supporting teachers in adopting these approaches through targeted training and resources is vital for ensuring their widespread and effective implementation.

Collaboration and community-building emerge as essential elements in fostering sustainable mathematics teaching. Teachers who are part of supportive professional networks and learning communities are more likely to stay motivated, share best practices, and continuously refine their skills. School leaders play a crucial role in cultivating these environments, highlighting the importance of leadership in sustaining teacher growth and success (Godino et al., 2007; Niss & Højgaard, 2019).

Finally, a commitment to equity must underpin all efforts in mathematics education. Ensuring that all students, regardless of their background or abilities, have access to high-quality instruction requires systemic changes that prioritize inclusivity, differentiated instruction, and culturally responsive teaching practices. Sustainable mathematics teaching is, at its core, equitable mathematics teaching (An & Wu, 2012; Godino et al., 2019).

In conclusion, sustainable and effective mathematics teaching is a collaborative endeavor that hinges on the well-being and professional capacity of teachers. By embracing holistic approaches, fostering innovation, and prioritizing equity, educators, leaders, and policymakers can work together to build a future where mathematics education thrives. These reflections serve as a call to action for continued investment in the development of mathematics teachers, ensuring that they are equipped to inspire and empower the next generation of learners.

REFERENCES

An, S., & Wu, Z. (2012). Enhancing mathematics teachers' knowledge of students' thinking from assessing and analyzing misconceptions. *International Journal of Science and Mathematics Education*, *10*(3), 717–753. DOI: 10.1007/s10763-011-9324-x

Barmby, P., Bolden, D., & Thompson, L. (2014). *Understanding and enriching problem solving in primary mathematics*. Critical Publishing.

Bell, C. A., Wilson, S. M., Higgins, T., & McCoach, D. B. (2010). Measuring the effects of professional development on teacher knowledge: The case of developing mathematical ideas. *Journal for Research in Mathematics Education*, *41*(5), 479–512. DOI: 10.5951/jresematheduc.41.5.0479

Blömeke, S., & Delaney, S. (2012). Assessment of teacher knowledge across countries: A review of the state of research. *ZDM Mathematics Education*, *44*(3), 223–247. DOI: 10.1007/s11858-012-0429-7

Felmer, P., Liljedahl, P., & Koichu, B. (Eds.). (2019). *Problem solving in mathematics instruction and teacher professional development*. Springer., DOI: 10.1007/978-3-030-29215-7

Felmer, P., Pehkonen, E., & Kilpatrick, J. (Eds.). (2016). *Posing and solving mathematical problems: Advances and new perspectives*. Springer., DOI: 10.1007/978-3-319-28023-3

Fernández, C., & Yoshida, M. (2004). *Lesson study: A Japanese approach to improving mathematics teaching and learning*. Lawrence Erlbaum., DOI: 10.4324/9781410610867

Godino, J. D., Batanero, C., & Font, V. (2007). The onto-semiotic approach to research in mathematics education. *ZDM Mathematics Education*, *39*(1), 127–135. DOI: 10.1007/s11858-006-0004-1

Godino, J. D., Giakomone, B., Batanero, C., & Font, V. (2017). Enfoque ontosemiótico de los conocimientos y competencias del profesor de matemáticas. *Bolema*, *31*(57), 90–113. DOI: 10.1590/1980-4415v31n57a05

Hill, H. C., Ball, D. L., & Schilling, S. G. (2008). Unpacking pedagogical content knowledge: Conceptualizing and measuring teachers' topic-specific knowledge of students. *Journal for Research in Mathematics Education*, *39*(4), 372–400. DOI: 10.5951/jresematheduc.39.4.0372

Korthagen, F. (2010). La práctica, la teoría y la persona en la formación del profesorado [Practice, theory, and person in teachers' training]. *Revista Interuniversitaria De Formación Del Profesorado*, *68*(24), 83–101.

Llinares, S. (2012). Construcción de conocimiento y desarrollo de una mirada profesional para la práctica de enseñar matemáticas en entornos en línea. [Building knowledge and developing a professional approach to teaching mathematics in online environments*]. Avances de Investigación en Educación Matemática*, *2*(2), 53–70. DOI: 10.35763/aiem.v1i2.18

Niss, M., & Højgaard, T. (2019). Mathematical competencies revisited. *Educational Studies in Mathematics*, *102*(1), 9–28. DOI: 10.1007/s10649-019-09903-9

Presmeg, N. (2014). Contemplating visualization as an epistemological learning tool in mathematics. *ZDM Mathematics Education*, *46*(1), 151–157. DOI: 10.1007/s11858-013-0561-z

Rowland, T., & Ruthven, K. (Eds.). (2011). *Mathematical knowledge in teaching*. Springer., DOI: 10.1007/978-90-481-9766-8

Sánchez, V., & García, M. (2014). Socio-mathematical and mathematical norms related to definition in pre-service primary teachers' discourse. *Educational Studies in Mathematics*, *85*(2), 305–320. DOI: 10.1007/s10649-013-9516-0

Shulman, L. S. (1987). Knowledge and teaching: Foundations of the new reform. *Harvard Educational Review*, *57*(1), 1–22. DOI: 10.17763/haer.57.1.j463w79r56455411

Stahnke, R., Schueler, S., & Roesken-Winter, B. (2016). Teachers' perception, interpretation, and decision-making: A systematic review of empirical mathematics education research. *ZDM Mathematics Education*, *48*(1–2), 1–27. DOI: 10.1007/s11858-016-0775-y

Watson, A., & Ohtani, M. (Eds.). (2015). *Task design in mathematics education*. Springer., DOI: 10.1007/978-3-319-09629-2

Chapter 9
Teaching With Adaptability:
Cognitive and Pedagogical Strategies for Geriatric Nursing

Tiago Manuel Horta Reis da Silva
https://orcid.org/0000-0001-5220-1718
King's College London, UK

ABSTRACT

This chapter explores the role of adaptability in teaching geriatric nursing, emphasising cognitive and pedagogical strategies that cater to diverse learning needs. It delves into the cognitive processes involved in learning complex geriatric care concepts and the pedagogical methods that enhance adaptability in teaching. The chapter discusses case studies and evidence-based approaches to integrating flexible teaching strategies, promoting critical thinking and problem-solving skills among nursing students. It also addresses the challenges and solutions in implementing adaptable teaching practices in geriatric nursing education. The chapter concludes with recommendations for educators to develop adaptable teaching frameworks that foster comprehensive learning in geriatric care.

INTRODUCTION

The field of geriatric nursing is increasingly recognised for its complexity and the unique challenges it presents (Reis da Silva, 2024a). As the global population ages, the demand for skilled nursing professionals who can adapt to the evolving needs of older adults has never been more critical (Fitzpatrick et al., 2023). Adaptability in teaching geriatric nursing is paramount, as it ensures that nursing educators can effectively prepare students to meet the diverse and dynamic challenges of caring for older adults (Reis da Silva, 2024a). This adaptability encompasses not only the ability to modify teaching methods in response to changing educational environments but also the capacity to integrate new knowledge and practices that reflect the latest advancements in geriatric care.

Adaptability in teaching geriatric nursing is essential for several reasons. First, the geriatric population presents a wide array of health issues, including chronic diseases, cognitive impairments, and complex psychosocial needs (Reis da Silva, 2024b, Reis da Silva, 2024c). Educators must be prepared to teach students how to assess and address these multifaceted challenges effectively. Moreover, the rapid pace of medical advancements necessitates that nursing curricula remain current and relevant. This

DOI: 10.4018/979-8-3373-0472-4.ch009

requires educators to be flexible in their teaching approaches, incorporating new research findings and evidence-based practices into their instruction (Reis da Silva and Mitchell, 2024a). Furthermore, the COVID-19 pandemic has underscored the importance of adaptability in nursing education. The shift to online learning and virtual clinical experiences has challenged traditional teaching methods, compelling educators to explore innovative pedagogical strategies that engage students and enhance their learning outcomes (Mukattil, 2024; Nashwan et al., 2020). As nursing education continues to evolve, the ability to adapt to new technologies and educational modalities will be crucial for preparing future nurses to deliver high-quality care to older adults (Reis da Silva and Mitchell, 2024a).

In addition to the necessity of adaptability, understanding cognitive and pedagogical strategies is vital for effective teaching in geriatric nursing. Cognitive strategies refer to the mental processes involved in learning, such as critical thinking, problem-solving, and decision-making. These strategies are essential for nursing students as they learn to navigate complex clinical scenarios and make informed decisions about patient care (Leffers et al., 2017; Reis da Silva, 2024a). Educators can foster these cognitive skills by employing active learning techniques, such as case studies, simulations, and group discussions, which encourage students to engage with the material and apply their knowledge in practical settings (Nurse-Clarke & Sockol, 2022).

Pedagogical strategies, on the other hand, encompass the methods and approaches used to facilitate learning. In geriatric nursing education, a variety of pedagogical strategies can be employed to enhance student engagement and learning outcomes. For instance, experiential learning opportunities, such as clinical placements, simulation with service users and community service projects, allow students to gain hands-on experience in geriatric care settings (Jana, 2023; Reis da Silva and Mitchell, 2024b). Additionally, the integration of technology in nursing education, such as online learning platforms and simulation tools, can provide students with flexible and interactive learning experiences that cater to diverse learning styles (Jung, 2023; Reis da Silva and Mitchell, 2024a).

The scope of this chapter will encompass a comprehensive exploration of the importance of adaptability in teaching geriatric nursing, as well as an overview of cognitive and pedagogical strategies that can enhance nursing education. The objectives of this chapter are threefold: first, to highlight the critical need for adaptability in nursing education, particularly in the context of geriatric care; second, to examine various cognitive strategies that can be employed to foster critical thinking and problem-solving skills among nursing students; and third, to explore innovative pedagogical approaches that can enhance student engagement and learning outcomes in geriatric nursing education.

COGNITIVE THEORIES IN GERIATRIC NURSING EDUCATION

Overview of Key Cognitive Theories

Cognitive theories of learning are foundational in understanding how individuals process information, particularly in the context of nursing education, which is crucial for preparing competent geriatric nurses. Among the most influential cognitive theorists are Jean Piaget, Lev Vygotsky, and Benjamin

Bloom, each of whom has contributed significantly to our understanding of learning processes that can be effectively applied to nursing education.

Jean Piaget's theory of cognitive development posits that individuals progress through distinct stages of cognitive growth, which influence their understanding and interaction with the world. Piaget identified four stages: sensorimotor stage (0-2 years old), preoperational stage (2-7 years old), concrete operational stage (7-11 years old), and formal operational stage (12 years and older). In nursing education, recognising these stages allows educators to tailor their teaching strategies to provide training to the formal operational stage, ensuring that complex concepts related to geriatric care are presented in a manner that aligns with their developmental stage. This means that it would be useful, not only to include hands-on experiences and practical applications of theoretical knowledge (concrete operational stage) but also promote experiences to engage in abstract reasoning and critical thinking about complex patient scenarios (formal operational stage) (Nascimento et al., 2021; Hidayat, 2023).

Lev Vygotsky's sociocultural theory emphasises the importance of social interaction and cultural context in cognitive development. Vygotsky introduced the concept of the Zone of Proximal Development (ZPD), which refers to the difference between what a learner can do independently and what they can achieve with guidance (Domínguez-González, 2023). This theory is particularly relevant in nursing education, where collaborative learning and mentorship are essential. By leveraging the ZPD, nursing educators can facilitate peer learning and collaborative problem-solving, allowing students to develop their skills in geriatric care through guided practice and social interaction. Vygotsky's emphasis on the role of language and communication in learning also highlights the importance of effective communication skills in nursing, especially when interacting with older patients who may have varying levels of cognitive function (Domínguez-González, 2023).

In exploring innovative teaching strategies for older adults, particularly in nursing education, it is essential to synthesise key cognitive learning theories, particularly those of Piaget and Vygotsky. These theories provide a foundation for developing effective geriatric nursing curricula that cater to the unique cognitive needs of mature learners (Erbil, 2020). Piaget's cognitive development theory emphasises the stages of maturity in thought processes, while Vygotsky's sociocultural theory highlights the importance of social interaction in cognitive development. Vygotsky's concepts of the Zone of Proximal Development (ZPD) and scaffolding play a crucial role in designing teaching methodologies that enable older adults to engage more effectively with educational material, fostering independence and adaptability in learning environments (Erbil, 2020; Semmar & Al-Thani, 2015). Vygotsky believed that cognitive growth is predominantly a social process, stating that learning occurs when individuals interact with more knowledgeable others (Erbil, 2020). This engagement allows learners to build upon their existing knowledge and skills, moving through their ZPD to gain independence (Stenberg et al., 2024). In the context of nursing education, where collaborative learning is prevalent, Vygotsky's theories are applied through practices such as peer-assisted learning and mentorship, which can be particularly beneficial for older learners who may feel isolated in traditional educational settings. Research indicates that the inclusion of these collaborative methods enhances cognitive outcomes and knowledge retention in nursing students (Higgins et al., 2020; Isaacs & Spencer, 2022).

Bloom's Taxonomy, developed by Benjamin Bloom in the 1950s, categorises educational objectives into a hierarchy of cognitive processes. The original taxonomy included six levels: knowledge, comprehension, application, analysis, synthesis, and evaluation. The revised version, which updates the terminology and structure, includes remembering, understanding, applying, analysing, evaluating, and creating. In geriatric nursing education, Bloom's Taxonomy can be utilised to design learning objectives and assessments that promote higher-order thinking skills. For example, educators can create scenarios that require students to analyse patient data, evaluate treatment options, and create care plans tailored to the unique needs of older population. By structuring learning experiences around these cognitive levels, educators can foster a deeper understanding of geriatric care and encourage critical thinking among nursing students (Ailani, 2023; Verenna et al., 2018).

Application of Cognitive Theories to Geriatric Nursing Education

The application of cognitive theories in geriatric nursing education can significantly enhance the learning experience for nursing students. By integrating Piaget's, Vygotsky's, and Bloom's theories into the curriculum, educators can create a more effective and engaging learning environment that prepares students for the complexities of geriatric care (Nascimento et al., 2021).

Incorporating Piaget's stages of cognitive development into nursing education involves recognising the varying cognitive abilities of students and adapting teaching methods accordingly. For instance, educators can use concrete examples and hands-on activities to help students grasp complex concepts related to ageing and health care. Simulations and role-playing exercises can provide students with opportunities to apply their knowledge in realistic scenarios, thereby reinforcing their understanding and retention of information (Reis da Silva and Mitchell, 2024a). Additionally, educators can encourage reflective practice, allowing students to analyse their experiences and integrate new knowledge into their existing cognitive frameworks (Nascimento et al., 2021; Hidayat, 2023).

Vygotsky's sociocultural theory can be applied in geriatric nursing education through collaborative learning strategies. Group discussions, peer teaching, and mentorship programs can facilitate social interaction and knowledge sharing among students. By working together, students can support each other's learning and develop a deeper understanding of geriatric care concepts. Furthermore, educators can use scaffolding techniques, providing support and guidance as students navigate complex topics related to ageing, such as dementia care or palliative care. This approach not only enhances cognitive development but also fosters a sense of community and collaboration among nursing students (Domínguez-González, 2023; Reis da Silva, 2024d).

Bloom's Taxonomy serves as a valuable framework for designing curriculum and assessments in geriatric nursing education. By aligning learning objectives with the various levels of the taxonomy, educators can ensure that students engage in higher-order thinking and develop critical skills necessary for effective patient care (Ailani, 2023). For example, assessments can be designed to evaluate students' ability to analyse case studies, synthesise information from multiple sources, and create comprehensive care plans for older patients (Reis da Silva, 2024e; Reis da Silva, 2024f). Additionally, educators can incorporate technology, such as online quizzes and interactive simulations, to assess students' understanding at different cognitive levels, providing immediate feedback and opportunities for improvement (Ailani, 2023; Verenna et al., 2018).

The advent of technology also necessitates revisiting these foundational theories. Digital platforms can provide socially interactive environments that promote collaborative learning among students, despite geographical constraints (Reis da Silva and Mitchell, 2024a). For instance, in nursing education, online simulation tools help bridge the gap between theoretical knowledge and practical application, allowing older adults to learn at their own pace while benefiting from interactions with peers and educators (Souza, 2023; Reis da Silva and Mitchell, 2024a). The use of multimedia and instructional designs tailored for older adults is particularly vital due to cognitive changes associated with aging, such as cognitive overload caused by sensory impairments (Nunneley et al., 2020). Incorporating elements of Mayer's Cognitive Theory of Multimedia Learning can alleviate these challenges, enhancing engagement and comprehension in geriatric nursing students (Erbil, 2020).

Fundamental to these innovative practices is the recognition of the growing global demand for effective geriatric care due to demographic shifts toward an ageing population (Reis da Silva, 2023a; Reis da Silva, 2023b). According to the World Health Organization, the number of people aged 60 years and older is expected to reach 2 billion by 2050, marking a significant increase that necessitates an expansion of the geriatric nursing workforce (Muñoz & Letouzé, 2022). This context highlights the crucial need for adaptable teaching strategies that respond not only to age-related learning styles but also to the diverse backgrounds and experiences of older students (Chen et al., 2019). Furthermore, the evolving landscape of nursing education must address various workforce challenges, such as the shortage of qualified geriatric nurses. Policy reforms at both national and international levels aim to enhance nursing graduates' capabilities through innovative curricula that incorporate social constructivist approaches rooted in Vygotskian principles (Erbil, 2020). These changes are vital to produce nursing professionals who are not only proficient in geriatric care but are also adaptable to the unique societal needs of an aging demographic.

In addition, integrating evidence-based best practices into nursing curricula that cater specifically to older adults is essential. Empirical studies indicate that the adoption of problem-based learning, simulation, and cooperative learning significantly benefits nursing students, particularly those engaging with geriatric populations (Ohreen et al., 2021; Shah, 2022; Reis da Silva and Mitchell, 2024a). For example, research has shown that cooperative learning strategies improve motivation and knowledge comprehension among nursing students in aged care settings, reinforcing Vygotsky's assertion that group learning enhances cognitive development (Freitas & Dias, 2023). Moreover, teaching strategies should also be flexible and dynamic, reflecting an understanding of adult learning principles, particularly the need for relevancy and practical application. By designing courses that include real-world scenarios and interprofessional collaboration, nursing educators can better serve their older students while ensuring they are adequately prepared for the complexities of modern geriatric nursing (Smalley et al., 2020; Siregar et al., 2024). This alignment of teaching practices with the cognitive and emotional realities of older adults reinforces a more inclusive educational environment that values experience and practical knowledge.

Importantly, the role of policy in influencing these educational structures cannot be overlooked. With officials recognising the need for specialised geriatric training, there is a growing emphasis on integrating gerontology content into nursing curricula (Ortíz et al., 2023). Educational institutions must respond to these policy changes by developing courses that explicitly prepare nurses for the complex needs of the elderly, thereby ensuring a competent workforce equipped to tackle the challenges posed by an ageing population.

The demographic transitions impacting geriatric nursing education require a proactive approach, embracing innovations and evidence-based practices designed to meet the cognitive needs of older learners while simultaneously addressing workforce development challenges. By implementing Vygotsky's social constructivist theory and Piaget's cognitive development principles, educators can create responsive and meaningful learning experiences that prepare nursing professionals to deliver high-quality care to an increasingly elderly population.

As this discourse advances, the synthesis of cognitive theories with practical nursing pedagogy illustrates the ongoing evolution of educational strategies aimed at fostering adaptability in older adult learners. With a collaborative framework rooted in cognitive and social theories, nursing education can significantly enhance its approach to preparing students for real-world geriatric care challenges.

Cognitive Challenges in Teaching Older Adult Care

Teaching geriatric nursing presents unique cognitive challenges that educators must address to effectively prepare students for the complexities of caring for older adults. One significant challenge is the variability in cognitive function among elderly patients, which can complicate communication and understanding (Hidayat, 2023). Nursing students must learn to navigate these differences and develop strategies for effectively interacting with patients who may have cognitive impairments, such as dementia or delirium. This requires educators to emphasise the importance of empathy, active listening, and adaptability in communication, ensuring that students are equipped to provide patient-centred care (Nascimento et al., 2021; Hidayat, 2023).

Emotional Intelligence (EI) plays a crucial role in addressing the cognitive challenges inherent in teaching geriatric nursing (Reis da Silva, 2025). As educators prepare students to care for older adults, they must cultivate EI skills such as empathy, self-awareness, and emotional regulation (da Silva, 2022; Reis da Silva et al., 2024; Reis da Silva, 2025a). These skills enable nursing students to better understand and respond to the diverse cognitive needs of elderly patients, particularly those with impairments like dementia or delirium. By integrating EI into the curriculum, educators can enhance students' ability to engage in patient-centred communication, fostering a compassionate and adaptable approach to care (Reis da Silva, 2024g; Reis da Silva, 2025b; Reis da Silva, 2025c; Reis da Silva, 2025d). This holistic development is essential for preparing nurses to manage the emotional and cognitive complexities of geriatric care effectively (Reis da Silva, 2024h; Reis da Silva, 2025e). Another cognitive challenge in geriatric nursing education is the need for students to integrate knowledge from various disciplines, including pharmacology, psychology, and sociology, to provide holistic care for older patients. The complexity of geriatric care necessitates a multidisciplinary approach, and students must learn to synthesise information from multiple sources to make informed clinical decisions. Educators can address this challenge by incorporating interdisciplinary case studies and collaborative projects into the curriculum, encouraging students to draw connections between different areas of knowledge and apply them to real-world scenarios (Domínguez-González, 2023).

Additionally, the rapid pace of advancements in healthcare technology and treatment options can overwhelm nursing students, making it difficult for them to stay current with best practices in geriatric care. Educators must emphasise the importance of lifelong learning and professional development, encouraging students to seek out continuing education opportunities and stay informed about emerging trends in geriatric nursing. By fostering a culture of inquiry and critical thinking, educators can help students develop the skills necessary to adapt to the ever-changing landscape of healthcare (Ailani, 2023). Finally, the emotional and psychological aspects of caring for elderly patients can pose cognitive challenges for nursing students. The realities of ageing, illness, and death can be difficult for students to confront, and they may experience feelings of anxiety or inadequacy when faced with these issues. Educators must provide support and resources to help students process these emotions and develop resilience. This can include incorporating discussions about ethical dilemmas in geriatric care, providing opportunities for reflective practice, and promoting self-care strategies to help students manage stress and maintain their well-being (Verenna et al., 2018).

PEDAGOGICAL PRACTICES IN GERIATRIC NURSING

Traditional vs. Innovative Pedagogical Methods

The landscape of nursing education, particularly in the field of geriatric nursing, has undergone significant transformation over the past few decades (Reis da Silva, 2024a; Reis da Silva, 2024c). Traditional pedagogical methods, which often rely on lecture-based instruction and rote memorisation, have been increasingly challenged by innovative approaches that emphasise active learning, critical thinking, and real-world application. Traditional methods typically focus on the transmission of knowledge from instructor to student, where students passively receive information without engaging deeply with the material. This approach can lead to superficial understanding and retention, particularly in a complex field like geriatric nursing, where nuanced understanding of patient care is essential (Thompson et al., 2020; Fox et al., 2015). In contrast, innovative pedagogical methods, such as case-based learning, simulation, and problem-based learning (PBL), encourage students to actively engage with the material, fostering deeper understanding and retention (Reis da Silva and Mitchell, 2024a). These methods promote critical thinking and problem-solving skills, which are crucial for nursing students who will face complex clinical scenarios in their future practice (Reis da Silva and Mitchell, 2024a). For instance, case-based learning allows students to analyse real-life patient cases, encouraging them to apply theoretical knowledge to practical situations. This approach not only enhances understanding but also prepares students for the realities of geriatric care, where they must make informed decisions based on patient needs and circumstances (Fox et al., 2015).

Moreover, innovative pedagogical practices often incorporate interprofessional education (IPE), which emphasises collaboration among students from different healthcare disciplines (Reis da Silva, 2025b). This approach is particularly relevant in geriatric care, where a multidisciplinary team is essential for providing comprehensive care to older adults. Research has shown that IPE enhances students' understanding of team dynamics and improves their ability to work collaboratively, ultimately leading to better patient outcomes (Masud et al., 2022a). By integrating IPE into the curriculum, nursing educators can prepare students to function effectively in diverse healthcare teams, a critical competency in geriatric nursing (Reis da Silva, 2025b). Collaboration among healthcare professionals — including nurses, physicians, and social workers — is vital in providing comprehensive care within interprofessional education (IPE). Effective teamwork enhances patient outcomes and addresses the complex healthcare needs of diverse populations. Studies show that when nursing students learn alongside other healthcare candidates, they develop a better understanding of different roles within a healthcare team, leading to improved communication and collaborative skills (Kleib et al., 2024; Shin et al., 2024; Abujaber et al., 2023). For example, coordinated efforts among team members help create tailored care plans that respect cultural sensitivities and individual patient needs, which is particularly important in geriatric care settings where patient backgrounds may vary significantly (Kleib et al., 2024; Abujaber et al., 2023). Implementing IPE competencies in nursing education prepares students for real-world interactions, fostering mutual respect and shared decision-making among professionals (Abujaber et al., 2023).

Case-Based Learning, Simulation, and Problem-Based Learning

Case-based learning (CBL) is an instructional strategy that uses real or simulated patient cases to facilitate learning. In geriatric nursing education, CBL allows students to explore complex scenarios that reflect the multifaceted nature of geriatric care (Reis da Silva, 2024a). For example, students might analyse a case involving a patient with multiple comorbidities, requiring them to consider various aspects of care, including medication management, psychosocial support, and family dynamics. This method encourages students to think critically and develop clinical reasoning skills, as they must evaluate information, identify problems, and propose appropriate interventions (Haske-Palomino et al., 2023; Nakrem et al., 2019).

The integration of technology in nursing education, particularly through the use of simulations, virtual labs, and AI-driven learning tools, offers novel ways to enhance the training and adaptability of nursing students (Reis da Silva and Mitchell, 2024a). Simulation is another powerful pedagogical tool in nursing education, particularly for geriatric care (Reis da Silva and Mitchell, 2024b). High-fidelity simulations provide students with realistic clinical experiences in a safe environment, allowing them to practice skills and decision-making without the risk of harming real patients (Reis da Silva and Mitchell, 2024a). Research indicates that simulation enhances students' confidence and competence in managing geriatric patients, as they can practice complex procedures and communication skills in a controlled setting (Heeren et al., 2019; Southerland et al., 2021). Additionally, simulation can be tailored to address specific learning objectives, such as managing delirium or conducting comprehensive geriatric assessments, thereby ensuring that students are well-prepared for clinical practice (Hesselink et al., 2019). Simulations, such as virtual clinical environments, allow students to practice clinical scenarios in a safe, controlled setting, facilitating experiential learning that is critical in developing clinical competencies (Makowski et al., 2021). For instance, such technologies enable students to perform procedures virtually, which in turn boosts their confidence and competence in real-life situations (Jung, 2023). Furthermore, the incorporation of adaptive learning platforms driven by AI can personalise educational content based on student performance, helping them bridge knowledge gaps effectively (Jung, 2023; Wynn et al., 2025). These innovations not only adapt to the cognitive needs of diverse learners but also stimulate engagement through interactive and experiential learning methodologies (Makowski et al., 2021).

Problem-based learning (PBL) is an instructional method that centres on students working collaboratively to solve complex, real-world problems. In geriatric nursing education, PBL can be particularly effective in fostering critical thinking and teamwork skills. Students are presented with a clinical problem related to geriatric care and must work together to research, analyse, and propose solutions. This collaborative approach not only enhances learning but also mirrors the interprofessional nature of geriatric care, where nurses must collaborate with other healthcare professionals to provide optimal patient care (Lin, 2024; Miguél et al., 2022). Studies have shown that PBL improves students' self-directed learning skills and their ability to apply knowledge in clinical settings, making it a valuable pedagogical approach in nursing education (Shinohara, 2024).

The significance of cultural competence in nursing education cannot be overstated, as nurses often work in culturally diverse environments where patient backgrounds can profoundly influence care delivery. Cultural competence encompasses the ability to recognise and adapt to varying beliefs, values, and practices among patient populations (Notarnicola et al., 2024; Vázquez-Sánchez et al., 2022). Research indicates that incorporating cultural training into nursing curricula enhances students' capabilities to engage respectfully with patients from different backgrounds, improving therapeutic relationships and health outcomes (Notarnicola et al., 2024).

Furthermore, culturally tailored educational interventions enable nursing students to apply knowledge of cultural practices to care planning, thus ensuring adaptability in their teaching and delivery methods (Notarnicola et al., 2024; Vázquez-Sánchez et al., 2022). Providing real-life scenarios that reflect diverse patient backgrounds helps learners develop the necessary skills to navigate complex cultural dynamics in clinical practice (Vázquez-Sánchez et al., 2022).

Effective assessment methods for evaluating adaptability in nursing education are critical to ensure that future nurses possess the competencies required to meet the evolving demands of healthcare. These assessments can range from direct measures, such as simulations and practical exams, to reflective practices where students analyse their learning experiences and gather feedback from peers (Abujaber et al., 2023; Nurse-Clarke & Sockol, 2022). A study highlighted the use of portfolio assessments as a means for students to document their learning journey, emphasising the importance of adaptability in various clinical situations and fostering critical thinking skills (Abujaber et al., 2023).

Additionally, formative assessments that focus on real-world scenarios and patient interactions can help educators evaluate students' ability to adapt learned concepts to clinical practice effectively (Nurse-Clarke & Sockol, 2022). By employing diverse assessment strategies, nursing educators can identify individual learning needs, thus enhancing instructional methods and curricular designs to better prepare students for the complexities of patient care (Abujaber et al., 2023).

As the demand for geriatric care rises due to increasing life expectancy and an aging population, the relevant trends in global health structures and nursing education are paramount. The current workforce is witnessing demographic shifts, emphasising the need for adaptable and culturally competent nursing practices (Makowski et al., 2021). Policies increasingly stress the importance of integrating interdisciplinary collaboration, technological proficiency, and cultural awareness into nursing curricula, thereby preparing future nurses to handle the complexities associated with aging patients and diverse populations (Kleib et al., 2024). These changes necessitate continuous professional development and adaptability among nursing educators themselves, ensuring that they remain responsive to emerging educational techniques and workforce health strategies (Kleib et al., 2024; Shin et al., 2024).

Role of Technology in Enhancing Pedagogy

The integration of technology in nursing education has revolutionised the way geriatric nursing is taught. Technological advancements have provided new opportunities for enhancing pedagogy, making learning more engaging and accessible. E-learning platforms, virtual simulations, and mobile applications are just a few examples of how technology can support nursing education and improve student outcomes (Reis da Silva, 2024g; Reis da Silva, 2024i).

E-learning platforms offer flexibility and accessibility, allowing students to engage with course materials at their own pace. This is particularly beneficial for adult learners, many of whom may be balancing work, family, and education. Online modules can include interactive elements, such as quizzes and discussion forums, which encourage active participation and facilitate peer learning (Hu et al., 2021; Reis da Silva, 2024a). Additionally, e-learning can be used to supplement traditional classroom instruction, providing students with additional resources and opportunities for self-directed learning.

Virtual simulations are another innovative use of technology in nursing education. These simulations can replicate real-life clinical scenarios, allowing students to practice their skills in a safe and controlled environment. For example, virtual reality (VR) simulations can immerse students in geriatric care situations, enabling them to interact with virtual patients and practice communication and assessment skills (Park & Lim, 2022; Farrell et al., 2018). Research has shown that VR simulations enhance students' learning experiences and improve their confidence in managing geriatric patients (Eklund et al., 2022; Huls et al., 2015).

Mobile applications also play a significant role in enhancing geriatric nursing education (Reis da Silva and Mitchell, 2024a). These apps can provide students with quick access to clinical guidelines, assessment tools, and educational resources, facilitating just-in-time learning. For instance, apps that focus on geriatric syndromes, medication management, and patient assessment can support students in their clinical practice, ensuring they have the necessary information at their fingertips (Mitchell et al., 2018; Reis da Silva, 2024j). Furthermore, mobile technology can facilitate communication and collaboration among students, allowing them to share resources and discuss cases in real-time (Reis da Silva and Mitchell, 2024a).

ADAPTABILITY IN TEACHING STRATEGIES

Definition and Importance of Adaptability in Education

Adaptability in education refers to the ability of educators to modify their teaching methods and strategies in response to the diverse needs of students, changing educational environments, and evolving subject matter (Thompson et al., 2020). This concept is particularly crucial in nursing education, especially in the context of geriatric nursing, where students must be prepared to address the complex and dynamic needs of older adults. The importance of adaptability in education can be underscored by several factors, including the diversity of student learning styles, the rapid advancements in healthcare knowledge, and the necessity for educators to respond to real-time challenges in clinical settings (Mahmud et al., 2020; Thompson et al., 2020).

In geriatric nursing education, adaptability is essential for several reasons. First, older adults present a unique set of health challenges that require nursing students to be flexible in their approach to care (Reis da Silva, 2024a). For instance, students must learn to communicate effectively with patients who may have cognitive impairments or sensory deficits. This necessitates a teaching approach that is responsive to the varying levels of understanding and engagement among students (Masud et al., 2022b). Second, the healthcare landscape is constantly evolving, with new research findings, technologies, and treatment modalities emerging regularly. Educators must be able to integrate these advancements into their curricula to ensure that nursing students are equipped with the most current knowledge and skills (Huh & Shin, 2021).

Furthermore, adaptability in teaching strategies fosters a culture of lifelong learning among nursing students. By modelling flexible teaching practices, educators encourage students to embrace change and seek continuous improvement in their professional development. This is particularly important in geriatric nursing, where ongoing education is vital for staying abreast of best practices and emerging trends in elder care (Fitzpatrick et al., 2023). Overall, adaptability in education not only enhances the learning experience for nursing students but also prepares them to become competent and compassionate caregivers in an increasingly complex healthcare environment.

Strategies for Developing Adaptable Teaching Methods

To cultivate adaptability in teaching methods, educators can employ several strategies that promote flexibility and responsiveness to student needs (Hesselink et al., 2019). One effective approach is the incorporation of differentiated instruction, which involves tailoring teaching methods to accommodate the diverse learning styles, preferences, and abilities of students. This can include offering a variety of instructional modalities, such as lectures, hands-on activities, group discussions, and online resources, to engage students and enhance their understanding of geriatric nursing concepts (Hesselink et al., 2019). Another strategy is the use of formative assessments to gauge student understanding and adjust teaching methods accordingly. By regularly assessing students' knowledge and skills through quizzes, reflections, and practical demonstrations, educators can identify areas where students may be struggling and modify their instructional approaches to address these gaps (Reis da Silva, 2024a). This ongoing feedback loop allows educators to be responsive to student needs and ensures that all learners are progressing in their understanding of geriatric care (Marsden et al., 2022).

Incorporating technology into teaching practices can also enhance adaptability. Online learning platforms, simulation tools, and interactive resources provide educators with the flexibility to deliver content in various formats and engage students in active learning. For example, virtual simulations can allow nursing students to practice their skills in a safe environment, enabling them to adapt their approaches based on real-time feedback and outcomes (Voorend et al., 2021; Reis da Silva and Mitchell, 2024a). Additionally, technology can facilitate collaboration among students, allowing them to share insights and learn from one another's experiences, further enhancing adaptability in the learning process (Abera, 2024).

Professional development opportunities for educators are also crucial for fostering adaptability in teaching methods (Reis da Silva, 2024a). By participating in workshops, conferences, and training sessions focused on innovative teaching strategies and advancements in geriatric care, educators can stay informed about best practices and emerging trends. This ongoing professional growth enables them to adapt their teaching approaches to meet the evolving needs of their students and the healthcare landscape (Benko & Peršolja, 2023).

Examples of Adaptive Teaching in Geriatric Nursing

Adaptive teaching strategies can be observed in various aspects of geriatric nursing education. One prominent example is the use of case-based learning, where students analyse real-life scenarios involving older adults. This method encourages students to apply their theoretical knowledge to practical situations, fostering critical thinking and problem-solving skills (Haske-Palomino et al., 2023; Nakrem et al., 2019). By adapting case studies to reflect current issues in geriatric care, educators can ensure that students are engaged and learning relevant content (Grootven et al., 2021).

Simulation-based learning is another effective adaptive teaching strategy in geriatric nursing. High-fidelity simulations allow students to practice their skills in a controlled environment, where they can receive immediate feedback on their performance (Reis da Silva and Mitchell, 2024a). For instance, nursing students may participate in simulations that involve managing complex patient scenarios, such as a geriatric patient experiencing a fall or a sudden change in mental status (Reis da Silva, 2023b;

Reis da Silva, 2024k). This hands-on experience enables students to adapt their clinical reasoning and decision-making skills in real-time, preparing them for the challenges they will face in clinical practice (Kim et al., 2021).

IPE is also a valuable approach to adaptive teaching in geriatric nursing. By collaborating with students from other healthcare disciplines, nursing students can gain a broader understanding of the multidisciplinary nature of geriatric care (Masud et al., 2022b). For example, joint learning experiences with medical, pharmacy, and social work students can help nursing students appreciate the importance of teamwork and communication in providing comprehensive care for older adults. This adaptability in learning environments prepares students to work effectively in diverse healthcare teams (Doležalová et al., 2022).

Additionally, educators can promote adaptability by incorporating reflective practice into the curriculum. Encouraging students to reflect on their experiences, challenges, and successes in geriatric nursing fosters a growth mindset and helps them develop the ability to adapt their approaches based on feedback and self-assessment. This reflective process is essential for cultivating lifelong learning and adaptability in nursing practice (Haske-Palomino et al., 2023; Reis da Silva, 2024d).

PROMOTING CRITICAL THINKING AND PROBLEM-SOLVING

Techniques to Foster Critical Thinking in Nursing Students

Critical thinking is a vital skill for nursing students, particularly in geriatric care, where complex patient needs and multifaceted health issues require sound clinical judgment. To foster critical thinking, educators can employ several techniques that encourage students to analyse, evaluate, and synthesize information effectively (Fox et al., 2015). One effective technique is the use of Socratic questioning, which involves posing open-ended questions that challenge students to think deeply about their responses (Fox et al., 2015). By encouraging students to articulate their reasoning and consider alternative perspectives, educators can promote a culture of inquiry and critical analysis. For example, educators might ask students to evaluate the potential outcomes of different treatment options for an elderly patient with multiple comorbidities, prompting them to consider the implications of their choices (Fox et al., 2015).

Another technique is the integration of reflective journaling into the curriculum. By encouraging students to document their thoughts, experiences, and learning processes, educators can help them develop metacognitive skills that enhance critical thinking (Reis da Silva, 2025a). Reflective journaling allows students to analyse their decision-making processes, identify areas for improvement, and connect theoretical knowledge to practical experiences in geriatric care (Li et al., 2022).

Simulation-based learning also plays a crucial role in fostering critical thinking among nursing students (Reis da Silva and Mitchell, 2024a). By engaging in realistic scenarios that mimic clinical situations, students are challenged to apply their knowledge and make decisions in real-time. For instance, a simulation involving a geriatric patient with acute confusion may require students to assess the patient's condition, consider potential causes, and develop an appropriate care plan (Reis da Silva, 2024e).

This experiential learning approach enhances critical thinking skills by requiring students to analyse complex information and make informed decisions under pressure (Detroyer et al., 2018). Group discussions and collaborative learning activities can further promote critical thinking. By working together to solve problems and analyse case studies, students can share diverse perspectives and learn from one another's insights. This collaborative approach not only enhances critical thinking but also fosters teamwork and communication skills, which are essential in geriatric nursing (Fahim, 2023).

Problem-Solving Approaches in Geriatric Care Education

Problem-solving is a critical competency for nursing students, particularly in geriatric care, where patients often present with complex and multifaceted health issues. Educators can employ several approaches to enhance problem-solving skills among nursing students (Haske-Palomino et al., 2023; Nakrem et al., 2019).

One effective approach is the use of problem-based learning (PBL), where students are presented with real-world clinical problems and work collaboratively to develop solutions. In geriatric nursing education, PBL can involve case studies that reflect common challenges faced by older adults, such as managing polypharmacy or addressing functional decline. By engaging in PBL, students learn to analyse patient data, consider multiple factors influencing health outcomes, and develop comprehensive care plans (Pavithra et al., 2023). Another approach is the use of clinical reasoning frameworks, which provide students with structured methods for approaching complex patient scenarios (Reis da Silva, 2024f).

For example, the "ABCDE" framework (Airway, Breathing, Circulation, Disability, Exposure) can guide students in assessing geriatric patients systematically or the Clinical Reasoning Levett-Jones cycle to encourage students to develop their problem-solving skills and enhance their ability to prioritise care based on patient needs (Switzer & Schroder, 2019; Reis da Silva, 2024f).

Educators can also incorporate case studies that require students to navigate ethical dilemmas in geriatric care. For instance, a case study involving a patient with advanced dementia and a family requesting aggressive treatment can prompt students to consider the ethical implications of their decisions. This approach encourages critical thinking and problem-solving by requiring students to weigh the benefits and burdens of different treatment options (Lucchetti et al., 2019).

Case Studies and Practical Applications

Case studies serve as valuable tools for promoting critical thinking and problem-solving in geriatric nursing education (Ghamgosar et al., 2021). By analysing real-life scenarios, students can apply their theoretical knowledge to practical situations, enhancing their understanding of complex patient care issues. For example, a case study involving a geriatric patient with chronic kidney disease may require students to assess the patient's symptoms, review their medication regimen, and develop a comprehensive care plan that addresses both medical and psychosocial needs (Reis da Silva, 2024e; Reis da Silva, 2024l). This practical application of knowledge encourages students to think critically about the patient's condition and consider the broader context of their care (Hakvoort et al., 2021).

Additionally, case studies can be used to facilitate interprofessional collaboration among nursing students and their peers from other healthcare disciplines. By working together to analyse a case, students can gain insights into the roles and responsibilities of different team members in geriatric care. This collaborative approach not only enhances problem-solving skills but also prepares students for the realities of working in multidisciplinary teams (Ghamgosar et al., 2021).

CHALLENGES IN IMPLEMENTING ADAPTABLE TEACHING

Institutional and Curricular Barriers

Implementing adaptable teaching strategies in nursing education, particularly in geriatric nursing, faces several institutional and curricular barriers (Reis da Silva, 2025f). One significant challenge is the rigidity of existing curricula, which often prioritise standardized content delivery over flexible teaching methods (Reis da Silva, 2024a). Many nursing programs are bound by accreditation requirements and established learning outcomes that may not easily accommodate innovative teaching approaches (Reis da Silva, 2024a). This rigidity can stifle creativity and limit educators' ability to tailor their instruction to meet the diverse needs of students and the complexities of geriatric care (Ahouah & Rothan-Tondeur, 2019).

Additionally, institutional policies and administrative structures can hinder the adoption of adaptable teaching methods. For instance, faculty may be required to adhere to specific teaching formats or assessment methods that do not allow for flexibility. This can create a disconnect between the teaching strategies employed and the actual learning needs of students, particularly in a field as dynamic as geriatric nursing, where patient care scenarios can vary widely (Sinvani et al., 2016; Thompson et al., 2020). Furthermore, limited resources, such as access to technology and training for faculty, can impede the implementation of innovative teaching practices that promote adaptability (Thompson et al., 2020).

Another barrier is the lack of interprofessional collaboration within nursing education. Geriatric care often requires a multidisciplinary approach, yet many nursing programs operate in silos, limiting opportunities for students to engage with peers from other healthcare disciplines. This lack of collaboration can restrict the development of adaptable teaching strategies that incorporate diverse perspectives and expertise, ultimately impacting the quality of education provided to nursing students (Thompson et al., 2020; Zambrotta et al., 2021).

Resistance to Change Among Educators

Resistance to change among educators is another significant challenge in implementing adaptable teaching strategies. Many nursing faculty members may be accustomed to traditional teaching methods and may feel uncomfortable or unprepared to adopt new approaches (Song & Tang, 2019). This resistance can stem from a lack of familiarity with innovative pedagogical practices, fear of the unknown, or concerns about the effectiveness of new methods (King et al., 2017; Song & Tang, 2019). Additionally, educators may feel overwhelmed by the demands of their roles, including teaching, research, and clinical practice, leaving little time or energy to explore and implement new teaching strategies (Marsden et al., 2022).

Moreover, the perception that adaptable teaching methods require more effort and resources can deter educators from making changes (Masud et al., 2022a). For instance, implementing simulation-based learning or case-based discussions may necessitate additional preparation time, training, and resources, which some educators may view as burdensome (Shen, 2024; Masud et al., 2022a). This reluctance to embrace change can perpetuate a cycle of traditional teaching practices, limiting the potential for innovation and improvement in nursing education.

Solutions and Best Practices

To overcome these challenges, several solutions and best practices can be implemented. First, institutions should foster a culture of continuous professional development for nursing educators (Fitzpatrick et al., 2023). Providing training and resources on innovative teaching methods, such as adaptable teaching strategies, can empower faculty to embrace change and enhance their instructional practices (Duan, 2024; Liao, 2024). Workshops, seminars, and peer mentoring programs can facilitate knowledge sharing and support educators in their efforts to implement new approaches.

Second, nursing programs should prioritise flexibility in their curricula. This can be achieved by allowing educators to modify course content and delivery methods based on student needs and emerging trends in geriatric care (Silva et al., 2021). Institutions can also encourage interdisciplinary collaboration by creating opportunities for nursing students to engage with peers from other healthcare disciplines, fostering a more holistic approach to geriatric education (Hesselink et al., 2019; Silva et al., 2021).

Finally, institutions should actively involve educators in the decision-making process regarding curriculum development and teaching practices. By soliciting input from faculty members, institutions can create a sense of ownership and investment in the implementation of adaptable teaching strategies (Gcawu & Rooyen, 2022). This collaborative approach can help address concerns and resistance to change, ultimately leading to a more dynamic and responsive educational environment (Gcawu & Rooyen, 2022).

CASE STUDIES OF ADAPTIVE TEACHING IN GERIATRIC NURSING

Analysis of Successful Adaptive Teaching Models

Several case studies illustrate the successful implementation of adaptive teaching models in geriatric nursing education. One notable example is the use of simulation-based learning in a nursing program that focuses on geriatric care (Reis da Silva and Mitchell, 2024a). In this program, students engage in high-fidelity simulations that replicate real-life scenarios involving older adults, such as managing chronic illnesses or addressing cognitive impairments. The simulations are designed to be adaptable, allowing educators to modify scenarios based on student performance and learning needs. This approach has been shown to enhance students' clinical reasoning skills and confidence in managing complex patient situations (Wissanji et al., 2022; Meiers & Eggenberger, 2018).

Another successful model involves the integration of interprofessional education (IPE) into the geriatric nursing curriculum (Saripella et al., 2021). In this case study, nursing students collaborate with students from other healthcare disciplines, such as medicine, pharmacy, and social work, to address the needs of older adults. Through joint case discussions and collaborative projects, students learn to appreciate the value of teamwork and communication in providing comprehensive care. This adaptive teaching model not only enhances students' understanding of geriatric care but also prepares them for real-world practice in multidisciplinary teams (Saripella et al., 2021).

Lessons Learned from Case Studies

The analysis of these case studies reveals several key lessons for nursing education. First, the importance of flexibility in teaching methods cannot be overstated (Engel et al., 2016; Zúñiga et al., 2019). Adaptive teaching models that allow for real-time modifications based on student performance and engagement lead to more effective learning outcomes. Educators should be encouraged to embrace flexibility and creativity in their instructional practices to meet the diverse needs of nursing students (Engel et al., 2016; Zúñiga et al., 2019). Second, the integration of interprofessional education is essential for preparing nursing students for the complexities of geriatric care. By collaborating with peers from other disciplines, nursing students gain valuable insights into the roles and responsibilities of various healthcare providers, enhancing their ability to work effectively in multidisciplinary teams (Cheng, 2024; Mariano, 2023). This collaborative approach not only improves students' understanding of geriatric care but also fosters a culture of teamwork and communication that is vital in healthcare settings (Engel et al., 2016; Zúñiga et al., 2019).

Finally, ongoing assessment and feedback are crucial components of successful adaptive teaching models. Regularly evaluating student performance and soliciting feedback on teaching methods can help educators identify areas for improvement and make necessary adjustments to their instructional practices (Ma et al., 2022). This continuous improvement process ensures that teaching strategies remain relevant and effective in meeting the evolving needs of nursing students (Ma et al., 2022).

Implications for Nursing Education

The implications of these case studies for nursing education are significant. As the demand for geriatric care continues to grow, nursing programs must adapt their curricula and teaching methods to prepare students for the unique challenges of this field (Muntinga et al., 2015). By embracing adaptive teaching strategies, nursing educators can enhance students' critical thinking, problem-solving, and clinical reasoning skills, ultimately leading to improved patient outcomes in geriatric care (Muntinga et al., 2015; Luo, 2023). Furthermore, the integration of interprofessional education into nursing curricula can foster a more holistic approach to geriatric care, preparing students to collaborate effectively with other healthcare providers. This collaborative mindset is essential for addressing the complex needs of older adults and ensuring that they receive comprehensive, patient-centred care (Zúñiga et al., 2022).

Future Trends in Nursing Education

Several trends are emerging in nursing education that may influence the future of adaptable teaching in geriatric nursing. One significant trend is the increasing emphasis on competency-based education, which focuses on developing specific skills and competencies required for effective nursing practice (Abera, 2024). This approach aligns well with adaptable teaching strategies, as it allows educators to tailor instruction to meet the unique needs of students and the demands of the healthcare environment (Marsden et al., 2022; Abera, 2024). Another trend is the growing integration of technology in nursing education. The use of virtual simulations, online learning platforms, and digital resources is becoming more prevalent, providing educators with new tools to enhance adaptability in teaching (Reis da Silva, 2025f). As technology continues to advance, nursing programs will need to explore innovative ways to incorporate these resources into their curricula (Wilson et al., 2021; Wu, 2024). Additionally, the focus on interprofessional education is likely to expand in nursing programs (Cheng, 2024). As healthcare becomes increasingly collaborative, nursing students will benefit from learning alongside peers from other disciplines. This trend will not only enhance adaptability but also prepare students for the complexities of geriatric care in multidisciplinary teams (Zhong et al., 2021; Cheng, 2024).

CONCLUSION

Summary of Key Points

In summary, adaptability in teaching is essential for preparing nursing students to meet the challenges of geriatric care. The importance of adaptable teaching strategies lies in their ability to address the diverse needs of students, respond to the complexities of geriatric care, and foster a culture of lifelong learning. By embracing flexible teaching methods, incorporating active learning, and using technology, nursing educators can create engaging and effective learning environments.

Furthermore, the challenges associated with implementing adaptable teaching, including institutional barriers and resistance to change among educators, must be addressed through targeted interventions and professional development. Research into the effectiveness of adaptive teaching strategies and the integration of interprofessional education will provide valuable insights for enhancing nursing education.

Final Thoughts on Enhancing Geriatric Nursing Education

Enhancing geriatric nursing education through adaptable teaching strategies is crucial for preparing the next generation of nursing professionals. By embracing flexibility, fostering collaboration, and integrating technology, educators can create dynamic learning environments that empower students to thrive in their future careers. As the demand for geriatric care continues to grow, it is imperative that nursing programs prioritise adaptability in their curricula, ensuring that graduates are well-prepared to meet the unique challenges of caring for older adults.

Practical Tips for Educators

To effectively implement adaptable teaching strategies in geriatric nursing education, educators can consider the following practical tips:

1. Embrace Flexibility: Be open to modifying lesson plans and teaching methods based on student feedback and performance. Flexibility allows educators to respond to the unique needs of their students and the dynamic nature of geriatric care (Hesselink et al., 2019).
2. Incorporate Active Learning: Use active learning techniques, such as case-based learning, simulations, and group discussions, to engage students and promote critical thinking. These methods encourage students to apply their knowledge in real-world scenarios, enhancing their adaptability (Reis da Silva and Mitchell, 2024a).
3. Foster a Collaborative Learning Environment: Create opportunities for students to work together and learn from one another. Collaborative learning not only enhances adaptability but also prepares students for the teamwork required in geriatric care (Fox et al., 2015).

4. Use Technology: Integrate technology into teaching practices to enhance engagement and accessibility. Online resources, simulations, and interactive platforms can provide students with diverse learning experiences that cater to different learning styles (Zúñiga et al., 2019).
5. Encourage Reflection: Incorporate reflective practices into the curriculum, allowing students to analyse their experiences and identify areas for improvement. Reflection fosters a growth mindset and encourages students to adapt their approaches based on feedback and self-assessment (Li et al., 2022).

REFERENCES

Abera, W., Daba, A. K., Gebre, B. B., Ashagere, M., Yeheyis, T., Addisu, D., Amlaku, T., Tadesse, F., & Samuel, T. (2024). Knowledge and attitude toward geriatric nursing care and associated factors among nurses working at hospitals in Hawassa city, Ethiopia, 2022. *Frontiers in Medicine*, *11*, 1284845. Advance online publication. DOI: 10.3389/fmed.2024.1284845 PMID: 38912339

Abujaber, A., Abd-Alrazaq, A., Al-Qudimat, A., & Nashwan, A. (2023). A strengths, weaknesses, opportunities, and threats (swot) analysis of ChatGPT integration in nursing education: A narrative review. *Cureus*. Advance online publication. DOI: 10.7759/cureus.48643 PMID: 38090452

Ahouah, M., & Rothan-Tondeur, M. (2019). End-users and caregivers' involvement in health interventional research carried out in geriatric facilities: A systematic review. *International Journal of Environmental Research and Public Health*, *16*(16), 2812. DOI: 10.3390/ijerph16162812 PMID: 31394752

Ailani, A. (2023). Examining the impact of bloom's taxonomy integration in entrepreneurship education on the competitiveness and performance of micro, small, and medium enterprises (MSMES). *Makreju*, *1*(3), 181–188. DOI: 10.55606/makreju.v1i3.2924

Benko, E., & Peršolja, M. (2023). Nursing students' views of the impact of geriatric role-play workshops on professional competencies: Survey. *BMC Nursing*, *22*(1), 203. Advance online publication. DOI: 10.1186/s12912-023-01373-y PMID: 37316872

Chen, W., Shah, U., & Brechtelsbauer, C. (2019). A framework for hands-on learning in chemical engineering education—Training students with the end goal in mind. *Education for Chemical Engineers*, *28*, 25–29. DOI: 10.1016/j.ece.2019.03.002

Cheng, Y., Sun, S., Hu, Y., Wang, J., Chen, W., Miao, Y., & Wang, H. (2024). Effects of different geriatric nursing teaching methods on nursing students' knowledge and attitude: Systematic review and network meta-analysis. *PLoS One*, *19*(5), e0300618. DOI: 10.1371/journal.pone.0300618 PMID: 38820259

da Silva, T. H. R. (2022). Emotional awareness and emotional intelligence. *British Journal of Community Nursing*, *27*(12), 573–574. DOI: 10.12968/bjcn.2022.27.12.573 PMID: 36519463

Detroyer, E., Dobbels, F., Teodorczuk, A., Depaifve, Y., Joosten, E., & Milisen, K. (2018). Effect of an interactive e-learning tool for delirium on patient and nursing outcomes in a geriatric hospital setting: Findings of a before-after study. *BMC Geriatrics*, *18*(1), 19. Advance online publication. DOI: 10.1186/s12877-018-0715-5 PMID: 29351772

Doležalová, J., Tóthová, V., & Neugebauer, J. (2022). Selected aspects of nursing practice related to the usage of evaluation tools for population aged over 60 years. *Kontakt*, *24*(2), 131–136. DOI: 10.32725/kont.2021.044

Domínguez-González, M., Hervás-Gómez, C., Díaz-Noguera, M. D., & Reina-Parrado, M. (2023). Attention to diversity from artificial intelligence. *The European Educational Researcher*, *6*(3), 101–115. DOI: 10.31757/euer.633

Duan, A., Jiang, F., Li, L., Li, Q., & Chen, W. (2024). Design and practice of blended teaching of internal medicine nursing based on o-amas effective teaching model. *BMC Medical Education*, *24*(1), 580. Advance online publication. DOI: 10.1186/s12909-024-05588-8 PMID: 38807140

Eklund, A., Carlson, E., Sundin-Andersson, C., & Olsson, C. (2022). Experiences of the clinical learning environment in peer learning: A mixed method study from bachelor students' and main preceptors' perspectives. *Nordic Journal of Nursing Research*, *43*(1), 20571585221126597. Advance online publication. DOI: 10.1177/20571585221126597

Engel, P., Spencer, J., Paul, T., & Boardman, J. (2016). The geriatrics in primary care demonstration: integrating comprehensive geriatric care into the medical home: preliminary data. *Journal of the American Geriatrics Society*, *64*(4), 875–879. DOI: 10.1111/jgs.14026 PMID: 27100583

Erbil, D. (2020). A review of flipped classroom and cooperative learning method within the context of vygotsky theory. *Frontiers in Psychology*, *11*, 1157. Advance online publication. DOI: 10.3389/fpsyg.2020.01157 PMID: 32655437

Fahim, R., Boughdady, A., & Meawad, E. (2023). Knowledge and attitude of nurses toward geriatric palliative care. *Mansoura Nursing Journal*, *10*(2), 307–311. DOI: 10.21608/mnj.2023.340395

Farrell, T., Luptak, M., Supiano, K., Pacala, J., & Lisser, R. (2018). State of the science: Interprofessional approaches to aging, dementia, and mental health. *Journal of the American Geriatrics Society*, *66*(S1). Advance online publication. DOI: 10.1111/jgs.15309 PMID: 29659009

Fitzpatrick, J. M., Bianchi, L. A., Hayes, N., Da Silva, T., & Harris, R. (2023). Professional development and career planning for nurses working in care homes for older people: A scoping review. *International Journal of Older People Nursing*, *18*(1), e12519. DOI: 10.1111/opn.12519 PMID: 36441621

Fox, M., Butler, J., Persaud, M., Tregunno, D., Sidani, S., & McCague, H. (2015). A multi-method study of the geriatric learning needs of acute care hospital nurses in Ontario, Canada. *Research in Nursing & Health*, *39*(1), 66–76. DOI: 10.1002/nur.21699 PMID: 26471253

Freitas, C., & Dias, R. (2023). Professional training in industrial maintenance: A social interactionist approach. *International Journal of Mechanical Engineering Education*, *52*(3), 288–314. DOI: 10.1177/03064190231188866

Gcawu, S., & Rooyen, D. (2022). Clinical teaching practices of nurse educators: An integrative literature review. *Health SA = SA Gesondheid*, *27*. Advance online publication. DOI: 10.4102/hsag.v27i0.1728 PMID: 36262921

Ghamgosar, A., Zarghani, M., & Nemati-Anaraki, L. (2021). Bibliometric analysis on geriatric nursing research in web of science (1900–2020). *Biomed Research International*, 2021(1). DOI: 10.1155/2021/8758161

Grootven, B., Jeuris, A., Jonckers, M., Devriendt, E., Casterlé, B., Dubois, C., & Flamaing, J. (2021). Geriatric co-management for cardiology patients in the hospital: A quasi-experimental study. *Journal of the American Geriatrics Society*, *69*(5), 1377–1387. DOI: 10.1111/jgs.17093 PMID: 33730373

Hakvoort, L., Dikken, J., Wel, M., Derks, C., & Schuurmans, M. (2021). Minimizing the knowledge-to-action gap; identification of interventions to change nurses' behavior regarding fall prevention, a mixed method study. *BMC Nursing*, *20*(1), 80. Advance online publication. DOI: 10.1186/s12912-021-00598-z PMID: 34016106

Haske-Palomino, M., Carlson, R., & Gatley, L. (2023). Elder veteran program: A description of a nurse practitioner–led inpatient geriatric consult model. *Journal of the American Association of Nurse Practitioners*, *35*(4), 229–234. DOI: 10.1097/JXX.0000000000000844 PMID: 36857530

Heeren, P., Devriendt, E., Fieuws, S., Wellens, N., Flamaing, J., Sabbé, M., & Milisen, K. (2019). Unplanned readmission prevention by a geriatric emergency network for transitional care (urgent): A prospective before-after study. *BMC Geriatrics*, *19*(1), 215. Advance online publication. DOI: 10.1186/s12877-019-1233-9 PMID: 31390994

Hesselink, G., Demirbas, M., Rikkert, M., & Schoon, Y. (2019). Geriatric education programs for emergency department professionals: A systematic review. *Journal of the American Geriatrics Society*, *67*(11), 2402–2409. DOI: 10.1111/jgs.16067 PMID: 31335964

Hidayat, Z. (2023). Prosumers of hybrid entertainment-educational communication in a media convergence environment. *Journal of Biomedical Research & Environmental Sciences*, *4*(9), 1281–1289. DOI: 10.37871/jbres1797

Higgins, D., Hayes, M., Taylor, J., & Wallace, J. (2020). How do we teach simulation-based dental education? time for an evidence-based, best-practice framework. *European Journal of Dental Education*, *24*(4), 815–821. DOI: 10.1111/eje.12551 PMID: 32521066

Hu, F., Lee, H., & Li, Y. (2021). Exploration of geriatric care competencies in registered nurses in hospitals. *The Journal of Nursing Research*, *29*(4), e159. DOI: 10.1097/JNR.0000000000000441 PMID: 34034305

Huh, A., & Shin, J. (2021). Person-centered care practice, patient safety competence, and patient safety nursing activities of nurses working in geriatric hospitals. *International Journal of Environmental Research and Public Health*, *18*(10), 5169. DOI: 10.3390/ijerph18105169 PMID: 34068125

Huls, M., Rooij, S., Diepstraten, A., Koopmans, R., & Helmich, E. (2015). Learning to care for older patients: Hospitals and nursing homes as learning environments. *Medical Education*, *49*(3), 332–339. DOI: 10.1111/medu.12646 PMID: 25693992

Isaacs, A., & Spencer, B. (2022). Factors affecting learning during internship: A qualitative study of junior doctors. *Australian Journal of General Practice*, *51*(1-2), 83–89. DOI: 10.31128/AJGP-02-21-5821 PMID: 35098280

Jana, V., Radizuan, T. H., & Maldisa, G. S. (2023). Qualitative perspectives of new graduate nurses to educational preparation for patient care. *Malahayati International Journal of Nursing and Health Science*, *6*(4), 330–337. DOI: 10.33024/minh.v6i4.12611

Jung, S. (2023). Challenges for future directions for artificial intelligence integrated nursing simulation education. *Korean Journal of Women Health Nursing*, *29*(3), 239–242. DOI: 10.4069/kjwhn.2023.09.06.1 PMID: 37813667

Kim, B., Lee, M., & Jang, S. (2021). Hospital nurses' empathy and moral sensitivity toward elderly care: A cross-sectional study. *Journal of Nursing Management, 30*(7), 2138–2146. DOI: 10.1111/jonm.13442 PMID: 34350659

King, A., Boyd, M., Dagley, L., & Raphael, D. (2017). Implementation of a gerontology nurse specialist role in primary health care: Health professional and older adult perspectives. *Journal of Clinical Nursing, 27*(3-4), 807–818. DOI: 10.1111/jocn.14110 PMID: 29052288

Kleib, M., Arnaert, A., Nagle, L., Sugars, R., & Costa, D. (2024). Newly qualified Canadian nurses' experiences with digital health in the workplace: Comparative qualitative analysis. *JMIR Medical Education, 10*, e53258. DOI: 10.2196/53258 PMID: 39159452

Leffers, J., McDermott-Levy, R., Nicholas, P., & Sweeney, C. (2017). Mandate for the nursing profession to address climate change through nursing education. *Journal of Nursing Scholarship, 49*(6), 679–687. DOI: 10.1111/jnu.12331 PMID: 28806483

Li, Q., Bu, Z., Xue, M., Huang, A., Tu, W., & Xu, G. (2022). Visual analysis of research hotspots in geriatric nursing education in China and abroad. *Journal of Healthcare Engineering, 2022*, 1–7. DOI: 10.1155/2022/7472188 PMID: 35480150

Liao, L., Bai, X., He, X., Tan, L., Yang, L., Long, H., Huang, S., Li, X., Han, Y., Peng, X., Chen, H., Yang, X., Li, H., Wang, S., & Liu, Y. (2024). Identification of formation mechanism and key elements of quality geriatric care behavior of nursing assistants in nursing homes: A grounded theory study. *Frontiers in Public Health, 12*, 1425883. Advance online publication. DOI: 10.3389/fpubh.2024.1425883 PMID: 38993701

Lin, P., Wung, S.-F., Lin, P.-C., Lin, Y.-C., Lin, C.-Y., & Huang, H.-L. (2024). Virtual reality-based simulation learning on geriatric oral health care for nursing students: A pilot study. *BMC Oral Health, 24*(1), 627. Advance online publication. DOI: 10.1186/s12903-024-04249-y PMID: 38807116

Lucchetti, A., Duarte, B., Assis, T., Laurindo, B., & Lucchetti, G. (2019). Is it possible to teach geriatric medicine in a stimulating way? measuring the effect of active learning activities in Brazilian medical students. *Australasian Journal on Ageing, 38*(2). Advance online publication. DOI: 10.1111/ajag.12626 PMID: 30761707

Luo, Y. (2023). Construction and application of an interdisciplinary team teaching model for foreign-related nursing English in higher vocational colleges from the perspective of ecolinguistics. *Journal of Contemporary Educational Research, 7*(11), 203–208. DOI: 10.26689/jcer.v7i11.5666

Ma, W., Zhang, L., Wang, C., & Xu, X. (2022). Effect of multidisciplinary team collaborative nursing model combined with mind mapping teaching method on postoperative complications and mental health of patients with advanced pancreatic cancer. *Iranian Journal of Public Health*. Advance online publication. DOI: 10.18502/ijph.v51i3.8929 PMID: 35865063

Mahmud, S., Begum, M., & Akhter, J. (2020). Knowledge and attitude of senior staff nurses regarding geriatrics health care in a tertiary public hospital. *Asian Journal of Medical and Biological Research, 6*(3), 431–439. DOI: 10.3329/ajmbr.v6i3.49791

Makowski, A., Hampton, D., Hardin-Pierce, M., Scott, L., Garritano, N., Tovar, E., & Melander, S. (2021). Adapting clinical experiences during a pandemic: One college of nursing's approach. *Journal of the American Association of Nurse Practitioners*, *34*(2), 254–260. DOI: 10.1097/JXX.0000000000000656 PMID: 34537798

Mariano, M. (2023). Assessment of learning style preferences of saudi nursing students. *Ibnosina Journal of Medicine and Biomedical Sciences*, 16(01), 010-016. DOI: 10.1055/s-0043-1776016

Marsden, E., Craswell, A., Taylor, A., Barnett, A., Wong, P., & Wallis, M. (2022). Translation of the geriatric emergency department intervention into other emergency departments: A post implementation evaluation of outcomes for older adults. *BMC Geriatrics*, *22*(1), 290. Advance online publication. DOI: 10.1186/s12877-022-02999-4 PMID: 35392828

Masud, T., Blundell, A., & Stuck, A. (2022a). Geriatric medicine learning objectives and entrustable professional activities in undergraduate medical curricula: A scoping review. *Age and Ageing*, *51*(5), afac100. Advance online publication. DOI: 10.1093/ageing/afac100 PMID: 35536879

Masud, T., Ogliari, G., Lunt, E., Blundell, A., Gordon, A., Roller-Wirnsberger, R., Vassallo, M., Mari, D., Kotsani, M., Singler, K., Romero-Ortuno, R., Cruz-Jentoft, A. J., & Stuck, A. (2022b). A scoping review of the changing landscape of geriatric medicine in undergraduate medical education: Curricula, topics and teaching methods. *European Geriatric Medicine*, *13*(3), 513–528. DOI: 10.1007/s41999-021-00595-0 PMID: 34973151

Meiers, S., Eggenberger, S., & Krumwiede, N. (2018). Development and implementation of a family-focused undergraduate nursing curriculum: Minnesota state university, Mankato. *Journal of Family Nursing*, *24*(3), 307–344. DOI: 10.1177/1074840718787274 PMID: 30101655

Miguél, M., Gómara, A., Izagirre-Otaegi, A., Elguea-Díaz, F., Badiola-Etxaburu, I., & Gómez-Gastiasoro, A. (2022). Development, application and evaluation of an active learning methodology for health science students, oriented towards equity and cultural diversity in the treatment and care of geriatric patients. *International Journal of Environmental Research and Public Health*, *19*(21), 14573. DOI: 10.3390/ijerph192114573 PMID: 36361451

Mitchell, H., Lucas, C., Charlton, K., & McMahon, A. (2018). Models of nutrition-focused continuing education programs for nurses: A systematic review of the evidence. *Australian Journal of Primary Health*, *24*(2), 101. DOI: 10.1071/PY17088 PMID: 29362024

Mukattil, N., & Pangandaman, H. (2024). Barriers and strategies in the implementation of virtual learning modalities amid covid-19 among students in nursing schools at sulu province, philippines. *Critical Medical and Surgical Nursing Journal*, *13*(1), 20–25. DOI: 10.20473/cmsnj.v13i1.51366

Muñoz, J., & Letouzé, P. (2022). Some considerations on the principles of the cognitive theory of multimedia learning for instructional video design for the elderly. *Research Society and Development*, *11*(10), e499111032333. DOI: 10.33448/rsd-v11i10.32333

Muntinga, M., Leeuwen, K., Schellevis, F., Nijpels, G., & Jansen, A. (2015). From concept to content: Assessing the implementation fidelity of a chronic care model for frail, older people who live at home. *BMC Health Services Research, 15*(1), 18. Advance online publication. DOI: 10.1186/s12913-014-0662-6 PMID: 25608876

Nakrem, S., Stensvik, G., Skjong, R., & Ostaszkiewicz, J. (2019). Staff experiences with implementing a case conferencing care model in nursing homes: A focus group study. *BMC Health Services Research, 19*(1), 191. Advance online publication. DOI: 10.1186/s12913-019-4034-0 PMID: 30917815

Nascimento, J., Siqueira, T., Oliveira, J., Alves, M., Regino, D., & Dalrí, M. (2021). Development of clinical competence in nursing in simulation: The perspective of bloom's taxonomy. *Revista Brasileira de Enfermagem, 74*(1), e20200135. Advance online publication. DOI: 10.1590/0034-7167-2020-0135 PMID: 33787788

Nashwan, A., Waghmare, J., & Ladd, E. (2020). How will the covid-19 pandemic reshape nursing education globally? *Open Journal of Nursing, 10*(10), 973–976. DOI: 10.4236/ojn.2020.1010068

Notarnicola, I., Duka, B., Lommi, M., Prendi, E., Cristofori, E., Mele, T., Ivziku, D., Rocco, G., & Stievano, A. (2024). Empowering nurse health education: Linguistic and cultural validation of the nurse health education competence instrument (NHECI) in the Italian context. *Health Care, 12*(14), 1445. DOI: 10.3390/healthcare12141445 PMID: 39057588

Nunneley, C., Fishman, M., Sundheim, K., Korus, R., Rosen, R., Streater, B., O'Donnell, K. A., Newman, L. R., & Marcus, C. (2020). Leading synchronous virtual teaching sessions. *The Clinical Teacher, 18*(3), 231–235. DOI: 10.1111/tct.13282 PMID: 33063455

Nurse-Clarke, N., & Sockol, L. (2022). An exploration of resiliency among nurse educators during the covid-19 pandemic. *Nursing Education Perspectives, 43*(5), 283–286. DOI: 10.1097/01.NEP.0000000000001014 PMID: 35947139

Ohreen, D., Sundararajan, B., Trifts, V., & Comber, S. (2021). Vygotskian business ethics: The influence of peers on moral reasoning in business ethics education. *Journal of Management Education, 46*(1), 70–105. DOI: 10.1177/1052562921996019 PMID: 35013659

Ortíz, J., Argudo, U., López, D., & Mejía, S. (2023). Impact of an educational intervention for healthy eating in older adults: A quasi-experimental study. *International Journal of Environmental Research and Public Health, 20*(19), 6820. DOI: 10.3390/ijerph20196820 PMID: 37835089

Park, H., & Lim, S. (2022). Awareness of pain assessment for elderly patients with dementia among geriatric care hospital nurses. *Asia-Pacific Journal of Convergent Research Interchange, 8*(1), 77–86. DOI: 10.47116/apjcri.2022.01.06

Pavithra, J., Gangadhar, K. R., & Nanjappan, D. (2023). Evaluate the effectiveness of video assisted teaching on pranayama for reducing stress among geriatric clients at selected old age homes in rural area, Bengaluru. *International Journal of Advanced Psychiatric Nursing, 5*(1), 73–86. DOI: 10.33545/26641348.2023.v5.i1a.113

Reis da Silva, T. H. (2023a). Ageing in place: Ageing at home and in the community. *British Journal of Community Nursing, 28*(5), 213–214. DOI: 10.12968/bjcn.2023.28.5.213

Reis da Silva, T. H. (2023b). Falls assessment and prevention in the nursing home and community. *British Journal of Community Nursing*, *28*(2), 68–72. DOI: 10.12968/bjcn.2023.28.2.68

Reis da Silva, T. H. (2024a). Integrating Effective Strategies for Teaching Nursing Care for Older Adults into the Curriculum. *International Journal of Biomedical Research*, *3*(5). Advance online publication. DOI: 10.31579/2834-5029/063

Reis da Silva, T. H (2024b). Oncology and Cancer Medicine: Understanding the complexities in Older Patients. Biomed J Sci & Tech Res 55(3)-2024. DOI: https://doi.org/10.26717/BJSTR.2024.55.008720

Reis da Silva, T. (2024c). The Evolution of Nursing for Older Adult: A Historical Perspective. Associative J Health Sci. 3(3). AJHS. 000561. 2024. DOI: . Link: https://crimsonpublishers.com/ajhs/pdf/AJHS .000561.pdfDOI: 10.31031/AJHS.2024.03.000561

Reis da Silva, T. H. (2024d). Death and Its Significance in Nursing Practice. Palliat Med Care Int J. 2024; 4(3): 555640. DOI: https://juniperpublishers.com/pmcij/pdf/PMCIJ.MS.ID.555640.pdf

Reis da Silva, T. H. (2024e). The Nursing Process in Geriatric Care: A Comprehensive Approach. *International Journal of Biomedical Research*, *3*(5). Advance online publication. DOI: 10.31579/2834-5029/064

Reis da Silva, T. H. (2024f). Enhancing Nursing Care Planning Through the Levett-Jones Clinical Reasoning Cycle: A Comprehensive Analysis. *International Journal of Biomedical Research*, *3*(5). Advance online publication. DOI: 10.31579/2834-5029/065

Reis da Silva, T. H. (2024g). Chapter 7 – Emotional Intelligence in Higher Education: Humanising Technology for Holistic Student Development. In Tikadar, S.; Liu, H.; Bhattacharya, P.; Bhattacharya, S. *Humanizing Technology With Emotional Intelligence*. Pp. 117-140. IGI Global Scientific Publishing. DOI: 10.4018/979-8-3693-7011-7.ch007

Reis da Silva, T. H. (2024h). Chapter 8 – Emotional Intelligence in Teaching Geriatric Nursing: Humanising Technology for Compassionate Care in Higher Education. In Tikadar, S.; Liu, H.; Bhattacharya, P.; Bhattacharya, S. *Humanizing Technology With Emotional Intelligence*. Pp. 141-160. IGI Global Scientific Publishing. DOI: 10.4018/979-8-3693-7011-7.ch008

Reis da Silva, T. H. (2024i). Chapter 12 – Emotional Intelligence in Integrative Medicine: Enhancing Patient-Centred Care Through Humanised Technology. In Tikadar, S.; Liu, H.; Bhattacharya, P.; Bhattacharya, S. *Humanizing Technology With Emotional Intelligence*. Pp. 263-280. IGI Global Scientific Publishing. DOI: 10.4018/979-8-3693-7011-7.ch012

Reis da Silva, T. H. (2024j). Pharmacokinetics in older people: An overview of prescribing practice. *Journal of Prescribing Practice*, *6*(9), 374–381. DOI: 10.12968/jprp.2024.6.9.374

Reis da Silva, T. H. (2024k). Falls prevention in older people and the role of nursing. *British Journal of Community Nursing*, *29*(7), 335–339. DOI: 10.12968/bjcn.2024.0005 PMID: 38963269

Reis da Silva, T. H. (2024l). Chronic kidney disease in older adults: Nursing implications for community nurses. *Journal of Kidney Care*, *9*(4), 174–179. DOI: 10.12968/jokc.2024.9.4.174

Reis da Silva, T. M. (2025a). Fostering Emotional Intelligence for Competency Development in Nursing Education. In Aldosemani, T., Lytras, M., & Ordóñez de Pablos, P. (Eds.), *Innovative Educational Frameworks for Future Skills and Competencies* (pp. 39–62). IGI Global Scientific Publishing., DOI: 10.4018/979-8-3693-7555-6.ch002

Reis da Silva, T. H. (2025b). Promoting Emotional Well-Being in Older Adults: The Role of AI in Supporting Emotional Intelligence and Mental Health. In Jermsittiparsert, K., & Marzo, R. R. (Eds.), *Physical Health, Mental Health, and Human Well-Being in the Age of AI* (pp. 137–167). IGI Global Scientific Publishing., https://doi.org/10.4018/979-8-3693-6190-0.ch8

Reis da Silva, T. H. (2025c). AI-Powered Emotional Intelligence in Nursing: Bridging the Gap in Older People Care. In Jermsittiparsert, K., & Marzo, R. R. (Eds.), *Physical Health, Mental Health, and Human Well-Being in the Age of AI*. IGI Global Scientific Publishing., DOI: 10.4018/979-8-3693-6190-0.ch9

Reis da Silva, T. M. H. (2025d). The Role of Emotional Intelligence in Promoting Mental Peace, Healthy Work Environments, and Emotional Well-Being Among Entrepreneurs. In Tunio, M. N. (Ed.), *Supporting Psychological and Emotional Wellbeing Among Entrepreneurs*. IGI Global Scientific Publishing., DOI: 10.4018/979-8-3693-3673-1.ch21

Reis da Silva, T. H. (2025e). Emotional Intelligence and AI in Geriatric Nursing: Bridging Technology and Compassionate Care. In Jermsittiparsert, K., & Marzo, R. R. (Eds.), *Physical Health, Mental Health, and Human Well-Being in the Age of AI*. IGI Global Scientific Publishing., https://doi.org/10.4018/979-8-3693-6190-0.ch10

Reis da Silva, T. M. (2025f). Gamification as a Pedagogical Tool in Geriatric Nursing: Engaging and Empowering Future Nurses. In Marcão, R., & Ribeiro Santos, V. (Eds.), *Enhancing Engagement With Gamification: Education, Business, and Healthcare Perspectives* (pp. 103–136). IGI Global Scientific Publishing., DOI: 10.4018/979-8-3693-8322-3.ch005

Reis da Silva, T. M., Hammett, R., & Low, G. (2025). Emotional Intelligence in Business: Enhancing Leadership, Collaboration, and Performance. In Sedky, A. (Ed.), *Building Business Knowledge for Complex Modern Business Environments* (pp. 149–178). IGI Global Scientific Publishing., https://doi.org/10.4018/979-8-3693-6447-5.ch006

Reis da Silva, T. H., & Mitchell, A. (2024a). Integrating Digital Transformation in Nursing Education: Best Practices and Challenges in Curriculum Development. In Lytras, M., Serban, A. C., Alkhaldi, A., Malik, S., & Aldosemani, T. (Eds.), *Digital Transformation in Higher Education, Part B Cases, Examples and Good Practices*. Emerald Publishing Limited.

Reis Da Silva, T. H., & Mitchell, A. (2024b). Simulation in nursing: The importance of involving service users. *British Journal of Nursing (Mark Allen Publishing)*, *33*(5), 262–265. DOI: 10.12968/bjon.2024.33.5.262

Saripella, A., Wasef, S., Nagappa, M., Riazi, S., Englesakis, M., Wong, J., & Chung, F. (2021). Effects of comprehensive geriatric care models on postoperative outcomes in geriatric surgical patients: A systematic review and meta-analysis. *BMC Anesthesiology*, *21*(1), 127. Advance online publication. DOI: 10.1186/s12871-021-01337-2 PMID: 33888071

Semmar, Y., & Al-Thani, T. (2015). Piagetian and Vygotskian approaches to cognitive development in the kindergarten classroom. *Journal of Educational and Developmental Psychology*, *5*(2). Advance online publication. DOI: 10.5539/jedp.v5n2p1

Shah, M. (2022). Teachers as reflective practitioners: From individualism to Vygotskian social constructivism. *The Alberta Journal of Educational Research*, *68*(3), 297–307. DOI: 10.55016/ojs/ajer.v68i3.68598

Shen, Y. (2024). The application of the BOPPPS model in the ward rounds of nurses' standardized training in southwest China: A mixed methods study. *Frontiers in Medicine*, *11*, 1276652. Advance online publication. DOI: 10.3389/fmed.2024.1276652 PMID: 38994347

Shin, S., Hong, E., Lee, I., & Kim, J. (2024). Effects of governmental support on clinical nurse educators in south Korea: A repeated cross-sectional study. *International Nursing Review*, *71*(4), 933–941. DOI: 10.1111/inr.12942 PMID: 38305725

Shinohara, A. (2024). Learning outcomes from preventive home visits practicum in an undergraduate nursing program: a qualitative study. *Journal of International Nursing Research*, 3(1), e2022-0035-e2022-0035. DOI: 10.53044/jinr.2022-0035

Silva, C., Toriyama, A., Claro, H., Borghi, C., Castro, T., & Salvador, P. (2021). Covid-19 pandemic, emergency remote teaching and nursing now: challenges for nursing education. *Revista Gaúcha De Enfermagem*, 42(spe). https://doi.org/DOI: 10.1590/1983-1447.2021.20200248

Sinvani, L., Kozikowski, A., Pekmezaris, R., Akerman, M., & Wolf-Klein, G. (2016). Delirium: A survey of healthcare professionals' knowledge, beliefs, and practices. *Journal of the American Geriatrics Society*, *64*(12). Advance online publication. DOI: 10.1111/jgs.14544 PMID: 27879983

Siregar, T., Luali, N., Vinalistyosari, R., Hanurawan, F., & Anggraini, A. E. (2024). Implementation of Vygotsky's constructivism learning theory through project-based learning (PJBL) in elementary science education. *Al Qalam Jurnal Ilmiah Keagamaan Dan Kemasyarakatan*, *18*(4), 2586. DOI: 10.35931/aq.v18i4.3620

Smalley, S., Solomonson, J., & Schramm, K. (2020). The role of collaboration throughout the agricultural education student teaching experience. *Journal of Agricultural Education*, *61*(4), 329–342. DOI: 10.5032/jae.2020.04329

Song, P., & Tang, W. (2019). The community-based integrated care system in Japan: Health care and nursing care challenges posed by super-aged society. *Bioscience Trends*, *13*(3), 279–281. DOI: 10.5582/bst.2019.01173 PMID: 31327797

Southerland, L., Hunold, K., Fossen, J., Caterino, J., Gulker, P., Bischof, J., & Mion, L. (2021). An implementation science approach to geriatric screening in an emergency department. *Journal of the American Geriatrics Society*, *70*(1), 178–187. DOI: 10.1111/jgs.17481 PMID: 34580860

Souza, B. (2023). Regular schools as a straitjacket for zone of proximal development. *Perspectives in Education*, *41*(4). Advance online publication. DOI: 10.38140/pie.v41i4.6880

Stenberg, M., Bengtsson, M., Mangrio, E., & Carlson, E. (2024). Supporting each other towards independence: A narrative analysis of first-year nursing students' collaborative process. *Nursing Inquiry*, *31*(3), e12627. Advance online publication. DOI: 10.1111/nin.12627 PMID: 38351462

Switzer, J., & Schroder, L. (2019). Mobile outreach: An innovative program for older orthopedic patients in care facilities. *Geriatric Orthopaedic Surgery & Rehabilitation*, *10*, 2151459319826476. Advance online publication. DOI: 10.1177/2151459319826476 PMID: 30886761

Thompson, S., Metcalfe, K., Boncey, K., Merriman, C., Flynn, L., Alg, G., Bothwell, H., Forde-Johnston, C., Puffett, E., Hardy, C., Wright, L., & Beale, J. (2020). Interprofessional education in geriatric medicine: Towards best practice. a controlled before–after study of medical and nursing students. *BMJ Open*, *10*(1), e018041. DOI: 10.1136/bmjopen-2017-018041 PMID: 31964659

Vázquez-Sánchez, M., Casals-Vázquez, A., Sánchez-Ojeda, M., García-Gámez, M., Papadopoulos, I., & Casals, C. (2022). Spanish adaptation and validation of the cultural competence assessment tool (CCATOOL) for undergraduate nursing students. *International Nursing Review*, *70*(1), 43–49. DOI: 10.1111/inr.12792 PMID: 35879822

Verenna, A., Noble, K., Pearson, H., & Miller, S. (2018). Role of comprehension on performance at higher levels of bloom's taxonomy: Findings from assessments of healthcare professional students. *Anatomical Sciences Education*, *11*(5), 433–444. DOI: 10.1002/ase.1768 PMID: 29346708

Voorend, C., Joosten, H., Berkhout-Byrne, N., Diepenbroek, A., Franssen, C., Bos, W., Van Buren, M., Mooijaart, S. P., van Alphen, A., Berkhout-Byrne, N., van Breda, F., van Buren, M., Boom, H., Bos, W. J., Diepenbroek, A., Emmelot-Vonk, M., Franssen, C., Gaillard, C. A. J. M., Groeneweg, N., & Stehouwer, C. (2021). Design of a consensus-based geriatric assessment tailored for older chronic kidney disease patients: Results of a pragmatic approach. *European Geriatric Medicine*, *12*(5), 931–942. DOI: 10.1007/s41999-021-00498-0 PMID: 33871790

Wilson, J., Hensley, A., Culp-Roche, A., Hampton, D., Hardin-Fanning, F., & Thaxton-Wiggins, A. (2021). Transitioning to teaching online during the covid-19 pandemic. *SAGE Open Nursing*, *7*, 23779608211026137. Advance online publication. DOI: 10.1177/23779608211026137 PMID: 34222655

Wissanji, T., Forget, M., Muscedere, J., Beaudin, D., Coveney, R., & Wang, H. (2022). Models of care in geriatric intensive care—A scoping review on the optimal structure of care for critically ill older adults admitted in an ICU. *Critical Care Explorations*, *4*(4), e0661. DOI: 10.1097/CCE.0000000000000661 PMID: 35382113

Wu, M., Chen, Z., Xu, Y., Zhao, L., Zhao, L., & Xia, L. (2024). A qualitative study of geriatric specialist nurses' experiences to navigate delirium in the elderly. *BMC Nursing*, *23*(1), 426. Advance online publication. DOI: 10.1186/s12912-024-02100-x PMID: 38918789

Wynn, M., Reis da Silva, T. H., & Pearson-Jenkins, J. (2025). Introduction to digital nursing and nursing theory. In Wynn, M. (Ed.), *Digital Nursing: Shaping Practice and Identity in the Age of Informatics*. Routledge. DOI: 10.4324/9781032714547-1

Zambrotta, M., Aylward, P., Roy, C., Piper-Vallillo, E., Pelletier, S., Honan, J., Heller, N., Ramani, S., & Shields, H. (2021). Nurse-doctor co-teaching: A pilot study of the design, development, and implementation of structured interprofessional co-teaching sessions. *Advances in Medical Education and Practice*, *12*, 339–348. DOI: 10.2147/AMEP.S300231 PMID: 33889044

Zhong, X., Liu, X., & Sheng, Y. (2021). The effect of the humanistic care teaching model on nurse patient conflict and nurse turnover intention in a pediatric outpatient department: Results of a randomized trial. *Translational Pediatrics*, *10*(8), 2016–2023. DOI: 10.21037/tp-21-214 PMID: 34584871

Zúñiga, F., Geest, S., Guerbaaï, R., Basinska, K., Nicca, D., Kressig, R., & Simon, M. (2019). Strengthening geriatric expertise in Swiss nursing homes: InterCare implementation study protocol. *Journal of the American Geriatrics Society*, *67*(10), 2145–2150. DOI: 10.1111/jgs.16074 PMID: 31317544

Zúñiga, F., Guerbaaï, R., Geest, S., Popejoy, L., Bartáková, J., Denhaerynck, K., & Simon, M. (2022). Positive effect of the InterCare nurse-led model on reducing nursing home transfers: A nonrandomized stepped-wedge design. *Journal of the American Geriatrics Society*, *70*(5), 1546–1557. DOI: 10.1111/jgs.17677 PMID: 35122238

Chapter 10
Investigating the Relationship Between Teacher Stress and Classroom Performance:
A Well-Being Perspective

A. K. Monisha
https://orcid.org/0009-0002-8293-9152
Jain University, India

Sandhya Reddy S.
https://orcid.org/0009-0001-2663-1615
Bishop Cotton Women's Christian College, India

Shree S. Kavya
https://orcid.org/0009-0007-8593-9146
Bishop Cotton Women's Christian College, India

M. Swetha
https://orcid.org/0009-0007-5539-0447
Bishop Cotton Women's Christian College, India

ABSTRACT

Abstract: Teaching profession is highly demanding, often exposing the teachers to high stress which is affecting their well-being and classroom effectiveness. The study explores the relationship between teacher stress and classroom performance, emphasizing the mediating role of teacher's well-being. Through thematic analysis of teacher's semi structured interviews, key stressors such as workload, administrative pressures and student behaviour were identified. Findings indicated that these stressors negatively influence teaching quality engagement, and student outcomes. The Study also highlights the effectiveness of well-being programs including mindfulness practices, emotional support systems and work-life balance strategies to decrease stress and enhance teacher's performance. Based on the findings, it is recommended that institutions to prioritize well-being programs to foster a healthier and resilient teaching environment. Implementing holistic interventions can support teachers' mental health while also contributing to improved classroom dynamics and student success.

DOI: 10.4018/979-8-3373-0472-4.ch010

INTRODUCTION

Teaching is a profession where an individual role as a teacher is to guide, instruct and educate students for different fields in Academic Progression. A teacher plays a very important role of imparting knowledge, fostering critical thinking, understanding concepts and social development. They help individuals to develop various skills like reading, learning, understanding, and analysing in various subjects.

Teachers takes many different forms as a coach, tutor, educator, motivator. They perform duties by creating experiential learning, participative learning, developing problem solving skills in students, assign grades &students' academic progress. Their roles can go beyond teaching, by helping the students to become better individuals for the society. The Role of a teacher in Graduate levels plays a significant role and their role's varies in curriculum by providing basic foundational knowledge under UG level (Undergraduate level) and Advance knowledge and specialization of the subject opted by the student under PG level (Post graduate level).

According to survey (All-India Survey of Higher Education (AISHE) conducted the total number of teachers that are reported in the survey year 2021-22 are 15,97,688. Out of which, 56.6% are male teachers and 43.4% are female teachers. With Reference to At All-India level, number of female teachers per 100 male teachers is 77. The percentage based on Designation Wise Distribution of Teachers: Among all teachers, 68% are Lecturers/Assistant Professors. Around 10% each are Associate Professors and Professor & Equivalent and 5.7% of the teachers are temporary teachers. (The All-India Survey of Higher Education (AISHE), 2021-22 was conducted with reference period as academic session 2021-22).

Stress: Teacher and Stressors

According to World Health Organization (WHO) Stress is defined as a state of worry or tension caused by a difficult situation. It affects both mind and body. Every individual needs a certain amount of stress which can be consider as good to make the individual perform daily activities. But extended period of stress can hinder the performance and affect the individual over all functioning which can be both physically and mentally. (WHO).

The role of teacher as a profession itself is challenging as they significantly impact the overall quality of education system. Teachers who play a key role in imparting knowledge, fostering critical thinking and help students to develop relevant skills in their fields. As teachers they face different challenging situations in their profession, the stressors that they face in their challenging role can be directly from administration such as excessive workloads, time pressure, target-based research contribution, intense course load, working hours, assignments, constant change in practices, technology integration without training & lack of growth and opportunities. Other stressors can be high demands placed by the institutions & parents, managing & guiding students with the fast-growing technology to build their competency level.

These stressors faced professionally can affect the non-working life personally due to lack of work life balance, teachers can develop personal issues not able to manage the responsibility at home efficiently, fear of losing job, no family time as work can be carried over to personal space which can lead to a conflict between the individual personality and profession. The challenges faced by a teacher in the profession can lead to stress which can affect them both physically and mentally.

As the stressors are constantly faced on daily basis and prolonged over period of time can also impact negatively both professionally and personally. There is lack of work life balance for teachers which can lead dissatisfaction in job, they are no more motivated, absenteeism, burnouts, lack of interest, lack of interaction with students, student engagement would be affected. All these can result in negative effect on the overall classroom performance which can hinder teacher – student relationships as well as quality of imparting Knowledge.

Figure 1. Teacher stressors

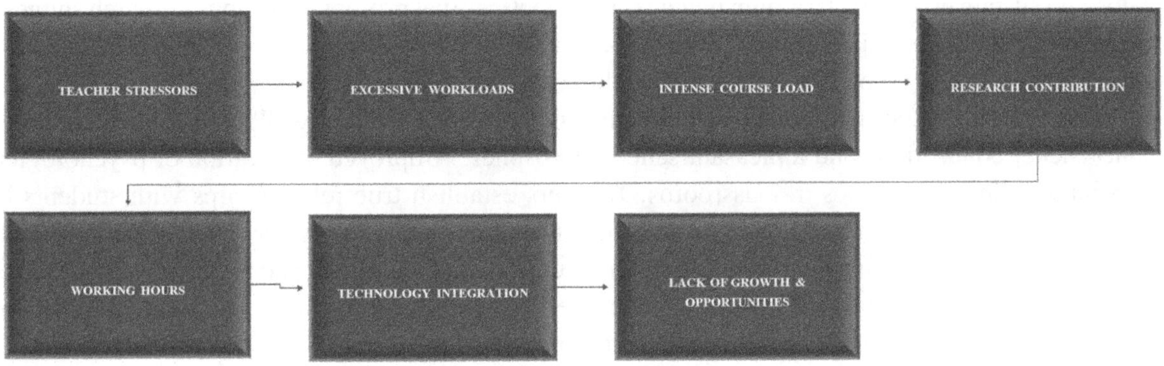

Well-Being

Well-being is state where an individual experiences joy, happiness, contentment in his / her life. They will be able to recognise their potential, optimal functioning abilities, develop a sense of purpose in their life. According to World Health Organization (WHO) defines positive mental health as "a state of well-being in which the individual realizes his or her own abilities, can cope with the normal stresses of life, can work productively and fruitfully, and is able to make a contribution to his or her community"

Teacher Well-Being

Teacher well-being is the sense of satisfaction and accomplishment that individuals feel in their professional lives. For teachers, this fulfilment can arise from various factors, such as their work environment, opportunities for professional development, benefits and perks offered by their organizations, a healthy work-life balance, and their motivation towards their work. Together, these elements can greatly enhance a teacher sense of satisfaction in their professional life, thus supporting their mental health. Blending both personal and professional life is effectively handling the competing needs of both positions while maintaining one's wellness, personal connections, and general well-being.

Figure 2. Teacher well-being factors

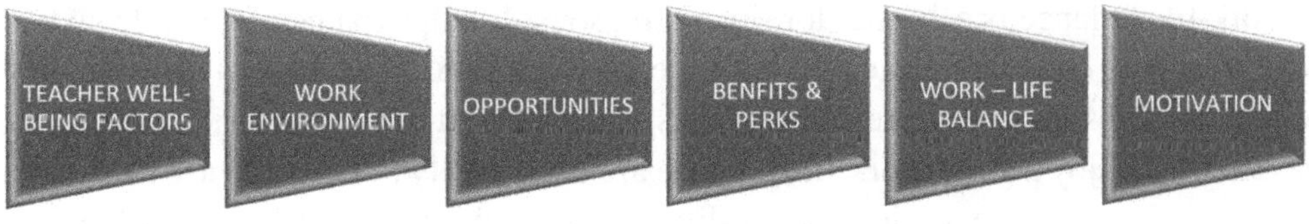

Teacher wellness is shown in a good perspective on teaching, which comes from helpful connections. Effective teaching requires strong relationships with colleagues and students, as well as a sense of personal and professional fulfilments. Teachers, mental health, psychological, and physical well-being is a shared duty that impacts student achievement and the wellness of the community. Teacher well-being important consequently to overcome the daily stress of teaching. Teachers with higher levels of dedication had a reduced likelihood of stress and destruction, as well as more positive attitudes towards teaching. Develop and sustain effective classroom management practices that promote learning. Research suggests that teacher wellbeing is linked to increased efficacy, improved knowledge of classroom dynamics and also for effective use of emotional emotions to enhance learning experiences.

Develop helpful and compassionate connections with students. Teacher wellbeing is linked to better student-teacher connections and a pleasant school atmosphere. Improved application of psychological and social learning approaches in classrooms. Teachers establish true relationships with students by understanding what they need. Positive partnerships need continual contacts and active participation. Observing and tolerance. Teachers who demonstrate compassion, support, and real concern can boost student's confidence and sense of worth. A healthy teacher-student connection involves pupils feeling noticed, understood, encouraged, and valued. Effective relationships among peers improve student welfare in the classroom. They promote enjoyment, emotional wellness, and togetherness. They enhance solving issues and interpersonal skills. Strong school involvement correlates with positive peer connections.

Collaboration amid teachers promotes student well-being in classrooms. High Impact Wellbeing Strategies (HIWS) assist classroom teachers increase student wellbeing. Everyone may implement these strategies. HIWS are most effective when used with other mental health programs include learning about society and emotions, the Health and Physical Education curriculum, and the Personal and Social Capabilities program. The HIWS should be adopted as part of a school-wide strategy to increasing student wellness.

Well-being strategies follow as: Management, Assessing, Aid and Resources, Participation, Methods for teaching and learning. Wellbeing is impacted by several variables, especially school-level ones. Teachers may promote well-being by giving. Wellbeing educational initiatives include cognitive and behavioural methods, as well as excellent role-modelling. HIWS tend to be successful whenever utilized to foster healthy peer interactions, student opinion, action, leadership, and student participation in learning.

Teachers' presence in schools and colleges is vital. Teachers have a crucial role in achieving educational objectives. Teachers performed various roles in achieving the national educational objective. Teachers have a crucial role in enhancing education in our society. It is genuine to suggest that instructors are significant academic assets since they possess knowledge, leading talents, a strong personality, and are dedicated to guaranteeing student greatness.

The phrase work life balance is frequently used to describe the difficult balance we strike between our professional and personal lives. In order the majority of us, striking the correct balance is an ongoing battle. We may feel pushed in different directions by the requirements of our jobs and families, or we might find it difficult to unplug with work when we are at home. Some days, we may feel as if we have to balance multiple balls, one of which is going to drop.

Teacher work-life balance is essential for several reasons. It helps to keep people from being burned out. Whenever teachers can strike a healthy balance between their professional and personal life, they are less likely to feel stressed by their responsibilities as teachers. It can increase job fulfilment. Teachers who believe that they enjoy a good work-life balance are more likely to be content with their professions and less likely to seek new career options. Work-life balance is defined as striking a balance among both personal and professional lives by managing, organizing, and prioritizing personal responsibilities and work routines (Fazal et al., 2019).

LITERATURE REVIEW

Many studies have shown the importance of teacher's well-being. Blasé (1986), in their qualitative study investigated the teacher's perception of work stress and its influence on their performance. The study highlighted a strong link between extended work stress and negative effects on teacher's instructional abilities. The data was discussed in terms of the Performance Adaptation Syndrome (PAS), the findings suggested that stress has a negative impact on the performance of the teacher. Qin (2024) examined the correlation between the stress levels and classroom performance of classroom instructors. The study identified that stress negatively impacts the quality of instruction. In order to assess the classroom performance, indicators like engagement of the student, achievement in academics and evaluations of teacher were used. The results emphasize a strong relationship between high stress levels and lower engagement of student, decreased academic achievement and negative teacher evaluations. A study by Shi (2024) investigated the impact of stress on classroom performance of the teachers. The stressors like teaching tasks, research demands and student management impacted teacher's stress levels and performance. The study reveals excessive stress led to emotional exhaustion, reduced classroom interaction and negative impact on teaching effectiveness and student satisfaction. Jõgi et al. (2022) examined the relationship between teacher's daily physiological stress and positive affect in classroom settings. Positive affect was assessed by measuring salivary cortisol level which was an indicator of physiological stress of 45 classroom teachers by using experience sampling. The findings revealed no significant relationship between teacher's stress and positive affect. The study suggests some kind of improvement in school resources and teacher's self -efficacy will be useful in stress reduction and improve affect in the classroom. Study by klusmann investigated the relationship between emotional exhaustion of teacher's and academic performance of student's, specifically in mathematics. The study was conducted on 1.102 German elementary school teachers and their students. The study found that the relationship between teacher's emotional exhaustion and student's math achievement was significantly negative, even when there was a control on various teacher and student characteristics. The study implies the significance of teachers' well-being for student learning outcomes.

Penttinen et al., investigated the relationship between teachers' occupational well-being and the quality of teacher-child interactions in Finnish kindergarten classrooms. The findings suggest that there is a negative impact of teaching-related stress on the quality of emotional support and classroom organization, while there was a strong correlation between teacher's work engagement and the quality of instructional support. Carroll et al., (2021) investigated the influence of teacher stress reduction programs on both teachers and their students. The results revealed that reductions in teacher's stress and burnout were linked with improved student outcomes, and includes better perceptions of teacher support and enhanced academic self-perception. Bidi et al., analysed the impact of stress factors affecting teachers on their impact on psychological well-being. The quantitative review included 60 studies, focusing on stressors such as workload, classroom management, and lack of support. The common issues found were emotional exhaustion, depersonalization, and reduced personal accomplishment. Coping strategies and mindfulness training were the most effective interventions in order to reduce teacher stress. The study recommends the need to address and manage these stressors to improve teacher's well-being. Nwoko et al. in their systematic review examined 38 studies to understand the affecting factors of occupational well-being of kindergarten, primary, and secondary school teachers. The study emphasizes the roles of self-efficacy, socioemotional competence, professional relationships, and organizational support in order to enhance teachers' well-being. The study revealed that teacher positive well-being was associated to improved teaching quality, healthier relationship between teacher-student, and better student engagement. This research recommends the need to integrate occupational well-being initiatives into teacher professional development plans and fostering collaboration among teachers, parents, colleagues, and school leadership in order to create supportive work environments. A systematic review by Ozturk et al. (2024) explored 61 studies to analyse the multidimensional role of well-being of teacher and evaluate methodological quality. Authors found that teacher well-being is often viewed from professionalism lens, in eighteen studies adopting this perspective exclusively, while fewer studies explored stress, stressors and burnout (negativity/deficiency approach). Only six studies integrated all three discourses: negativity/deficiency, positivity/flourishing, and professionalism. This research paper emphasized the importance for a holistic understanding of teacher well-being in order to notify the interventions and strategies that enhance teacher well-being, improve student outcomes, and enhance the overall quality of education. McLean and Connor (2015), studied the relationship between teacher teaching in third grade, they studied depressive symptoms, classroom learning environment (CLE) quality, and student achievement in mathematics and literacy. Findings revealed that Depressive symptoms of the teachers were negatively predicted student's spring math performance, especially for students with weaker fall math skills. In addition, depressive symptoms were linked to lower CLE quality, which facilitated the correlation between mental health of the teacher's and the outcome of. The results highlight the need for supporting mental health of the teachers to enhance classroom quality and student success. Flook et al., (2013) analysed the impact of a modified Mindfulness-Based Stress Reduction (MBSR) course tailored for teachers. The randomized controlled trial revealed substantial benefits for participants, including reduced psychological symptoms and burnout, enhanced classroom organization, and increased self-compassion. Furthermore, there was a positive relationship between mindfulness changes with improved psychological outcomes and sustained attention. Conversely, control group participants showed a decline in the cortisol functioning and marginal increases in burnout. The findings highlight the potential of mindfulness-based interventions and assessments in order to support teacher well-being and classroom effectiveness.

In the study by Klassen and chiu, explored the relationships between teacher characteristics such as gender and years of experience, self-efficacy domains such as instructional strategies, management of the classroom, and engagement of the student, job stress, and job satisfaction. The study found that there was an increase in early to mid-career of teacher's self-efficacy but decreased afterward. Many Female teachers reported higher workload and stress in classroom and managing the class. Teachers teaching younger children revealed an increase in self-efficacy in managing classroom and engaging the students. The study also emphasized that higher self-efficacy in classroom management and instructional strategies was associated with greater job satisfaction, while higher workload and classroom stress negatively impacted both self-efficacy and job satisfaction. Pagán-Castaño et al. (2021), in their research they found out the association between human resource management (HRM), teacher well-being, and performance in secondary schools. The study involved 315 teachers and used structural equation modelling to substantiate the positive effect of leadership by example and information management on HRM and teacher performance. The results highlight the significant impact of HRM on educational performance, both directly and through its effect on teacher well-being. Furthermore, the findings emphasize the need innovation in schools and classrooms for enhancing performance and sustaining educational quality. The findings suggest that human resource policies focused on teacher well-being and school innovation are the keys in order to enhance both teacher and student outcomes. Karjalainen et al. (2020), did a pilot study which explored the association between teachers' well-being and classroom acoustics, emphasizing on factors such as vocal health, stress, burnout, and self-efficacy. The study involved 23 primary school teachers, which explored the impact of classroom acoustics, including reverberation time, speech clarity, and ventilation system noise. The study highlighted and recommend a strong relationship between burnout and higher levels of ventilation system noise (VSN), as well as a connection between voice symptoms and VSN, with teachers in lower grades reporting more voice symptoms due to higher noise levels. The study also highlights the potential impact of classroom noise on teachers' well-being, particularly in relation to burnout and vocal health. Gustems-Carnicer et al., analysed the correlation between stress, coping strategies, and academic performance in 334 university students. The study suggested that there is frequent usage of neglecting coping strategies among students who experienced stress. The study highlights that those students under higher stress showed poorer performance and students who employed problem focused coping strategies and lower stress levels achieved better academic performance. Furthermore, the study suggests that age moderated the impact of stress on achieving academics, with older students being less affected by stress. Mudin et al., in their systematic literature review explored 121 studies to analyse the association between teacher stress and well-being. There was a strong negative association between teacher stress and well-being, both directly and indirectly. Teacher stress was shown to adversely affecting effectiveness in teaching, outcomes of the students, and overall educational governance. The study prioritized the importance of teacher stress to improve teacher well-being as well as the quality of education.

Neuenschwander et al. (2017), examined how teacher stress affects student's executive function (EF), an important aspect of self-regulation. The authors and revealed significant negative relationship between teacher stress and child EF, though classroom quality did not mediate this effect. In addition, school-level poverty measured the relationship: in low-poverty schools, higher teacher stress significantly impacted with smaller gains in child EF, while in high-poverty schools, teacher stress did not significantly impact child EF. The results suggest the need to address teacher stress, particularly in low-poverty schools, to support child development. Liao et al. (2023 examined the impact of work stress affecting teacher's well-being, a focus on conflict in family-work and self-transcendent meaning in life. The study involving 562 teachers, found that work stress negatively predicted teachers' well-being, and family-work conflict mediated this relationship. Furthermore, a self-transcendent meaning of life moderated the effect of work stress on well-being. The study suggests the need for addressing both work stress and family-work conflict to support teachers' well-being, with the potential role of self-transcendence in mitigating stress. Arifin et al. (2024) in their research examined how environment of the classroom and professional development of teacher's influence the academic performance of the students in secondary education. Using a mixed-methods approach, the study found strong positive correlation between both classroom environment and teacher professional development with student academic performance. The results highlight the need to create inclusive and nurturing classroom spaces and providing comprehensive professional development programs for teachers to enhance student outcomes. Jeon et al. (2018), investigated the factors influencing early childhood teachers' psychological well-being. The study recommends that teachers' self-efficacy and work environment were significant predictors of their psychological well-being, including depression, stress, and emotional exhaustion, beyond their personal and professional backgrounds. The study suggests that improving teachers' teaching competence, efficacy, and creating positive work climates are key strategies to support their well-being and reduce psychological burdens.

McCarthy et al. (2016) reviewed the Classroom Appraisal of Resources and Demands (CARD) as a tool for measuring teachers' appraisals of classroom demands and resources. The study's meta-analysis of 18 studies revealed moderate effects between CARD scores and variables such as job satisfaction, occupational commitment, burnout symptoms, stress prevention resources, and challenging student demands. The study highlights the utility of CARD in assessing teacher stress and its implications for educational policies aimed at reducing teacher turnover and improving welfare. Hao (2022), studied "Analysis of the Effect of Work Stress and Organizational Culture on Teacher Performance," explained how work stress and organizational culture influence teacher performance at The Global Prima School. Using descriptive quantitative and explanatory research methods, the study suggests that both work stress and organizational culture had a positive correlation and strong effect on performance of the teachers. The statistical analysis showed significant results (work stress: t-count = 5.045, organizational culture: t-count = 5.221), with a coefficient of determination (R^2) of 74.6%, indicating that these factors strongly contribute to teacher performance. The study recommends that other unexamined variables, such as supervision and leadership, account for the remaining 25.4% of teacher performance variance. Gonzales (2024) studied the association between stress related to work and performance of teachers in San Francisco, Cluster IV. The study highlights that teacher experienced moderate work-related stress, with significant differences based on age, teaching experience, and grade level. Despite the moderate stress levels, the majority of teachers received outstanding or very satisfactory performance ratings. Notably, finding shows a negative association of workload stress on teaching performance, suggesting that while managing workloads is important, it may not directly affect teaching quality.

METHODOLOGY

Research Objective

The primary aim of this study was to investigate the relationship between teacher stress and classroom performance, with a specific focus on how well-being initiatives may mediate this relationship. The research was designed to explore four key objectives: (1) to assess the relationship between teacher stress and classroom performance; (2) to evaluate the impact of stress on teaching effectiveness; (3) to examine the role of well-being initiatives in managing teacher stress and improving performance; and (4) to assess how teacher stress influences classroom dynamics, student engagement, and instructional quality. These objectives guided the development of the study's hypotheses, interview framework, and thematic analysis approach. Based on the research objectives, four hypotheses were formulated to structure the investigation.

H1: "There is a significant relationship between teacher stress and classroom performance"
H2: "Increased teacher stress negatively impacts teaching effectiveness"
H3: "Well-being initiatives significantly reduce teacher stress and enhance teaching performance"
H4: "Higher levels of teacher stress lead to disruptions in classroom dynamics, lower student engagement and reduced instructional quality"

Variables

The variables were clearly defined for consistency and analytical precision. The independent variable, teacher stress, was measured through multiple factors, including administrative pressures, workload, and student behavior. The dependent variable, classroom performance, was assessed in terms of teacher engagement, instructional quality, and student learning outcomes. Additionally, teacher well-being functioned as a mediating variable and was explored through mindfulness practices, work-life balance strategies, and access to emotional support systems. This three-variable framework allowed for a more comprehensive understanding of how stress and well-being interplay to impact teaching performance.

Procedure

Data for this study was collected through one-on-one interviews with the participants, ensuring a focused and personal approach to understanding their experiences. The researchers employed thematic analysis to examine the collected data, enabling them to identify common patterns, themes, and insights across the interviews. A semi-structured interview format was used to guide the conversation while allowing flexibility for participants to elaborate on their thoughts and reflections. The interview was designed to cover five core areas relevant to the research objectives.

The first section, titled *Understanding Stressors*, included five questions aimed at uncovering teachers' perceptions of their workload, administrative responsibilities, and challenges with student behavior. The second section, *Impact of Stress on Performance*, featured four questions that explored how stress influences teaching quality, teacher engagement, and student outcomes. The third section, *Exploring Well-being*, also contained five questions and focused on how teachers perceive and manage their emotional and psychological well-being. The fourth area, *Role of Well-being Initiatives*, included four questions assessing the effectiveness of strategies such as mindfulness practices, emotional support systems, and work-life balance. Finally, a *Personal Reflections* section with three questions encouraged teachers to share their subjective experiences and offer suggestions for improving well-being in educational settings.

The study utilized clearly defined inclusion criteria to ensure participant relevance and the reliability of collected data. Only teachers currently employed in either undergraduate (UG) or postgraduate (PG) academic programs were considered for participation. Additionally, all participants were required to have a minimum of **two years of teaching experience**, ensuring they had adequate exposure to the teaching environment and related stressors. Participation was voluntary, and only those who **provided informed consent** were included in the final sample. These criteria ensured that the study focused on experienced educators capable of reflecting meaningfully on their professional and personal challenges.

To maintain the integrity and focus of the study, several exclusion criteria were applied. Teachers with **less than two years of teaching experience** were excluded to avoid perspectives that may not be fully developed or reflective of long-term professional engagement. In addition, **part-time teachers and guest faculty** were not included, as their working conditions and responsibilities often differ significantly from full-time educators. Participants who were unavailable for interviews or **did not provide consent** were also excluded to uphold ethical standards. Finally, teachers who were not actively teaching during the time of the study were excluded, as their experiences might not accurately represent current stress levels or teaching dynamics. These exclusions helped ensure the study maintained a focused and relevant participant pool.

Case Study

Case Study on teacher Well-being Initiative Through Mindfulness in college. Faculty staff at a college in Bengaluru, India, were found to be under a lot of stress as a result of personal struggles, administrative duties, and academic demands. The college created a Mindfulness-Based Stress Reduction (MBSR) curriculum specifically for teachers in order to address this.

Intervention: An 8-week mindfulness training course was part of the curriculum; weekly sessions covered body scans, mindful breathing, and meditation. Teachers were urged to include mindfulness activities into their lessons and to practice them every day. In order to promote a sense of community and exchange experiences, peer support groups were established.

Results: Participants reported feeling better overall and experiencing a considerable decrease in stress. Teachers reported being more patient and focused when interacting with students and co-workers. With more cooperation and lower absenteeism, the college noticed a good change in the workplace.

Research Gap

The study focused on teacher stress, well-being, and classroom performance into unified framework which focused on teachers. The study emphasised on stressors and their impact on teaching quality of the teachers and exploring the role of well-being interventions in decreasing these effects. The study fills the gap in existing literature by contributing different perspective which considers contextual factors and personal experiences, eventually aiming to practical, evidence based recommendations for strengthen teacher well-being and classroom dynamics. The study also highlights the need for a holistic understanding of how well-being programs can transform the teaching environment and enhance teacher and student success.

THEMATIC ANALYSIS

Theme 01: Engagement in Well-Being Initiatives

Sub-Theme 1.1: Participation in Mindfulness and Stress Management

Teachers who engaged in mindfulness and stress management programs reported better coping mechanisms, improved their emotional regulation and sustained energy levels, which positively influenced their teaching performance.

Participate 03- "I developed better coping strategies like time management and mindfulness. It really helped me stay calm even during stressful days"

Participant 06- "Over the years, I've learned to recognize when I'm reaching my limits and developed better self-care habits, which has greatly improved my ability to handle stress."

Participate 10- "I've become more proactive in maintaining my well-being as my career has evolved. I now incorporate relaxation techniques and set clear boundaries to avoid burnout."

Participant 04- "I used to feel guilty for taking time for myself, but as my career progressed, I realized that maintaining my well-being is essential for my effectiveness as a teacher."

Participant 09- "At the start, I was constantly feeling burned out, but with experience, I've found ways to manage stress, like seeking support from colleagues and utilizing stress-reduction techniques."

Sub-Theme 1.2: Practical Benefits of Well-Being Programs

Teachers reported tangible benefits of well-being initiatives such as reduced burnout, increased job satisfaction and enhanced classroom performance.

Participant 05- "At first, I didn't pay much attention to my mental health, but I've grown to understand that managing stress is key to sustaining a long and successful teaching career."

Participant 07- "In the beginning, I was overwhelmed by the demands of teaching, but now I've learned how to delegate tasks and use resources to better manage stress."

Participant 04- "I used to feel guilty for taking time for myself, but as my career progressed, I realized that maintaining my well-being is essential for my effectiveness as a teacher."

Participant 01- "In the beginning, I didn't recognize the signs of stress, but over time, I've learned to set boundaries and prioritize self-care to protect my well-being."

Theme 02: Institutional Support for Teacher Well-Being

Sub-Theme 2.1: Flexible Work Policies

Teachers emphasised the role of flexible work policies in reducing stress and improving teaching effectiveness.

Participant 06- "It would feature flexible work policies, like the option to work from home when needed, and an understanding of the importance of personal time for mental health."

Participant 08- "A supportive leadership team would encourage transparency and open communication about stress and challenges, offering guidance and support without judgment."

Participant 02- "It would offer reasonable class sizes and manageable workloads to allow teachers to focus on delivering quality lessons and maintain their well-being."

Participant 10- "Teachers would have access to an effective support system for handling student behaviour issues, administrative tasks, and personal concerns, ensuring that their workload and emotional demands are manageable."

Sub-Theme 2.2: Professional Development for Stress Management

Teachers expressed a need for professional training in stress management and stress management and self-care practices.

Participant 10- "Seek out professional development opportunities focused on stress management, mindfulness, and self-care, as these can provide valuable tools to maintain your well-being."

Participant 07- "In an ideal setting, teachers would be given the resources and time to focus on professional development, ensuring their growth without feeling overwhelmed by expectations."

Participate 06- "Workshops on self-care and time management would go a long way in helping teachers"

Participant 04- "When institutions prioritize teacher well-bring, performance and student outcomes improve"
Participate 03- "Mandatory well-being training should be implemented to quip teachers with stress management strategies"

Theme 03: Structural and Policy Recommendations for Stress Reduction

Sub-Theme 3.1: Workload Management

Teachers emphasized that reducing excessive workloads would enhance classroom performance.
Participate 02- "if class sizes were more manageable, teachers could focus on more on quality lessons"
Participate 10- "Access to better administrative support would significantly reduce our stress levels"
Participate 09- "Managing workload efficiently directly affects how well we engage with students"
Participant 07- "When overwhelmed, my patience and creativity in the classroom diminish"
Participant 04- "An institution that prioritizes workload balance would result in better teacher performance"

Sub-Theme 3.2: Clarity in Communication and Expectations

Clear Communication reduces stress and improves classroom management.
Participate 08- "A leadership team that encourages transparency helps alleviate stress"
Participate 06- "Knowing expectations ahead of time makes it easier to plan lessons"
Participate 04- "Unclear guidelines increase stress and affect how I engage with students"
Participate 09- "The more structured the admin is; the better teachers can focus on their job"
Participate 07- "Good communication from management reduces unnecessary stress"

Theme 04: Impact of stress on Classroom Performance

Sub-Theme 4.1: Impact on Student Engagement

Teachers observed that high stress levels negatively affect student engagement and learning outcomes.
Participate 05- "When I'm stressed, I notice that my students are less engaged"
Participate 10- "Lower energy levels directly impact how actively I interact with students"
Participate 02- "Managing classroom disruptions becomes harder when overwhelmed"
Participant 07- "A relaxed teacher fosters a more positive learning environment"
Participate 09- "There is direct correlation between my stress and overall classroom atmosphere"

Sub-Theme 4.2: Burnout and Job Satisfaction

Teachers who managed stress well reported good job satisfaction
Participant 01- "I learned to set boundaries to void burnout
Participant 07- "Finding work-life balance early helped me avoid stress-related burnout"
Participant 05- "When I'm well rested, I'm more patient with students"
Participant 08- "job satisfaction increases when stress is managed effectively"

The findings from the thematic analysis revealed four major themes: Engagement in Well-Being Initiatives, Institutional Support for Teacher Well-Being, Structural and Policy Recommendations for Stress Reduction, and the Impact of Well-Being Initiatives on Teaching and Student Outcomes. Each of these themes, supported by participant responses and structured interview data, contributes to a comprehensive understanding of how teacher stress interacts with institutional factors and personal well-being practices to influence classroom performance and student engagement.

The first major theme, Engagement in Well-Being Initiatives, encompassed two sub-themes: *Participation in Mindfulness and Stress Management* and *Practical Benefits of Well-Being Programs*. Participants who actively engaged in mindfulness practices and stress-reduction techniques reported significant improvements in emotional regulation, energy levels, and patience in the classroom. Teachers shared how such practices helped them stay composed during high-pressure teaching moments and contributed to a more focused and calm demeanor when interacting with students. Additionally, the practical advantages of structured well-being programs were emphasized. Teachers indicated that these initiatives not only reduced burnout but also enhanced their job satisfaction and teaching effectiveness. Many noted that participating in regular mindfulness sessions or stress-management training helped them feel more in control of their work-life balance, thereby fostering long-term professional sustainability.

The second theme, Institutional Support for Teacher Well-Being, focused on the role of school policies and administrative practices in managing teacher stress. Under the sub-theme *Flexible Work Policies*, participants highlighted the importance of work environments that accommodate personal needs and offer flexibility in scheduling and workload distribution. Teachers emphasized that institutions with adaptable policies were better able to prevent burnout and promote a healthier work-life balance. The second sub-theme, *Professional Development for Stress Management*, revealed that teachers valued opportunities for learning how to cope with stress. Workshops on self-care, time management, and emotional resilience were cited as especially beneficial. Furthermore, *Access to Mental Health Resources* emerged as a critical component of institutional support. Teachers from institutions that offered counseling services, wellness seminars, or informal peer support systems reported experiencing reduced stress levels and greater emotional stability. The availability of such resources created a culture of care and understanding, which not only supported teacher well-being but also improved the overall workplace environment.

The third theme, Structural and Policy Recommendations for Stress Reduction, addressed broader institutional factors affecting stress levels. Under the sub-theme *Workload Management*, teachers described how excessive responsibilities, such as large class sizes and heavy administrative tasks, undermined their ability to deliver high-quality instruction. Reducing these burdens was seen as essential for maintaining focus on student learning and instructional design. Participants stated that managing workload efficiently directly affected their energy levels, patience, and overall classroom presence. The second sub-theme, *Clarity in Communication and Expectations*, underlined the need for transparent and consistent communication from school leadership. Teachers expressed that vague directives and sudden policy changes increased stress and disrupted their planning. In contrast, clear and timely communication helped educators feel more secure in their roles and better prepared to meet institutional goals. Institutions that encouraged open dialogue and clearly outlined teacher responsibilities were associated with a more positive and productive work atmosphere.

The fourth and final theme, Impact of Well-Being Initiatives on Teaching and Student Outcomes, focused on the downstream effects of stress reduction on classroom dynamics. Under *Improved Teacher Effectiveness*, teachers shared that when their stress was well managed, they could engage more actively with students, adapt their instructional methods, and maintain consistency in lesson planning. This in turn improved learning outcomes and student satisfaction. The sub-theme *Positive Influence on Students* reinforced the importance of teacher well-being in fostering a supportive learning environment. A balanced and mentally healthy teacher was described as more patient, empathetic, and creative, traits that significantly impact student motivation and behavior. Participants consistently reported that when they were less stressed, students responded more positively in class, resulting in better participation, reduced behavioral issues, and a more collaborative classroom culture.

In summary, the results of the thematic analysis provide strong evidence that teacher well-being is deeply interconnected with institutional support systems, stress-management strategies, and overall teaching quality. Teachers who were empowered through mindfulness practices, supported by flexible policies, and given access to mental health resources reported not only improved personal well-being but also better student engagement and academic performance. The findings emphasize the importance of systemic, sustained efforts to support teachers, both individually and institutionally. By implementing structured well-being programs, managing workloads, and fostering open communication, educational institutions can play a pivotal role in enhancing teacher performance and, ultimately, student success.

DISCUSSION

The findings of thematic analysis emphasized the relationship between teacher stress, well-being initiatives and classroom performance. The data collected from participants suggested that stress significantly impacts teaching effectiveness, but institutional support and well-being initiatives can play an important role in mitigating its adverse effects.

Teacher Stress and Classroom Performance: The themes related to workload management and clarity in communication and expectations supports H1 that "There is a significant relationship between teacher stress and classroom performance" and H2 that "Increased teacher stress negatively impacts teaching effectiveness".

Participants highlighted that excessive workloads, unclear expectations from administration and lack of structural support contribute to high stress levels, which will result in decreased teaching efficiency and disruptions in classroom dynamics. For example, teachers who struggled with unmanageable workloads reported feeling of overwhelmed, which negatively affected their instructional quality. Additionally, when leadership provided clear expectations and structured support, teachers reported feeling more in place, which improved their classroom performance.

Well-Being Initiatives and Stress Reduction: The themes of Engagement in Well-Being Initiatives and Institutional Support for Teacher Well-Being strongly supports H3 that is "Well-being initiatives significantly reduce teacher stress and enhance teaching performance." Teachers who inculcated mindfulness practices, relaxation techniques and self-care routines reported that these strategies improved their emotional regulation and patience in the classroom. Moreover, access to mental health resources and professional development in stress management was prioritized as an essential factor in reducing burnout and improve job satisfaction. Institutions that invested in teacher well-being programs or workshops resulted in a positive work culture, allowing teachers to manage stress more effectively and maintain high levels of teaching performance.

Impact of Teacher Stress on Student Engagement: The themes of impact of Well-Being Initiatives on Teaching and Student Outcomes supports H4 that is "Higher levels of teacher stress lead to disruptions in classroom dynamics, lower student engagement and reduced instructional quality." Teachers repeatedly reported on stress negatively influencing their classroom interactions. When most of the teachers were stressed, they found out that it was harder to engage students, maintain patience and create a positive learning environment. However, teachers who adopted Well-Being initiatives and stress management strategies observed a significant improvement in student behaviour and engagement. A well-balanced teacher fostered a calm and supportive atmosphere, enhancing student participation and learning outcomes.

Thematic analysis provided strong evidence that addressing teacher stress through structured well-being initiatives and institutional support leads to improved teaching effectiveness and positive classroom dynamics. The acceptance of H1, H2, H3 AND H4 suggest that reducing stress and promoting well-being among teachers in very important for maintain high quality education. Colleges should prioritize teacher well-being programs, workload management strategies and clear institutional communication to ensure a healthier and more effective teaching environment

Discussion of Participants Responses

Understanding stressors: Participants responses emphasised the multifaceted nature of stress in the teaching profession, specially its impact on daily responsibilities and overall well-being. *Participant 01- "Balancing administrative duties with lesson planning and student engagement", Participant 02- "Managing diverse learning needs in the classroom, especially in inclusive environments" and Participant 03- "The emotional toll of addressing students' personal struggles, including mental health issues."* These challenges may often lead to high pressure and decreased capacity to maintain quality instruction.

Teachers workload was an important factor affecting teacher's routine and classroom performance. *Participant 01- "My workload often extends into the evenings and weekends, leaving little time for personal life" and Participant 02- "The sheer volume of grading and paperwork often means that I have to rush through lesson planning and preparation."* Participates reported that their workload often extended beyond college hours, leaving a very little room for their personal life or adequate lesson preparation. *Participant 04- "Workload affects my energy levels, making me feel drained and less enthusiastic about teaching", Participant 05- "It can lead to procrastination and stress, as I try to manage competing priorities with limited time" and Participant 09- "The workload increases during certain periods (e.g., grading seasons), which can leave me feeling overwhelmed and underprepared." Teacher* Stress was linked to reduce energy level in teachers, procrastination and feeling of being unprepared during class hours

Administrative tasks were noted as significant contributing factor of stress, consuming a lot of time and energy of the teachers other than dedicated to teaching. In semi structured interview participants expressed frustration over paperwork and meetings, which often lead to lack of student interaction and added to teacher's mental load. *Participant 03- "I feel stressed because administrative tasks are often seen as more important than the teaching itself" and Participant 09- "It can feel like an ongoing juggling act between teaching duties and administrative responsibilities, leading to burnout."* The juggling act between teaching and administrative work often lead to feelings of burnout and reduced in productivity

Teachers dealing with difficult student behaviour, most of the teachers used strategies like positive reinforcement, conflict resolution techniques, restorative justice practices. *Participant 01- "I try to remain calm and use positive reinforcement to address behaviour while maintaining firm boundaries", Participant 05- "I use conflict resolution strategies, like mediation or role-playing, to help students navigate disagreements" and Participant 08- "In challenging situations, I prioritize restorative justice practices to rebuild relationships and prevent future issues."* All these techniques were effective but they require emotional resilience and more effort, which can be paired with existing stressors.

Support from college was seen as inconsistent. Many participants appreciated access to counselling services and flexibility. *Participant 01- "I feel supported, as the administration provides access to counselling services and professional development opportunities" and Participant 04- "The administration is supportive in providing flexibility when stress levels become too high, which I appreciate."* Others participants felt that teacher well-being was not much prioritized, were there was excessive focus on performance metrics and administrative tasks. *Participant 02- "I don't always feel supported because administrative tasks often take priority over teacher well-being", Participant 07- "Sometimes, the administration doesn't seem to understand the challenges teachers face, which makes me feel isolated in managing stress" and Participant 09- "The support from administration is inconsistent, and it often feels like more focus is placed on performance metrics than on teacher well-being."* Lack of sufficient resources for mental health and work life balance left many teachers feeling unsupported in managing stress (Participate 01).

By this we understand that the need for colleges to implement holistic well-being initiatives that shall address workload management, provides teachers emotional support, and foster a supportive environment. By prioritizing teacher well-being, colleges can enhance classroom performance and can create a more sustainable ecosystem.

Impact of stress on performance: By observing the responses from the participate its very clear and direct impact of teacher stress on classroom performance and dynamics. Stress can vanish teacher's patience, their creativity, and their emotional engagement, which can lead to challenges in maintaining effective classroom management and delivering quality the content what they have prepared for. For instance, *Participant 01 reported that "Stress makes me more irritable, which can affect my tone and interaction with students, impacting classroom dynamics" and participate 05 "High stress levels make me more prone to distractions, leading to less effective classroom management and reduced student attention."* Stress also reduced teacher's ability to plan and adapt, which lead some teachers to rush their lessons and less one on one attention to their students *Participant 03- "Stress can lead me to rush through lessons, which reduces the quality of student engagement and learning" and Participant 08- "Stress affects my ability to adapt to students' individual needs, leading to less personalized support and feedback."*

During high stress engaging with students becomes harder. Teachers described withdrawing emotionally and focusing on completing tasks, which resulted in diminished rapport and interaction with students. *Participant 01- "Yes, I tend to withdraw more and don't engage with students as much when I'm feeling stressed", Participant 03- "Stress makes me more likely to focus on the task at hand rather than fostering meaningful discussions or student involvement" and Participant 07- "When I'm stressed, I tend to rely more on lectures and less on student-centred activities or discussions."* This lack of engagement affected the flow of meaningful conversations and can reduces student motivation and participation. For instance, *Participant 10 reported that "I find that when stressed, I become less open to students' ideas and contributions, limiting the flow of conversation in the classroom."*

The effects of teacher stress on classroom dynamics and student outcomes are very much concerning. Participate 01 reported high stress can create a tense atmosphere that will impact student focus and confidence. *Participant 01- "High stress can lead to a tense atmosphere, making students anxious, which in turn affects their ability to concentrate and learn."* If we observe that participate 05,06 and 08 spoke about teacher's inability to provide consistent feedback or manage disruption effectively can lead to confusion, decreased motivation and lower academic performance. *Participant 05- "High stress levels can lead to inconsistency in my feedback, which might confuse students and affect their motivation", Participant 06- "Stress can cause me to become less flexible, meaning students have fewer opportunities to explore topics in depth or in creative ways" and Participant 08- "If I'm stressed, I may not manage classroom disruptions as effectively, leading to negative behaviour from students."* Stress also vanishes teacher's enthusiasm, which might impact the classroom environment and overall quality of education (participate 09). *Participant 09- "Stress can diminish my enthusiasm, making it harder to inspire students and leading to a less motivating classroom environment."*

Many participants shared how stress can manifest in immediate challenges. For example, *Participant 01 reported that "Yes, during an exam week, I was so stressed with grading and meeting deadlines that I didn't manage to explain the exam material clearly, leaving some students confused."*

Other participants reported forgetting important tasks such as *Participant 02 reported "I recall a time when I was overwhelmed by personal issues, and it caused me to forget about a scheduled group activity, which disappointed the students." Participant 07 reported "Once, during a particularly stressful teaching period, I forgot to assign important homework, which led to confusion and frustration among my students."*

By observing the responses of the participants we can observe there is an immediate requirement for the colleges to support to help teachers manage stress. Its high time to prioritize teacher well-being through mindfulness training, peer support groups, and workload management strategies, college can talk about the effects of stress, leading to improved classroom dynamics and better student outcomes.

Exploring Well-Being: Talking about teachers well-being is a holistic concept which includes emotional, physical and mental health which is defined by our participates. It includes achieving a good balance between professional life and personal life, managing stress effectively and maintain a sense of fulfilment. Participant *01- "Teacher well-being is the overall physical, emotional, and mental health that allows a teacher to perform their job effectively while maintaining a sense of fulfilment and balance in life", Participant 03- "Teacher well-being involves a balance of personal self-care, professional satisfaction, and positive relationships, enabling me to be my best in both my personal and professional life" and Participant 09- "It's about emotional resilience, physical health, and mental clarity, ensuring that I can stay present for my students while maintaining a sense of happiness and personal fulfilment."*

Participants 08 and 10 emphasized the importance of being valued, having manageable workloads and access to recourses to balance in both personal and professional life. *Participant 08- "I define teacher well-being as a holistic state where a teacher feels valued, supported, and has the resources and time needed to take care of themselves both professionally and personally" and Participant 10- "Teacher well-being to me means feeling respected and valued in the workplace, having a manageable workload, and having access to the tools and support needed to thrive in both my personal and professional life."*

Many teachers to maintain their well-being they completely depend on strategies such as mindfulness, exercise and self-care. *Participant 01- "I practice mindfulness and meditation regularly to manage stress and clear my mind after a busy day at work", Participant 02- "I make sure to carve out time for physical exercise, whether it's yoga, walking, or just stretching, to help relieve tension" and Participant 05- "I make self-care a priority, whether it's indulging in a hobby, spending quality time with loved ones, or simply having quiet time to myself."*

Setting boundaries after work such as limiting work tasks after the college hours and they engaged in creative or relaxing activities were also effective approaches to our participate 04 and 07. *Participant 04- "I set boundaries by limiting work-related tasks outside of school hours and ensuring I have time to relax in the evenings" and Participant 07- "I engage in creative outlets, like reading, writing, or drawing, to help me disconnect from work and engage my mind in different ways."* Participates also spoke about the importance of having a healthier support system from family, colleagues, and friends (Participate 06 and 09) *Participant 06- "I lean on a support system of friends, family, and colleagues to share my challenges and seek advice when needed" and Participant 09- "I prioritize sleep and nutrition, making sure I get enough rest and eat well to keep my energy levels up."*

Irrespective of the individual efforts it's important that the colleges support for well-being, but unfortunately it's not inconsistent. Most of the teachers reported that their colleges lack well-being programs, leaving teachers to manage their stress independently (Participants 01, 03, and 06). *Participant 01- "No, my college does not have any formal well-being programs. I think such initiatives would make a big difference in supporting teachers", Participant 03- "No, we don't have any well-being initiatives, and it feels like teacher well-being is not a priority for the administration" and Participant 06- "My college does not offer any well-being initiatives. I believe such programs could help us better manage stress and improve our work environment."*

While few colleges offer wellness weeks, mindfulness workshops or peer support groups, their infrequency diminishes their impact (Participants 07, 08, and 10). *Participant 07: "Yes, my school organizes mindfulness workshops, but they happen infrequently, and it would be better if they were more regular", Participant 08*: *"We have a wellness program with stress management seminars and yoga sessions, which I find helpful for reducing stress" and Participant 10*: *"Our school holds an annual wellness week with activities like meditation and fitness challenges. It's effective but could be more consistent throughout the year."* A strong light must be throwed on regular and structured programs such as counselling services or stress management seminars were also suggested as a way to support teachers.

Work-life balance is recognized universally for preventing burnout and maintaining effectiveness. Few teachers such as participant maintained this balance by prioritizing their personal time, planning ahead and avoiding over commitment (Participants 02, 06, and 08). *Participant 02- "It's crucial for my well-being, so I ensure I prioritize personal time, especially on weekends, to recharge and disconnect from work" Participant 06- "It's essential to my mental health, so I maintain work-life balance by keeping a strict boundary between work and personal time" and Participant 08- "Work-life balance is crucial for avoiding burnout, and I manage it by setting realistic expectations and saying no when needed to preserve my personal time."* To add on teachers with clear boundaries and time management strategies found it easier to recharge and remain passionate about their teaching profession.

Peers supporting emotional is an important factor to manage stress. Participants often depended on their colleagues for advice, encouragement and camaraderie (Participants 01, 03, and 09). *Participant 01- "Yes, I feel emotionally supported by my peers, as we have a strong network where we can share challenges and offer advice", Participant 03- "My peers are always there to lend a helping hand and provide emotional support, which makes a big difference in how I manage stress" and Participant 09- "I feel emotionally supported by my colleagues, especially during tough times, and I think our collaborative spirit really helps with stress management."* Moreover, emotional support from leadership was seen inconsistent with some teachers wising for more proactive check-ins and acknowledgment of their challenges (Participants 02, 06, and 08). *Participant 02- "I feel somewhat supported by my leadership team, but I believe more regular check-ins would help strengthen the emotional support provided", Participant 06- "Sometimes, I don't feel enough emotional support from the leadership team, as they tend to focus more on administrative tasks than teacher well-being" and Participant 08- "I wish I felt more emotionally supported by my leadership team, as I believe they could do more to check in on teachers' well-being."*

By observing the responses, we can understand that the colleges need to adopt for well-being initiatives, which must include regular workshop, easily accessible mental health resources and good communication from leadership. By fostering a supportive environment, colleges can help teachers maintain their well-being, ensuing a good motivation and better outcomes for students.

Role of Well-Being Initiatives: Responses from the participates shows us role of wellbeing initiatives plays an important role in reducing teacher stress and it enhances classroom performance. Teachers who participated in stress management, mindfulness, and self-care programs reported reduced anxiety, improved concentration, and greater emotional resilience

(Participants 01, 03, and 09). *Participant 01-"Yes, I participated in a mindfulness program that focused on deep breathing and meditation. It helped me feel more centred and less reactive to stress", Participant 03- "Yes, I took part in a self-care retreat that emphasized physical health, mental relaxation, and emotional resilience. It gave me a fresh perspective on managing stress" and Participant 09- "Yes, I participated in a group stress management program that helped me understand the importance of emotional regulation. It has improved my classroom patience."* Teachers remained calmed during challenging situations because of mindfulness sessions they have attended, while stress management workshops helped teachers with practical tools like time management and relaxation requires continuous effort (Participates 06 and 10). *Participant 06- "Yes, I participated in a stress reduction seminar. The strategies were useful, but I found I needed to apply them more consistently to see lasting effects" And Participant 10- "I engaged in a mindfulness training program and learned to approach stressful situations more calmly. However, integrating it into my daily routine takes ongoing effort."*

Teachers suggested that to support teacher well-being, it is important college starts implementing flexible working conditions, such as work from home options for administrative tasks and mental health days (Participants 01 and 10). *Participant 01- "The college could offer more flexible working hours or work-from-home options, especially for grading or administrative tasks, to help reduce stress" and Participant 10-"The college could create a more open environment for discussing mental health, encouraging teachers to seek help without fear of judgment or stigma."*

Teachers suggested certain strategies for mitigating stress such as reducing class size, providing administrative support, and addressing behavioural issues (Participants 01, 03, and 05). A good access from college to use professional development programs which focused on stress management and mental health resources were also emphasized (Participants 06 and 08). Peer support was also highlighted as a way to coping strategies (Participant 09).

Well-being initiatives have a direct impact on teacher's performance and student outcome. Decrease stress enables teachers to remain focused, patient, and creative, which enhances student engagement and fostering a positive classroom environment (Participants 01, 02, and 05). *Participant 01- "I would like to see policies that encourage reasonable class sizes and manageable teaching loads to reduce the stress of balancing multiple responsibilities", Participant 03- "More administrative support for grading, paperwork, and other tasks could free up teachers' time and reduce stress from having to juggle everything" and Participant 05 "A clear and structured plan for handling student behavioural issues in the classroom, with support from the administration, could alleviate stress for teachers."* Teachers who are supported and valued by their colleagues and family are more consistent in their teaching approach, and which can contribute to student motivation and better learning outcomes Therefore, effective well- being programs or workshop can improve teacher retention, and their continuity and stability in education

It's very important to prioritize teacher well-being through consistent initiatives not only for the professional and personal lives of the teachers but to also create a healthier college culture which can benefit the students. Colleges must opt for holistic approaches, inculcating flexible policies, mental health accessibility to build a resilient and effective teaching workspace.

Personal Reflections: Personal Reflections of participants gives us an outline in understanding and addressing stress and well-being throughout their teaching careers. By observing we can understand that most of the teachers struggled to manage stress which resulted in neglecting self-care and over committing to their professional responsibilities (Participants 01, 02, and 05). *Participant 01- "In the beginning, I didn't recognize the signs of stress, but over time, I've learned to set boundaries and prioritize self-care to protect my well-being", Participant 02- "Early in my career, I used to overcommit and neglect my own needs. Now, I've learned the importance of taking breaks and saying no when necessary" and Participant 05- "At first, I didn't pay much attention to my mental health, but I've grown to understand that managing stress is key to sustaining a long and successful teaching career."* However, we can observe that they developed strategies such as setting boundaries, practicing mindfulness, and seeking support to maintain their work and personal life (Participants 03, 06, and 09). *Participant 03- "Initially, I struggled with the workload and emotional demands of teaching, but with time, I developed better coping strategies like time management and mindfulness", Participant 06- "Over the years, I've learned to recognize when I'm reaching my limits and developed better self-care habits, which has greatly improved my ability to handle stress" and Participant 09- "At the start, I was constantly feeling burned out, but with experience, I've found ways to manage stress*, like seeking support from colleagues and utilizing stress-reduction techniques." This emphasized the importance of experience and self-care in opting to the demands of teaching.

Experienced Teachers advice for new teachers on managing stress and prioritizing wellbeing in their very initial of their careers. They also suggested that it is very important to set a clear boundary, to avoid over commitment, and to maintain a good support environment (Participants 01, 03, and 07). *Participant 01-"Start by setting clear boundaries with your time, and don't feel guilty for taking breaks or saying no to additional tasks if it affects your well-being", Participant 03- "Establish a strong support system with colleagues, mentors, and friends who understand the challenges of teaching and can offer advice and encouragement" and Participant 07-"Find a work-life balance early in your career to avoid burnout. Set aside time for yourself and your loved ones, and don't let work consume your entire life."*Teachers also emphasized the importance of physical health, mental health, relaxation and self-care (Participants 04, 06, and 08). *Participant 04- "Take care of your body by exercising, eating well, and getting enough rest, your physical health is crucial for managing stress", Participant 06- "Make self-care a priority by scheduling regular breaks, relaxation time, and hobbies that bring you joy outside of teaching" and Participant 08- "Be kind to yourself, teaching is challenging, and it's okay to make mistakes. Give yourself permission to learn and grow without putting too much pressure on yourself."*

It is very important to have organizational skills and professional development in stress management to help upcoming teachers to navigate the challenges of their profession (Participants 09 and 10). *Participant 09- "The ideal teaching environment would have a strong focus on professional and emotional well-being, with dedicated time for relaxation and reflection to recharge during the workday" and Participant 10- "Teachers would have access to an effective support system for handling student behaviour issues, administrative tasks, and personal concerns, ensuring that their workload and emotional demands are manageable."*

The participates vision of an ideal teaching environment underscores the need for colleges support to foster teacher well-being and performance. For instances collaborative work culture where teachers can share emotional support, workloads, and a decent class size (Participants 01, 02, and 05). *Participant 01- "An ideal teaching environment would have a collaborative culture where teachers support one another, share resources, and offer emotional support to handle stress together", Participant 02- "It would offer reasonable class sizes and manageable workloads to allow teachers to focus on delivering quality lessons and maintain their well-being", and Participant 05- "The ideal environment would include a balance of autonomy and collaboration, where teachers have the freedom to manage their classrooms but also the support of a strong network of colleagues."*

An easy access to wellness workshop, mental health resources and a flexible working culture were also emphasized as important component of a supportive working environment (Participants 03, 04, and 06*). Participant 03- "Teachers would have access to wellness programs, including stress management workshops and mental health resources, to help them cope with the demands of teaching", Participant 04- "In my ideal environment, leadership would prioritize teacher well-being by providing regular check-ins, offering mental health days, and ensuring teachers feel supported" and Participant 06- "It would feature flexible work policies, like the option to work from home when needed, and an understanding of the importance of personal time for mental health."*

Leadership plays an important role in creating a good environment, were participants advising for transparent communication, regular check-ins and open discussion on stress and challenges (Participants 04 and 08). *Participant 04- "In my ideal environment, leadership would prioritize teacher well-being by providing regular check-ins, offering mental health days, and ensuring teachers feel supported" and Participant 08- "A supportive leadership team would encourage transparency and open communication about stress and challenges, offering guidance and support without judgment."* A time for relaxation, reflection and professional development opportunities were also identified as an important factor for teacher's growth and emotional well-being (Participants 07 and 09). *Participant 07- "In an ideal setting, teachers would be given the resources and time to focus on professional development, ensuring their growth without feeling overwhelmed by expectations" and Participant 09- "The ideal teaching environment would have a strong focus on professional and emotional well-being, with dedicated time for relaxation and reflection to recharge during the workday."*

By this we can understand that it is important for both personal strategies and college initiatives in addressing teacher stress and promoting well-being. By having a supportive and balanced teaching environment, colleges can help teachers thrive, ensuring both their personal satisfaction and professional effectiveness.

CONCLUSION

The study emphasised on the impact of stress on teacher's well-being and classroom performance, disclosing the multiple challenges they face in balancing their personal lives and their professional lives. Teachers experience high stress because of administrative work, workload, managing a diverse classroom and addressing student's emotional needs. All this stress directly affects teacher's patience, creativity, and their ability to engage with students effectively which might lead to decrease in classroom dynamics and learning outcomes of the students.

The findings emphasised on the immediate need for the colleges to prioritize teacher's well-being. Colleges must identify the important role teacher's play and implement continuous support system to decrease their stress. Colleges must encourage work-life balance, fostering emotional support, and making sure that the colleges provides mental health resources and stress management for maintaining teacher satisfaction and effectiveness.

Teaching environment must be supportive by clear communication, manageable workloads and regular well-being programs, which can decrease stress and enhance teacher's retention. By prioritizing teacher's well-being colleges not only improve the quality of the teachers personal and professional lives but also foster a positive learning environment for students.

Recommendations

Based on the study, it is recommended that the colleges start implementing well-being programs, such as mindfulness workshops, stress management workshops and counselling services to identify teachers stress effectively. Reworking on administrative works, providing flexible working hours and assigning proper resources for classroom management, can slowly reduce workload pressures. Leadership should work on transparent communication, providing professional development programs on stress management and creative teaching strategies, and conducting regular check-ins. Colleges should create a collaborative work culture through team building activities, peer support groups and colleagues sharing responsibilities can enhance emotional resilience. Colleges must support and encourage work-life balance, time management and most importantly identifying and recognizing teacher's efforts can improve teacher's satisfaction, eventually contributing to better outcomes and positive educational environment, contributing to the overall success of the education system.

Limitations of the Study

While this study provides valuable insights into the relationship between teacher stress, institutional well-being initiatives, and classroom performance, it is important to acknowledge several limitations that may influence the interpretation and applicability of the findings. First, the study relies solely on qualitative data gathered from semi-structured interviews. While this approach allows for in-depth exploration of personal experiences, the responses may be influenced by individual perceptions and biases, which could affect the objectivity of the findings. As such, the results may reflect the subjective views of the participants and may not be fully generalizable to the broader population of educators.

Secondly, the sample size is limited to ten teachers from a single Indian institution, which constrains the scope of the study. The lack of diversity in the participant pool—both in terms of geography and educational context—limits the extent to which the results can be applied to educators working in different countries, school systems, or socio-cultural environments. A larger and more diverse sample would have enhanced the external validity of the findings and offered a more comprehensive understanding of how teacher stress manifests across various educational settings.

Another key limitation is the cross-sectional nature of the study. Because the data was collected at a single point in time, it does not capture the long-term impact of well-being initiatives on teacher performance or psychological health. A longitudinal design, involving repeated observations over an extended period, would have allowed for a more accurate assessment of whether well-being strategies produce sustained improvements in teaching effectiveness and classroom engagement. Without this temporal perspective, the study cannot conclusively determine the lasting impact of the interventions.

Additionally, while the study focuses on institutional stressors such as workload, administrative pressure, and access to mental health resources, it does not consider external influences like family responsibilities, socio-economic conditions, or cultural expectations. These external factors may significantly affect teacher stress and well-being but fall outside the current study's framework. Including such variables in future research could provide a more holistic view of the sources and effects of teacher stress.

Finally, the study does not incorporate a comparative analysis between teachers who actively engage in well-being programs and those who do not. Such a comparison could have offered stronger evidence regarding the effectiveness of the interventions discussed. Without a control group or contrasting experiences, it is difficult to isolate the specific impact of the well-being initiatives from other variables influencing teacher performance. Future studies that incorporate a comparative or mixed-methods approach could help strengthen the validity of these findings and offer deeper insights into best practices for reducing teacher stress.

REFERENCES

Aydin, B., & Kaya, A. (2016). Sources of stress for teachers working in private elementary schools and methods of coping with stress. *Universal Journal of Educational Research*, *4*(12A), 186–195. DOI: 10.13189/ujer.2016.041324

Blase, J. (1986). A qualitative analysis of sources of teacher stress: Consequences for performance. *American Educational Research Journal*, *23*(1), 13–40. DOI: 10.3102/00028312023001013

Carroll, A., York, A., Fynes-Clinton, S., Sanders-O'Connor, E., Flynn, L., Bower, J., Forrest, K., & Ziaei, M. (2021). The downstream effects of teacher well-being programs: Improvements in teachers' stress, cognition, and well-being benefit their students. *Frontiers in Psychology*, *12*, 689628. Advance online publication. DOI: 10.3389/fpsyg.2021.689628 PMID: 34276519

Center for Mindfulness in Medicine, Health Care, and Society, University of Massachusetts Medical School. (2017). *Mindfulness-Based Stress Reduction (MBSR) curriculum guide 2017*.https://mbsr.website/sites/default/files/docs/mbsr-curriculum-guide 2017.pdf?form=MG0AV3

Collie, R., Shapka, J., & Perry, N. (2012). School climate and social-emotional learning: Predicting teacher stress, job satisfaction, and teaching efficacy. *Journal of Educational Psychology*, *104*(4), 1189–1204. DOI: 10.1037/a0029356

Department of Education and Training. Victoria. (n.d.). High-impact well-being strategies. Retrieved from https://www.education.vic.gov.au/Documents/school/teachers/teachingresources/practice/High_Impact_Wellbeing_Strategies.pdf

Doherty, S. (2024, April 11). Teacher stress - Teachers UK. *Teachers UK*.https://www.teachers-uk.co.uk/teacher-stress/

Eloor, A. G., & Menon, S. D. (2024). The work-life balance of teachers: A review. *International Journal of Research in Management*, *6*(1), 52–56. DOI: 10.33545/26648792.2024.v6.i1a.125

Flook, L., Goldberg, S., Pinger, L., Bonus, K., & Davidson, R. (2013). Mindfulness for teachers: A pilot study to assess effects on stress, burnout, and teaching efficacy. *Mind, Brain and Education : the Official Journal of the International Mind, Brain, and Education Society*, *7*(3), 182–195. DOI: 10.1111/mbe.12026 PMID: 24324528

Franco, L., Picinin, C., Pilatti, L., & Franco, A. (2021). Work-life balance in higher education: A systematic review of the impact on the well-being of teachers. *Ensaio: Avaliação e Políticas Públicas em Educação, 29*(111). https://doi.org/DOI: 10.1590/s0104-403620210002903021

Gonzales, A. (2024). Work-related stress and teaching performance of public elementary school teachers of San Francisco Cluster IV. *International Journal of Innovation and Scientific Research*, *11*(5), 203–210. DOI: 10.51244/IJRSI.2024.1105003

Government of India. Ministry of Education, & Department of Higher Education. (2021). *All India Survey on Higher Education 2021-22*. Retrieved from https://cdnbbsr.s3waas.gov.in

Gustems-Carnicer, J., Calderón, C., & Calderón-Garrido, D. (2019). Stress, coping strategies, and academic achievement in teacher education students. *European Journal of Teacher Education*, *42*(3), 375–390. DOI: 10.1080/02619768.2019.1576629

Hao, L. (2022). Analysis of the effect of work stress and organizational culture on teacher performance. *International Journal of Progressive Sciences and Technologies*, *34*(1), 45–50. DOI: 10.52155/ijpsat.v34.1.4503

Jeon, L., Buettner, C., & Grant, A. (2018). Early childhood teachers' psychological well-being: Exploring potential predictors of depression, stress, and emotional exhaustion. *Early Education and Development*, *29*(1), 53–69. DOI: 10.1080/10409289.2017.1341806

Jõgi, A., Aulén, A., Pakarinen, E., & Lerkkanen, M. (2022). Teachers' daily physiological stress and positive affect in relation to their general occupational well-being. *The British Journal of Educational Psychology*, *93*(2), 368–385. DOI: 10.1111/bjep.12561 PMID: 36336902

Karjalainen, S., Brännström, J., Christensson, J., Sahlén, B., & Lyberg-Åhlander, V. (2020). A pilot study on the relationship between primary-school teachers' well-being and the acoustics of their classrooms. *International Journal of Environmental Research and Public Health*, *17*(6), 2083. DOI: 10.3390/ijerph17062083 PMID: 32245168

Klassen, R., & Chiu, M. (2010). Effects on teachers' self-efficacy and job satisfaction: Teacher gender, years of experience, and job stress. *Journal of Educational Psychology*, *102*(3), 741–756. DOI: 10.1037/a0019237

Klusmann, U., Richter, D., & Lüdtke, O. (2016). Teachers' emotional exhaustion is negatively related to students' achievement: Evidence from a large-scale assessment study. *Journal of Educational Psychology*, *108*(8), 1193–1203. DOI: 10.1037/edu0000125

Kukalo, E. (2024, September 25). 5 Secrets to Achieving a Perfect Teacher Work-Life Balance. *actiTIME*. https://www.actitime.com/school-management-software/teacher-work-life-balance

Liao, J., Wang, X., & Wang, X. (2023). The effect of work stress on the well-being of primary and secondary school teachers in China. *International Journal of Environmental Research and Public Health*, *20*(2), 154. DOI: 10.3390/ijerph20021154 PMID: 36673909

McCarthy, C., Lambert, R., Lineback, S., Fitchett, P., & Baddouh, P. (2016). Assessing teacher appraisals and stress in the classroom: Review of the Classroom Appraisal of Resources and Demands. *Educational Psychology Review*, *28*(3), 577–603. DOI: 10.1007/s10648-015-9322-6

McLean, L., & Connor, C. (2015). Depressive symptoms in third-grade teachers: Relations to classroom quality and student achievement. *Child Development*, *86*(3), 945–954. DOI: 10.1111/cdev.12344 PMID: 25676719

Mudin, N., Ahmad, A., & Kee, P. (2023). Association of stress on well-being among teachers: A systematic literature review. *International Journal of Academic Research in Progressive Education and Development*, *12*(3), 75–85. DOI: 10.6007/IJARPED/v12-i3/17936

Neuenschwander, R., Friedman-Krauss, A., Raver, C., & Blair, C. (2017). Teacher stress predicts child executive function: Moderation by school poverty. *Early Education and Development*, *28*(7), 880–900. DOI: 10.1080/10409289.2017.1287993

Nwoko, J., Emeto, T., Malau-Aduli, A., & Malau-Aduli, B. (2023). A systematic review of the factors that influence teachers' occupational well-being. *International Journal of Environmental Research and Public Health*, *20*(12), 1–20. DOI: 10.3390/ijerph20126070 PMID: 37372657

Ozturk, M., Wigelsworth, M., & Squires, G. (2024). A systematic review of primary school teachers' wellbeing: Room for a holistic approach. *Frontiers in Psychology*, *15*, 1358424. Advance online publication. DOI: 10.3389/fpsyg.2024.1358424 PMID: 38915431

Pagán-Castaño, E., Sánchez-García, J., Garrigós-Simón, F., & Guijarro-García, M. (2021). The influence of management on teacher well-being and the development of sustainable schools. *Sustainability (Basel)*, *13*(5), 2909. DOI: 10.3390/su13052909

Penttinen, V., Pakarinen, E., Von Suchodoletz, A., & Lerkkanen, M. (2020). Relations between kindergarten teachers' occupational well-being and the quality of teacher-child interactions. *Early Education and Development*, *31*(7), 994–1010. DOI: 10.1080/10409289.2020.1785265

Qin, H. (2024). Stress level of classroom instructors and its influence on their classroom performance. *World Journal of Educational Research (Los Angeles, Calif.)*, *11*(1), p29. Advance online publication. DOI: 10.22158/wjer.v11n1p29

Ruggeri, K., Garcia-Garzon, E., Maguire, Á., Matz, S., & Huppert, F. A. (2020). Well-being is more than happiness and life satisfaction: A multidimensional analysis of 21 countries. *Health and Quality of Life Outcomes*, *18*(1), 192. Advance online publication. DOI: 10.1186/s12955-020-01423-y PMID: 32560725

Shi, J. (2024). Stress level of full-time university classroom instructors affecting their classroom performance. *Journal of Education and Educational Research*. https://doi.org/DOI: 10.54097/q70x5r98

Stress. (2022, June 17). *World Health Organization*. https://www.who.int/news-room/questions-and-answers/item/stress

Suri, S., & Sharma, A. (2024). Occupational stress and its effect on teaching performance among school teachers. *International Journal of Applied Educational Studies*, *13*(1), 34–45. DOI: 10.5281/zenodo.8237695

Wettstein, A., Schneider, S., Holtforth, M. G., & La Marca, R. (2021). Teacher stress: A psychobiological approach to stressful interactions in the classroom. *Frontiers in Education*, *6*, 681258. Advance online publication. DOI: 10.3389/feduc.2021.681258

Yang, W., Liu, W., & Liu, X. (2023). Work-related stress among teachers: Interventions and effectiveness. *Educational Psychology*, *43*(4), 783–799. DOI: 10.1080/01443410.2022.2046752

Zembylas, M., & Papanastasiou, E. (2006). Sources of teacher job satisfaction and dissatisfaction in Cyprus. *Compare: A Journal of Comparative Education*, *36*(2), 229–247. DOI: 10.1080/03057920600741289

Zhao, Y., & Li, X. (2023). Mindfulness and teacher resilience: An intervention study in China. *International Journal of Psychology*, *58*(2), 229–241. DOI: 10.1002/ijop.12956

Chapter 11
Spiritual Wellbeing and Self-Efficacy as Catalysts for Culturally Responsive Teaching in Diverse Classrooms

Vijendra Nath Pathak
https://orcid.org/0000-0003-0731-7003
Parul Institute of Liberal Arts, Parul University, India

Mamta
https://orcid.org/0009-0009-7242-7437
School of Social Science and Language, Lovely Professional University, India

ABSTRACT

Integrating spiritual wellbeing and self-efficacy equips educators with the emotional resilience and confidence needed to navigate diverse classrooms effectively. Spiritual wellbeing fosters empathy, patience, and emotional balance, enabling teachers to manage stress, form meaningful connections, and create inclusive environments where students feel valued and motivated. Self-efficacy enhances teachers' belief in their ability to positively impact student learning, driving reflective practices, adaptability, and innovative strategies. Together, these elements empower educators to address diverse needs, foster equity, and create compassionate, adaptive learning environments, benefiting both students and the broader educational system.

INTRODUCTION

Today's classrooms are vibrant and diverse, bringing together students from different cultural, linguistic, and socioeconomic backgrounds. This diversity enriches education by fostering creativity, cultural awareness, and a more global perspective. However, it also presents challenges for educators, who must meet the needs of all students, address biases, and create an inclusive environment where everyone feels valued (Gay, 2018). Traditional teaching methods often take a one-size-fits-all approach and may struggle to accommodate these differences. This is where culturally responsive teaching plays a crucial

DOI: 10.4018/979-8-3373-0472-4.ch011

role, incorporating students' backgrounds into lessons and classroom management to ensure that every learner's identity is recognized and respected (Ladson-Billings, 1995). Culturally responsive teaching is not just about adjusting teaching methods, it requires a shift in perspective. Educators must self-reflect to recognize their biases, build meaningful connections with students, and foster an atmosphere of equity and mutual respect (Hammond, 2015). While this approach is rewarding, it can also be demanding, particularly when balancing academic pressures and limited resources. To successfully navigate these challenges, teachers need internal strengths that promote resilience, adaptability, and confidence.

Two key psychological factors that support educators in diverse classrooms are spiritual well-being and self-efficacy. Spiritual well-being, which encompasses purpose, balance, and connection, helps teachers maintain emotional stability and make ethical decisions. It nurtures empathy and compassion, creating a supportive atmosphere where students feel safe and valued (Nelson, 2009). When educators align their values with their professional roles, they become more responsive to the needs of their students (Tisdell, 2003).

Self-efficacy, or confidence in one's success, is equally vital. Teachers with high self-efficacy are more likely to embrace innovative strategies, persist through challenges, and proactively address diverse learning needs (Bandura, 1997). This belief in their effectiveness fosters continuous growth, allowing educators to create meaningful learning experiences that support academic achievement and emotional well-being (Tschannen-Moran & Hoy, 2001).

Spiritual well-being and self-efficacy create a strong foundation for culturally responsive teaching. Spiritual well-being strengthens emotional resilience and empathy, while self-efficacy fuels the motivation for meaningful change. Educators can develop these qualities through self-reflection, mindfulness, and professional training in cultural competency (Gay, 2018). Collaboration with colleagues, mentorship, and peer support are crucial for building resilience and fostering professional growth (Hammond, 2015).

By cultivating these strengths, teachers can create learning environments that are inclusive, supportive, and equitable, ensuring that all students, regardless of background, have the opportunity to succeed. This chapter explores how spiritual well-being and self-efficacy shape culturally responsive teaching, offering research-based insights and practical strategies to help educators confidently navigate diverse classrooms. By adopting these approaches, teachers can foster an environment where students feel recognized, supported, and empowered to reach their full potential.

UNDERSTANDING KEY CONCEPTS

Spiritual Wellbeing

Spiritual well-being is a vital aspect of human life, focused on finding meaning, purpose, and balance within oneself and with the world. It plays a crucial role in personal growth, resilience, and overall fulfillment, positively influencing mental, emotional, and physical health (Pargament, 1997). This form of well-being is not limited to any particular religion or belief system but embraces a universal sense of connection and unity across diverse experiences (Koenig, 2012).

Core Aspects of Spiritual Wellbeing

Finding Meaning and Purpose

At its core, spiritual well-being involves understanding what gives life meaning and purpose. This includes identifying personal values, recognizing one's place in the broader context of life, and living intentionally. A strong sense of purpose helps provide clarity and strength, especially during difficult times.

Living Authentically

Living in harmony with one's core values and beliefs enhances spiritual well-being. This alignment fosters authenticity, reduces inner conflict, and promotes a deep sense of personal peace and integrity.

Cultivating Connections

A significant component of spiritual well-being is the ability to connect deeply with oneself, others, and the world. These connections help foster self-awareness, empathy, and an appreciation for life's interconnectedness.

Experiencing Inner Peace and Transcendence

Spiritual well-being often involves moments where individuals feel a part of something greater than themselves. These experiences, whether through meditation, creativity, acts of kindness, or time spent in nature, bring tranquillity and fulfillment.

Strength in Adversity

Spirituality offers a profound framework for managing life's difficulties, providing individuals with resilience, hope, and perspective in adversity. It empowers people to approach challenges confidently and with composure, fostering inner strength to cope with stress, loss, and uncertainty. Whether one follows religious or secular practices, spirituality fosters a sense of connection to something greater, offering emotional stability and solace during turbulent times (Pargament, 1997).

For those who adhere to a religious path, prayer, meditation, and rituals deepen spiritual growth by establishing a connection to a higher power, bringing peace and comfort. These practices help individuals stay true to their core values, providing moral clarity during challenging moments. However, spiritual well-being is not confined to religious practices alone. Non-religious activities like mindfulness, creative expression, outdoor spending, or exploring philosophical concepts can also enhance spiritual well-being by encouraging introspection, self-awareness, and emotional balance (Koenig, 2012).

Community involvement and acts of service are also integral to nurturing spiritual health. Being part of a supportive network and contributing to the welfare of others helps individuals feel a more profound sense of purpose and connection. Acts of kindness or volunteer work enrich the spirit by fostering a sense of fulfillment and reinforcing the belief that one's life has meaning. Additionally, personal reflection, whether through journaling or quiet contemplation, allows individuals to deepen their understanding of themselves and align their actions with their values, offering greater clarity and purpose (Wong, 2016).

Relevance to Teaching

Spiritual well-being is vital in teaching, helping educators approach their work with resilience, patience, and empathy. Teachers who cultivate spiritual well-being are likelier to possess the emotional and psychological tools to engage with students from diverse backgrounds. This quality is essential for understanding students' varied perspectives and needs, particularly in multicultural or socioeconomically diverse classrooms (Palmer, 1998).

Spiritual well-being encourages teachers to adopt a non-judgmental and open attitude, allowing them to create inclusive environments where all students feel valued. It promotes a deeper connection with students, enabling educators to appreciate the cultural and individual experiences that each learner brings to the classroom. This connection fosters a classroom climate that prioritizes respect, equity, and the celebration of diversity (Tisdell, 2003).

Additionally, spiritual well-being supports emotional stability, allowing teachers to remain calm, centered, and adaptable when dealing with challenges. Whether addressing conflicts or adjusting to marginalized students' needs, teachers with intense spiritual well-being can maintain a positive and supportive learning environment. They inspire similar student qualities by modeling patience, empathy, and ethical behavior, cultivating a culture of mutual respect (Miller, 2000).

Ultimately, spiritual well-being empowers educators to create meaningful and inclusive learning experiences, fostering classrooms where diversity is embraced, and every student has the opportunity to succeed.

Self-Efficacy

As psychologist Albert Bandura conceptualized, self-efficacy reflects an individual's belief in their ability to carry out tasks and achieve specific goals. In education, self-efficacy represents a teacher's confidence in their skills to engage students, manage the classroom environment, and deliver compelling learning experiences despite challenges. This belief profoundly impacts how educators approach their profession, influencing their resilience, adaptability, and overall effectiveness in fostering academic and personal growth among students (Bandura, 1997).

Key Aspects of Teacher Self-Efficacy

Encouraging Active Learning

Teachers with strong self-efficacy demonstrate a belief in their capacity to make learning engaging and inclusive for all students. They utilize creative approaches to address different learning styles, ensuring every student feels motivated to participate and achieve their potential.

Building a Supportive Classroom Environment

A cornerstone of self-efficacy is confidence in classroom management. Educators with high self-efficacy maintain order and foster respect while addressing behavioral challenges constructively. They create spaces where students feel valued and safe, promoting academic and personal development.

Achieving Meaningful Educational Outcomes

Teachers with a robust sense of self-efficacy are goal-oriented and solution-focused. They approach challenges as opportunities to refine their strategies, adapt their methods to diverse needs, and inspire students to reach their academic goals.

Role in Culturally Responsive Teaching

Self-efficacy plays a crucial role in the successful implementation of culturally responsive teaching. Educators with high self-efficacy are more likely to view diversity as an asset, embracing their students' unique needs and strengths. They are also more open to trying new teaching strategies, modifying their approaches, and integrating content that resonates with their learners' cultural backgrounds (Gay, 2018).

Teachers with strong self-efficacy are also more proactive in creating inclusive lessons that reflect diverse perspectives and learning styles. They will likely engage in continuous professional development to refine their understanding of cultural differences and improve their teaching strategies. This proactive mindset encourages teachers to build classrooms where all students feel understood, respected, and empowered to succeed (Siwatu, 2011).

In addition, self-efficacy fosters perseverance and flexibility. Teachers with high self-efficacy are more likely to overcome obstacles and remain committed to their student's success, even when faced with limited resources or institutional challenges. Their resilience helps them focus on creating equitable learning opportunities, which is critical in diverse educational settings (Bandura, 1997).

Self-efficacy, when combined with spiritual well-being, provides educators with emotional fortitude and practical skills. This balance allows teachers to engage in culturally responsive teaching with compassion and confidence, ensuring their classrooms are places where all students feel valued and supported.

THEORETICAL FOUNDATIONS

Bandura's Social Cognitive Theory

Albert Bandura's Social Cognitive Theory (SCT) highlights the interconnected influence of cognitive processes, behaviors, and environmental factors on learning and adaptation. At the core of this theory is self-efficacy, or an individual's belief in their ability to accomplish tasks and achieve goals. This belief significantly shapes motivation, persistence, and overall performance, particularly in dynamic and challenging environments like education.

Self-Efficacy in the Educational Context

Teaching is as much about confidence as it is about knowledge. When educators believe in their ability to make a difference, it shapes everything they do, from teaching to connecting with their students. Teachers with a strong sense of self-efficacy trust in their skills to create meaningful, inclusive learning environments where every student, regardless of their background or abilities, feels seen and supported (Bandura, 1997). They are not afraid to experiment with new teaching methods, whether it is through interactive activities, personalized learning plans, or culturally responsive lessons. Their confidence allows them to adapt and meet students where they are, making learning a more engaging and enriching experience for everyone (Tschannen-Moran & Hoy, 2001).

Beyond teaching strategies, self-efficacy also plays a significant role in classroom management. Teachers who believe in maintaining a positive learning space handle challenges with patience and clarity. They approach disruptions not with frustration but with understanding, ensuring that their classrooms remain places of respect and open communication. This sense of assurance creates an environment where students feel comfortable asking questions, sharing ideas, and truly participating in their education (Hoy & Spero, 2005).

Resilience is another hallmark of teachers with strong self-efficacy. They see opportunities rather than setbacks when faced with obstacles like limited resources or difficult classroom situations. Instead of feeling defeated, they tap into their creativity and problem-solving skills, setting an example for their students about perseverance and growth. Their adaptability allows them to incorporate diverse teaching practices, collaborative projects, differentiated instruction, and innovative learning techniques designed to support every student's unique needs (Skaalvik & Skaalvik, 2010).

On the other hand, when teachers struggle with self-doubt, it can make the job feel overwhelming. They may hesitate to step outside their comfort zones, sticking to rigid teaching methods that do not always work for today's diverse classrooms. This lack of confidence can lead to frustration, burnout, and disconnect from their students, making it harder to create a genuinely engaging and supportive learning environment (Klassen & Chiu, 2010).

Ultimately, a teacher's belief in themselves has a profound ripple effect, not just on their sense of fulfillment but also on their students' success. When educators feel empowered, they empower others, shaping classrooms where curiosity, resilience, and growth thrive.

For example, Ms. Sharma is a passionate and resourceful science teacher who believes in her ability to make learning meaningful for her students. She keeps her lessons engaging by incorporating hands-on activities, real-world examples, and personalized support. When challenges arise, whether a student is struggling with a concept or a classroom distraction, she remains calm and adaptable, finding creative solutions to keep students motivated and involved.

On the other hand, Mr. Verma often doubts his effectiveness as a teacher. He sticks to traditional lectures and is hesitant to try new approaches. When students struggle, he assumes they are uninterested rather than adjusting his methods to support them better. Over time, his lack of confidence creates a disengaged classroom where learning feels rigid and uninspiring.

This contrast shows how teachers' belief in themselves shapes their teaching style, the classroom environment, and, ultimately, their students' success.

Building Self-Efficacy in Educators

Fostering self-efficacy in educators is a gradual process that requires deliberate effort and a nurturing environment, as outlined in Bandura's research. One powerful strategy for building self-efficacy is to provide success-driven experiences. Encouraging teachers to engage in tasks where they can achieve small, incremental successes boosts their confidence and strengthens their skills over time. Furthermore, learning from others is a vital component in developing self-efficacy. Teachers can gain valuable perspectives and inspiration to enhance their practices by collaborating with seasoned educators or observing effective teaching strategies. Providing supportive feedback is equally important. Offering constructive criticism that highlights both strengths and areas for growth enables teachers to refine their techniques and build greater self-assurance. Another key factor is professional development, particularly training focused on inclusive teaching methods and culturally responsive strategies. These opportunities equip educators with essential tools and deepen their understanding of effectively addressing the needs of diverse classrooms. Finally, prioritizing emotional well-being through initiatives like mindfulness programs, peer support groups, and stress management resources is crucial. These initiatives help educators develop resilience and prepare them to handle challenges more confidently. These strategies create an environment where educators can continually grow, adapt, and cultivate a strong sense of self-efficacy in their professional roles.

For example, Ms. Priya Mehta, a newly appointed elementary school teacher, initially struggled with classroom management and catering to the diverse learning styles of her students. However, she gradually gained confidence through mentorship and professional development opportunities and refined her teaching methods. Observing experienced educators helped her adopt effective strategies, while small achievements, like successfully engaging a quiet student, reinforced her belief in her abilities. Supportive feedback from colleagues allowed her to fine-tune her approach, and participating in mindfulness and stress management programs helped her navigate the emotional demands of teaching. Over time, Priya became more adaptable, integrating inclusive teaching practices and fostering a nurturing learning environment. Her journey highlights how guidance, skill-building, and emotional well-being can empower educators, ultimately enhancing their effectiveness and enriching students' learning experiences.

Positive Psychology in Education: Nurturing Spiritual Wellbeing to Empower Educators

Positive psychology, conceptualized by Martin Seligman, centers on strengthening individual abilities, fostering resilience, and enhancing well-being. In the educational landscape, these principles bring transformative benefits, enriching the experiences of both students and educators. Central to this approach is spiritual well-being, which supports teachers in managing challenges, building meaningful relationships, and cultivating inclusive learning spaces through a sense of emotional balance, purpose, and interconnectedness.

THE ROLE OF SPIRITUAL WELLBEING IN EFFECTIVE TEACHING

Teaching in diverse classrooms comes with its share of challenges, but spiritual well-being helps educators navigate them with emotional strength and balance. Simple practices like mindfulness and self-reflection allow teachers to manage stress, stay composed, and respond to challenging situations with patience and clarity (Schonert-Reichl, 2017). When educators feel emotionally grounded, they create a stable and supportive environment where students can thrive, no matter what external pressures exist. Beyond emotional resilience, spiritual well-being fosters empathy, allowing teachers to understand their students' academic and personal struggles deeply. When educators tune in to their students' needs, they build genuine connections that make students feel valued and supported. This sense of trust encourages students to open up, engage more in class, and develop confidence in their learning abilities. Classrooms that nurture empathy and understanding lead to better academic outcomes and a stronger sense of belonging for all students (Jennings & Greenberg, 2009).

At the same time, a strong sense of purpose helps teachers stay committed, even in the face of daily challenges. Educators who connect their values and personal beliefs with their teaching feel more fulfilled and motivated (Sharma & Sharma, 2016). When teachers find meaning in their work, they are more likely to persevere through obstacles, inspire their students, and continuously improve their teaching methods (Palmer, 2007). Their passion becomes contagious, encouraging students to approach learning with the same enthusiasm and dedication. Perhaps most importantly, spiritual well-being strengthens relationships between teachers and students. Educators who invest in their emotional and spiritual health create safe, trusting environments where students feel comfortable expressing themselves and taking academic risks (Roeser et al., 2012). These bonds transform the classroom into a mutual respect and encouragement space where teachers and students grow together. In such an environment, learning becomes about more than academics, it becomes a personal and emotional development journey, equipping students with the confidence and resilience they need to succeed.

Integrating Positive Psychology to Strengthen Spiritual Wellbeing

Positive psychology provides a robust framework for nurturing spiritual well-being, with a particular focus on personal growth, resilience, and purposeful action. These principles are especially valuable for educators, helping them create fulfilling and impactful professional and personal lives (Seligman, 2011).

One of the foundational concepts of positive psychology is the emphasis on recognizing and utilizing personal strengths. It encourages individuals to identify and harness their unique talents meaningfully to maximize their potential (Peterson & Seligman, 2004). For teachers, this means leveraging their creativity, problem-solving skills, and emotional intelligence strengths when designing engaging lessons and addressing diverse student needs. By focusing on their natural abilities, educators are more likely to experience a sense of empowerment and fulfillment, directly translating into a more positive and effective learning environment for students (Lyubomirsky, King, & Diener, 2005).

Resilience is another core aspect of positive psychology crucial to personal and professional success. Teachers face numerous challenges, from managing student behavior to adapting to evolving educational demands. Cultivating resilience enables educators to confront these challenges positively (Fredrickson, 2001). Strategies like practicing gratitude, self-reflection, and pursuing professional development can help educators build their resilience (Reivich & Shatté, 2002). By adopting these practices, teachers can maintain optimism, recover from setbacks, and cultivate greater self-efficacy. Resilient educators enhance their well-being and serve as role models for their students, demonstrating how to handle adversity with confidence and perseverance (Masten, 2001).

Another important principle of positive psychology is the search for meaning and purpose in one's work. Teachers who find purpose in their teaching are more motivated and engaged, positively impacting their overall well-being (Dweck, 2006). By aligning their professional roles with their values, educators can experience a more profound satisfaction and fulfillment in their work. Volunteering, supporting community initiatives, or engaging in projects that resonate with their core beliefs can give teachers a sense of purpose (Steger, Frazier, Oishi, & Kaler, 2006). This alignment between personal values and professional goals sustains their motivation and enhances their teaching practices, leading to more meaningful and effective interactions with students.

Furthermore, positive psychology highlights the importance of supportive relationships in fostering overall well-being. Teachers who build strong, positive relationships with colleagues, students, and the wider community are better equipped to navigate the stresses and demands of their profession (Ryff & Singer, 2000). These connections provide emotional support, encourage collaboration, and offer personal and professional growth opportunities. When educators feel supported and connected, their job satisfaction increases, and they are more effective in their roles (Huppert, 2009). This creates a positive cycle where strong relationships contribute to a more supportive and thriving classroom environment for teachers and students.

For example, Ms. Kapoor, an elementary school teacher, strives to make learning meaningful by tailoring her lessons to reflect her students' diverse backgrounds. She uses her creativity and emotional intelligence to design engaging activities that help every child feel seen and valued. When faced with challenges like classroom disruptions or evolving curriculum demands, she relies on self-reflection and gratitude to stay resilient, focusing on small successes rather than setbacks. Teaching is not just a job for her, it is a calling. She incorporates service-learning projects to instill empathy and a sense of purpose in her students, making education more than academics. Beyond the classroom, she actively collaborates with colleagues, exchanging ideas and finding encouragement in a supportive professional network. By balancing personal strengths, resilience, purpose, and meaningful relationships, Ms. Kapoor creates a positive learning environment where she and her students thrive.

Transformative Impact on Students

When educators embrace spiritual well-being and the principles of positive psychology, they inspire their students to adopt similar values. By modeling qualities such as resilience, empathy, and purpose, teachers create classrooms that promote mutual respect and empower students to reach their full potential. These classrooms become environments where academic, emotional, and social development are equally prioritized.

Culturally Responsive Pedagogy (Gay, 2010)

Geneva Gay's Culturally Responsive Pedagogy (CRP) provides a framework for understanding how teachers can integrate students' cultural identities into their teaching practices. CRP emphasizes that an effective teacher acknowledges and actively incorporates students' diverse cultural backgrounds and experiences into the curriculum, instructional methods, and classroom environment.

CRP encourages educators to use culturally relevant materials, recognize the diverse ways students learn, and foster an inclusive classroom culture where all students feel respected and understood. Teachers who practice CRP build strong relationships with students, maintain high expectations, and create a learning space where cultural differences are celebrated. This approach helps ensure that all students, regardless of their background, feel a sense of belonging, enhancing engagement, motivation, and academic success. Culturally responsive pedagogy aligns with the goals of spiritual well-being and self-efficacy, as it encourages teachers to be both compassionate and confident in addressing the diverse needs of their students.

Core Principles of CRP

Using Culturally Relevant Content

Teachers practicing CRP ensure that their materials reflect their students' cultural backgrounds. This could involve selecting books, historical examples, and learning resources that align with students' experiences, fostering stronger engagement and deeper understanding.

Adapting to Different Learning Styles

Students learn in ways that often reflect their cultural contexts. CRP emphasizes the importance of flexible teaching strategies, such as storytelling, group work, and project-based learning, that adapt to these varied approaches, ensuring that all students can access and benefit from the curriculum.

Creating an Inclusive Classroom Atmosphere

Inclusivity in CRP means cultivating a classroom culture that respects and celebrates diversity. Teachers foster this environment by encouraging open discussions, challenging stereotypes, and ensuring students from all backgrounds feel seen and valued.

Maintaining High Expectations

A foundational belief of CRP is that all students can succeed academically. Teachers communicate this belief by setting clear, ambitious goals and supporting students in achieving them, ensuring equity in opportunities and outcomes.

Building Strong Relationships

Understanding students' cultural contexts allows teachers to build trust and meaningful relationships. These connections strengthen the classroom community and motivate students to engage more fully with their education.

Professional development programs focusing on Culturally Responsive Pedagogy (CRP) have significantly enhanced teachers' confidence and effectiveness in diverse classrooms. A study by Liberty University (2021) found that educators who underwent CRP training were better at incorporating inclusive teaching strategies, leading to higher student engagement and improved academic outcomes. This highlights the importance of equipping teachers with the tools to create learning environments that embrace cultural diversity and foster student success.

Future educators who feel prepared to address diverse student needs tend to be more flexible and effective in their teaching approaches. Siwatu (2007) examined the confidence levels of pre-service teachers in delivering culturally responsive instruction and found that those with higher self-efficacy were more proactive in adapting their methods. This underscores the need for teacher preparation programs to emphasize CRP principles, ensuring that educators enter the profession with the skills and confidence to support students from all backgrounds.

Educators who integrate students' cultural backgrounds into their lessons often see a noticeable improvement in engagement and participation. Research by Villegas and Lucas (2002) highlights that teachers who use culturally relevant materials and interactive teaching styles, such as storytelling, group discussions, and real-world applications, help students connect more deeply with the content. When students see their identities reflected in their learning, they are more motivated, leading to better comprehension and academic performance.

Bringing cultural values and traditions into the classroom profoundly impacts student motivation and well-being. A study by Kana'iaupuni, Ledward, and Jensen (2010) on Native Hawaiian students revealed that incorporating cultural identity into education strengthened students' sense of belonging, boosted confidence, and enhanced academic achievement. This research reinforces that honoring students' cultural backgrounds is not just about representation, it directly contributes to their personal and academic growth.

Siwatu (2011) explored how teachers' confidence in implementing CRP affects student outcomes. The study found that educators with strong self-efficacy were more likely to embrace inclusive teaching methods, creating a more supportive and equitable learning environment. Their students, in turn, experienced higher levels of academic success and engagement. This suggests that when teachers believe in their ability to meet the needs of diverse learners, they are more effective in fostering positive educational experiences.

These studies collectively emphasize the vital role of Culturally Responsive Pedagogy in shaping teaching practices and student success. Teacher self-efficacy plays a key role in effectively implementing CRP, making it essential for professional development programs and teacher education curricula to prioritize culturally inclusive strategies. Schools can create classrooms where every student feels valued, engaged, and empowered to succeed by equipping educators with the knowledge and confidence to support diverse learners.

The Interplay of CRP with Broader Educational Theories

CRP works effectively alongside other educational frameworks, such as Bandura's Social Cognitive Theory and Positive Psychology, to promote a well-rounded approach to teaching:

- Bandura's Social Cognitive Theory emphasizes learning through observation and the importance of self-efficacy. Teachers who model cultural awareness and competence encourage students to develop these traits, fostering mutual respect and resilience.
- Positive Psychology focuses on strengths, well-being, and resilience. CRP aligns with this by helping students recognize and celebrate their abilities, boosting confidence and a sense of purpose.

These theoretical frameworks, Bandura's Social Cognitive Theory, Positive Psychology, and Culturally Responsive Pedagogy, interact in ways that strengthen teaching practices. Teachers confident in their abilities (self-efficacy), spiritually grounded, and culturally responsive are better equipped to create inclusive, supportive, and effective learning environments. By integrating these frameworks, educators can improve their teaching strategies and foster an educational atmosphere where every student can thrive, regardless of their background.

Spiritual Wellbeing and Cultural Responsiveness

Empathy and Connection

Educators with strong spiritual well-being are exceptionally skilled at forming meaningful, empathetic connections with their students. Spiritual practices help teachers cultivate a deep sense of compassion and understanding, allowing them to relate to the unique cultural experiences and perspectives that each student brings to the classroom (Palmer, 1998). This empathy fosters an inclusive atmosphere where students feel recognized and valued for who they are rather than simply for their academic performance (Tisdell, 2003).

A teacher grounded in spiritual well-being approaches each student with an open and nonjudgmental mindset, which helps minimize the influence of biases. Rather than making assumptions based on stereotypes, they strive to truly understand each student's needs, creating a classroom where diverse perspectives are acknowledged and respected (Zajonc, 2013). This approach enhances students' sense of belonging, encourages them to engage more fully, and fosters an environment that celebrates differences (Miller, 2000).

Ethical Decision-Making

Spiritual well-being is key in shaping how educators make decisions about fairness, equity, and inclusivity in the classroom. Spiritually connected teachers are likelier to base their decisions on principles of justice, ensuring that all students have access to the resources and opportunities they need to succeed. Their sense of ethics, rooted in personal values, promotes fairness and guides them in fostering a supportive classroom environment where every student feels respected and valued.

Teachers who maintain a strong spiritual foundation approach their work holistically, considering academic outcomes and students' social and emotional needs. This commitment to fairness and inclusivity drives them to create a classroom culture where all students are given equal opportunities to thrive regardless of their cultural background. By prioritizing these values, spiritually well educators help create an atmosphere of respect, equity, and trust, essential for student success in a diverse environment.

Stress Reduction

Navigating a culturally diverse classroom can be highly demanding, with teachers often facing challenges related to different learning styles, cultural expectations, and social dynamics. Spiritual well-being provides a valuable tool for managing the emotional pressures that come with these responsibilities. By incorporating mindfulness, meditation, or reflective practices, educators can achieve inner calm and emotional balance, enabling them to handle stress more effectively.

Teachers who focus on their spiritual well-being can maintain emotional stability, which helps them remain composed in classroom challenges. This stability enhances their ability to manage complex classroom dynamics and helps prevent burnout, allowing them to sustain their energy and enthusiasm over time. In addition, reducing stress helps teachers stay engaged and present in their work, leading to a more positive and nurturing environment for students and educators. Teachers can better support their students in meaningful and lasting ways by caring for their emotional well-being.

Self-Efficacy as a Catalyst for Change

Enhancing Instructional Confidence

Educators with strong self-efficacy are more confident in implementing culturally relevant and adaptive teaching strategies. This belief in their capabilities motivates teachers to engage in creative, student-centered methods that acknowledge and respect their students' cultural identities. Such confidence helps teachers build an inclusive learning environment where students from various backgrounds feel seen and understood, fostering deeper engagement and connection to the material.

Additionally, self-efficacy allows teachers to approach common classroom challenges, such as language barriers or diverse learning styles, with creative and resourceful solutions. Teachers with high self-efficacy are not easily discouraged by obstacles; instead, they actively seek ways to modify their teaching to accommodate all learners. This approach ensures that cultural diversity is embraced in the classroom, enhancing both student outcomes and the richness of the learning experience.

Overcoming Challenges with Resilience

A key strength of self-efficacy is its role in fostering resilience. Teachers who believe in their ability to influence student success are better equipped to tackle challenges, such as resistance to cultural inclusion or limited resources. These educators view challenges not as insurmountable barriers but as opportunities to explore new strategies and solutions, allowing them to persist in creating equitable learning environments.

Resilient educators with high self-efficacy continue to prioritize diversity and inclusion, even in the face of opposition. They remain committed to making their classrooms welcoming for all students, consistently adapting their teaching practices to reflect the needs of diverse learners. This unwavering dedication to inclusivity is critical in creating classrooms where cultural differences are respected and every student is encouraged to succeed.

Strengthening Professional Growth

Ongoing professional development is crucial for enhancing teachers' self-efficacy, particularly when integrating cultural responsiveness into their teaching methods. Effective training programs equip educators with the knowledge, tools, and strategies needed to address the complexities of a multicultural classroom. These programs help teachers develop a deeper understanding of engaging and supporting students from diverse backgrounds, building their confidence in their ability to manage diverse learning environments.

As teachers develop new skills and gain confidence in their ability to navigate the challenges of a culturally diverse classroom, their self-efficacy grows. This strengthened belief in their capabilities leads them to feel more empowered in their teaching, allowing them to make a lasting impact on their students' learning experiences. By prioritizing professional development, schools ensure that educators are well-prepared to meet the diverse needs of their students, leading to positive outcomes for both teachers and learners.

INTEGRATING SPIRITUAL WELLBEING AND SELF-EFFICACY

A Holistic Framework

Integrating spiritual well-being with self-efficacy offers a well-rounded approach to culturally responsive teaching. Spiritual well-being provides the emotional and inner resources needed to foster empathy and resilience, while self-efficacy enables teachers to confidently take practical actions in the classroom. Together, these aspects equip educators to address the diverse needs of their students with compassion, confidence, and adaptability.

This integrated model emphasizes both personal reflection and professional development. Spiritual well-being encourages mindfulness, self-awareness, and emotional regulation, enabling educators to respond to challenges with patience and understanding. Self-efficacy, on the other hand, motivates educators to implement inclusive teaching practices that actively promote diversity and equity. By combining inner strength with outward action, educators can create an environment where all students feel supported and valued.

Strategies for Educators

- **Mindfulness Practices**: Meditation, breathing exercises, or mindful reflection help educators strengthen their spiritual well-being while managing stress. These activities foster emotional stability, essential for maintaining a calm and focused mindset in the classroom. Mindfulness helps educators better understand their emotional responses and enhances their ability to empathize with students, especially when addressing their diverse needs. It promotes self-awareness and the ability to stay present, ensuring that teachers can respond thoughtfully and appropriately to their students' concerns.
- **Cultural Immersion**: Engaging directly with different cultures, whether through community activities, educational events, or travel, allows educators to broaden their understanding of their students' backgrounds and experiences. This exposure helps teachers build stronger, more authentic connections with their students and creates a classroom culture that values diversity. Cultural immersion also provides insights that inform teaching practices, allowing educators to adapt their methods to meet the needs of all learners better, fostering a more inclusive educational experience.
- **Collaborative Learning Communities**: Creating peer networks and collaborative learning environments within schools enables teachers to share insights, challenges, and solutions related to culturally responsive teaching. These communities allow educators to learn from each other, exchanging strategies and experiences that enhance their collective ability to support diverse student populations. Collaboration nurtures a sense of shared purpose and support, strengthening teachers' confidence and efficacy. Educators can develop new approaches and refine existing ones by working together, making their teaching practices more inclusive and effective.

  for example, Ms. Ananya Verma, a passionate middle school teacher in a diverse urban school, embraced spiritual well-being and self-efficacy to create a supportive and engaging learning space. To stay grounded and manage the daily challenges of teaching, she practiced mindfulness, using deep breathing and reflective journaling to maintain patience and emotional balance. Understanding the importance of cultural awareness, she actively participated in community events and incorporated diverse perspectives into her lessons, ensuring every student felt valued. She also built collaborative relationships with fellow educators, exchanging ideas and strategies to improve inclusive teaching. Her efforts made students more engaged, confident, and connected to their learning. Ananya's story highlights how inner resilience and self-belief enable teachers to cultivate classrooms where diversity is celebrated, learning is meaningful, and every student has the opportunity to thrive.

By integrating spiritual well-being and self-efficacy, educators are empowered to cultivate a classroom environment that values diversity and actively seeks to embrace it. This dual focus on personal development and professional practice provides a solid foundation for creating inclusive, supportive, and culturally responsive classrooms.

Implications for Policy and Practice

For Educators

Integrating spiritual well-being and self-efficacy into educational frameworks can significantly improve teaching strategies, classroom interactions, and student success. Educators who focus on their spiritual growth and develop confidence in their ability to make a difference are better prepared to address the challenges of contemporary classrooms. Engaging in mindfulness and reflective practices enhances emotional awareness, allowing teachers to connect with students' diverse cultural, social, and emotional needs with greater sensitivity. This approach fosters an environment of respect and inclusivity where every student feels valued and understood.

Designing lessons incorporating culturally relevant material increases student interest and ensures the learning process is more inclusive. Lessons that align with students' backgrounds make education more relatable, fostering a stronger connection between students and the content. Teachers with strong self-efficacy are likelier to implement creative and adaptive teaching strategies, such as differentiated learning and hands-on activities, to cater to various learning preferences and abilities.

Encouraging students to participate in collaborative tasks and share diverse perspectives promotes mutual respect and understanding. These interactions help students appreciate different cultures and viewpoints, preparing them for meaningful engagement in a global society. Educators who demonstrate spiritual wellbeing and self-assurance inspire their students, showcasing qualities such as resilience, empathy, and the ability to adapt to challenges.

Incorporating spiritual wellbeing and self-efficacy into educational practices not only supports academic achievement but also contributes to the overall development of both teachers and students. This holistic approach builds a supportive and dynamic learning environment that empowers individuals to excel academically while fostering emotional and social growth.

For Administrators

School administrators have a crucial role in shaping a supportive environment for educators, promoting both their spiritual wellbeing and professional growth. Administrators should design professional development programs that integrate these concepts, including training on mindfulness, emotional resilience, and culturally inclusive pedagogy. These programs help educators build the skills needed to navigate the complexities of teaching in a diverse environment while enhancing their emotional well-being and effectiveness.

Administrators should also ensure that teachers have the resources and time needed to prioritize self-care, which supports their overall well-being and strengthens their ability to be effective in the classroom. A collaborative school culture that encourages peer support and shared learning is essential for building confidence among educators. Administrators can foster a sense of collective efficacy by providing platforms for teachers to exchange ideas and experiences, ultimately benefiting the entire school community.

Comprehensive Professional Development

Comprehensive professional development for educators should encompass both technical skills and the emotional and practical challenges teachers face in today's classrooms. One key aspect of this development is integrating mindfulness practices, such as meditation and reflective exercises, which help teachers manage stress, maintain focus, and build emotional resilience. These techniques enable educators to stay calm and composed, even in stressful situations. Regular mindfulness practices support teachers' well-being and enhance their capacity to create a positive and nurturing classroom environment. This emotional balance allows teachers to approach their work clearly, reducing burnout and promoting a healthy, balanced approach to their teaching practice.

In addition to mindfulness, resilience training is critical for ensuring that teachers can thrive in the face of the profession's challenges. Teaching can be emotionally demanding, especially when navigating difficult situations or managing a diverse classroom. Resilience-focused training equips educators with strategies to cope with stress, recover from setbacks, and remain motivated throughout their careers. This training includes practical techniques for managing conflicts, reducing stress, and sustaining enthusiasm for teaching, even in adversity. Emphasizing work-life balance is also key to resilience development, helping teachers maintain a healthy equilibrium and avoid burnout.

Moreover, professional development programs must also emphasize inclusive teaching practices. As classrooms become increasingly diverse, educators must be prepared to meet the varying needs of their students. Culturally responsive teaching equips teachers with the knowledge to recognize and respect cultural differences, adjust their teaching strategies, and foster an inclusive atmosphere where every student feels valued. This training encourages teachers to design lessons that connect with students from diverse backgrounds and to manage classrooms in inclusive and culturally sensitive ways. By cultivating an understanding of diversity and inclusion, teachers can create equitable learning environments that support all students' academic and emotional needs.

Comprehensive professional development enables educators to succeed in today's complex and diverse classrooms by addressing the emotional, practical, and cultural challenges that teachers face. Focusing on mindfulness, resilience, and inclusive practices, such development not only improves teachers' technical skills but also strengthens their ability to support students and manage their well-being. This holistic approach to professional growth ultimately enhances teacher satisfaction and student success, fostering a more positive and effective educational environment.

Promoting Collaboration

A collaborative culture within a school is vital for strengthening professional relationships and improving teaching effectiveness. Administrators can encourage collaboration by establishing support systems such as mentorship and peer coaching programs. These initiatives allow teachers to connect, exchange ideas, and offer guidance, helping to create a unified and supportive community. Peer coaching and mentorship contribute to individual professional development and promote a collaborative environment where educators feel valued and empowered. Such systems foster collective growth and mutual respect, enhancing teaching practices and improving student outcomes.

Along with these support systems, administrators can further promote collaboration by facilitating the creation of professional learning communities (PLCs). These groups provide teachers a dedicated space to engage in meaningful discussions, share knowledge, and explore new teaching strategies. Regular group meetings allow educators to address challenges, collaborate on solutions, and continuously refine their practices. By encouraging ongoing dialogue, administrators help teachers stay informed about the latest educational trends, share best practices, and build stronger professional networks, ultimately enriching the entire school community.

Recognizing and celebrating the achievements of educators is another key strategy for reinforcing a collaborative culture. Administrators should make a concerted effort to acknowledge teachers' hard work and successes through formal recognition events or personal gestures of appreciation. These acknowledgments boost morale and create a positive atmosphere that reinforces the value of collaboration. When teachers feel their contributions are celebrated, it fosters a sense of accomplishment. It encourages continued engagement with their professional growth and commitment to the school's success. Administrators can cultivate a dynamic and effective teaching environment that benefits educators and students by prioritizing collaboration, support, and recognition.

Encouraging Collective Confidence

Administrators have a crucial role in cultivating a sense of collective efficacy among educators, which refers to the shared belief in their ability to achieve success together. One key approach to fostering this unity is establishing clear, unified goals that direct the efforts of all teachers toward a common purpose. This alignment enhances collaboration and reinforces a sense of shared responsibility, helping educators feel connected to a broader mission. Additionally, by involving teachers in leadership roles and decision-making, administrators empower them to take ownership of school policies and initiatives, fostering a culture of trust and collective accountability.

Furthermore, creating opportunities for teachers to share ideas and strategies is vital for building a strong sense of collective efficacy. Administrators can facilitate platforms for collaboration, allowing educators to exchange innovative teaching practices and learn from each other's experiences. This knowledge exchange encourages professional growth, builds confidence, and strengthens the sense of community among teachers. By setting shared goals, involving educators in leadership, and providing spaces for collaboration, administrators can cultivate an environment where collective efficacy thrives, resulting in a more supportive and effective learning environment for teachers and students.

Enhancing the School Community

When administrators prioritize teachers' holistic development, they create a supportive, collaborative, and dynamic school culture. Teachers who feel valued and empowered are likelier to approach their work with enthusiasm, creativity, and dedication, leading to better student engagement and learning outcomes. A positive and supportive environment also helps attract and retain talented educators, ensuring the school's long-term success.

Administrators focusing on educators' well-being and professional growth lay the foundation for a thriving, inclusive, high-performing educational community. Through thoughtful leadership, they create an atmosphere where teachers and students alike can reach their full potential.

For Policymakers

Policymakers are important in creating educational frameworks prioritizing teacher wellbeing and continuous professional development. They can advocate for funding to support programs promoting mindfulness, emotional resilience, and self-efficacy, ensuring teachers have access to the resources they need. Policies that focus on ongoing professional learning, particularly in areas like diversity and culturally responsive teaching, are crucial for preparing educators to address the varied needs of a diverse student population.

Additionally, policymakers should support initiatives that allow teachers to engage in mentorship programs, peer collaboration, and professional networks, which provide opportunities for personal and professional growth. By creating policies that promote teacher wellbeing and continuous development, policymakers can help ensure that educators are equipped to create inclusive and responsive learning environments. This ultimately leads to better educational outcomes for students, regardless of their cultural or socioeconomic background.

Securing Funding and Essential Resources

A fundamental responsibility of policymakers is to secure adequate funding to support professional development programs for teachers. Financial support is crucial in creating and maintaining initiatives that enhance emotional resilience, mindfulness, and self-efficacy, all essential for teachers' personal and professional growth. These programs provide teachers with the skills to manage stress, stay focused, and boost their confidence in the classroom.

In addition to well-being-focused programs, policymakers should ensure teachers have access to resources like mental health services, wellness initiatives, and continuing education opportunities. A holistic approach to supporting teachers can reduce burnout, improve retention, and help educators feel more satisfied and engaged. Well-resourced teachers are more likely to be effective, leading to a positive impact on student outcomes.

Fostering Ongoing Professional Development

Policymakers must advocate for policies that prioritize continuous professional growth for teachers. This includes ensuring that professional development is not a one-time event but a sustained process that evolves to meet the challenges of a changing educational landscape. It is crucial to emphasize areas such as diversity, inclusive teaching, and culturally responsive pedagogy to prepare educators for diverse classrooms.

Continuous professional learning enables teachers to stay current with new pedagogical methods and technologies, which enhances their effectiveness in meeting students' diverse needs. Training that focuses on cultural awareness and inclusion empowers educators to address the unique challenges of students from various racial, cultural, and socioeconomic backgrounds, creating a more equitable and effective learning environment for all students.

Encouraging Teacher Collaboration and Mentorship

Policymakers can further enhance teacher development by promoting policies that foster collaboration and mentorship among educators. Establishing opportunities for teachers to work together, exchange ideas, and mentor one another contributes to their professional growth and improves teaching practices. These collaborative programs allow teachers to share best practices, learn from each other's experiences, and strengthen their skills.

Encouraging mentorship and peer collaboration also combats teachers' feelings of isolation. By providing structured platforms for teachers to connect with colleagues, policymakers help build a supportive network that fosters professional growth and enhances teaching effectiveness.

Promoting Inclusive and Responsive Classrooms

Policymakers are instrumental in ensuring educators have the necessary training and resources to build inclusive and responsive classrooms. By creating policies that prioritize teacher well-being and professional development, they enable teachers to respond effectively to their students' diverse needs. Professional development in areas such as social-emotional learning, inclusive teaching strategies, and equity-driven practices helps educators create classrooms where all students feel valued and supported, which leads to higher engagement and better academic results.

Enhancing Student Success Through Teacher Support

When teachers feel supported in their well-being and professional growth, their effectiveness in the classroom improves, leading to better student outcomes. Educators who are confident, motivated, and equipped with the necessary tools to address their students' diverse needs are more likely to foster an environment that encourages student engagement and success.

Furthermore, investing in teacher wellbeing and ongoing professional development contributes to teacher retention. When teachers receive the support they need, they are more likely to stay in the profession, ensuring school stability and consistent, high-quality teaching for students.

Case Study

Ms. Aditi Sharma, a dedicated high school teacher in a diverse urban school, faces daily challenges such as managing classroom disruptions, addressing varied learning needs, and working with limited resources. Despite these obstacles, she creates a warm and engaging learning environment, drawing strength from her spiritual well-being and self-efficacy. She continuously adapts her teaching methods, incorporating creative strategies like project-based learning and storytelling to cater to different learning styles. Beyond academics, Aditi prioritizes building trust and meaningful relationships with her students, ensuring they feel valued and motivated. She fosters an inclusive atmosphere by approaching challenges with patience, empathy, and resilience, often using mindfulness and self-reflection to stay composed. Her ability and confidence allow her to embrace difficulties as learning opportunities, maintain high expectations for her students, and create a structured yet supportive classroom.

As a result of her approach, student engagement and academic performance improve while behavioral issues decrease. More importantly, she finds deep professional fulfillment, knowing she is making a lasting impact on her students' lives. Aditi's story highlights how spiritual well-being and self-efficacy empower educators to navigate challenges with compassion, adaptability, and innovation, ultimately creating a nurturing and effective learning environment for all.

CONCLUSION

Great teaching transcends the delivery of content; it encompasses the creation of a supportive, empathetic, and inspiring learning environment. A teacher's spiritual well-being and self-efficacy are central to cultivating such spaces where students not only learn but also feel seen, valued, and empowered. Educators who are grounded in their inner values and who believe in their ability to effect change are better equipped to navigate the challenges of a diverse and dynamic classroom. Spiritual well-being helps teachers develop emotional balance, patience, and ethical clarity, allowing them to build inclusive spaces that honor student diversity. Simultaneously, self-efficacy, the confidence in one's ability to influence outcomes, encourages teachers to be adaptable, proactive, and innovative in their approach, enabling them to respond constructively to the varied needs of their students.

In today's increasingly multicultural classrooms, the importance of these internal strengths cannot be overstated. Culturally responsive teaching demands emotional resilience, reflective practice, and an ability to engage meaningfully with students from diverse backgrounds. Teachers with high levels of spiritual well-being are more likely to approach this responsibility with compassion and empathy, forming deep connections that transcend academic content. This empathetic engagement creates a sense of safety and belonging for students, encouraging greater participation and authentic self-expression. Moreover, educators who embrace their own cultural humility are more likely to model inclusive behaviors and challenge systemic biases, making their classrooms catalysts for social justice and equity.

Self-efficacy enhances this process by giving educators the belief and determination to take meaningful action. Confident teachers do not shy away from complexity; instead, they experiment with differentiated instruction, integrate culturally relevant content, and foster collaborative learning environments. When educators believe they can meet their students' needs, they are more likely to persevere in the face of challenges such as language barriers, behavioral issues, or limited institutional resources. These teachers adapt their methods to honor diverse learning styles and identities, ultimately contributing to stronger academic outcomes and deeper student engagement. In essence, self-efficacy empowers educators to take initiative, remain resilient, and serve as transformative agents in their classrooms.

The interplay between spiritual well-being and self-efficacy also enhances educators' capacity to manage stress and avoid burnout. Teaching is an emotionally demanding profession, particularly in under-resourced or high-pressure environments. Spiritual practices such as mindfulness, reflection, and connection to personal values help teachers maintain perspective, enabling them to approach difficult situations with calm and compassion. At the same time, a strong sense of self-efficacy helps them stay solution-focused and optimistic. This balance not only benefits teachers' mental health but also enhances their presence in the classroom, allowing them to model emotional regulation and perseverance for their students. A resilient teacher is more likely to foster a resilient student.

Furthermore, when teachers embody both spiritual wellness and high self-efficacy, they become more effective advocates for inclusive education. They engage with professional development not just as a requirement, but as an opportunity for personal and pedagogical growth. These educators seek out culturally responsive teaching strategies, participate in collaborative networks, and continuously refine their practice. Their classrooms reflect an intentional commitment to justice, equity, and holistic learning. Students in such environments benefit from personalized instruction, meaningful relationships, and a sense of cultural affirmation. These are the classrooms where academic achievement and social-emotional growth go hand-in-hand, where students are not only prepared for tests but for life.

The broader implications of these findings extend to school leadership and educational policy. School administrators should recognize that teacher well-being is not a personal matter but a professional imperative. Institutions that foster spiritual growth and support teachers' belief in their abilities create a culture of trust, motivation, and collective efficacy. Providing opportunities for mindfulness training, emotional support systems, and culturally responsive professional development can greatly enhance teacher satisfaction and retention. Moreover, policies that allow teacher voice in decision-making processes reinforce their sense of agency, thereby increasing their investment in student success. When educators feel supported, respected, and empowered, the entire school community flourishes.

From a policy perspective, investing in teacher well-being and self-efficacy is an investment in educational equity. As the student population becomes increasingly diverse, the demand for culturally competent, emotionally grounded, and confident educators grows. Teacher preparation programs must incorporate training in self-awareness, cultural humility, and resilience alongside instructional strategies. By embedding these elements into curricula, teacher education can produce educators who are not only academically prepared but also personally equipped to navigate the realities of diverse classrooms. This proactive approach can close opportunity gaps, reduce teacher attrition, and foster environments that truly reflect the democratic ideals of education.

The conclusions drawn from this study also have significant relevance for researchers and scholars in the field of education. While much attention has been given to curriculum design and student performance, there remains a critical need to explore the internal resources that empower teachers to bring these designs to life. Future research should continue to investigate how constructs like spiritual well-being and self-efficacy influence teacher performance, especially in relation to inclusive and culturally sustaining pedagogies. Longitudinal studies could provide deeper insights into how these traits evolve over time and in response to various professional development interventions. By broadening our understanding of teacher identity and inner development, the field of education can create more holistic and sustainable models of teaching and learning.

At the classroom level, the integration of spiritual well-being and self-efficacy transforms not only teaching practices but also student experiences. Teachers who embody these qualities provide safe, nurturing, and intellectually stimulating environments that validate students' identities and promote deep engagement. These educators serve as role models, demonstrating how to navigate life with integrity, compassion, and purpose. Students in such classrooms learn more than academic content, they learn how to think critically, collaborate with others, and develop resilience in the face of adversity. They are empowered to take ownership of their learning and to see themselves as capable contributors to society.

In conclusion, the union of spiritual well-being and self-efficacy forms the foundation of truly transformative teaching. Together, these attributes empower educators to meet the complex demands of culturally responsive education with confidence, empathy, and innovation. They support the personal resilience of teachers while enhancing their ability to foster inclusive, high-performing learning environments. As schools and education systems strive to prepare students for a complex and interconnected world, the role of the teacher, as guide, role model, and change-maker, becomes even more vital. Investing in teachers' internal growth is not only a matter of individual wellness; it is essential to achieving educational equity and excellence for all learners. By prioritizing both spiritual and psychological empowerment, we can move closer to realizing the full potential of education as a force for human flourishing.

REFERENCES

Bandura, A. (1997). *Self-efficacy: The exercise of control.* W. H. Freeman and Company.

Banks, J. A., & Banks, C. A. M. (Eds.). (2020). *Multicultural education: Issues and perspectives* (10th ed.). Wiley.

Brookfield, S. D. (2017). *Becoming a critically reflective teacher* (2nd ed.). Jossey-Bass.

Conner, L. A., & Pope, J. R. (2013). Cultivating culturally responsive teaching through spirituality: Implications for educators. *Journal of Education and Social Justice*, *3*(1), 5–22.

Cranton, P. (2006). *Understanding and promoting transformative learning: A guide for educators of adults* (2nd ed.). Jossey-Bass.

Darling-Hammond, L., & Bransford, J. (2005). *Preparing teachers for a changing world: What teachers should learn and be able to do.* Jossey-Bass.

Dweck, C. S. (2006). *Mindset: The new psychology of success.* Random House.

Emmons, R. A. (2003). Personal goals, life meaning, and virtue: Wellsprings of a positive life. In Keyes, C. L. M., & Haidt, J. (Eds.), *Flourishing: Positive psychology and the life well-lived* (pp. 105–128). American Psychological Association. DOI: 10.1037/10594-005

Fredrickson, B. L. (2001). The role of positive emotions in positive psychology: The broaden-and-build theory of positive emotions. *The American Psychologist, 56*(3), 218–226. DOI: 10.1037/0003-066X.56.3.218 PMID: 11315248

Fullan, M. (2007). *The new meaning of educational change* (4th ed.). Teachers College Press.

Gay, G. (2010). *Culturally Responsive Teaching: Theory, Research, and Practice.* Teachers College Press.

Gay, G. (2018). *Culturally responsive teaching: Theory, research, and practice* (3rd ed.). Teachers College Press.

Hammond, Z. (2015). *Culturally responsive teaching and the brain: Promoting authentic engagement and rigor among culturally and linguistically diverse students.* Corwin Press.

Hattie, J., & Timperley, H. (2007). The power of feedback. *Review of Educational Research, 77*(1), 81–112. DOI: 10.3102/003465430298487

Hodge, D. R. (2013). Spirituality and social work: A cultural and contextual perspective. *Journal of Social Work : JSW, 13*(3), 254–271.

Hooks, B. (1994). *Teaching to transgress: Education as the practice of freedom.* Routledge.

Hoy, A. W., & Spero, R. B. (2005). Changes in teacher efficacy during the early years of teaching: A comparison of four measures. *Teaching and Teacher Education, 21*(4), 343–356. DOI: 10.1016/j.tate.2005.01.007

Huppert, F. A. (2009). Psychological well-being: Evidence regarding its causes and consequences. *Applied Psychology. Health and Well-Being, 1*(2), 137–164. DOI: 10.1111/j.1758-0854.2009.01008.x

Jennings, P. A., & Greenberg, M. T. (2009). The prosocial classroom: Teacher social and emotional competence in relation to student and classroom outcomes. *Review of Educational Research, 79*(1), 491–525. DOI: 10.3102/0034654308325693

Johnson, S. L., & Johnson, R. T. (2007). The role of collaboration in building professional efficacy. *Teaching and Teacher Education, 23*(6), 740–752.

Klassen, R. M., & Chiu, M. M. (2010). Effects on teachers' self-efficacy and job satisfaction: Teacher gender, years of experience, and job stress. *Journal of Educational Psychology, 102*(3), 741–756. DOI: 10.1037/a0019237

Koenig, H. G. (2012). Religion, spirituality, and health: The research and clinical implications. *ISRN Psychiatry, 278730*, 1–33. Advance online publication. DOI: 10.5402/2012/278730 PMID: 23762764

Koenig, H. G. (2012). *The Handbook of Religion and Health*. Oxford University Press.

Ladson-Billings, G. (1995). Toward a theory of culturally relevant pedagogy. *American Educational Research Journal, 32*(3), 465–491. DOI: 10.3102/00028312032003465

Ladson-Billings, G. (2021). *Critical race theory in education: A scholar's journey*. Teachers College Press.

Liberty University. (2021). *The impact of professional development on culturally responsive teaching and teacher self-efficacy*. Liberty University Press.

Lyubomirsky, S., King, L., & Diener, E. (2005). The benefits of frequent positive affect: Does happiness lead to success? *Psychological Bulletin, 131*(6), 803–855. DOI: 10.1037/0033-2909.131.6.803 PMID: 16351326

Masten, A. S. (2001). Ordinary magic: Resilience processes in development. *The American Psychologist, 56*(3), 227–238. DOI: 10.1037/0003-066X.56.3.227 PMID: 11315249

Mezirow, J., & Taylor, E. W. (Eds.). (2009). *Transformative learning in practice: Insights from community, workplace, and higher education*. Jossey-Bass.

Miller, J. P. (2000). *Education and the soul: Toward a spiritual curriculum*. State University of New York Press.

Nieto, S. (2002). *Affirming diversity: The sociopolitical context of multicultural education* (3rd ed.). Longman.

Palmer, P. J. (1998). *The courage to teach: Exploring the inner landscape of a teacher's life*. Jossey-Bass.

Palmer, P. J. (2000). *Let your life speak: Listening for the voice of vocation*. Jossey-Bass.

Palmer, P. J. (2007). *The courage to teach: Exploring the inner landscape of a teacher's life*. Jossey-Bass.

Pargament, K. I. (1997). *The psychology of religion and coping: Theory, research, practice*. Guilford Press.

Pargament, K. I. (2011). *Spiritually integrated psychotherapy: Understanding and addressing the sacred*. Guilford Press.

Paris, D., & Alim, H. S. (2017). *Culturally sustaining pedagogies: Teaching and learning for justice in a changing world*. Teachers College Press.

Peterson, C., & Seligman, M. E. P. (2004). *Character strengths and virtues: A handbook and classification*. Oxford University Press.

Reivich, K., & Shatté, A. (2002). *The resilience factor: 7 essential skills for overcoming life's inevitable obstacles*. Broadway Books.

Roeser, R. W., Skinner, E., Beers, J., & Jennings, P. A. (2012). Mindfulness training and teachers' professional development: An emerging area of research and practice. *Child Development Perspectives, 6*(2), 167–173. DOI: 10.1111/j.1750-8606.2012.00238.x

Ryan, R. M., & Deci, E. L. (2000). Self-determination theory and the facilitation of intrinsic motivation, social development, and well-being. *The American Psychologist, 55*(1), 68–78. DOI: 10.1037/0003-066X.55.1.68 PMID: 11392867

Ryff, C. D., & Singer, B. (2000). Interpersonal flourishing: A positive health agenda for the new millennium. *Personality and Social Psychology Review, 4*(1), 30–44. DOI: 10.1207/S15327957PSPR0401_4

Schon, D. A. (1987). *Educating the reflective practitioner*. Jossey-Bass.

Schonert-Reichl, K. A. (2017). Social and emotional learning and teachers. *The Future of Children, 27*(1), 137–155. DOI: 10.1353/foc.2017.0007

Seligman, M. E. P. (2011). *Flourish: A visionary new understanding of happiness and well-being*. Free Press.

Seligman, M. E. P., & Csikszentmihalyi, M. (2000). Positive psychology: An introduction. *The American Psychologist, 55*(1), 5–14. DOI: 10.1037/0003-066X.55.1.5 PMID: 11392865

Shapiro, S. L., & Carlson, L. E. (2009). *The art and science of mindfulness: Integrating mindfulness into psychology and the helping professions*. American Psychological Association. DOI: 10.1037/11885-000

Sharma, M., & Rush, S. E. (2014). Mindfulness-based stress reduction as a stress management intervention for healthy individuals: A systematic review. *Journal of Evidence-Based Complementary & Alternative Medicine, 19*(4), 271–286. DOI: 10.1177/2156587214543143 PMID: 25053754

Shor, I. (1992). *Empowering education: Critical teaching for social change*. University of Chicago Press. DOI: 10.7208/chicago/9780226147864.001.0001

Siwatu, K. O. (2011). Preservice teachers' self-efficacy beliefs in implementing culturally responsive teaching techniques. *The Urban Review, 43*(4), 467–487. DOI: 10.1007/s11256-010-0166-3

Skaalvik, E. M., & Skaalvik, S. (2010). Teacher self-efficacy and teacher burnout: A study of relations. *Teaching and Teacher Education, 26*(4), 1059–1069. DOI: 10.1016/j.tate.2009.11.001

Steger, M. F., Frazier, P., Oishi, S., & Kaler, M. (2006). The meaning in life questionnaire: Assessing the presence of and search for meaning in life. *Journal of Counseling Psychology, 53*(1), 80–93. DOI: 10.1037/0022-0167.53.1.80

Tisdell, E. J. (2003). *Exploring spirituality and culture in adult and higher education*. John Wiley & Sons.

Tomlinson, C. A., & Imbeau, M. B. (2010). *Leading and managing a differentiated classroom*. ASCD.

Tschannen-Moran, M., & Hoy, A. W. (2001). Teacher efficacy: Capturing an elusive construct. *Teaching and Teacher Education, 17*(7), 783–805. DOI: 10.1016/S0742-051X(01)00036-1

Villegas, A. M., & Lucas, T. (2002). Preparing culturally responsive teachers: Rethinking the curriculum. *Journal of Teacher Education, 53*(1), 20–32. DOI: 10.1177/0022487102053001003

Wong, P. T. P. (2012). Toward a dual-systems model of what makes life worth living. In Wong, P. T. P. (Ed.), *The Human Quest for Meaning* (2nd ed., pp. 3–22). Routledge.

Zajonc, A. (2013). *Meditation as contemplative inquiry: When knowing becomes love*. Lindisfarne Books.

Zaslow, M., & Martinez-Beck, I. (2006). *Self-efficacy and teachers' beliefs about their ability to teach diverse students*. Teachers College Press.

Zins, J. E., Weissberg, R. P., Wang, M. C., & Walberg, H. J. (Eds.). (2004). *Building Academic Success on Social and Emotional Learning: What Does the Research Say?* Teachers College Press.

Chapter 12
Digital Pathways to Excellence for Bridging Gaps and Building Competencies:
The Role of Educational Technology in Transforming Teacher Professional Development

Seema Yadav
https://orcid.org/0000-0002-8477-5895
Bhopal School of Social Sciences, India

ABSTRACT

This chapter explores the integration of educational technology into Continuing Professional Development (CPD) for educators, highlighting the transformative potential of advanced digital tools and strategies. It examines how technologies such as Artificial Intelligence (AI), Virtual Reality (VR), Augmented Reality (AR), and microlearning can create personalized, adaptive, and immersive learning experiences, enhancing the effectiveness of CPD programs. It emphasizes the importance of peer-to-peer learning, mentorship, and global collaboration in promoting knowledge-sharing and professional development. Challenges such as ensuring digital accessibility, addressing attitudes towards technology, and providing ongoing support are also explored. The chapter concludes by stressing the need for inclusive, equitable CPD programs that prioritize teacher agency and hands-on learning, ensuring that educators are equipped with the skills and confidence to navigate the evolving digital landscape and provide high-quality, student-centered education.

INTRODUCTION

Sustainable Development Goal (SDG) 4, which seeks to "ensure inclusive and equitable quality education and promote lifelong learning opportunities for all," highlights the need to promote high-quality education as a key component of attaining sustainable development on a global scale. A comprehensive understanding of educational quality necessitates taking into account a number of variables, including

DOI: 10.4018/979-8-3373-0472-4.ch012

student characteristics, teaching and learning procedures, learning outcomes, and contextual elements. Significantly increasing the supply of trained teachers is one of the main goals to accomplish SDG 4, including through international collaboration for teacher training in many nations (S. Li et al., 2019). In any nation, teachers are a key player on the agenda. The dissemination of knowledge and the development of skilled human resources are two crucial roles of educators. A teacher is a professional, noble, and superior person who uses his knowledge, wisdom, enlightenment, and authority to guide others and develop good character. Thus, teachers are not only responsible for instructing; they are also responsible for transporting students (Nurtanto et al., 2022). In an era characterized by rapid technological advancements and evolving educational paradigms, the role of continuous professional development (CPD) for teachers has gained renewed significance (M. Li et al., 2022). As educators navigate increasingly dynamic and diverse learning environments, equipping themselves with the skills to integrate technology into their pedagogy has become paramount. Educational technology not only facilitates innovative teaching and learning practices but also serves as a powerful enabler of ongoing professional growth (Yureva et al., 2020). In the digital age of the twenty-first century, professional development is consistently a crucial factor in determining how meaningfully (language) teachers engage with technology integration (Liu & Kleinsasser, 2023). The sustainability of high-quality education is significantly influenced by the professional development of teachers. It is challenging to provide sustained professional development programs, nevertheless, due to a number of issues, including a lack of resources, access, and training opportunities, as well as a shortage of skilled staff. Educational institutions presently support the adoption and use of contemporary educational technologies for the professional development of instructors (Dahri et al., 2022). Teachers are encouraged to create successful learning programs that are tailored to the individual needs of each student by encouraging personalized planning and organization of learning processes made possible by technology. Given the difficulties, preschool teachers must improve their professional competence and pick up the skills they need to teach (H. T. Nguyen et al., 2023). Since technology is developing so quickly, digital literacy has emerged as a crucial element of professional development for teachers (A. T. Nguyen & Habók, 2024).

To enhance teaching and learning in schools, educators are increasingly utilizing educational technologies (ET) such as tablets, laptops, and dependable internet in the classroom. It is thought that the Fourth Industrial Revolution (4IR) will have a positive effect on how teachers present the curriculum if these tools are used (Kgosi et al., 2023). The most recent national educational technology plan in the United States described how technology can give students a more relevant and engaging experience, incorporate more real-world elements into the classroom, and engage students in a variety of activities that extend beyond the traditional classroom and let them follow their interests (Xie et al., 2023). A comprehensive set of interrelated structural elements make up future teachers' digital competency as a personal education: cognitive (creation of personally acquired knowledge of contemporary digital learning technologies, methods of professional activity based on information and communication technologies), motivational (increasing awareness of the necessity of using digital technologies in professional activities, focus on collective interaction, and the desire to succeed through information technology), activity (a collection of skills and abilities related to the use of digital technologies for the organisation of educational activities and solving professional problems), and personal (ability to self-study, professional self-development in a digital society) (Fursykova et al., 2022).

Digital literacy is now a crucial part of educators' professional development because of the speed at which technology is developing (Charania et al., 2023). The extent to which digital technology has been incorporated into university instruction has altered our perceptions of teacher competency and preparedness. The concept of teacher digital competence (TDC) has recently surfaced in response to the digitization of higher education (HE) (Peters et al., 2022). The primary areas that comprise professional activity, including teachers' understanding of professional activity, how the educational process is organized, and their own personal growth, all reflect their professional development. This, in turn, necessitates that the teacher knows exactly how to teach his subject and apply it, that he or she acquire the fundamental skills required to plan, organize, and assess the educational process, and that the teaching and educational process be intensively organized based on cutting-edge scientific discoveries in the field of education, contemporary pedagogical and information and communication technologies, and ongoing and methodical self-development (Shodiyeva, 2022). In the twenty-first century, digital technologies are evolving, altering, and proliferating at a rapid pace. We now use digital technologies in many aspects of our lives and consider them to be essential. Education and training are two of the most important domains where 21st-century digital skills are implemented (Yeni & Can, 2022). Teachers must take advantage of opportunities to incorporate new tools into their pedagogy as technology continues to offer new teaching alternatives in the classroom. One way to encourage teachers to use new technologies in the classroom is through professional development, which links them to the resources they need, creates continuous training that increases teachers' technological knowledge and proficiency, and develops site-based leaders for technology use through professional development and professional learning communities (Love et al., 2020).

Using technology in the classroom can improve student learning and increase the effectiveness of teachers' instruction. Since educational technology is still widely employed in classroom instruction, technology-infused learning is a powerful tool in the modern era (Zhou et al., 2022). Technology, content, and pedagogy are the three "core" aspects of education. Technological pedagogical content knowledge (TPACK) is knowledge that goes beyond these traditional divisions. Not all instructional philosophies, subjects, or classrooms can use the same technological resources. Teachers in all fields should take part in professional development programs to learn about and apply cutting-edge methods for teaching documentation, assessment, and material delivery. Enhancing education using technology is a difficult, organized process. No one method is better than another when it comes to using technology in the classroom (Adipat et al., 2023). Teachers must be ready to include technology in the classroom in a meaningful way with high-quality training if they want to fully realize its potential. One crucial way to get in-service teachers ready to incorporate technology into their lessons is through professional development or PD (Fütterer et al., 2023). To encourage teachers to engage in technology-related professional development, it is still important to educate them about the benefits of using technology in the classroom (Fütterer et al., 2023). Enhancing knowledge, abilities, motivation, creativity, networking possibilities, and learning materials are just a few benefits of teachers using technology (Ikhwan et al., 2023).

Technology's role in education is becoming more and more crucial as it develops further, influencing how educators interact with students, content, and pedagogical approaches. They investigate the possible effects of ICT, Smart Classrooms, and the Internet of Things (IoT) on elementary school teachers' professional growth. IoT integration in educational settings offers a special chance to improve teaching methods, expedite administrative duties, and promote a creative and cooperative learning environment as technology develops further (Mihir Kumar Jena et al., 2023). Since teacher-centered learning gave way to student-centered learning, most educators have been motivated to be more selective and innovative in how they manage the teaching and learning process. Support is given under the stipulation that, because of the pandemic, all instruction must take place online. It is necessary to improve instructors' knowledge of how to use technology in the classroom (Puspandari, 2023). Professional development programs for teachers are crucial for preparing, assisting, and equipping pre-service and in-service educators with the most recent technology abilities. These initiatives must serve as sources or catalysts for changing educators' perspectives by ensuring that technology use in the classroom is required. In the classroom, technology plays a significant role in teaching academic performance, general skills, socioemotional skills, metacognition and societal readiness, and creative development (Abel et al., 2022). Continuing Professional Development (CPD) has been reshaped by technology, which has made learning more interactive, personalized, and accessible. Digital solutions such as Learning Management Systems (LMS), mobile apps, and online platforms like Coursera and LinkedIn Learning allow workers to upskill at their own pace and convenience. While augmented reality (AR) and virtual reality (VR) offer immersive simulations for practical skill development, artificial intelligence (AI) and other technologies allow adaptive learning by customizing content to each learner's needs. Peer-to-peer learning and networking are encouraged via virtual communities and collaborative tools, which strengthens the social component of CPD. Furthermore, frameworks such as TPACK and the SAMR Model guarantee the successful use of technology in the planning and execution of CPD.

Technology enables professionals to participate in lifelong learning and maintain their competitiveness in their fields by dismantling the conventional boundaries of time, place, and expense. The necessity of using technology in the classroom might have provided teachers with instant feedback on their level of expertise. Teachers may have a more accurate assessment of their technological knowledge and proficiency, and some who had a tendency to overestimate themselves may now be more inclined to engage in technology-related professional development (Fütterer et al., 2023). Teachers play a vital role in knowledge dissemination and skill development, requiring continuous professional development (CPD) to adapt to evolving educational paradigms and technological advancements. Integrating technology into teaching enhances pedagogy, fosters digital literacy, and supports innovative learning practices. Tools like Learning Management Systems, AR/VR, and AI-driven platforms personalize and enrich CPD experiences, enabling lifelong learning. However, challenges such as resource limitations and skill gaps persist, necessitating well-structured programs. Frameworks like TPACK and the SAMR Model ensure effective technology integration, empowering teachers to create dynamic, student-centered learning environments while staying competitive in their profession.

FRAMEWORKS FOR TECHNOLOGY-ENABLED CPD: AN ESSENTIAL ELEMENT FOR INTEGRATION OF TECHNOLOGY IN PROFESSIONAL DEVELOPMENT OF TEACHERS

Using digital tools and platforms, technology-enabled continuing professional development (CPD) offers professionals individualized, flexible, and accessible learning opportunities. Through interactive simulations, adaptive learning, and gamified content, CPD programs boost skill development and increase engagement by incorporating cutting-edge technologies like Learning Management Systems (LMS), Artificial Intelligence (AI), Augmented Reality (AR), and Virtual Reality (VR). CPD is made more interesting and scalable by platforms like Coursera, LinkedIn Learning, and Zoom, which combine live webinars, self-paced modules, and group projects. Technology-enhanced CPD is designed and delivered using frameworks like TPACK, SAMR, and the ADDIE Model, which guarantee a balance between pedagogy, technology, and content. A useful paradigm for understanding how to integrate technology into various theoretical contexts, including EFL lessons, is Technological Pedagogical Content Knowledge (TPACK). This project began as an academic framework to enhance the body of knowledge that enables educators to use technology to educate more effectively (Su, 2023).

These programs support social learning through virtual communities, offer content in a variety of formats, and use AI-powered tools to provide real-time feedback and assessment in order to meet the different learning needs of their students. Although issues like the digital divide and self-directed learning motivation still exist, technology-enabled CPD enables professionals to stay current in their industries, develop their competencies, and promote ongoing development in a world that is becoming more and more digital. Excellent technology-based education is dependent on three pillars: available technology tools, teaching strategies, and subject content. These three interconnected criteria account for all of the variations in the amount and quality of technology used in classrooms among schools. As a result, educators need to understand the advantages and disadvantages of different learning resources. The most effective teachers are able to see beyond the obvious uses of technology and find new ways to use it for learning. TPK programs should inspire kids to think creatively, look ahead, and be receptive to the ways that technology might enhance their education. In an effort to improve students' educational experiences, schools are progressively integrating technology into their curricula (Adipat et al., 2023).

Technology-enabled Continuing Professional Development (CPD) offers personalized, flexible, and accessible learning through tools like Learning Management Systems (LMS), Artificial Intelligence (AI), Augmented Reality (AR), and Virtual Reality (VR). Platforms such as Coursera and Zoom integrate live webinars, self-paced modules, and group projects, while frameworks like TPACK, SAMR, and ADDIE ensure balanced and effective design. These programs foster social learning, real-time feedback, and creative engagement, empowering educators to integrate technology innovatively. Despite challenges like the digital divide, technology-enabled CPD equips professionals with skills to adapt in a digital era, enhancing teaching strategies, content delivery, and overall educational experiences.

DIGITAL TOOLS AND PLATFORMS FOR PERSONALIZED PROFESSIONAL DEVELOPMENT

The term "continuous professional development" (CPD) describes the process of improving, monitoring, and documenting the abilities, information, and experience gained both formally and informally through employment after initial training. It documents what one encounters, acquires, and subsequently uses. Eventually, this process of learning alters teachers' professional practices and perspectives about those practices (Lee et al., 2023). Technology-related skills are becoming increasingly important not only in professional contexts but also in nearly every aspect of life due to the growing availability of new technologies in an increasingly digitalized environment. Therefore, it is the duty of schools to teach these skills to their students and, by extension, to future generations of professionals. Teachers play a crucial part in this process since their proficiency with new technologies is a necessary precondition for the successful application of these abilities (Drossel & Eickelmann, 2017). The incorporation of smart classrooms is essential to elementary school teachers' professional growth. Teachers can improve student engagement, improve their teaching methods, and keep up with changing educational trends by embracing technology. Elementary school teachers' continued development in the twenty-first century will be greatly aided by the ongoing advancement of smart classrooms and the dedication to resolving related issues. The use of ICT in professional development programs has the potential to completely transform the way that primary school teachers teach. Teachers can be empowered to adjust to the demands of 21st-century education by embracing the potential that ICT presents and tackling the related obstacles (Mihir Kumar Jena et al., 2023). (Zhou et al., 2022) came to the conclusion that, in contrast to the conventional wisdom that more seasoned teachers possess superior teaching skills, primary EFL teachers with the most experience had the least technological expertise. In order to improve their teaching abilities, the study recommends that EFL teachers in China, including those with varying years of teaching experience, be provided with technology-based professional development training. Digital resources and technological tools are varied, and there is growing discussion on how to use them as mediating tools in the classroom to help students build their scientific knowledge. It is widely believed that the introduction of digital technologies into professional development programs for teachers to enhance science instruction is a potential innovation in science education (Fernandes et al., 2020). The ongoing use of technology in the classroom was influenced by a confluence of institutional, professional development, technological, and individual variables (Kafyulilo et al., 2016).

Professional development has been revolutionized by digital tools and platforms that allow for a highly customized and flexible approach to learning. This allows professionals to stay current, gain new skills, and accomplish their career goals in ways that are specific to their requirements and preferences. To personalize learning, these solutions make use of cutting-edge technologies like data analytics, artificial intelligence (AI), and adaptive learning algorithms. Large course libraries in a variety of subject areas are offered by platforms like LinkedIn Learning, Coursera, and Udemy, enabling users to choose learning trajectories that complement their career goals. In order to provide tailored information and make sure students concentrate on areas that need attention or interest, many of these platforms use artificial intelligence (AI) to evaluate user behavior, preferences, and progress. Furthermore, learning on the go is made possible by mobile learning programs like Duolingo, Edmodo, and Khan Academy, which remove time and location constraints while providing interactive content in a variety of formats like videos, tests, and interactive activities.

Digital tools also improve social learning and collaboration. Professionals can participate in peer-to-peer knowledge sharing, group discussions, and virtual workshops using platforms like Zoom, Slack, and Microsoft Teams, which create dynamic, rich learning environments. Professionals may practice skills in risk-free, realistic environments thanks to immersive simulations offered by virtual reality (VR) and augmented reality (AR) solutions like zSpace and Mursion. For instance, teachers can create lesson plans using virtual reality classrooms, and doctors can practice difficult procedures by simulating them. Personalized professional development is further supported by learning management systems (LMS) like Moodle, Blackboard, and Canvas, which offer a centralized platform for self-paced learning modules, resource access, and progress monitoring. Organizations may also track employee progress, design customized training programs, and award micro-credentials or badges for skill mastery thanks to these platforms. Additionally, the incorporation of gamification components into several digital platforms enhances the fun and engagement of professional development. Leaderboards, medals, and awards are used by platforms such as Kahoot and Quizlet to encourage students, while AI-powered evaluation tools give users immediate feedback so they can monitor their progress. Professionals can record and display their accomplishments, abilities, and development with e-portfolios, such as those offered by Seesaw or Pathbrite. However, overcoming obstacles like digital accessibility, guaranteeing content quality, and encouraging self-discipline in students are necessary for these technologies to be effective. By removing these obstacles, digital tools and platforms can facilitate highly individualized, significant professional development that enables people to succeed in their jobs and adjust to a constantly changing work environment.

The integration of smart classrooms and digital tools is pivotal for the professional growth of elementary school teachers, enabling them to enhance student engagement, modernize teaching methods, and adapt to 21st-century educational demands. Platforms like LinkedIn Learning, Coursera, and mobile apps such as Duolingo and Khan Academy offer personalized, flexible learning through adaptive algorithms and diverse formats, while tools like VR and AR provide immersive, risk-free environments for skill development. Social learning is supported by platforms like Zoom and Slack, fostering collaboration and peer-to-peer knowledge sharing. Gamification, micro-credentials, and AI-powered feedback further boost engagement and track progress. Overcoming challenges like accessibility and self-discipline is crucial to realizing the full potential of these technologies in creating impactful professional development opportunities.

BENEFITS OF TECHNOLOGY IN TEACHER CPD

There are many advantages to incorporating technology into Teacher Continuing Professional Development (CPD), which changes how teachers develop their abilities, keep abreast of educational developments, and enhance classroom procedures. Flexibility and accessibility rank among the biggest benefits. Teachers can take part in CPD programs whenever it's convenient for them thanks to technology, which eliminates time and location restrictions. Teachers can learn at any time and from any location with the help of online learning platforms like Coursera, Moodle, and Edmodo, which offer top-notch materials, webinars given by experts, and self-paced learning modules. Teachers in rural areas who might not otherwise have access to professional training can particularly benefit from this. Personalization is another important advantage since technology makes it possible to customize CPD programs to meet the needs of specific teachers. By using data analytics and artificial intelligence to suggest pertinent courses and resources, adaptive learning systems make sure teachers concentrate on topics that support their career objectives or deal with particular difficulties. Learning is made more efficient by platforms such as Khan Academy and LinkedIn Learning, which provide curated content that changes according to the learner's progress and pace. Additionally, technology promotes collaborative learning by bringing educators together from around the world through online communities, discussion boards, and social media sites like Facebook Groups and Twitter (with hashtags like #EdChat). By exchanging best practices, asking for guidance, and working together on creative teaching methods, these virtual networks give educators access to a wealth of collective knowledge. Through immersive technologies like virtual reality (VR) and augmented reality (AR), technology also makes experiential and active learning easier. Teachers may experience classroom simulations, create efficient teaching methods, and experiment with cutting-edge pedagogies in a risk-free setting with the use of tools like zSpace and Nearpod VR. Additionally, by adding features like leaderboards, prizes, and quizzes, gamified platforms like Kahoot and Quizizz make CPD interesting. AI-driven tests also give teachers real-time feedback, allowing them to monitor student progress and modify their lesson plans as necessary. Finally, technology improves scalability and cost-effectiveness. By using online tools and platforms, educational institutions may execute extensive training programs with less funding. This makes CPD more economical and ecological by lowering the need for travel, printed materials, and physical infrastructure. Teachers gain the digital skills required for classrooms in the twenty-first century and stay current with the newest trends in education by incorporating technology into their continuing professional development. This leads to an improvement in student results and a promotion of innovation in education.

Incorporating technology into Teacher Continuing Professional Development (CPD) offers numerous benefits, transforming how educators enhance their skills and classroom practices. Technology provides flexibility and accessibility, enabling teachers to engage in CPD anytime and anywhere through platforms like Coursera, Moodle, and Edmodo, particularly benefiting those in remote areas. Personalization through adaptive learning systems and AI ensures tailored content that aligns with individual career goals. Collaborative learning is enhanced by global online communities and social media networks like #EdChat, fostering knowledge exchange and innovation. Immersive technologies such as VR and AR facilitate experiential learning, while gamified platforms like Kahoot make CPD engaging. Additionally, technology improves scalability and cost-effectiveness, reducing costs and environmental impact while equipping teachers with 21st-century digital skills, ultimately enhancing student outcomes and educational innovation.

CHALLENGES IN TECHNOLOGY-DRIVEN CPD FOR TEACHERS

The fourth technological revolution has brought up both new challenges and benefits for all facets of social life, but most significantly, it has changed how nations educate their citizens. One tool to support, improve, boost productivity, and enrich educational content is the fourth technological revolution (H. T. Nguyen et al., 2023). (Kgosi et al., 2023) showed that there was insufficient professional development for using the educational tools that were available in schools. Technology is seen as a tool that supports and upholds formalized and crystallized patterns of cooperation, frameworks, frames, etc. Because technology is exclusively regarded through the lens of organized forms and institutional arrangements that it may or may not properly serve, it is therefore perceived as straightforward and unambiguous, and its role aligns with its intended use (Eradze et al., 2021). STEM (science, technology, engineering, and mathematics) is a meta-discipline that uses problem-focused, active methods like game-based learning. STEM skills are a crucial component of the educational response to the changes brought about by the fourth industrial revolution, which is being driven by the convergence of several exponential technologies. An essential component of any technology-enhanced learning innovation is the attitude of the teachers (Mystakidis & Christopoulos, 2022). Since early 2020, in order to ensure that everyone has access to school, educational systems around the world have had to rely on technology. The epidemic has made social injustice, inequality, and the digital divide worse, necessitating special and targeted responses based on compassion and understanding, open educational viewpoints, and serious worries about ethics, monitoring methods, and data privacy (Rodés et al., 2021).

Technology-driven continuing professional development (CPD) has many benefits, but it also has drawbacks that must be resolved to guarantee its efficacy and inclusion. The digital gap, which results in differences in access to technology and internet connectivity, is one of the biggest problems. Teachers encounter difficulties because of inadequate funding, support, infrastructure, and access. These educators are also lacking in preparedness, competency, security, and moral knowledge. Among the challenges, barriers, and hazards that educators face are network outages, malware, viruses, spam, phishing, hacking, cyberbullying, plagiarism, and more. The difficulties in using technology for the professional development of teachers include issues related to infrastructure, ethics, skills, and attitudes. As a result, educators need to be sufficiently skilled and prepared to use technology in their professional growth (Ikhwan et al., 2023). (Melash et al., 2020) found that general digital competencies align with the educational system's context. The context of integration, preparing future educators for the chance to use technology in their future instruction of students, critically evaluating the use of digital devices in the classroom, and teaching children are the main focusses of competency in integrating technologies into teaching practice.

Programs for professional development that are based on technology must offer teachers a variety of activities, including brainstorming sessions, discussions, practical exercises, and just-in-time coaching. Despite the fact that there are several instances of teachers using applications in various fields, there may be some issues. One of these issues is the dearth of technical assistance and the inadequate or non-existent hardware needed to integrate the apps into the classroom. Teachers benefit from the professional development program's technology-based approach by learning how to employ technology for enrichment activities (Elmali & Balkan Kiyici, 2022). Technology-specific experience gave them a variety of technological skills, self-control to pick up new tools, resilience to overcome technical problems, and confidence when using technology. Age, career background, technology abilities, and views about technology self-efficacy were all contributing personal factors that influenced beliefs and practice (Rowston et al., 2022).

Teachers in underdeveloped or rural locations might not have the technology or infrastructure needed to take part in online continuing professional development programs, which would limit their possibilities for professional development. For many educators and institutions, the expense of gadgets, software licensing, and dependable internet connections can be a deterrent, even in areas with good connectivity. The use of technology in teacher education has become increasingly widespread, giving teachers more options for developing innovative teaching methods. However, the practical ramifications of using technology have continued to stimulate controversial discussions about the support systems required for the smooth integration of technology into teaching methods (Theodorio, 2024). Proficiency with technology is another difficulty. To use digital tools and platforms efficiently, many teachers need training, and not all of them are tech-savvy. Insufficient assistance may cause educators to feel overburdened or disheartened, which may impede their engagement in technology-based professional development. Furthermore, some applications' lack of personalization can seem generic or unimportant, failing to cater to the particular requirements of different teachers or their particular classroom environments. The necessity of flexible and context-sensitive CPD programs that complement instructors' subject areas and professional objectives is therefore highlighted.

In self-paced, tech-driven CPD, motivation and engagement can also be problems. Without direct supervision or in-person interactions, instructors could find it difficult to prioritize professional growth or maintain motivation in the face of their hectic schedules. This problem is made worse by the possibility of information overload, as there are so many online courses and resources available that it might be hard for teachers to find relevant, high-quality content. Additionally, there are issues with online CPD programs' quality assurance because not all digital platforms offer verified or evidence-based content, which may jeopardize the learning objectives. Concerns about privacy and security add still another level of complexity. The likelihood of data breaches and the misuse of personal information has increased due to the increased reliance on internet platforms. It is imperative for educators and educational institutions to guarantee that CPD tools adhere to data protection laws and offer safe learning environments. Last but not least, incorporating technology-driven CPD into institutional timetables and procedures might be difficult. In order to support digital learning projects, administrators may encounter opposition from educators who favor more conventional approaches or who find it difficult to manage their time and resources efficiently. As educators encounter more research and pressure to help students achieve higher degrees, the quality of teacher professional development is becoming a crucial academic issue. Teachers now have unrestricted access to professional development and learning without regard to time or location because to technological advancements. Using technology to help teachers develop creates affinity spaces that are different from traditional settings (Zhang, 2022). Infrastructure investment, teacher training and technical assistance, and accessible, pertinent, and interesting CPD programs are all essential to overcoming these obstacles. Technology-driven CPD has the potential to be a potent instrument for empowering teachers and improving instruction in a variety of educational environments by removing these obstacles. The roles of facilitators, motivators, inspirers, mentors, cultivators of imagination and creativity, cultivators of character values, promoters of teamwork, and providers of social empathy cannot be fulfilled by technology, so teachers cannot be replaced. Additionally, instructors' role in molding pupils' character cannot be replaced by technology. However, in order to use technology to raise the standard of education in Indonesia, educators are expected to keep improving their skills (Ikhwan et al., 2023).

The fourth technological revolution has transformed education, introducing both opportunities and challenges for teacher professional development (CPD). Technology-driven CPD offers flexibility, personalization, and innovative teaching practices through platforms like Moodle and Coursera, while tools such as VR and game-based learning foster active, experiential learning. However, barriers such as digital divides, inadequate infrastructure, lack of technical support, and teachers' varying technology competencies hinder effective implementation. Additional challenges include high costs, privacy concerns, motivation issues in self-paced programs, and difficulty ensuring quality content. Addressing these requires investment in infrastructure, training, technical assistance, and the development of engaging context-sensitive CPD programs. Although technology enhances professional growth, it cannot replace teachers' critical roles as mentors, motivators, and character builders. With proper support, technology-driven CPD has the potential to empower educators, improve teaching methods, and elevate education quality across diverse contexts.

STRATEGIES FOR EFFECTIVE INTEGRATION OF EDUCATIONAL TECHNOLOGY IN CPD

It is becoming increasingly clear that teachers must use technology to enhance their instruction and students' learning. Professional development (PD) is one of the most crucial strategies to assist educators in using technology more successfully. In order to improve teachers' knowledge, tactics, and other relevant teacher attributes that could affect their instruction, professional development (PD) programs are provided to in-service teachers (Bowman et al., 2022). For educators to have meaningful and lasting learning experiences, integrating educational technology into Continuing Professional Development (CPD) effectively necessitates strategic planning and a learner-centered approach. Aligning CPD activities with particular teaching needs and goals is an important tactic. To determine the skills and knowledge gaps among instructors, a requirements analysis must be carried out prior to technology integration. This aids in customizing CPD programs to fill these gaps using the best available technology. For example, adding virtual laboratories, simulation tools, or coding platforms like Scratch to CPD might give instructors practical, pertinent experiences if they need to improve their comprehension of STEM teaching. In these times of accelerated innovation and growing academic duties, continuing professional development (CPD) is essential for academics. Gaining new skills is necessary to perform the necessary roles in leadership, administration, and the fields of knowledge and instruction (Spilker et al., 2020). Technology use in

CPD directly promotes teaching techniques and enhances student learning when clear objectives are set and linked to classroom outcomes. Teachers can be better equipped to meet the needs of education in the twenty-first century by embracing the potential that ICT presents and tackling related issues. A dedication to thorough ICT-based professional development is essential as we continue to traverse the digital age to make sure primary school teachers are prepared to give their pupils a top-notch education (Mihir Kumar Jena et al., 2023). Therefore, capacity building should concentrate on elevating the quality of topic teaching, providing resources (particularly with high-quality digital content), and encouraging change in classroom practices rather than just incorporating technology. We believe that in the SSA setting, where technology is now scarce, developing pedagogical ability necessitates programs designed to make use of technology resources when they are accessible without becoming reliant on them. These programs establish educational environments that facilitate the effective use of digital technologies as they become accessible (Hennessy et al., 2015). When it is viewed as a problem that only affects the micro level, technology adoption becomes challenging. Teachers' hesitancies about implementing technology in the reorganisation of teaching and learning should not be attributed to a lack of knowledge or skills or a personal attitude. Even when they are expressed as insecurities, doubts are real manifestations of unresolved matters pertaining to the usage of technology (Eradze et al., 2021).

Giving instructors continual assistance and training to increase their self-assurance and technological proficiency is another crucial tactic. To help instructors adjust to technology-driven CPD, workshops, tutorials, and peer mentorship programs can be offered as many may not be comfortable with the newest digital tools. For instance, schools can provide "train-the-trainer" events in which tech-savvy educators assist their peers in using Moodle or Google Workspace for Education. Offering a hybrid learning strategy, which combines online modules with in-person meetings, can also facilitate the shift to fully digital CPD programs. To reduce interruptions and promote a productive learning environment, it is also critical to give teachers access to technical assistance and troubleshooting tools. More ongoing assistance and interventions are required for instructors to adopt comprehensive and long-term practice adjustments (Kul, 2018). For the online educational course "Simultaneous and asynchronous translation in a digital environment," (Wang, 2023) developed the pedagogical concept by using a competency-based approach to interpreter training and taking into account the need to develop the abilities, knowledge, and skills necessary for a successful professional translation activity. The high degree of integration between information technology and educational development has given rise to a new model of educational development called "smart education," which gives educational reform endless life. To confront the problems of the modern world, educators must adopt a new way of thinking about education and develop the idea of lifelong learning (Huang, 2023).

To create a high-quality educational system, preschool teachers' professional competence needs to be strengthened through training and capacity-building courses. In addition to creating courses and training programs to improve teachers' professional ability, the teacher training management system is crucial in establishing clear strategies and goals (H. T. Nguyen et al., 2023).Encouragement of teamwork and community development is another important tactic. By enabling educators to exchange ideas, resources, and best practices, virtual communities of practice facilitated by platforms such as Microsoft Teams, Slack, or Edmodo promote peer-to-peer learning and support. Co-creating lesson plans or taking part in group simulations are examples of collaborative initiatives that can improve motivation and engagement. These cooperative learning activities support the growth of a continuous learning culture inside schools or districts in addition to assisting educators in learning from one another. Additionally, using gamification features on sites like Kahoot or Quizizz, such as leaderboards, badges, or awards, can enhance CPD by making it more engaging and fun, which will increase participation and retention.

Leveraging data and feedback is essential for refining and personalizing CPD programs. Learning Management Systems (LMS) like Canvas or Blackboard can track educators' progress, provide analytics on their learning behavior, and identify areas for improvement. Using this data, program designers can customize CPD content to cater to individual needs, ensuring a more personalized experience. For instance, adaptive learning technologies powered by Artificial Intelligence (AI) can suggest modules or activities based on a teacher's performance, ensuring that CPD remains relevant and effective. Additionally, collecting feedback through surveys or focus groups helps in evaluating the success of the program and identifying areas for enhancement. The need for professional development is emphasised here since it is still helpful in raising teachers' technology self-efficacy and equipping them with the skills they need to successfully incorporate technology into their lessons. PD programs, however, function best when they are directed by suggested cautions based on research. Teachers will be better positioned and able to increase their levels of technology integration self-efficacy to facilitate twenty-first-century teaching, whether in-person or remotely, through professional development intervention, flexible knowledge acquisition, and a willingness to take instructional risks with technology integration (Gomez et al., 2022).

In conclusion, the effective incorporation of educational technology into CPD depends on guaranteeing accessibility and inclusivity. Programs must take into account instructors' varied demands, particularly those who work in remote locations or have little experience with technology. This can entail making resources accessible offline, delivering free or reasonably priced tech tools, and making sure that different devices are compatible. The principles of Universal Design for Learning (UDL) can direct the development of inclusive CPD programs by providing a variety of interaction, representation, and expression options that accommodate varying learning styles and skill levels. By using these tactics, educational institutions can successfully incorporate educational technology into continuing professional development, enabling teachers to prosper in a quickly changing digital environment and eventually improving student learning. Effective integration of educational technology into Continuing Professional Development (CPD) requires strategic planning, accessibility, and a learner-centered approach. It begins with a needs analysis to align CPD programs with teachers' skill gaps, leveraging tools like virtual labs, simulation platforms, and coding tools for practical learning. Providing ongoing support through workshops, mentorship, and hybrid learning ensures smooth adoption while fostering collaboration through virtual communities and gamified platforms enhances engagement. Data-driven insights from Learning Management Systems (LMS) and feedback loops enable personalization, ensuring relevance and effectiveness. Addressing barriers such as digital divides, technical challenges, and inclusivity is crucial, incorporating offline resources, affordable tools, and Universal Design for Learning (UDL) principles to cater to diverse teacher needs. These strategies empower educators with the skills and confidence to navigate the digital age, ultimately enhancing teaching practices and improving student outcomes.

FUTURE DIRECTIONS FOR EFFECTIVE INTEGRATION OF EDUCATIONAL TECHNOLOGY IN CPD

Utilizing subject-specific software/applications, blended learning, social messaging, virtual coaching, and video-stimulated reflection were all promising, regionally contextualized types of technology-mediated TPD. We discuss the uneven efficacy of programs and the disregard for marginalized populations. The role of facilitators or expert peers is crucial, but frequently overlooked, and the interpersonal aspect of teacher learning must be preserved in order to maximize the efficiency of technology-enhanced TPD. Researchers, legislators, educators, and teacher educators are all given advice (Hennessy et al., 2022). Despite obstacles related to technology, administration, or evaluation, educators who hold student-centered views have a tendency to implement student-centered curricula. The greatest influence on students' success was thought to be the teachers' own attitudes and ideas regarding the value of technology in the classroom. Their behaviors were significantly shaped by both internal (such as a love of technology and a problem-solving mindset) and external (administrators and personal learning networks) elements. Teachers reported that their present attitudes and views about technology, together with their current levels of knowledge and skills, were the biggest obstacles keeping other teachers from utilizing technology. Refocusing professional development initiatives on methods for promoting shifts in teachers' attitudes and beliefs is necessary (Ertmer et al., 2012). The ability of teachers to integrate technology into their lessons is a prerequisite for the professional development program. Teachers' continued use of technology in the classroom is also dependent on their access to and proficiency with it, as well as the support they receive from school administration (Kafyulilo et al., 2016). Supporting teachers' use of technology in the classroom requires professional development. Research on professional development must keep looking for proof of the illusive relationships between teachers' involvement in professional development, their choices and implementation of inquiry in science classes, and the effect on student accomplishment. Though there is still much to learn about how teachers can utilize technology more skilfully in inquiry-based instruction, it is a potent tool that can be used to scaffold students' comprehension (Higgins & Spitulnik, 2008).

Digital tool developments, shifting educational demands, and shifting pedagogical methods will all have a major impact on how educational technology is integrated into continuing professional development (CPD) in the future. Growing personalized, adaptable learning is one exciting avenue. As artificial intelligence (AI) and machine learning become more prevalent, continuing professional development (CPD) platforms will be able to offer increasingly more personalized learning experiences, adjusting the pace and material to each educator's unique requirements, interests, and past expertise. With the ability to modify learning routes in real-time, this adaptive technology has the potential to completely transform continuing professional development by giving every teacher the tools they need to advance their craft. In the context of both formal and informal education, individuals with functional diversity (FD) can learn through information and communication technology (ICT). The full growth of this population depends on having qualified experts in the field of educational technology, as these digital solutions promote their inclusion, participation, and motivation (Gallardo-Montes et al., 2024). Furthermore, incorporating data analytics will be essential for monitoring teacher development, giving feedback, and spotting new patterns in professional development requirements. This will enable more accurate, fact-based choices to be made when creating CPD programs.

The rise of immersive learning technologies like virtual reality (VR) and augmented reality (AR), which have enormous potential to improve experiential learning in CPD, is another significant development. Without requiring resources or physical presence, educators can engage in virtual classroom simulations, investigate various pedagogical approaches, and receive practical training. VR might, for instance, replicate difficult classroom management scenarios or provide immersive learning environments for history, science, and art classes where teachers can engage with the material in dynamic ways. There are instructional uses for mobile augmented reality games (MARGs). Since the technologies that facilitate MARGs' use in education are becoming more and more common, it is time for teacher practices to embrace them. Teachers' personal and professional needs were addressed in the workshop, which had an effect on their practices regarding the use of games and mobile and augmented reality technology. Teachers found that creating a new game during training and making it publicly accessible to others was a highly fulfilling experience and a useful habit to adopt (Pombo & Marques, 2021). The proposal of a set of recommendations that will be helpful to academics and teacher trainers worldwide who want to create and implement ongoing professional development programs for the integration of MARGs into teaching practices seemed feasible (Pombo & Marques, 2021). Likewise, augmented reality (AR) has the potential to assist educators in visualizing intricate ideas and facilitating interactive learning that improves comprehension and memory. By encouraging active learning and critical thinking, the integration of these technologies into CPD will give educators more impactful and captivating professional development experiences.

Another emerging trend that will influence the direction of technology-driven professional development in the future is the combination of bite-sized CPD content with microlearning. Short, targeted learning modules that teachers can do in a few minutes possibly during a break or in between lessons will be beneficial to them because they frequently have hectic schedules. Mobile apps can be used to deliver microlearning, giving users access to content like brief courses, podcasts, and video clips while they're on the go. Teachers will be able to include professional development into their daily schedules thanks to this approach, which will make CPD more adaptable, practical, and pertinent. When combined with gamification, which turns minor accomplishments into more substantial rewards, this strategy will improve motivation and long-term involvement. Teachers' perceptions of educational technology not only affect how they use new resources, but they also have an impact on their professional growth. A sustainable approach to academic development is to educate university instructors about the range of viewpoints on teaching and learning while cooperating to enhance instruction in university courses (van der Rijst et al., 2019).

Future CPD models will also progressively emphasize peer-to-peer learning and mentorship driven by AI. Through the use of clever matching algorithms based on professional interests, teaching philosophies, or areas of expertise, AI can help teachers connect with seasoned mentors around the world to support virtual mentoring. Teachers can work together to solve problems in the classroom, share knowledge, and give feedback in personalized learning communities that are facilitated by this technology. By encouraging instructors to share best practices, learn from one another, and work together to solve problems, peer-to-peer learning facilitated by collaborative platforms like Slack or Google Workspace—will foster an ongoing loop of knowledge sharing and support. The integration of global and interdisciplinary learning networks is another crucial avenue. Geographical and institutional obstacles will be removed by the use of technology in CPD, allowing educators to work together and gain knowledge from different viewpoints throughout the world. Teachers from various places and experiences can communicate through platforms like Edmodo, Twitter, or specialized forums, which can facilitate debates on global educational concerns and share a variety of teaching and learning methodologies. The sharing of creative concepts, research results, and instructional strategies from around the globe will be encouraged by this networked learning environment, enhancing the CPD experience and expanding the professional horizons of educators. Alternative approaches to providing our present and future educators with successful professional development must be the subject of more research (Schrum, 1999). Teachers are prepared and eager to embrace new approaches and tools that will enable them to implement student-centered and active blended learning in the post-pandemic period. To thrive in the Industry 4.0 era, education leaders and policymakers should accelerate systematic professional development actions that will unleash teachers' communities of practice and creativity, improving learning quality, student motivation, and engagement with STEM. This is encouraged by the limited number of teacher professional development projects, but also by their encouraging outcomes (Mystakidis & Christopoulos, 2022).

Finally, a primary priority of educational technology in CPD will be equity and inclusion. Every teacher will have access to top-notch CPD programs, regardless of their socioeconomic status, geographic location, or level of technological proficiency. It will be necessary to design technology-enabled CPD with inclusivity in mind, providing a variety of content delivery methods, multilingual support, and mobile-friendly formats to suit all learners. Prioritizing accessibility can help close the digital divide and ensure that all educators have equitable access to the resources and tools they need to further their careers. Teacher agency must be attained in order for digital technology to be integrated into the knowledge base. This ambition necessitates complete participation in professional learning activities within an environment that acknowledges teachers as actors and the owners of the authority to ask questions in a process where time is also regarded as a critical component (Magnusson, 2022). In conclusion, AI, VR/AR, microlearning, and international cooperation will support more individualized, adaptable, and immersive learning experiences that define the future of educational technology in CPD. As technology advances, it will give educators more creative opportunities to hone their craft, advance their practice, and ultimately improve their students' educational experiences. For these developments to be fair and effective for all educators, it will be essential to address issues like digital accessibility, data protection, and teacher support. Academic developers ought to be cognizant of the variety of learning demands and preferences among teachers. Despite being accessible, formal learning activities like workshops, training, and hands-on guides are not the most popular methods of learning how to use educational technologies. Teachers at universities learn by carrying out their duties, trying new things, and thinking back on what they have taught. In our capacity as academic developers, we must support teacher learning methods far beyond creating new workshops and training sessions. For instance, we must collaborate with teaching staff and academic developers to co-create student activities and co-design assessment techniques. These cooperative organizations or professional practice communities give academic developers a place to offer prompt, on-the-spot guidance and feedback (van der Rijst et al., 2019).

The future of technology-driven Continuing Professional Development (CPD) emphasizes personalization, equity, and inclusivity, integrating tools like AI, VR/AR, microlearning, and global networks to create adaptive, immersive, and collaborative learning experiences for educators. Emerging trends include personalized learning paths, data analytics for progress monitoring, gamification, and peer-to-peer mentorship facilitated by AI. Tools such as mobile apps and immersive technologies like VR/AR offer flexible and experiential learning opportunities, while platforms enable global collaboration and sharing of diverse teaching practices. Addressing barriers like accessibility, digital divides, and teacher support is crucial for equitable CPD. Teachers' attitudes, collaboration, and hands-on learning play pivotal roles in technology adoption, highlighting the need for CPD that values teacher agency and integrates professional learning into daily practices. By fostering inclusive, innovative, and practical approaches, CPD can empower educators to enhance their skills, embrace student-centered teaching, and improve learning outcomes in the rapidly evolving digital landscape.

CONCLUSION

In conclusion, the integration of educational technology into Continuing Professional Development (CPD) presents transformative opportunities to empower educators and improve teaching practices in an ever-evolving digital era. The use of advanced tools such as AI, VR/AR, and microlearning ensures that CPD programs are personalized, adaptive, and engaging, catering to the diverse needs and preferences of educators. By leveraging data analytics, educators' progress can be closely monitored, enabling targeted interventions and fostering continual professional growth. Immersive technologies allow educators to engage in experiential learning, while global networks and peer-to-peer mentoring enhance collaboration, knowledge-sharing, and exposure to innovative teaching strategies. However, these advancements also bring critical challenges that require attention, such as ensuring digital accessibility, bridging the digital divide, and providing continuous support to educators, particularly in underserved regions. Moreover, fostering teacher agency, valuing hands-on and reflective learning methods, and addressing attitudes and beliefs about technology are essential to maximizing the effectiveness of CPD. Collaborative efforts between policymakers, academic developers, and educators are vital to designing CPD programs that not only enhance technical proficiency but also promote equity, inclusivity, and sustainable professional growth. By embracing these innovations and addressing the associated challenges, educational institutions can ensure that CPD remains a powerful tool for equipping educators to deliver high-quality, student-centered learning experiences, ultimately transforming education for the 21st century.

REFERENCES

Abel, V. R., Tondeur, J., & Sang, G. (2022). Teacher Perceptions about ICT Integration into Classroom Instruction. In *Education Sciences* (Vol. 12, Issue 9). DOI: 10.3390/educsci12090609

Adipat, S., Chotikapanich, R., Laksana, K., Busayanon, K., Piatanom, P., Ausawasowan, A., & Elbasouni, I. (2023). Technological Pedagogical Content Knowledge for Professional Teacher Development. *Academic Journal of Interdisciplinary Studies, 12*(1), 173–182. DOI: 10.36941/ajis-2023-0015

Bowman, M. A., Vongkulluksn, V. W., Jiang, Z., & Xie, K. (2022). Teachers' exposure to professional development and the quality of their instructional technology use: The mediating role of teachers' value and ability beliefs. *Journal of Research on Technology in Education, 54*(2), 1–18. DOI: 10.1080/15391523.2020.1830895

Charania, A., Paltiwale, S., Sen, S., Sarkar, D., & Bakshani, U. (2023). Leading Edge Use of Technology for Teacher Professional Development in Indian Schools. *Education Sciences, 13*(4), 1–20. DOI: 10.3390/educsci13040386

Dahri, N. A., Vighio, M. S., Al-Rahmi, W. M., & Alismaiel, O. A.Muhammad Saleem VighioWaleed Mugahed Al-RahmiOmar A. Alismaiel. (2022). Usability Evaluation of Mobile App for the Sustainable Professional Development of Teachers. *International Journal of Interactive Mobile Technologies, 16*(16), 4–30. DOI: 10.3991/ijim.v16i16.32015

Drossel, K., & Eickelmann, B. (2017). Teachers' participation in professional development concerning the implementation of new technologies in class: A latent class analysis of teachers and the relationship with the use of computers, ICT self-efficacy and emphasis on teaching ICT skills. *Large-Scale Assessments in Education, 5*(1), 1–13. DOI: 10.1186/s40536-017-0053-7

Elmali, Ş., & Balkan Kiyici, F. (2022). Technology-based professional development program: Experiences of science teachers. *Journal of Educational Technology and Online Learning, 5*(2), 297–315. Advance online publication. DOI: 10.31681/jetol.1081367

Eradze, M., Bardone, E., & Dipace, A. (2021). Theorising on covid-19 educational emergency: Magnifying glasses for the field of educational technology. *Learning, Media and Technology, 46*(4), 404–419. DOI: 10.1080/17439884.2021.1961802

Ertmer, P. A., Ottenbreit-Leftwich, A. T., Sadik, O., Sendurur, E., & Sendurur, P. (2012). Teacher beliefs and technology integration practices: A critical relationship. *Computers & Education, 59*(2), 423–435. DOI: 10.1016/j.compedu.2012.02.001

Fernandes, G. W. R., Rodrigues, A. M., & Ferreira, C. A. (2020). Professional Development and Use of Digital Technologies by Science Teachers: a Review of Theoretical Frameworks. In *Research in Science Education* (Vol. 50, Issue 2, pp. 673–708). DOI: 10.1007/s11165-018-9707-x

Fursykova, T., Habelko, O., & Chernii, V. (2022). The Development of Digital Competence of Future Teachers in the Process of Distance Learning. *International Journal of Emerging Technologies in Learning, 17*(10), 85–98. DOI: 10.3991/ijet.v17i10.28973

Fütterer, T., Scherer, R., Scheiter, K., Stürmer, K., & Lachner, A. (2023). Will, skills, or conscientiousness: What predicts teachers' intentions to participate in technology-related professional development? *Computers & Education*, *198*, 1–17. DOI: 10.1016/j.compedu.2023.104756

Gallardo-Montes, C. del P., Caurcel Cara, M. J., Rodríguez Fuentes, A., & Capperucci, D. (2024). Opinions, training and requirements regarding ICT of educators in Florence and Granada for students with functional diversity. *Universal Access in the Information Society*, *23*(2), 889–899. Advance online publication. DOI: 10.1007/s10209-023-00977-0 PMID: 36852016

Gomez, F. C.Jr, Trespalacios, J., Hsu, Y. C., & Yang, D. (2022). Exploring Teachers' Technology Integration Self-Efficacy through the 2017 ISTE Standards. *TechTrends*, *66*(2), 159–171. DOI: 10.1007/s11528-021-00639-z PMID: 34378007

Hennessy, S., D'Angelo, S., McIntyre, N., Koomar, S., Kreimeia, A., Cao, L., Brugha, M., & Zubairi, A. (2022). Technology Use for Teacher Professional Development in Low- and Middle-Income Countries: A systematic review. *Computers and Education Open*, *3*, 1–32. DOI: 10.1016/j.caeo.2022.100080

Hennessy, S., Haßler, B., & Hofmann, R. (2015). Challenges and opportunities for teacher professional development in interactive use of technology in African schools. *Technology, Pedagogy and Education*, *24*(5), 1–28. Advance online publication. DOI: 10.1080/1475939X.2015.1092466

Higgins, T. E., & Spitulnik, M. W. (2008). Supporting teachers' use of technology in science instruction through professional development: A literature review. In *Journal of Science Education and Technology* (Vol. 17, Issue 5, pp. 511–521). DOI: 10.1007/s10956-008-9118-2

Huang, R. (2023). Research on the Path of Teacher Professional Development in the Context of Smart Education. *Open Journal of Social Sciences*, *11*(03), 293–303. DOI: 10.4236/jss.2023.113021

Ikhwan, M., Fuadi, M., Mailizar, M., & Jannah, M. (2023). The Utilization of Information Technology for the Professional Development of Islamic Education Teachers in Indonesia. *Progresiva : Jurnal Pemikiran Dan Pendidikan Islam*, *12*(02), 209–222. Advance online publication. DOI: 10.22219/progresiva.v12i02.31169

Jena, M. K., Pandey, S. K., & Barad, S. (2023). Integrating Technology on Professional Development of Elementary School Teachers' in the 21st century. *International Journal of Multidisciplinary Research in Arts. Science and Technology*, *1*(5), 22–28. DOI: 10.61778/ijmrast.v1i5.27

Kafyulilo, A., Fisser, P., & Voogt, J. (2016). Factors affecting teachers' continuation of technology use in teaching. *Education and Information Technologies*, *21*(6), 1535–1554. DOI: 10.1007/s10639-015-9398-0

Kgosi, M. K., Makgato, M., & Skosana, N. M. (2023). Teachers' Views on the Application of Educational Technologies in the Classroom: A Case of Selected Tshwane West Secondary Schools in Gauteng. *Journal of Curriculum Studies Research*, *5*(2), 151–166. DOI: 10.46303/jcsr.2023.23

Kul, U. (2018). Influences of technology integrated professional development course on mathematics teachers *. *European Journal of Educational Research*, *7*(2), 233–243. DOI: 10.12973/eu-jer.7.2.233

Lee, J., Topping, K., & Lakin, E. (2023). Technology-Facilitated Continuous Professional Development During a Pandemic: A Hong Kong Primary School Case Study. *RELC Journal*, *54*(2), 376–393. DOI: 10.1177/00336882231175239

Li, M., Jotikasthira, N., & Pu, R. (2022). Digitalization, knowledge sharing and higher education for sustainable development. *World Journal on Educational Technology: Current Issues*, *14*(5), 1452–1468. DOI: 10.18844/wjet.v14i5.7713

Li, S., Yamaguchi, S., Sukhbaatar, J., & Takada, J. I. (2019). The influence of teachers' professional development activities on the factors promoting ICT integration in Primary Schools in Mongolia. *Education Sciences*, *9*(2), 1–18. DOI: 10.3390/educsci9020078

Liu, M. H., & Kleinsasser, R. (2023). Multiple Perspectives with Online EFL Practicum Technology Learning: Case Study on a Cloud. *TechTrends*, *67*(2), 342–353. DOI: 10.1007/s11528-022-00813-x PMID: 36466739

Love, M. L., Simpson, L. A., Golloher, A., Gadus, B., & Dorwin, J. (2020). Professional Development to Increase Teacher Capacity for the Use of New Technologies. *Intervention in School and Clinic*, *56*(2), 115–118. DOI: 10.1177/1053451220914886

Magnusson, P. (2022). UNDERSTANDING L1 AND L2 TEACHERS' TALK ABOUT 'DIGITALISATION': Discourses during the process of integrating digital technology in the educational practice. *L1-Educational Studies in Language and Literature*, *22*, 1–27. DOI: 10.21248/l1esll.2022.22.1.361

Melash, V. D., Molodychenko, V. V., Huz, V. V., Varenychenko, A. B., & Kirsanova, S. S. (2020). Modernization of education programs and formation of digital competences of future primary school teachers. *International Journal of Higher Education*, *9*(7), 377–386. DOI: 10.5430/ijhe.v9n7p377

Mystakidis, S., & Christopoulos, A. (2022). Teacher Perceptions on Virtual Reality Escape Rooms for STEM Education. *Information (Basel)*, *13*(3), 1–13. DOI: 10.3390/info13030136

Nguyen, H. T., Tran, V.-T., Nguyen, S. T., & Van Trinh, T. (2023). A Literature Review on the Management of Preschool Teacher's Professional Competence Development in the 4.0 Industrial Revolution. *Journal of Law and Sustainable Development*, *11*(9), 1–20. DOI: 10.55908/sdgs.v11i9.1060

Nguyen, L. A. T., & Habók, A. (2024). Tools for assessing teacher digital literacy: A review. *Journal of Computers in Education*, *11*(1), 305–346. DOI: 10.1007/s40692-022-00257-5

Nurtanto, M., Sudira, P., Sofyan, H., Kholifah, N., & Triyanto, T. (2022). Professional Identity of Vocational Teachers in the 21 Century in Indonesia. *Journal of Engineering Education Transformations*, *35*(3), 30–36. DOI: 10.16920/jeet/2022/v35i3/22085

Peters, M., Elasri-Ejjaberi, A., Martínez-Argüelles, M. J., & Fàbregues, S. (2022). Teacher digital competence development in higher education: Overview of systematic reviews. *Australasian Journal of Educational Technology*, *38*(3). Advance online publication. DOI: 10.14742/ajet.7543

Pombo, L., & Marques, M. M. (2021). Guidelines for teacher training in mobile augmented reality games: Hearing the teachers' voices. *Education Sciences*, *11*(10), 1–12. DOI: 10.3390/educsci11100597

Puspandari, L. E. (2023). Exploring pedagogical potentials of technology: Professional development for English teachers of vocational higher education. *Journal of Applied Studies in Language*, *7*(1), 107–116. DOI: 10.31940/jasl.v7i1.107-116

Rodés, V., Porta, M., Garófalo, L., & Enríquez, C. R. (2021). Teacher education in the emergency: A mooc-inspired teacher professional development strategy grounded in critical digital pedagogy and pedagogy of care. *Journal of Interactive Media in Education*, *2021*(1), 1–14. DOI: 10.5334/jime.657

Rowston, K., Bower, M., & Woodcock, S. (2022). The impact of prior occupations and initial teacher education on post-graduate pre-service teachers' conceptualization and realization of technology integration. In *International Journal of Technology and Design Education* (Vol. 32, Issue 5). Springer Netherlands. DOI: 10.1007/s10798-021-09710-5

Schrum, L. (1999). Technology professional development for teachers. *Educational Technology Research and Development*, *47*(4), 83–90. DOI: 10.1007/BF02299599

Shodiyeva, M. (2022). Technologies for Continuous Professional Development of Primary Class Teachers. *The American Journal of Interdisciplinary Innovations and Research*, *04*(11), 40–48. DOI: 10.37547/tajiir/Volume04Issue11-07

Spilker, M., Prinsen, F., & Kalz, M. (2020). Valuing technology-enhanced academic conferences for continuing professional development. A systematic literature review. *Professional Development in Education*, *46*(3), 482–499. DOI: 10.1080/19415257.2019.1629614

Su, Y. (2023). Delving into EFL teachers' digital literacy and professional identity in the pandemic era: Technological Pedagogical Content Knowledge (TPACK) framework. In *Heliyon* (Vol. 9, Issue 6, pp. 1–7). DOI: 10.1016/j.heliyon.2023.e16361

Theodorio, A. O. (2024). Examining the support required by educators for successful technology integration in teacher professional development program. *Cogent Education*, *11*(1), 1–18. DOI: 10.1080/2331186X.2023.2298607

van der Rijst, R., Baggen, Y., & Sjoer, E. (2019). University teachers' learning paths during technological innovation in education. *The International Journal for Academic Development*, *24*(1), 7–20. Advance online publication. DOI: 10.1080/1360144X.2018.1500916

Wang, Y. (2023). Artificial Intelligence Technologies in College English Translation Teaching. *Journal of Psycholinguistic Research*, *52*(5), 1525–1544. DOI: 10.1007/s10936-023-09960-5 PMID: 37100967

Xie, K., Nelson, M. J., Cheng, S. L., & Jiang, Z. (2023). Examining changes in teachers' perceptions of external and internal barriers in their integration of educational digital resources in K-12 classrooms. *Journal of Research on Technology in Education*, *55*(2), 281–306. DOI: 10.1080/15391523.2021.1951404

Yeni, G., & Can, T. (2022). The Effect of 21st Century Skills Training on Foreign Language Teachers' Perceptions Regarding Their Educational Technology and Materials Development Competencies. *Bartın Üniversitesi Eğitim Fakültesi Dergisi*, *11*(1), 118–136. DOI: 10.14686/buefad.777974

Yureva, O. V., Burganova, L. A., Kukushkina, O. Y., Myagkov, G. P., & Syradoev, D. V. (2020). Digital transformation and its risks in higher education: Students' and teachers' attitude. *Universal Journal of Educational Research, 8*(11 B), 5965–5971. DOI: 10.13189/ujer.2020.082232

Zhang, W. (2022). The Role of Technology-Based Education and Teacher Professional Development in English as a Foreign Language Classes. In *Frontiers in Psychology* (Vol. 13, pp. 1–7). DOI: 10.3389/fpsyg.2022.910315

Zhou, X., Padron, Y., & Waxman, H. (2022). Exploring the Relationship Between Professional Development Experience and Skills in Educational Technology Integration Among Primary EFL Teacher. *Contemporary Educational Technology, 14*(1), 1–13. DOI: 10.30935/cedtech/11365

KEY TERMS AND DEFINITIONS

Continuing Professional Development (CPD): A lifelong learning process aimed at enhancing the knowledge, skills, and competencies of educators to improve their teaching practices and adapt to changing educational demands. CPD can include formal workshops, training sessions, and informal learning methods.

Educational Technology: The use of digital tools and resources such as software, apps, and online platforms to enhance teaching and learning experiences. It includes technologies like Artificial Intelligence (AI), Virtual Reality (VR), and Augmented Reality (AR) that offer interactive and immersive learning opportunities.

Personalized Learning: An approach that tailors education to the individual needs, interests, and abilities of each learner. In CPD, this involves using adaptive technologies that adjust learning materials and pace according to the educator's unique learning profile.

Microlearning: A method of delivering small, focused learning modules that can be consumed in short, manageable segments. It is designed to fit into busy schedules, allowing educators to learn during brief periods throughout the day, such as during breaks or between lessons.

Peer-to-Peer Learning: A collaborative learning approach where educators share knowledge, resources, and best practices with one another. This fosters a culture of continuous learning and mutual support among peers, often facilitated through platforms like Google Workspace or Slack.

Virtual Reality (VR): A technology that creates immersive, computer-generated environments where educators can simulate real-world scenarios for teaching and training purposes. VR can enhance experiential learning by enabling simulations of classroom management or subject-specific environments.

Augmented Reality (AR): A technology that overlays digital content on the real world, providing interactive learning experiences. AR can assist educators in visualizing complex concepts and creating dynamic learning environments that enhance comprehension and memory.

Inclusive CPD: The design and implementation of CPD programs that cater to the diverse needs of all educators, ensuring accessibility and equity. This includes offering resources in various formats, multilingual support, and mobile-friendly content to accommodate different learning styles and technological proficiencies.

Compilation of References

(PDF) Equity and Inclusion in Physical Education and Sport. Second Edition. (n.d.). ResearchGate. https://www.researchgate.net/publication/263971685_Equity_and_Inclusion_in_Physical_Education_and_Sport_Second_Edition

(PDF) Leadership in Education. Decision-Making in Education. (n.d.). ResearchGate. https://www.researchgate.net/publication/339235695_Leadership_in_Education_Decision-Making_in_Education

Abel, V. R., Tondeur, J., & Sang, G. (2022). Teacher Perceptions about ICT Integration into Classroom Instruction. In *Education Sciences* (Vol. 12, Issue 9). DOI: 10.3390/educsci12090609

Abera, W., Daba, A. K., Gebre, B. B., Ashagere, M., Yeheyis, T., Addisu, D., Amlaku, T., Tadesse, F., & Samuel, T. (2024). Knowledge and attitude toward geriatric nursing care and associated factors among nurses working at hospitals in Hawassa city, Ethiopia, 2022. *Frontiers in Medicine*, *11*, 1284845. Advance online publication. DOI: 10.3389/fmed.2024.1284845 PMID: 38912339

Abujaber, A., Abd-Alrazaq, A., Al-Qudimat, A., & Nashwan, A. (2023). A strengths, weaknesses, opportunities, and threats (swot) analysis of ChatGPT integration in nursing education: A narrative review. *Cureus*. Advance online publication. DOI: 10.7759/cureus.48643 PMID: 38090452

Abulibdeh, A., Zaidan, E., & Abulibdeh, R. (2024). Navigating the confluence of artificial intelligence and education for sustainable development in the era of industry 4.0: Challenges, opportunities, and ethical dimensions. *Journal of Cleaner Production*, *437*, 140527. DOI: 10.1016/j.jclepro.2023.140527

Adaba, H. (2017). Exploring the practice of teacher-student classroom interaction in EFL to develop the learners' speaking skills in Tullu Sangota Primary School Grade Eight students in focus. *Arts and Social Sciences Journal*, *8*(4). Advance online publication. DOI: 10.4172/2151-6200.1000295

Adaptability and Resilience. (n.d.). Www.linkedin.com. https://www.linkedin.com/pulse/adaptability-resilience-ariel-matius-surbakti

Adaptive Learning AI Technology in Education. (2021, August 13). Belitsoft. https://belitsoft.com/custom-elearning-development/ai-in-education/adaptive-learning-ai

Adeyemo, D. A., & Chukwudi, A. R. (2014). Emotional intelligence and teacher efficacy as predictors of teacher effectiveness among pre-service teachers in some Nigerian universities. *International Journal of Evaluation and research in Education, 3*(2), 85-90.

Adipat, S., Chotikapanich, R., Laksana, K., Busayanon, K., Piatanom, P., Ausawasowan, A., & Elbasouni, I. (2023). Technological Pedagogical Content Knowledge for Professional Teacher Development. *Academic Journal of Interdisciplinary Studies*, *12*(1), 173–182. DOI: 10.36941/ajis-2023-0015

Agboola Sogunro, O. (2004). Efficacy of role-playing pedagogy in training leaders: Some reflections. *Journal of Management Development, 23*(4), 355–371. DOI: 10.1108/02621710410529802

Ahmad, K., Iqbal, W., El-Hassan, A., Qadir, J., Benhaddou, D., Ayyash, M., & Al-Fuqaha, A. (2023). Data-driven artificial intelligence in education: A comprehensive review. *IEEE Transactions on Learning Technologies, 17*, 12–31. DOI: 10.1109/TLT.2023.3314610

Ahouah, M., & Rothan-Tondeur, M. (2019). End-users and caregivers' involvement in health interventional research carried out in geriatric facilities: A systematic review. *International Journal of Environmental Research and Public Health, 16*(16), 2812. DOI: 10.3390/ijerph16162812 PMID: 31394752

Ailani, A. (2023). Examining the impact of bloom's taxonomy integration in entrepreneurship education on the competitiveness and performance of micro, small, and medium enterprises (MSMES). *Makreju, 1*(3), 181–188. DOI: 10.55606/makreju.v1i3.2924

Ainscow, M. (2020). Promoting inclusion and equity in education: lessons from international experiences. *Nordic journal of studies in educational policy, 6*(1), 7-16.

Ainscow, M. (2020). Promoting Inclusion and Equity in education: Lessons from International Experiences. *Nordic Journal of Studies in Educational Policy, 6*(1), 7–16. DOI: 10.1080/20020317.2020.1729587

Aksoy, N., Aksoy, E., & Usta, E. (2022). Metaphors Developed by Teachers for the Gamification Approach in Education. *Journal of Teacher Education and Lifelong Learning*.

Al-Abdullatif, A. M. (2024). Modeling teachers' acceptance of generative artificial intelligence use in higher education: The role of ai literacy, intelligent tpack, and perceived trust. *Education Sciences, 14*(11), 1209. DOI: 10.3390/educsci14111209

Al-Ahmadi, H. (2009). Factors affecting performance of hospital nurses in Riyadh Region, Saudi Arabia. *International Journal of Health Care Quality Assurance, 22*(1), 40–54. DOI: 10.1108/09526860910927943 PMID: 19284170

Almubarak, A., Alhalabi, W., Albidewi, I., & Alharbi, E. (2024). An analytical approach for an AI-based teacher performance evaluation system in Saudi Arabia's schools. *Discover Applied Sciences, 6*(8), 406. Advance online publication. DOI: 10.1007/s42452-024-06117-4

Alquraini, T. (2011). Special Education in Saudi Arabia: Challenges, Perspectives, Future Possibilities. *International Journal of Special Education, 26*(2), 149–159.

An, S., & Wu, Z. (2012). Enhancing mathematics teachers' knowledge of students' thinking from assessing and analyzing misconceptions. *International Journal of Science and Mathematics Education, 10*(3), 717–753. DOI: 10.1007/s10763-011-9324-x

Antonak, R. F., & Larrivee, B. (1995). Psychometric analysis and revision of the Opinions Relative to Integration of Students with Disabilities (ORI) Scale. *Journal of Developmental and Physical Disabilities, 7*(2), 137–146. DOI: 10.1007/BF02071408

Aptara. (2024, December 9). *Simulation-Based Learning: A Game Changer for Developing Corporate Leaders*. Aptara Corp - Home. https://www.aptaracorp.com/simulation-based-learning-a-game-changer-for-developing-corporate-leaders/

Arifin, A., Suryaningsih, S., & Arifudin, O. (2024). The relationship between classroom environment, teacher professional development, and student academic performance in secondary education. *International Education Trend Issues*, *2*(2), 151–159. DOI: 10.56442/ieti.v2i2.467

Arlin, C. (2023). Information processing theory in psychology. https://www.verywellmind.com/information-processing-theory-in-psychology-7503601

Armstrong, P. (2010). Bloom's taxonomy. Vanderbilt University Center for Teaching. Retrieved [31/12/24] from https://cft.vanderbilt.edu/guides-sub-pages/blooms-taxonomy/

Arnold, E. (1989). Artificial Intelligence in the UK: The Alvey Intelligent Knowledge-Based Systems (IKBS) Programme. *AI Communications*, *2*(1), 18–23. DOI: 10.3233/AIC-1989-2102

Arroyo, M., Quinn, L. F., Paretti, L., & McCarthy, J. (2020). A Content Analysis of Coaching Novice Teachers. *Educational Research: Theory and Practice*, *31*(1), 22–25.

Austin, D. B., & Brown, H. L., Jr. (1970). Report of the assistant principalship. Volume 3: The study of the secondary school principalship. National Association of Secondary School Principals.

Avcı, Ü., & Candan, Ö. (2023). Öğretmenlerin bilgi teknolojileri okuryazarlık düzeylerine göre endüstri 4.0 farkındalıklarının incelenmesi [Investigation of teachers' awareness of industry 4.0 according to their information technology literacy levels]. *Necmettin Erbakan Üniversitesi Ereğli Eğitim Fakültesi Dergisi*, *5*(1), 160–178.

Avdor, S. (2008). The Required Changes in Student-Teachers' Qualifications, Training and Professional Development. In Kfir, D., & Ariav, T. (Eds.), *The Crisis in Teacher Education: Reasons, Problems and Possible Solutions* [in Hebrew]. (pp. 145–175). VanLeer Institute.

Avramidis, E., & Norwich, B. (2002). Teachers' attitudes towards integration/inclusion: A review of the literature. *European Journal of Special Needs Education*, *17*(2), 129–147. DOI: 10.1080/08856250210129056

Avramidis, E., Toulia, A., Tsihouridis, C., & Strogilos, V. (2019). Teachers' attitudes towards inclusion and their self-efficacy for inclusive practices as predictors of willingness to implement peer tutoring. *Journal of Research in Special Educational Needs*, *19*(S1), 49–59. DOI: 10.1111/1471-3802.12477

Ayanwale, M. A., Adelana, O. P., Molefi, R. R., Adeeko, O., & Ishola, A. M. (2024). Examining artificial intelligence literacy among pre-service teachers for future classrooms. *Computers and Education Open*, *6*(100179), 1–15. DOI: 10.1016/j.caeo.2024.100179

Aydin, B., & Kaya, A. (2016). Sources of stress for teachers working in private elementary schools and methods of coping with stress. *Universal Journal of Educational Research*, *4*(12A), 186–195. DOI: 10.13189/ujer.2016.041324

Bahroun, Z., Anane, C., Ahmed, V., & Zacca, A. (2023). Transforming education: A comprehensive review of generative artificial intelligence in educational settings through bibliometric and content analysis. *Sustainability (Basel)*, *15*(17), 12983. DOI: 10.3390/su151712983

Baker, T., & Smith, L. (2019). *Educ-AI-tion Rebooted? Exploring the future of artificial intelligence in schools and colleges*. https://media.nesta.org.uk/documents/Future_of_AI_and_education_v5_WEB.pdf

Baker, R. S., & Inventado, P. S. (2016). Educational data mining and learning analytics. In Rupp, A. A., & Leighton, J. P. (Eds.), *The Cambridge handbook of the learning sciences* (Vol. 2, pp. 589–607). John Wiley.

Bala, R. (2017). Teacher effectiveness of secondary school teachers in relation to their emotional intelligence. *International Journal of Indian Psychology*, *4*(4), 72–78.

Balderson, K. (2023). AI In Education statistics you should know. https://mspoweruser.com/ai-in-education-statistics

Baltezarević, R., & Baltezarević, I. (2024). Students' attitudes on the role of artificial intelligence (AI) in personalized learning. *International Journal of Cognitive Research in Science* [IJCRSEE]. *Engineering and Education*, *12*(2), 387–397. DOI: 10.23947/2334-8496-2024-12-2-387-397

Banaz, E., & Demirel, O. (2024). Türkçe öğretmen adaylarının yapay zekâ okuryazarlıklarının farklı değişkenlere göre incelenmesi [Investigation of Turkish pre-service teachers' artificial intelligence literacy according to different variables]. *Dokuz Eylül Üniversitesi Buca Eğitim Fakültesi Dergisi*, (60), 1516–1529. DOI: 10.53444/deubefd.1461048

Bandura, A. (1991). Social cognitive theory of self-regulation. *Organizational Behavior and Human Decision Processes*, *50*(2), 248–287. DOI: 10.1016/0749-5978(91)90022-L

Bandura, A. (1997). *Self-efficacy: The exercise of control*. W. H. Freeman and Company.

Banks, J. A., & Banks, C. A. M. (Eds.). (2020). *Multicultural education: Issues and perspectives* (10th ed.). Wiley.

Baraishuk, D. (2021). *Adaptive Learning AI Technology in Education*. Belitsoft. https://belitsoft.com/custom-elearning-development/ai-in-education/adaptive-learning-ai

Barczak, G., Lassk, F., & Mulki, J. (2010). Antecedents of team creativity: An examination of team emotional intelligence, team trust, and collaborative culture. *Creativity and Innovation Management*, *19*(4), 332–345. DOI: 10.1111/j.1467-8691.2010.00574.x

Bardach, L., & Klassen, R. M. (2020). Smart teachers, successful students? A systematic review of the literature on teachers' cognitive abilities and teacher effectiveness. *Educational Research Review*, *30*, 100–312. DOI: 10.1016/j.edurev.2020.100312

Barmby, P., Bolden, D., & Thompson, L. (2014). *Understanding and enriching problem solving in primary mathematics*. Critical Publishing.

Bar-On, R. (2006). The Bar-On model of Emotional-Social Intelligence (ESI). *Psicothema*, *18*, 13–25. PMID: 17295953

Barrat, J. (2023). *Our final invention: Artificial intelligence and the end of the human era*. Hachette UK.

Baskonus, H. M., & Ozturk, O. M. (2023). Öğretmen adaylarının kaynaştırma/bütünleştirme eğitimine ilişkin duygu, tutum ve kaygı düzeylerinin incelenmesi. [Examining the emotions, attitudes, and anxiety levels of teacher candidates regarding inclusive education]. *Ahi Evran Üniversitesi Sosyal Bilimler Enstitüsü Dergisi*, *9*(1), 170–184. DOI: 10.31592/aeusbed.1212256

Bearman, M., & Ajjawi, R. (2023). Learning to work with the black box: Pedagogy for a world with artificial intelligence. *British Journal of Educational Technology*, *54*(5), 1160–1173. DOI: 10.1111/bjet.13337

Bell, C. A., Wilson, S. M., Higgins, T., & McCoach, D. B. (2010). Measuring the effects of professional development on teacher knowledge: The case of developing mathematical ideas. *Journal for Research in Mathematics Education*, *41*(5), 479–512. DOI: 10.5951/jresematheduc.41.5.0479

Belpaeme, T., Kennedy, J., Ramachandran, A., Scassellati, B., & Tanaka, F. (2018). Social robots for education: A review. *Science Robotics*, *3*(21), eaat5954. Advance online publication. DOI: 10.1126/scirobotics.aat5954 PMID: 33141719

Benko, E., & Peršolja, M. (2023). Nursing students' views of the impact of geriatric role-play workshops on professional competencies: Survey. *BMC Nursing*, *22*(1), 203. Advance online publication. DOI: 10.1186/s12912-023-01373-y PMID: 37316872

Berendt, B., Littlejohn, A., & Blakemore, M. (2020). AI in education: Learner choice and fundamental rights. *Learning, Media and Technology*, *45*(3), 312–324. DOI: 10.1080/17439884.2020.1786399

Bertolin, M., & Da Rin, D. (2020). Chatbots for K-12 education: A review of recent advances. Opportunities. and Challenges. *Education Sciences*, *10*(7), 181–191.

Best CBSE school in Nagpur. Delhi Public School Kamptee Road Nagpur. DPS Nagpur. (n.d.). Delhi Public School Kamptee Road Nagpur. https://dpsnagpur.edu.in/sports-shape-leadership-skills-build-character

Bhandari, S. (2021, October 25). *Benefits of Sports for Mental Health.* WebMD; WebMD. https://www.webmd.com/mental-health/benefits-of-sports-for-mental-health

Bhatnagar, N., & Das, A. (2014). Regular School Teachers' Concerns and Perceived Barriers to Implement Inclusive Education in New Delhi, India. *International Journal of Instruction*, *7*(2), 89–102.

Blackley, C., Redmond, P., & Peel, K. (2021). Teacher decision-making in the classroom: The influence of cognitive load and teacher affect. *Journal of Education for Teaching*, *47*(4), 1–14. DOI: 10.1080/02607476.2021.1902748

Blase, J. (1986). A qualitative analysis of sources of teacher stress: Consequences for performance. *American Educational Research Journal*, *23*(1), 13–40. DOI: 10.3102/00028312023001013

Block, M. E. (1995). Development and validation of the children's attitudes toward integrated physical education–revised (CAIPE–R) inventory. *Adapted Physical Activity Quarterly; APAQ*, *12*(1), 60–77. DOI: 10.1123/apaq.12.1.60

Blömeke, S., & Delaney, S. (2012). Assessment of teacher knowledge across countries: A review of the state of research. *ZDM Mathematics Education*, *44*(3), 223–247. DOI: 10.1007/s11858-012-0429-7

Bloom, B., Englehart, M., Furst, E., Hill, W., & Krathwohl, D. (1956). *Taxonomy of educational objectives.* Handbook One.

Bolman, L. G., & Deal, T. E. (2021). *Reframing organizations: Artistry, choice, and leadership* (7th ed.). Jossey-Bass.

Bond, N. (2021). Teacher leadership in an elementary school: A case study of instructional support specialists. *Education, 141*(4), 201–213.

Borko, H., Jacobs, J., & Koellner, K. (2010). Contemporary approaches to teacher professional development. *International encyclopedia of education. 7*(2), 548-556.

Bourdeau, J., & Grandbastien, M. (2010). Modeling tutoring knowledge. In *Advances in intelligent tutoring systems* (pp. 123–143). Springer Berlin Heidelberg. DOI: 10.1007/978-3-642-14363-2_7

Bowman, M. A., Vongkulluksn, V. W., Jiang, Z., & Xie, K. (2022). Teachers' exposure to professional development and the quality of their instructional technology use: The mediating role of teachers' value and ability beliefs. *Journal of Research on Technology in Education, 54*(2), 1–18. DOI: 10.1080/15391523.2020.1830895

Boyd, M. A. (2005). *The emotional intelligence of teachers and students' perceptions of their teachers' behaviour in the classroom.* Indiana University of Pennsylvania.

Broche-Pérez, Y., Herrera Jiménez, L. F., & Omar-Martínez, E. (2016). Neural substances of decision-making. [English Edition]. *Neurología : Publicación Oficial de la Sociedad Española de Neurología, 31*(5), 319–325. DOI: 10.1016/j.nrleng.2015.03.009 PMID: 25976946

Brookfield, S. D. (2017). *Becoming a critically reflective teacher* (2nd ed.). Jossey-Bass.

Brynjolfsson, E., & McAfee, A. (2014). *The second machine age: Work, progress, and prosperity in a time of brilliant technologies.* WW Norton & company.

Butakor, P. K., Ampadu, E., & Suleiman, S. J. (2020). Analysis of Ghanaian teachers' attitudes toward inclusive education. *International Journal of Inclusive Education, 24*(11), 1237–1252. DOI: 10.1080/13603116.2018.1512661

Büyüköztürk, Ş., Kılıç-Çakmak, E., Akgün, Ö. E., Karadeniz, Ş., & Demirel, F. (2024). *Bilimsel araştırma yöntemleri* [Scientific research methods]. Pegem Akademi.

Cambridge Dictionary (n.d). Intelligence. https://dictionary.cambridge.org/dictionary/english/intelligence

Campbell, M., Hoane, A. J.Jr, & Hsu, F. (2002). Deep Blue. *Artificial Intelligence, 134*(1-2), 57–83. DOI: 10.1016/S0004-3702(01)00129-1

Carmichael ST, Price J.L. (1994) Architectonic subdivision of the orbital and medial prefrontal cortex in the macaque monkey. *J Comp Neurol. 346*, 366, 402.

Carpenter, T. P., Fennema, E., Franke, M. L., & Levi, L. (2001). Capturing teachers' generative change: A follow-up study of professional development in mathematics. *American Educational Research Journal, 38*(3), 653–689. DOI: 10.3102/00028312038003653

Carroll, A., York, A., Fynes-Clinton, S., Sanders-O'Connor, E., Flynn, L., Bower, J., Forrest, K., & Ziaei, M. (2021). The downstream effects of teacher well-being programs: Improvements in teachers' stress, cognition, and well-being benefit their students. *Frontiers in Psychology, 12*, 689628. Advance online publication. DOI: 10.3389/fpsyg.2021.689628 PMID: 34276519

Carroll, J. B. (1997). The three-stratum theory of cognitive abilities. In Flanagan, D. P., Genshaft, J. L., & Harrison, P. L. (Eds.), *Contemporary intellectual assessment: Theories, tests, and issues* (pp. 122–130). Guilford Press.

Cattell, R. B. (1957). *Personality and motivation structure and measurement*. World Book.

Cattell, R. B. (1971). *Abilities: Their structure, growth, and action*. Houghton Mifflin.

Cefai, C., Ferrario, E., Cavioni, V., Carter, A., & Grech, T. (2013). Circle time for social and emotional learning in primary school. *Pastoral Care in Education*, *32*(2), 116–130. DOI: 10.1080/02643944.2013.861506

Center for Mindfulness in Medicine, Health Care, and Society, University of Massachusetts Medical School. (2017). *Mindfulness-Based Stress Reduction (MBSR) curriculum guide 2017*.https://mbsr.website/sites/default/files/docs/mbsr-curriculum-guide 2017.pdf?form=MG0AV3

Charania, A., Paltiwale, S., Sen, S., Sarkar, D., & Bakshani, U. (2023). Leading Edge Use of Technology for Teacher Professional Development in Indian Schools. *Education Sciences*, *13*(4), 1–20. DOI: 10.3390/educsci13040386

Chechi, K. V. (2012). Emotional intelligence and teaching. *International Journal of Research in Economics & Social Sciences*, *2*(2), 297–304.

Cheng, Y., Sun, S., Hu, Y., Wang, J., Chen, W., Miao, Y., & Wang, H. (2024). Effects of different geriatric nursing teaching methods on nursing students' knowledge and attitude: Systematic review and network meta-analysis. *PLoS One*, *19*(5), e0300618. DOI: 10.1371/journal.pone.0300618 PMID: 38820259

Chen, J., & Chi-Kin Lee, J. (2022). Teacher resilience matters: A buffering and boosting effect between job driving factors and their well-being and job performance. *Teachers and Teaching*, *28*(7), 1–18. DOI: 10.1080/13540602.2022.2116574

Chen, W., Shah, U., & Brechtelsbauer, C. (2019). A framework for hands-on learning in chemical engineering education—Training students with the end goal in mind. *Education for Chemical Engineers*, *28*, 25–29. DOI: 10.1016/j.ece.2019.03.002

Chen, X., Xie, H., Zou, D., & Hwang, G. J. (2020). Application and theory gaps during the rise of artificial intelligence in education. *Computers and Education: Artificial Intelligence*, *1*, 100002. DOI: 10.1016/j.caeai.2020.100002

Chen, X., Zou, D., Xie, H., Cheng, G., & Liu, C. (2022). Two decades of artificial intelligence in education. *Journal of Educational Technology & Society*, *25*(1), 28–47.

Chiu, T. K., Ahmad, Z., Ismailov, M., & Sanusi, I. T. (2024). What are artificial intelligence literacy and competency? A comprehensive framework to support them. *Computers and Education Open*, *6*(100171), 1–9. DOI: 10.1016/j.caeo.2024.100171

Chiu, T. K., Xia, Q., Zhou, X., Chai, C. S., & Cheng, M. (2023). Systematic literature review on opportunities. challenges. and future research recommendations of artificial intelligence in education. *Computers and Education: Artificial Intelligence*, *4*, 100–118. DOI: 10.1016/j.caeai.2022.100118

Chrzanowska, I. (2019). Inclusive education in the opinion of the teachers from special pre-schools, regarding the chances of success of the inclusive actions towards particular groups of students, education participants, and teachers' seniority. *Interdisciplinary Context of Special Pedagogy*, *25*(1), 127–149. DOI: 10.14746/ikps.2019.25.06

Clack, L., & Ellison, R. (2018). Innovative approaches to management education. *Journal of Management Policies and Practices*, *6*(1), 6–9. DOI: 10.15640/jmpp.v6n1a2

CMOE. (2018, May 15). *Strategic Thinking in Leadership [The Biggest Gap for New Leaders]*. CMOE. https://cmoe.com/blog/the-biggest-gap-for-new-leaders-at-any-level-is-thinking-strategically/

CoachMePlus. (2020, August 11). *Key Principles of Effective Collaboration in Sport*. CoachMePlus. https://coachmeplus.com/principles-team-staff-collaboration-sport/

Çoban, Ö. (2021). Liderlik ve sorumluluk. [Leadership and responsibility] In Karataş, K. (Ed.), *Eğitim ve 21.yy. becerileri* [[Education and 21st-*century skills]*]. (pp. 258–276). Nobel.

Çoban, Ö., & Gün, F. (2024). Eğitimde mükemmelliğe giden yolda öğretmen liderliği. [Teacher leadership on the road to excellence in education] In Atasoy, R., Çoban, Ö., & Yalçın, M. T. (Eds.), *Eğitimde mükemmellik haritası: Liderlik, beceriler ve başarı* [*Map of excellence in education: leadership, skills and success*]. (pp. 45–60). Eğitim.

Collaborative for Academic, Social, and Emotional Learning (CASEL). (2020). *Core SEL competencies*. https://casel.org/core-competencies/

Collie, R., Shapka, J., & Perry, N. (2012). School climate and social-emotional learning: Predicting teacher stress, job satisfaction, and teaching efficacy. *Journal of Educational Psychology*, *104*(4), 1189–1204. DOI: 10.1037/a0029356

Conflict Resolution in Leadership | In Professional Development. (2021, October 4). Www.inpd.co.uk. https://www.inpd.co.uk/blog/conflict-resolution-in-leadership

Conner, L. A., & Pope, J. R. (2013). Cultivating culturally responsive teaching through spirituality: Implications for educators. *Journal of Education and Social Justice*, *3*(1), 5–22.

Cooper, K. M., Downing, V. R., & Brownell, S. E. (2018). The influence of active learning practices on student anxiety in large-enrollment college science classrooms. *International Journal of STEM Education*, *5*(1), 23. Advance online publication. DOI: 10.1186/s40594-018-0123-6 PMID: 30631713

Council of Europe. (2012). *Culture and Sport*. Council of Europe. https://www.coe.int/en/web/compass/culture-and-sport

Cox, J. S., & Mullen, C. A. (2023). Impacting student achievement: Principals' instructional leadership practice in two Title I rural schools. *Journal of School Leadership*, *33*(1), 3–25. DOI: 10.1177/10526846221133996

Cranton, P. (2006). *Understanding and promoting transformative learning: A guide for educators of adults* (2nd ed.). Jossey-Bass.

Creswell, J. (2013). Qualitative inquiry & research design: Choosing among five approaches.

Creswell, J. W. (2013). *Qualitative inquiry & research design: Choosing among five approaches* (3rd ed.). SAGE Publications.

Creswell, J. W. (2014). *Research design: Qualitative, quantitative, and mixed methods approaches* (4th ed.). SAGE Publications.

Creswell, J. W., Hanson, W. E., Clark Plano, V. L., & Morales, A. (2007). Qualitative research designs: Selection and implementation. *The Counseling Psychologist*, *35*(2), 236–264.

Creswell, J. W., & Poth, C. N. (2018). *Qualitative inquiry and research design: Choosing among five approaches* (4th ed.). Sage Publications.

Crevier, D. (1993). *AI: The tumultuous search for artificial intelligence*. Basic Books.

Cross-disciplinary learning: the key to a successful future. (n.d.). Www.robowunderkind.com. https://www.robowunderkind.com/blog/cross-disciplinary-learning-the-key-to-a-successful-future

D'Alessio, S. (2012). *Inclusive education in Italy* (Vol. 10). Springer Science & Business Media.

da Silva, T. H. R. (2022). Emotional awareness and emotional intelligence. *British Journal of Community Nursing*, *27*(12), 573–574. DOI: 10.12968/bjcn.2022.27.12.573 PMID: 36519463

Dagli, G., Muhtaroglu, M. B., Bastas, M., Altinay, F., & Altinay, Z. (2023). Evaluation of primary school managers' duties in digital transformation. *GeSec: Revista de Gestão e Secretariado*, *14*(9), 15227–15249. DOI: 10.7769/gesec.v14i9.2524

Dahri, N. A., Vighio, M. S., Al-Rahmi, W. M., & Alismaiel, O. A.Muhammad Saleem VighioWaleed Mugahed Al-RahmiOmar A. Alismaiel. (2022). Usability Evaluation of Mobile App for the Sustainable Professional Development of Teachers. *International Journal of Interactive Mobile Technologies*, *16*(16), 4–30. DOI: 10.3991/ijim.v16i16.32015

Danielson, C. (2006). *Teacher leadership that strengthens professional practice*. Association for Supervision and Curriculum Development.

Darling-Hammond, L., & Bransford, J. (2005). *Preparing teachers for a changing world: What teachers should learn and be able to do*. Jossey-Bass.

Day, C., Sammons, P., Stobart, G., Kington, A., & Gu, Q. (2007). *Teachers matter: Connecting work, lives and effectiveness*. Open University Press.

Deiorio, N. M., Moore, M., Santen, S. A., Gazelle, G., Dalrymple, J. L., & Hammoud, M. (2022). Coaching models, theories, and structures: An overview for teaching faculty in the emergency department and educators in the offices. *AEM Education and Training*, *6*(5), e10801. DOI: 10.1002/aet2.10801 PMID: 36189456

Demir, K. (2020). Öğretmen liderliği. [Teacher leadership] In Yılmaz, K. (Ed.), *Liderlik: Kuram. araştırma. uygulama [[Leadership: Theory, research and implication]]*. (pp. 558–570). Pegem.

Deniz, E., & Coban, A. (2019). Kaynaştirma eğitimine ilişkin öğretmen görüşleri [Teachers views regarding inclusive education]. *Elektronik Sosyal Bilimler Dergisi*, *18*(70), 734–761. DOI: 10.17755/esosder.448379

Department of Education and Training. Victoria. (n.d.). High-impact well-being strategies. Retrieved from https://www.education.vic.gov.au/Documents/school/teachers/teachingresources/practice/High_Impact_Wellbeing_Strategies.pdf

Detroyer, E., Dobbels, F., Teodorczuk, A., Depaifve, Y., Joosten, E., & Milisen, K. (2018). Effect of an interactive e-learning tool for delirium on patient and nursing outcomes in a geriatric hospital setting: Findings of a before-after study. *BMC Geriatrics*, *18*(1), 19. Advance online publication. DOI: 10.1186/s12877-018-0715-5 PMID: 29351772

Doherty, S. (2024, April 11). Teacher stress - Teachers UK. *Teachers UK*.https://www.teachers-uk.co.uk/teacher-stress/

Doležalová, J., Tóthová, V., & Neugebauer, J. (2022). Selected aspects of nursing practice related to the usage of evaluation tools for population aged over 60 years. *Kontakt*, *24*(2), 131–136. DOI: 10.32725/kont.2021.044

Domagała-Zyśk, E., & Knopik, T. (2022). Attitudes Toward Inclusion Among Polish Primary School Teachers: A Strategic Factor in Implementing Inclusive Education. *Multidisciplinary Journal of School Education*, *11*(1 (21)), 211–233. DOI: 10.35765/mjse.2022.1121.11

Domínguez-González, M., Hervás-Gómez, C., Díaz-Noguera, M. D., & Reina-Parrado, M. (2023). Attention to diversity from artificial intelligence. *The European Educational Researcher*, *6*(3), 101–115. DOI: 10.31757/euer.633

Donker, M. H., Erisman, M. C., van Gog, T., & Mainhard, T. (2020). Teachers' Emotional Exhaustion: Associations With Their Typical Use of and Implicit Attitudes Toward Emotion Regulation Strategies. *Frontiers in Psychology*, *11*, 867. Advance online publication. DOI: 10.3389/fpsyg.2020.00867 PMID: 32547437

Dorji, R., Bailey, J., Paterson, D., Graham, L., & Miller, J. (2021). Bhutanese teachers' attitudes towards inclusive education. *International Journal of Inclusive Education*, *25*(5), 545–564. DOI: 10.1080/13603116.2018.1563645

Driscoll, M. P. (1994). *Psychology of learning for instruction*. Allyn & Bacon.

Drossel, K., & Eickelmann, B. (2017). Teachers' participation in professional development concerning the implementation of new technologies in class: A latent class analysis of teachers and the relationship with the use of computers, ICT self-efficacy and emphasis on teaching ICT skills. *Large-Scale Assessments in Education*, *5*(1), 1–13. DOI: 10.1186/s40536-017-0053-7

Duan, A., Jiang, F., Li, L., Li, Q., & Chen, W. (2024). Design and practice of blended teaching of internal medicine nursing based on o-amas effective teaching model. *BMC Medical Education*, *24*(1), 580. Advance online publication. DOI: 10.1186/s12909-024-05588-8 PMID: 38807140

Durlak, J. A., Weissberg, R. P., Dymnicki, A. B., Taylor, R. D., & Schellinger, K. B. (2011). The impact of enhancing students' social and emotional learning: A meta-analysis of school-based universal interventions. *Child Development*, *82*(1), 405–432. DOI: 10.1111/j.1467-8624.2010.01564.x PMID: 21291449

Durnalı, M., & Çoban, Ö. (2025). Artificial intelligence in higher education management: Innovations. opportunities and challenges. In González Lezcano, R. A., & Orakçı, Ş. (Eds.), *Innovation trends and educational technology in higher education* (pp. 501–516). Information Age Publishing.

Dweck, C. S. (2006). *Mindset: The new psychology of success*. Random House.

Ebubechukwu, R. (2021). Psychological education and mental well-being: Strategies for promoting emotional intelligence. *Interdisciplinary Journal Papier Human Review*, *2*(2), 1–4. DOI: 10.47667/ijphr.v2i2.218

Eklund, A., Carlson, E., Sundin-Andersson, C., & Olsson, C. (2022). Experiences of the clinical learning environment in peer learning: A mixed method study from bachelor students' and main preceptors' perspectives. *Nordic Journal of Nursing Research*, *43*(1), 20571585221126597. Advance online publication. DOI: 10.1177/20571585221126597

Elmali, Ş., & Balkan Kiyici, F. (2022). Technology-based professional development program: Experiences of science teachers. *Journal of Educational Technology and Online Learning*, *5*(2), 297–315. Advance online publication. DOI: 10.31681/jetol.1081367

Eloor, A. G., & Menon, S. D. (2024). The work-life balance of teachers: A review. *International Journal of Research in Management*, *6*(1), 52–56. DOI: 10.33545/26648792.2024.v6.i1a.125

Emmons, R. A. (2003). Personal goals, life meaning, and virtue: Wellsprings of a positive life. In Keyes, C. L. M., & Haidt, J. (Eds.), *Flourishing: Positive psychology and the life well-lived* (pp. 105–128). American Psychological Association. DOI: 10.1037/10594-005

Engelbrecht, P., Savolainen, H., Nel, M., & Malinen, O. P. (2013). How cultural histories shape South African and Finnish teachers' attitudes towards inclusive education: A comparative analysis. *European Journal of Special Needs Education*, *28*(3), 305–318. DOI: 10.1080/08856257.2013.777529

Engel, P., Spencer, J., Paul, T., & Boardman, J. (2016). The geriatrics in primary care demonstration: integrating comprehensive geriatric care into the medical home: preliminary data. *Journal of the American Geriatrics Society*, *64*(4), 875–879. DOI: 10.1111/jgs.14026 PMID: 27100583

Engreini, S., Amini, R., & Alfana, Y. D. (2025). Effectiveness of using artificial intelligence chatbox integrated case method on science learning outcomes of basic education students. *Jurnal Penelitian Pendidikan IPA*, *11*(1), 1206–1214.

Eng, T., Eng, L., Liew, C., Daneil, I., & Ting, G. (2023). An application of multiple linear regression analysis to establish the relationship between higher order thinking skills and affective domain. *International Journal of Academic Research in Business & Social Sciences*, *13*(5). Advance online publication. DOI: 10.6007/IJARBSS/v13-i5/16907

Equity and Inclusion in Physical Education and Sport, Second Edition. (n.d.). ResearchGate. https://www.researchgate.net/publication/263971685_Equity_and_Inclusion_in_Physical_Education_and_Sport_Second_Edition

Eradze, M., Bardone, E., & Dipace, A. (2021). Theorising on covid-19 educational emergency: Magnifying glasses for the field of educational technology. *Learning, Media and Technology*, *46*(4), 404–419. DOI: 10.1080/17439884.2021.1961802

Erbil, D. (2020). A review of flipped classroom and cooperative learning method within the context of vygotsky theory. *Frontiers in Psychology*, *11*, 1157. Advance online publication. DOI: 10.3389/fpsyg.2020.01157 PMID: 32655437

Erdoğan, M., & Kırılmaz, H. (2020). Hasta merkezlilik ve hasta merkezli bakım [Patient centerization and patient-centered care]. *İnsan ve İnsan*, *7*(24), 97-126.

Erdoğan, T. E., & Ekşioğlu, S. (2024). Yapay Zekâ Okuryazarlığı Ölçeği'nin Türkçeye uyarlanması [Adaptation of Artificial Intelligence Literacy Scale into Turkish]. *Türk Eğitim Bilimleri Dergisi.*, *22*(2), 1196–1211.

Ergen, M. (2019). What is artificial intelligence? Technical considerations and future perception. *The Anatolian Journal of Cardiology*, *22*(2), 5–7. DOI: 10.14744/AnatolJCardiol.2019.79091 PMID: 31670719

Ertmer, P. A., Ottenbreit-Leftwich, A. T., Sadik, O., Sendurur, E., & Sendurur, P. (2012). Teacher beliefs and technology integration practices: A critical relationship. *Computers & Education*, *59*(2), 423–435. DOI: 10.1016/j.compedu.2012.02.001

European Commission. (2014). *The Teaching and Learning International Survey (TALIS) 2013: Main findings from the survey and implications for education and training policies in Europe*. Publications Office of the European Union.

Every Student Succeeds Act of 2015, Pub. L. No. 114-95, 129 Stat. 1802 (2015). https://www.congress.gov/114/plaws/publ95/PLAW-114publ95.pdf

Ewing, L.-A. (2021). Mentoring novice teachers. *Mentoring & Tutoring*, *29*(2), 164–186. DOI: 10.1080/13611267.2021.1899585

Fahim, R., Boughdady, A., & Meawad, E. (2023). Knowledge and attitude of nurses toward geriatric palliative care. *Mansoura Nursing Journal*, *10*(2), 307–311. DOI: 10.21608/mnj.2023.340395

Farrell Buzbee Little, P. (2005). Peer coaching as a support to collaborative teaching. *Mentoring & Tutoring*, *13*(1), 83–94. DOI: 10.1080/13611260500040351

Farrell, T., Luptak, M., Supiano, K., Pacala, J., & Lisser, R. (2018). State of the science: Interprofessional approaches to aging, dementia, and mental health. *Journal of the American Geriatrics Society*, *66*(S1). Advance online publication. DOI: 10.1111/jgs.15309 PMID: 29659009

Fathima Anjila, P. K. (2021). Artificial intelligence. In J. Karthikeyan, T.S. Hie, & Ng. Y. Jin (Eds.). *Learning outcomes of classroom research.* (pp. 65-73). L'Ordine Nuovo Pub.

Felmer, P., Liljedahl, P., & Koichu, B. (Eds.). (2019). *Problem solving in mathematics instruction and teacher professional development.* Springer., DOI: 10.1007/978-3-030-29215-7

Felmer, P., Pehkonen, E., & Kilpatrick, J. (Eds.). (2016). *Posing and solving mathematical problems: Advances and new perspectives.* Springer., DOI: 10.1007/978-3-319-28023-3

Fernandes, G. W. R., Rodrigues, A. M., & Ferreira, C. A. (2020). Professional Development and Use of Digital Technologies by Science Teachers: a Review of Theoretical Frameworks. In *Research in Science Education* (Vol. 50, Issue 2, pp. 673–708). DOI: 10.1007/s11165-018-9707-x

Fernández, C., & Yoshida, M. (2004). *Lesson study: A Japanese approach to improving mathematics teaching and learning*. Lawrence Erlbaum., DOI: 10.4324/9781410610867

Finkelstein, S. (2018). *The Best Leaders Are Great Teachers*. Harvard Business Review. https://hbr.org/2018/01/the-best-leaders-are-great-teachers

Finlay, L. (2014). Engaging phenomenological analysis. *Qualitative Research in Psychology, 11*(2), 121–141. DOI: 10.1080/14780887.2013.807899

Fitzpatrick, J. M., Bianchi, L. A., Hayes, N., Da Silva, T., & Harris, R. (2023). Professional development and career planning for nurses working in care homes for older people: A scoping review. *International Journal of Older People Nursing, 18*(1), e12519. DOI: 10.1111/opn.12519 PMID: 36441621

Flanagan, D. P., & Dixon, S. G. (2014). The Cattell-Horn-Carroll theory of cognitive abilities. https://onlinelibrary.wiley.com/doi/pdf/10.1002/9781118660584.ese0431#:~:text=The%20Cattell%2DHorn%2D-Carroll%20(,&%20Alfonso%2C%20in%20press)

Flook, L., Goldberg, S. B., Pinger, L., Bonus, K., & Davidson, R. J. (2013). Mindfulness for teachers: A pilot study to assess effects on stress, burnout, and teaching efficacy. *Mind, Brain and Education : the Official Journal of the International Mind, Brain, and Education Society, 7*(3), 182–195. DOI: 10.1111/mbe.12026 PMID: 24324528

Fox, M., Butler, J., Persaud, M., Tregunno, D., Sidani, S., & McCague, H. (2015). A multi-method study of the geriatric learning needs of acute care hospital nurses in Ontario, Canada. *Research in Nursing & Health, 39*(1), 66–76. DOI: 10.1002/nur.21699 PMID: 26471253

Franco, L., Picinin, C., Pilatti, L., & Franco, A. (2021). Work-life balance in higher education: A systematic review of the impact on the well-being of teachers. *Ensaio: Avaliação e Políticas Públicas em Educação, 29*(111). https://doi.org/DOI: 10.1590/s0104-403620210002903021

Frasier, A. S. (2022). What makes classroom observation feedback useful? The perceptions of secondary math and English teachers. *Voices of Reform, 5*(1), 40–58. DOI: 10.32623/5.00004

Fredrickson, B. L. (2001). The role of positive emotions in positive psychology: The broaden-and-build theory of positive emotions. *The American Psychologist, 56*(3), 218–226. DOI: 10.1037/0003-066X.56.3.218 PMID: 11315248

Freitas, C., & Dias, R. (2023). Professional training in industrial maintenance: A social interactionist approach. *International Journal of Mechanical Engineering Education, 52*(3), 288–314. DOI: 10.1177/03064190231188866

Frenzel, A. C., Becker-Kurz, B., Pekrun, R., Goetz, T., & Lüdtke, O. (2018). Emotion transmission in the classroom revisited: A reciprocal effects model of teacher and student enjoyment. *Journal of Educational Psychology, 110*(5), 628–639. DOI: 10.1037/edu0000228

Fullan, M. (2007). *The new meaning of educational change* (4th ed.). Teachers College Press.

Furman, P. B. (2020). *Perceptions of assistant principals of curriculum and instruction regarding the effectiveness of their role* (Publication No. 1677) [Doctoral dissertation, University of Alabama at Birmingham]. UAB Digital Collections.

Fursykova, T., Habelko, O., & Chernii, V. (2022). The Development of Digital Competence of Future Teachers in the Process of Distance Learning. *International Journal of Emerging Technologies in Learning, 17*(10), 85–98. DOI: 10.3991/ijet.v17i10.28973

Fütterer, T., Scherer, R., Scheiter, K., Stürmer, K., & Lachner, A. (2023). Will, skills, or conscientiousness: What predicts teachers' intentions to participate in technology-related professional development? *Computers & Education, 198*, 1–17. DOI: 10.1016/j.compedu.2023.104756

Gallardo-Montes, C. del P., Caurcel Cara, M. J., Rodríguez Fuentes, A., & Capperucci, D. (2024). Opinions, training and requirements regarding ICT of educators in Florence and Granada for students with functional diversity. *Universal Access in the Information Society, 23*(2), 889–899. Advance online publication. DOI: 10.1007/s10209-023-00977-0 PMID: 36852016

Ganon-Shilon, S., & Schechter, C. (2017). Making sense while steering through the fog: Principals' metaphors within a national reform implementation. *Education Policy Analysis Archives, 25*(105), 105. Advance online publication. DOI: 10.14507/epaa.25.2942

Gardner, H. (1983). *Frames of mind: The theory of multiple intelligences.* Basic Books.

Gaurav, D., Kaushik, Y., Supraja, S., Yadav, M., Gupta, M. P., & Chaturvedi, M. (2022). Empirical study of adaptive serious games in enhancing learning outcome. *International Journal of Serious Games, 9*(2), 2. Advance online publication. DOI: 10.17083/ijsg.v9i2.486

Gay, G. (2010). *Culturally Responsive Teaching: Theory, Research, and Practice*. Teachers College Press.

Gay, G. (2018). *Culturally responsive teaching: Theory, research, and practice* (3rd ed.). Teachers College Press.

Gcawu, S., & Rooyen, D. (2022). Clinical teaching practices of nurse educators: An integrative literature review. *Health SA = SA Gesondheid, 27*. Advance online publication. DOI: 10.4102/hsag.v27i0.1728 PMID: 36262921

George, D., & Mallery, M. (2010). *SPSS for windows step by step: A simple guide and reference*. Pearson.

Ghahremani, M., & Ostovar, Z. (2017). The contribution of emotional intelligence and achievement motivation on psychological well-being. *Global Journal of Health Science, 9*(3), 88. DOI: 10.5539/gjhs.v9n3p88

Ghamgosar, A., Zarghani, M., & Nemati-Anaraki, L. (2021). Bibliometric analysis on geriatric nursing research in web of science (1900–2020). *Biomed Research International*, 2021(1). DOI: 10.1155/2021/8758161

Ghamrawi, N., Shal, T., & Ghamrawi, N. A. (2024). Exploring the impact of AI on teacher leadership: Regressing or expanding? *Education and Information Technologies, 29*(7), 8415–8433. DOI: 10.1007/s10639-023-12174-w

Gharaei, N., Thijs, J., & Verkuyten, M. (2018). Ethnic identity in diverse schools: Preadolescents' private regard and introjection in relation to classroom norms and composition. *Journal of Youth and Adolescence*, *48*(1), 132–144. DOI: 10.1007/s10964-018-0881-y PMID: 29922870

Glanz, J. (2004). The assistant principal's handbook: Strategies for success. *Sage (Atlanta, Ga.)*.

Gocen, A., & Aydemir, F. (2020). Artificial intelligence in education and schools. *Research on Education and Media*, *12*(1), 13–21. DOI: 10.2478/rem-2020-0003

Godino, J. D., Batanero, C., & Font, V. (2007). The onto-semiotic approach to research in mathematics education. *ZDM Mathematics Education*, *39*(1), 127–135. DOI: 10.1007/s11858-006-0004-1

Godino, J. D., Giakomone, B., Batanero, C., & Font, V. (2017). Enfoque ontosemiótico de los conocimientos y competencias del profesor de matemáticas. *Bolema*, *31*(57), 90–113. DOI: 10.1590/1980-4415v31n57a05

Goldring, E., Rubin, M., & Herrmann, M. (2021). *The role of assistant principals: Evidence and insights for advancing school leadership*. The Wallace Foundation. DOI: 10.59656/EL-SB1829.001

Goleman, D. (1996). *Emotional intelligence: Why it can matter more than IQ*. Bloomsbury Publishing.

Goleman, D. (1998). *Working with emotional intelligence*. Bantam Books.

Gomez, F. C.Jr, Trespalacios, J., Hsu, Y. C., & Yang, D. (2022). Exploring Teachers' Technology Integration Self-Efficacy through the 2017 ISTE Standards. *TechTrends*, *66*(2), 159–171. DOI: 10.1007/s11528-021-00639-z PMID: 34378007

Gondal, K. M. (2018). Artificial intelligence and educational leadership. *King Edward Medicine Üniversity Annuals*, *24*(4), 1–2.

Gong, Y. (2021). Application of virtual reality teaching method and artificial intelligence technology in digital media art creation. *Ecological Informatics*, *63*, 101304. DOI: 10.1016/j.ecoinf.2021.101304

Gonzales, A. (2024). Work-related stress and teaching performance of public elementary school teachers of San Francisco Cluster IV. *International Journal of Innovation and Scientific Research*, *11*(5), 203–210. DOI: 10.51244/IJRSI.2024.1105003

Gonzales, M. M., & Roberts, M. (2022). The impact of a franchise model school framework on the leadership development of assistant principals. *Journal of School Leadership*, *32*(5), 537–561. DOI: 10.1177/10526846211067636

Good, D. (2012). *Cognitive Agility: Adapting to Real-time Decision Making at Work. 44*(2), 13–17. https://www.researchgate.net/publication/265963826_Cognitive_Agility_Adapting_to_Real-time_Decision_Making_at_Work

Government of India. Ministry of Education, & Department of Higher Education. (2021). *All India Survey on Higher Education 2021-22*. Retrieved from https://cdnbbsr.s3waas.gov.in

Grade 4 Cross curricular Lesson Plan: A World of Sports. (2015). Touchableearth.org. https://www.touchableearth.org/blog/grade-4-cross-curricular-lesson-plan-a-world-of-sports

Grootven, B., Jeuris, A., Jonckers, M., Devriendt, E., Casterlé, B., Dubois, C., & Flamaing, J. (2021). Geriatric co-management for cardiology patients in the hospital: A quasi-experimental study. *Journal of the American Geriatrics Society, 69*(5), 1377–1387. DOI: 10.1111/jgs.17093 PMID: 33730373

Gross, J. J. (2014). Emotion regulation: Conceptual and empirical foundations. In Gross, J. J. (Ed.), *Handbook of emotion regulation* (2nd ed., pp. 3–20). Guilford Press.

Guan, C., Mou, J., & Jiang, Z. (2020). Artificial intelligence innovation in education: A twenty-year data-driven historical analysis. *International Journal of Innovation Studies, 4*(4), 134–147. DOI: 10.1016/j.ijis.2020.09.001

Güneş, F. (2019). Okuryazarlık yaklaşımları. *The Journal of Limitless Education and Research., 4*(3), 224–246.

Gupta, S., & McCarthy, P. J. (2022). The sporting resilience model: A systematic review of resilience in sport performers. *Frontiers in Psychology, 13*(1), 1003053. DOI: 10.3389/fpsyg.2022.1003053 PMID: 36619099

Gurley, D. K., Peters, G. B., Fifolt, M., Collins, L., & McNeese, R. (2015). Addressing the multiplicity of roles of the assistant principal: Insights from the field. *Journal of Behavioral and Social Sciences (Cedarville, Ohio), 3*(3), 135–147.

Guskey, T. R. (2002). Professional development and teacher change. *Teachers and Teaching, 8*(3), 381–391. DOI: 10.1080/135406002100000512

Gustems-Carnicer, J., Calderón, C., & Calderón-Garrido, D. (2019). Stress, coping strategies, and academic achievement in teacher education students. *European Journal of Teacher Education, 42*(3), 375–390. DOI: 10.1080/02619768.2019.1576629

Hakvoort, L., Dikken, J., Wel, M., Derks, C., & Schuurmans, M. (2021). Minimizing the knowledge-to-action gap; identification of interventions to change nurses' behavior regarding fall prevention, a mixed method study. *BMC Nursing, 20*(1), 80. Advance online publication. DOI: 10.1186/s12912-021-00598-z PMID: 34016106

Hall, R. A. (2011). Affective assessment: The missing piece of the educational reform puzzle. *Delta Kappa Gamma Bulletin, 77*(2), 7.

Halpin, M. (2024, February 26). *Equality and Equity in Sports*. Recite Me. https://reciteme.com/news/equality-and-equity-in-sports/

Hammond, Z. (2015). *Culturally responsive teaching and the brain: Promoting authentic engagement and rigor among culturally and linguistically diverse students*. Corwin Press.

Hanselman, P., Grigg, J., Bruch, S. K., & Gamoran, A. (2016). The consequences of principal and teacher turnover for school social resources. In *Family environments, school resources, and educational outcomes* (pp. 49–89). Emerald Group Publishing Limited., DOI: 10.1108/S1479-353920150000019004

Hao, L. (2022). Analysis of the effect of work stress and organizational culture on teacher performance. *International Journal of Progressive Sciences and Technologies, 34*(1), 45–50. DOI: 10.52155/ijpsat.v34.1.4503

Hargreaves, A. (1998). The emotional politics of teaching and teacher development: With implications for educational leadership. *International Journal of Leadership in Education, 1*(4), 315–336. DOI: 10.1080/1360312980010401

Hargreaves, A., & Fullan, M. (2012). *Professional capital: Transforming teaching in every school.* Teachers College Press.

Harper-collins-admin. (2023, August 25). *Embracing Sports Integration: A Vital Component of School Curriculum - HarperCollins.* HarperCollins. https://collins.in/embracing-sports-integration-a-vital-component-of-school-curriculum/

Harris, M. L. (2025). *Emerging Dialogues - Applying Sports Analogies to Motivate Assessment.* Aalhe.org. https://www.aalhe.org/emerging-dialogues--applying-sports-analogies-to-motivate-assessment

Harris, A. (2005). Teacher leadership: More than just a feel-good factor. *Leadership and Policy in Schools, 4*(3), 201–219. DOI: 10.1080/15700760500244777

Harris, A., Jones, M., & Ismail, N. (2022). Distributed leadership: Taking a retrospective and contemporary view of the evidence base. *School Leadership & Management, 42*(5), 438–456. DOI: 10.1080/13632434.2022.2109620

Harris, A., & Muijs, D. (2003). *Teacher leadership: Principles and practices.* National College for School Leadership.

Hasan, H. A., Adi, S., Hariyanto, E., Khongrungchok, A., Oktaviani, H. I., & Fajarianto, O.Humaid Ali Hasan. (2023). Physical Activity with Play and Game Model to Improve the Cognitive of Elementary School Students. *Edcomtech Jurnal Kajian Teknologi Pendidikan, 8*(1), 49–49. DOI: 10.17977/um039v8i12023p49

Hashim, S., Omar, M. K., Ab Jalil, H., & Sharef, N. M. (2022). Trends on technologies and artificial intelligence in education for personalized learning: Systematic literature. *Journal of Academic Research in Progressive Education and Development, 12*(1), 884–903. DOI: 10.6007/IJARPED/v11-i1/12230

Haske-Palomino, M., Carlson, R., & Gatley, L. (2023). Elder veteran program: A description of a nurse practitioner–led inpatient geriatric consult model. *Journal of the American Association of Nurse Practitioners, 35*(4), 229–234. DOI: 10.1097/JXX.0000000000000844 PMID: 36857530

Hattie, J. (2009). The black box of tertiary assessment: An impending revolution. In L. H. Meyer, S. Davidson, H. Anderson, R. Fletcher, P. M. Johnston, & M. Rees (Eds.), *Tertiary assessment & higher education student outcomes: Policy, practice & research* (pp. 259–275). Ako Aotearoa.

Hattie, J., & Timperley, H. (2007). The power of feedback. *Review of Educational Research, 77*(1), 81–112. DOI: 10.3102/003465430298487

Hausfather, S. J. (1996). Vygotsky and schooling: Creating a social contest for learning. *Action in Teacher Education, 18*(2), 1–10. DOI: 10.1080/01626620.1996.10462828

Health Direct. (2018, July 2). *Developing life skills through sports.* Healthdirect.gov.au;Australia, H. https://www.healthdirect.gov.au/developing-life-skills-through-sports

Heeren, P., Devriendt, E., Fieuws, S., Wellens, N., Flamaing, J., Sabbé, M., & Milisen, K. (2019). Unplanned readmission prevention by a geriatric emergency network for transitional care (urgent): A prospective before-after study. *BMC Geriatrics*, *19*(1), 215. Advance online publication. DOI: 10.1186/s12877-019-1233-9 PMID: 31390994

Heffernan, N. T., & Heffernan, C. L. (2014). The ASSISTments ecosystem: Building a platform that brings scientists and teachers together for minimally invasive research on human learning and teaching. *International Journal of Artificial Intelligence in Education*, *24*(4), 470–497. DOI: 10.1007/s40593-014-0024-x

Hehir, T., Grindal, T., Freeman, B., Lamoreau, R., Borquaye, Y., & Burke, S. (2016). *A summary of the evidence on inclusive education*. Abt Associates.

Hennessy, S., D'Angelo, S., McIntyre, N., Koomar, S., Kreimeia, A., Cao, L., Brugha, M., & Zubairi, A. (2022). Technology Use for Teacher Professional Development in Low- and Middle-Income Countries: A systematic review. *Computers and Education Open*, *3*, 1–32. DOI: 10.1016/j.caeo.2022.100080

Hennessy, S., Haßler, B., & Hofmann, R. (2015). Challenges and opportunities for teacher professional development in interactive use of technology in African schools. *Technology, Pedagogy and Education*, *24*(5), 1–28. Advance online publication. DOI: 10.1080/1475939X.2015.1092466

Henry, W. (2023, November 25). *Strategy Games: Engaging the Mind and Building Skills*. Medium. https://watson-henry.medium.com/strategy-games-engaging-the-mind-and-building-skills-dd09c219fca0

Hernández-Mendo, A., Reigal, R. E., López-Walle, J. M., Serpa, S., Samdal, O., Morales-Sánchez, V., Juárez-Ruiz de Mier, R., Tristán-Rodríguez, J. L., Rosado, A. F., & Falco, C. (2019). Physical Activity, Sports Practice, and Cognitive Functioning: The Current Research Status. *Frontiers in Psychology*, *10*, 2658. Advance online publication. DOI: 10.3389/fpsyg.2019.02658 PMID: 31866893

Herranz-Hernández, P., Fernández-Hernández, J., & Segovia-Torres, L. (2023). Emotional metaphors for an inclusive classroom climate. *Environment and Social Psychology*, *9*(1). Advance online publication. DOI: 10.54517/esp.v9i1.1945

Hesselink, G., Demirbas, M., Rikkert, M., & Schoon, Y. (2019). Geriatric education programs for emergency department professionals: A systematic review. *Journal of the American Geriatrics Society*, *67*(11), 2402–2409. DOI: 10.1111/jgs.16067 PMID: 31335964

Hidayat, Z. (2023). Prosumers of hybrid entertainment-educational communication in a media convergence environment. *Journal of Biomedical Research & Environmental Sciences*, *4*(9), 1281–1289. DOI: 10.37871/jbres1797

Hiebert, J., Gallimore, R., & Stigler, J. W. (2002). A knowledge base for the teaching profession: What would it look like and how can we get one? *Educational Researcher*, *31*(5), 3–15. DOI: 10.3102/0013189X031005003

Higgins, T. E., & Spitulnik, M. W. (2008). Supporting teachers' use of technology in science instruction through professional development: A literature review. In *Journal of Science Education and Technology* (Vol. 17, Issue 5, pp. 511–521). DOI: 10.1007/s10956-008-9118-2

Higgins, D., Hayes, M., Taylor, J., & Wallace, J. (2020). How do we teach simulation-based dental education? time for an evidence-based, best-practice framework. *European Journal of Dental Education, 24*(4), 815–821. DOI: 10.1111/eje.12551 PMID: 32521066

Higher Education Council. 2023. Pedagojik Formasyona Yönelik Bilgilendirme [Decision of the Higher Education Council regarding pedagogical formation education programme], Turkish Higher Education Council. https://www.yok.gov.tr/HaberBelgeleri/BasinAciklamasi/2023/pedagojik-formasyona-yonelik-bilgilendirme.pdf

Hill, H. C., Ball, D. L., & Schilling, S. G. (2008). Unpacking pedagogical content knowledge: Conceptualizing and measuring teachers' topic-specific knowledge of students. *Journal for Research in Mathematics Education, 39*(4), 372–400. DOI: 10.5951/jresematheduc.39.4.0372

Hodge, D. R. (2013). Spirituality and social work: A cultural and contextual perspective. *Journal of Social Work : JSW, 13*(3), 254–271.

Ho, J., Shaari, I., & Kang, T. (2023). Vice-principals as leaders: Role ambiguity and role conflicts faced by vice-principals in Singapore. *Educational Management Administration & Leadership, 51*(3), 575–593. DOI: 10.1177/17411432211002527

Hojat, M., & Gonnella, J. (2017). What matters more about the interpersonal reactivity index and the Jefferson Scale of Empathy? Their underlying constructs or their relationships with pertinent measures of clinical competence and patient outcomes? *Academic Medicine, 92*(6), 743–745. DOI: 10.1097/ACM.0000000000001424 PMID: 28557931

Holmes, W., Bialik, M., & Fadel, C. (2019). *Artificial intelligence in education: Promises and implications for teaching and learning*. Centre for Curriculum Redesign.

Hooks, B. (1994). *Teaching to transgress: Education as the practice of freedom*. Routledge.

Horn, J. L. (1968). Organization of abilities and the development of intelligence. *Psychological Review, 75*(3), 242–259. DOI: 10.1037/h0025662 PMID: 4875815

Hoy, A. W., & Spero, R. B. (2005). Changes in teacher efficacy during the early years of teaching: A comparison of four measures. *Teaching and Teacher Education, 21*(4), 343–356. DOI: 10.1016/j.tate.2005.01.007

Hsu, T. C., Hsu, T. P., & Lin, Y. T. (2023). The artificial intelligence learning anxiety and self-efficacy of in-service teachers taking AI training courses. *2023 International Conference on Artificial Intelligence and Education (ICAIE)*, Kobe, Japan, pp. 97-101. DOI: 10.1109/ICAIE56796.2023.00034

Huang, J., Saleh, S., & Liu, Y. (2021). A review on artificial intelligence in education. *Academic Journal of Interdisciplinary Studies, 10*(3), 206–217. DOI: 10.36941/ajis-2021-0077

Huang, R. (2023). Research on the Path of Teacher Professional Development in the Context of Smart Education. *Open Journal of Social Sciences, 11*(03), 293–303. DOI: 10.4236/jss.2023.113021

Huang, X. (2021). Aims for cultivating students' key competencies based on artificial intelligence education in China. *Education and Information Technologies, 26*(5), 5127–5147. DOI: 10.1007/s10639-021-10530-2

Hu, F., Lee, H., & Li, Y. (2021). Exploration of geriatric care competencies in registered nurses in hospitals. *The Journal of Nursing Research, 29*(4), e159. DOI: 10.1097/JNR.0000000000000441 PMID: 34034305

Huh, A., & Shin, J. (2021). Person-centered care practice, patient safety competence, and patient safety nursing activities of nurses working in geriatric hospitals. *International Journal of Environmental Research and Public Health, 18*(10), 5169. DOI: 10.3390/ijerph18105169 PMID: 34068125

Hu, L., & Bentler, P. M. (1999). Cutoff criteria for fit indexes in covariance structure analysis: Conventional criteria versus new alternatives. *Structural Equation Modeling, 6*(1), 1–55. DOI: 10.1080/10705519909540118

Huls, M., Rooij, S., Diepstraten, A., Koopmans, R., & Helmich, E. (2015). Learning to care for older patients: Hospitals and nursing homes as learning environments. *Medical Education, 49*(3), 332–339. DOI: 10.1111/medu.12646 PMID: 25693992

Humphries, M. L., Strickland, J., & Keenan, K. (2014). African American preschoolers' social and emotional competence at school: The influence of teachers and mothers. *Journal of Education, 194*(1), 11–18. DOI: 10.1177/002205741419400104

Huppert, F. A. (2009). Psychological well-being: Evidence regarding its causes and consequences. *Applied Psychology. Health and Well-Being, 1*(2), 137–164. DOI: 10.1111/j.1758-0854.2009.01008.x

Hussain, S., Lindh, J., & Shukur, G. (2024). The effect of LEGO training on pupils' school performance in mathematics, problem solving ability and attitude: Swedish data. *Journal of Educational Technology & Society, 9*(3), 182–194.

Hutchison, M. (2016). The empathy project: Using a project-based learning assignment to increase first-year college students' comfort with interdisciplinarity. *The Interdisciplinary Journal of Problem-Based Learning, 10*(1). Advance online publication. DOI: 10.7771/1541-5015.1580

Hwang, G. J., Xie, H., Wah, B. W., & Gašević, D. (2020). Vision. challenges. roles and research issues of Artificial Intelligence in Education. *Computers and Education: Artificial Intelligence, 1*(100001), 1–5. DOI: 10.1016/j.caeai.2020.100001

Ikhwan, M., Fuadi, M., Mailizar, M., & Jannah, M. (2023). The Utilization of Information Technology for the Professional Development of Islamic Education Teachers in Indonesia. *Progresiva : Jurnal Pemikiran Dan Pendidikan Islam, 12*(02), 209–222. Advance online publication. DOI: 10.22219/progresiva.v12i02.31169

Immordino-Yang, M. H., & Damasio, A. (2007). We feel, therefore we learn: The relevance of affective and social neuroscience to education. *Mind, Brain and Education : the Official Journal of the International Mind, Brain, and Education Society, 1*(1), 3–10. DOI: 10.1111/j.1751-228X.2007.00004.x

In what ways can we use sports to promote diversity and inclusion in our communities? (2019). Quora. https://www.quora.com/In-what-ways-can-we-use-sports-to-promote-diversity-and-inclusion-in-our-communities

Irvine, J. (2022). Practice and theory: Ten lessons that I have learned about being a vice principal. *Journal of Practical Studies in Education*, *3*(3), 7–16. DOI: 10.46809/jpse.v3i3.46

Isaacs, A., & Spencer, B. (2022). Factors affecting learning during internship: A qualitative study of junior doctors. *Australian Journal of General Practice*, *51*(1-2), 83–89. DOI: 10.31128/AJGP-02-21-5821 PMID: 35098280

Istiarsyah, I., Dawi, A. H., & Ahmad, N. A. (2019). The influence of special education training on teachers' attitudes towards inclusive education: Case study in Aceh Province, Indonesia. *International Journal of Academic Research in Progressive Education and Development*, *8*(4), 1016–1027.

Jackson, T., McBrayer, J. S., Pannell, S., Cleveland, R. E., Miller, S. B., & Fallon, K. (2021). The effect of school tasks on principals' and assistant principals' leadership self-efficacy. *School Leadership Review, 16*(1), Article 7. https://scholarworks.sfasu.edu/slr/vol16/iss1/7

Jana, V., Radizuan, T. H., & Maldisa, G. S. (2023). Qualitative perspectives of new graduate nurses to educational preparation for patient care. *Malahayati International Journal of Nursing and Health Science*, *6*(4), 330–337. DOI: 10.33024/minh.v6i4.12611

Järvelä, S., Nguyen, A., & Hadwin, A. (2023). Human and artificial intelligence collaboration for socially shared regulation in learning. *British Journal of Educational Technology*, *54*(5), 1057–1076. DOI: 10.1111/bjet.13325

Jena, M. K., Pandey, S. K., & Barad, S. (2023). Integrating Technology on Professional Development of Elementary School Teachers' in the 21st century. *International Journal of Multidisciplinary Research in Arts. Science and Technology*, *1*(5), 22–28. DOI: 10.61778/ijmrast.v1i5.27

Jennings, P. A., Brown, J. L., Frank, J. L., Doyle, S., Oh, Y., Davis, R., Rasheed, D., DeWeese, A., DeMauro, A. A., Cham, H., & Greenberg, M. T. (2017). Impacts of the CARE for Teachers program on teachers' social and emotional competence and classroom interactions. *Journal of Educational Psychology*, *109*(7), 1010–1028. DOI: 10.1037/edu0000187

Jennings, P. A., & Greenberg, M. T. (2009). The prosocial classroom: Teacher social and emotional competence in relation to student and classroom outcomes. *Review of Educational Research*, *79*(1), 491–525. DOI: 10.3102/0034654308325693

Jeon, L., Buettner, C., & Grant, A. (2018). Early childhood teachers' psychological well-being: Exploring potential predictors of depression, stress, and emotional exhaustion. *Early Education and Development*, *29*(1), 53–69. DOI: 10.1080/10409289.2017.1341806

Jiang, Y., Li, X., Luo, H., Yin, S., & Kaynak, O. (2022). Quo vadis artificial intelligence? *Discover Artificial Intelligence*, *2*(1), 4. DOI: 10.1007/s44163-022-00022-8

Jõgi, A., Aulén, A., Pakarinen, E., & Lerkkanen, M. (2022). Teachers' daily physiological stress and positive affect in relation to their general occupational well-being. *The British Journal of Educational Psychology*, *93*(2), 368–385. DOI: 10.1111/bjep.12561 PMID: 36336902

Johnson, S. L., & Johnson, R. T. (2007). The role of collaboration in building professional efficacy. *Teaching and Teacher Education*, *23*(6), 740–752.

Jones, K. M. L., Rubel, A., & LeClere, E. (2020). A matter of trust: Higher education institutions as information fiduciaries in an age of educational data mining and learning analytics. *Journal of the Association for Information Science and Technology*, *71*(10), 1227–1241. DOI: 10.1002/asi.24327

Jordan, A., Schwartz, E., & McGhie-Richmond, D. (2009). Preparing teachers for inclusive classrooms. *Teaching and Teacher Education*, *25*(4), 535–542. DOI: 10.1016/j.tate.2009.02.010

Jordan, M. I., & Mitchell, T. M. (2015). Machine learning: Trends, perspectives, and prospects. *Science*, *349*(6245), 255–260. DOI: 10.1126/science.aaa8415 PMID: 26185243

Jung, S. (2023). Challenges for future directions for artificial intelligence integrated nursing simulation education. *Korean Journal of Women Health Nursing*, *29*(3), 239–242. DOI: 10.4069/kjwhn.2023.09.06.1 PMID: 37813667

Kafyulilo, A., Fisser, P., & Voogt, J. (2016). Factors affecting teachers' continuation of technology use in teaching. *Education and Information Technologies*, *21*(6), 1535–1554. DOI: 10.1007/s10639-015-9398-0

Kamalov, F., Calonge, D. S., & Gurrib, I. (2023). New era of Artificial intelligence in education: Towards a sustainable multifaceted revolution. *Sustainability (Basel)*, *15*(16), 1–27. DOI: 10.3390/su151612451

Kandlhofer, M., Steinbauer, G., Hirschmugl-Gaisch, S., & Huber, P. (2016). *Artificial intelligence and computer science in education: From kindergarten to university*. In 2016 IEEE Frontiers in Education Conference (FIE) (pp. 1-9). NY: IEEE. DOI: 10.1109/FIE.2016.7757570

Kang, X., Meng, Q., & Su, C.-H. (2024). School-Based Team Sports as Catalysts for Holistic Student Wellness: A Narrative Review. *Behavioral Sciences (Basel, Switzerland)*, *14*(7), 528. DOI: 10.3390/bs14070528 PMID: 39062351

Kaplan, J. (2015). *Humans need not apply: A guide to wealth and work in the age of artificial intelligence*. Yale University Press.

Karataş, F., Eriçok, B., & Tanrikulu, L. (2024). Reshaping curriculum adaptation in the age of artificial intelligence: Mapping teachers' AI-driven curriculum adaptation patterns. *British Educational Research Journal*, *1*(0), 1–27.

Karjalainen, S., Brännström, J., Christensson, J., Sahlén, B., & Lyberg-Åhlander, V. (2020). A pilot study on the relationship between primary-school teachers' well-being and the acoustics of their classrooms. *International Journal of Environmental Research and Public Health*, *17*(6), 2083. DOI: 10.3390/ijerph17062083 PMID: 32245168

Kark, R. (2011). Games Managers Play: Play as a Form of Leadership Development. *Academy of Management Learning & Education*, *10*(3), 507–527. DOI: 10.5465/amle.2010.0048

Katzenmeyer, M., & Moller, G. (2013). *Uyuyan devi uyandırmak [Awaking the sleeping giant]*. Servet Özdemir (Trans. Ed.). Nobel.

Kearl, B. (n.d.). *5 Useful Performance Management Takeaways From Sports*. Www.bamboohr.com. https://www.bamboohr.com/blog/5-useful-performance-management-takeaways-sports

Kedzior, M., & Fifield, S. (2004). Teacher professional development. Delaware education research & development center. Learning matters (n.d). The importance of teacher training programs. https://learningmatters.ai/blog/the-importance-of-teacher-training-programs#:~:text=A%20teacher%20training%20program%20is%20a%20program%20that%20equips%20teachers,are%20conducted%20online%20or%20offline

Kgosi, M. K., Makgato, M., & Skosana, N. M. (2023). Teachers' Views on the Application of Educational Technologies in the Classroom: A Case of Selected Tshwane West Secondary Schools in Gauteng. *Journal of Curriculum Studies Research*, *5*(2), 151–166. DOI: 10.46303/jcsr.2023.23

Khushboo, P. (2021). Educating the mind without educating the heart is no education at all. https://timesofindia.indiatimes.com/readersblog/newvibe/educating-the-mind-without-educating-the-heart-is-no-education-at-all-40011/

Killoran, I., Woronko, D., & Zaretsky, H. (2014). Exploring preservice teachers' attitudes towards inclusion. *International Journal of Inclusive Education*, *18*(4), 427–442. DOI: 10.1080/13603116.2013.784367

Kim, B., Lee, M., & Jang, S. (2021). Hospital nurses' empathy and moral sensitivity toward elderly care: A cross-sectional study. *Journal of Nursing Management*, *30*(7), 2138–2146. DOI: 10.1111/jonm.13442 PMID: 34350659

Kim, K. (2020). Project-based learning approach to increase medical student empathy. *Medical Education Online*, *25*(1), 1742965. Advance online publication. DOI: 10.1080/10872981.2020.1742965 PMID: 32197574

Kim, S., Raza, M., & Seidman, E. (2019). Improving 21st-century teaching skills: The key to effective 21st-century learners. [sagepub.]. *Research in Comparative and International Education*, *14*(1), 99–117. DOI: 10.1177/1745499919829214

King, A., Boyd, M., Dagley, L., & Raphael, D. (2017). Implementation of a gerontology nurse specialist role in primary health care: Health professional and older adult perspectives. *Journal of Clinical Nursing*, *27*(3-4), 807–818. DOI: 10.1111/jocn.14110 PMID: 29052288

Kırksekiz, A., Yıldız, M., Kıyıcı, M., & Yıldız, M. (2024). Adaptation of artificial intelligence literacy scale: Latent profile analysis. *Sakarya University Journal of Education*, *14*(3), 581–596. DOI: 10.19126/suje.1479294

Klassen, R., & Chiu, M. (2010). Effects on teachers' self-efficacy and job satisfaction: Teacher gender, years of experience, and job stress. *Journal of Educational Psychology*, *102*(3), 741–756. DOI: 10.1037/a0019237

Kleib, M., Arnaert, A., Nagle, L., Sugars, R., & Costa, D. (2024). Newly qualified Canadian nurses' experiences with digital health in the workplace: Comparative qualitative analysis. *JMIR Medical Education*, *10*, e53258. DOI: 10.2196/53258 PMID: 39159452

Klusmann, U., Richter, D., & Lüdtke, O. (2016). Teachers' emotional exhaustion is negatively related to students' achievement: Evidence from a large-scale assessment study. *Journal of Educational Psychology*, *108*(8), 1193–1203. DOI: 10.1037/edu0000125

Knudsen, J. V., Elkrog-Hansen, L. M., & Christiansen, L. B. (2020). Life Skills Through School Sport: A Participatory Teacher Development Program. *Advances in Physical Education, 10*(03), 293–310. DOI: 10.4236/ape.2020.103024

Koenig, H. G. (2012). Religion, spirituality, and health: The research and clinical implications. *ISRN Psychiatry, 278730*, 1–33. Advance online publication. DOI: 10.5402/2012/278730 PMID: 23762764

Koenig, H. G. (2012). *The Handbook of Religion and Health*. Oxford University Press.

Komalavalli, K., Hemalatha, R., & Dhanalakshmi, S. (2020). A survey of artificial intelligence in smart phones and its applications among the students of higher education in and around Chennai City. *Shanlax International Journal of Education, 8*(3), 89–95. DOI: 10.34293/education.v8i3.2379

Korthagen, F. (2010). La práctica, la teoría y la persona en la formación del profesorado [Practice, theory, and person in teachers' training]. *Revista Interuniversitaria De Formación Del Profesorado, 68*(24), 83–101.

Kraft, M. A. (2022, February 24). Better teacher feedback can happen outside of the evaluation process. Brookings Institution. https://www.brookings.edu/articles/better-teacher-feedback-can-happen-outside-of-the-evaluation-process/

Krathwohl, D. R. (2002). A Revision of Bloom's Taxonomy: An Overview. *Theory into Practice, 41*(4), 212–218. DOI: 10.1207/s15430421tip4104_2

Kukalo, E. (2024, September 25). 5 Secrets to Achieving a Perfect Teacher Work-Life Balance. *actiTIME*. https://www.actitime.com/school-management-software/teacher-work-life-balance

Kul, U. (2018). Influences of technology integrated professional development course on mathematics teachers *. *European Journal of Educational Research, 7*(2), 233–243. DOI: 10.12973/eu-jer.7.2.233

Kuriscak, E., Marsalek, P., Stroffek, J., & Toth, P. G. (2015). Biological context of Hebb learning in artificial neural networks, a review. *Neurocomputing, 152*, 27–35. DOI: 10.1016/j.neucom.2014.11.022

Ladson-Billings, G. (1995). Toward a theory of culturally relevant pedagogy. *American Educational Research Journal, 32*(3), 465–491. DOI: 10.3102/00028312032003465

Ladson-Billings, G. (2021). *Critical race theory in education: A scholar's journey*. Teachers College Press.

Leadership and management in sports: A framework for success. (2025, November 12). KU SOE; School of Education and Human Sciences, Department of Health, Sport, and Exercise Sciences, University of Kansas. https://onlinesportmanagement.ku.edu/community/leadership-and-management-in-sports

Leadership in Education: Decision-Making in Education. (n.d.). ResearchGate. https://www.researchgate.net/publication/339235695_Leadership_in_Education_Decision-Making_in_Education

Lee, J., Topping, K., & Lakin, E. (2023). Technology-Facilitated Continuous Professional Development During a Pandemic: A Hong Kong Primary School Case Study. *RELC Journal, 54*(2), 376–393. DOI: 10.1177/00336882231175239

Leffers, J., McDermott-Levy, R., Nicholas, P., & Sweeney, C. (2017). Mandate for the nursing profession to address climate change through nursing education. *Journal of Nursing Scholarship*, *49*(6), 679–687. DOI: 10.1111/jnu.12331 PMID: 28806483

Leithwood, K., Leonard, L., & Sharratt, L. (1998). Conditions fostering organizational learning in schools. *Educational Administration Quarterly*, *34*(2), 243–276. DOI: 10.1177/0013161X98034002005

Leithwood, K., Sun, J., & Schumacker, R. (2020). How school leadership influences student learning: A test of "the four paths model.". *Educational Administration Quarterly*, *56*(4), 570–599. DOI: 10.1177/0013161X19878772

Leonard, N. M., & Smyth, S. (2020). Does training matter? Exploring teachers' attitudes towards the inclusion of children with autism spectrum disorder in mainstream education in Ireland. *International Journal of Inclusive Education*, •••, 1–15. DOI: 10.1080/13603116.2020.1718221

Lérias, E., Guerra, C., & Ferreira, P. (2024). Literacy in artificial intelligence as a challenge for teaching in higher education: A case study At Portalegre Polytechnic University. *Information (Basel)*, *15*(4), 205. DOI: 10.3390/info15040205

Levin, J. A., Fox, J. A., & Forde, D. R. (2010). *Elementary statistics in social research*. Allyn & Bacon Pearson.

Liao, J., Wang, X., & Wang, X. (2023). The effect of work stress on the well-being of primary and secondary school teachers in China. *International Journal of Environmental Research and Public Health*, *20*(2), 154. DOI: 10.3390/ijerph20021154 PMID: 36673909

Liao, L., Bai, X., He, X., Tan, L., Yang, L., Long, H., Huang, S., Li, X., Han, Y., Peng, X., Chen, H., Yang, X., Li, H., Wang, S., & Liu, Y. (2024). Identification of formation mechanism and key elements of quality geriatric care behavior of nursing assistants in nursing homes: A grounded theory study. *Frontiers in Public Health*, *12*, 1425883. Advance online publication. DOI: 10.3389/fpubh.2024.1425883 PMID: 38993701

Liberty University. (2021). *The impact of professional development on culturally responsive teaching and teacher self-efficacy*. Liberty University Press.

Li, M., Jotikasthira, N., & Pu, R. (2022). Digitalization, knowledge sharing and higher education for sustainable development. *World Journal on Educational Technology: Current Issues*, *14*(5), 1452–1468. DOI: 10.18844/wjet.v14i5.7713

Lin, P., Wung, S.-F., Lin, P.-C., Lin, Y.-C., Lin, C.-Y., & Huang, H.-L. (2024). Virtual reality-based simulation learning on geriatric oral health care for nursing students: A pilot study. *BMC Oral Health*, *24*(1), 627. Advance online publication. DOI: 10.1186/s12903-024-04249-y PMID: 38807116

Li, Q., Bu, Z., Xue, M., Huang, A., Tu, W., & Xu, G. (2022). Visual analysis of research hotspots in geriatric nursing education in China and abroad. *Journal of Healthcare Engineering*, *2022*, 1–7. DOI: 10.1155/2022/7472188 PMID: 35480150

Li, S., Yamaguchi, S., Sukhbaatar, J., & Takada, J. I. (2019). The influence of teachers' professional development activities on the factors promoting ICT integration in Primary Schools in Mongolia. *Education Sciences*, *9*(2), 1–18. DOI: 10.3390/educsci9020078

Liu, M. H., & Kleinsasser, R. (2023). Multiple Perspectives with Online EFL Practicum Technology Learning: Case Study on a Cloud. *TechTrends*, *67*(2), 342–353. DOI: 10.1007/s11528-022-00813-x PMID: 36466739

Liu, M., Ren, Y., Nyagoga, L. M., Stonier, F., Wu, Z., & Yu, L. (2023). Future of education in the era of generative artificial intelligence: Consensus among Chinese scholars on applications of ChatGPT in schools. *Future in Educational Research*, *1*(1), 72–101. DOI: 10.1002/fer3.10

Llinares, S. (2012). Construcción de conocimiento y desarrollo de una mirada profesional para la práctica de enseñar matemáticas en entornos en línea. [Building knowledge and developing a professional approach to teaching mathematics in online environments*]. Avances de Investigación en Educación Matemática*, *2*(2), 53–70. DOI: 10.35763/aiem.v1i2.18

Loeb Leadership. (2023, July 25). *The Importance of Adaptability in Successful Leadership*. Loeb Leadership. https://www.loebleadership.com/insights/importance-adaptability-successful-leadership

Logan, B. E., & Wimer, G. (2013). Tracing Inclusion: Determining Teacher Attitudes. *Research in Higher Education*, •••, 20.

Lokesh, B. (2024). Deciphering the science of decision making - insights from the brain. *Aster*. https://www.asterhospitals.in/blogs-events-news/aster-cmi-bangalore/deciphering-science-of-decision-making-insights-brain

Lortie, D. C. (1975). *Schoolteacher: A sociological study*. University of Chicago Press.

Louisiana Department of Education. (2019). *Bulletin 125—Standards for educational leaders in Louisiana*. Louisiana Administrative Code, Title 28, Part CXXXVII. https://www.doa.la.gov/media/glpah3zg/28v137.pdf

Love, M. L., Simpson, L. A., Golloher, A., Gadus, B., & Dorwin, J. (2020). Professional Development to Increase Teacher Capacity for the Use of New Technologies. *Intervention in School and Clinic*, *56*(2), 115–118. DOI: 10.1177/1053451220914886

Lucchetti, A., Duarte, B., Assis, T., Laurindo, B., & Lucchetti, G. (2019). Is it possible to teach geriatric medicine in a stimulating way? measuring the effect of active learning activities in Brazilian medical students. *Australasian Journal on Ageing*, *38*(2). Advance online publication. DOI: 10.1111/ajag.12626 PMID: 30761707

Lungu, M. (2019). Emotional intelligence and vocational schools. *Educa ia Plus*, *25*(2), 43–49.

Luo, X., Tong, S., Fang, Z., & Qu, Z. (2019). Machines versus humans: the impact of AI chatbot disclosure on customer purchases. *Nanyang Business School Research Paper, 20*(33), 1-30.

Luo, Y. (2023). Construction and application of an interdisciplinary team teaching model for foreign-related nursing English in higher vocational colleges from the perspective of ecolinguistics. *Journal of Contemporary Educational Research*, *7*(11), 203–208. DOI: 10.26689/jcer.v7i11.5666

Lyubomirsky, S., King, L., & Diener, E. (2005). The benefits of frequent positive affect: Does happiness lead to success? *Psychological Bulletin, 131*(6), 803–855. DOI: 10.1037/0033-2909.131.6.803 PMID: 16351326

Magazine, T. (2013, August 8). *Using Sports Analogies to Motivate Students*. Education for Today and Tomorrow | L'Education Aujourd'hui et Demain. https://teachmag.com/archives/6455

Magnusson, P. (2022). UNDERSTANDING L1 AND L2 TEACHERS' TALK ABOUT 'DIGITALISATION': Discourses during the process of integrating digital technology in the educational practice. *L1-Educational Studies in Language and Literature, 22*, 1–27. DOI: 10.21248/l1esll.2022.22.1.361

Mahindru, A., Patil, P., & Agrawal, V. (2023). Role of Physical Activity on Mental Health and Well-Being: A Review. *Cureus, 15*(1). Advance online publication. DOI: 10.7759/cureus.33475 PMID: 36756008

Mahmud, S., Begum, M., & Akhter, J. (2020). Knowledge and attitude of senior staff nurses regarding geriatrics health care in a tertiary public hospital. *Asian Journal of Medical and Biological Research, 6*(3), 431–439. DOI: 10.3329/ajmbr.v6i3.49791

Mahoney, S., Sladek, R., & Neild, T. (2016). A longitudinal study of empathy in pre-clinical and clinical medical students and clinical supervisors. *BMC Medical Education, 16*(1), 270. Advance online publication. DOI: 10.1186/s12909-016-0777-z PMID: 27756379

Maitlis, S., & Christianson, M. (2014). Sensemaking in organizations: Taking stock and moving forward. *The Academy of Management Annals, 8*(1), 57–125. DOI: 10.5465/19416520.2014.873177

Mai, Y. (2024). Unveiling the role of classroom climate in Chinese EFL teachers' perceptions of their emotional exhaustion and attrition. *Perceptual and Motor Skills, 132*(1), 169–194. DOI: 10.1177/00315125241291088 PMID: 39558200

Makowski, A., Hampton, D., Hardin-Pierce, M., Scott, L., Garritano, N., Tovar, E., & Melander, S. (2021). Adapting clinical experiences during a pandemic: One college of nursing's approach. *Journal of the American Association of Nurse Practitioners, 34*(2), 254–260. DOI: 10.1097/JXX.0000000000000656 PMID: 34537798

Malinen, O. P., Väisänen, P., & Savolainen, H. (2012). Teacher education in Finland: A review of a national effort for preparing teachers for the future. *Curriculum Journal, 23*(4), 567–584. DOI: 10.1080/09585176.2012.731011

Mandal, M. K. (2004). *Emotion: Basic issues and current trends*. Affiliated East-West Press.

Mariano, M. (2023). Assessment of learning style preferences of saudi nursing students. *Ibnosina Journal of Medicine and Biomedical Sciences*, 16(01), 010-016. DOI: 10.1055/s-0043-1776016

Markauskaite, L., Marrone, R., Poquet, O., Knight, S., Martinez-Maldonado, R., Howard, S., Tondeur, J., De Laat, M., Buckingham Shum, S., Gašević, D., & Siemens, G. (2022). Rethinking the entwinement between artificial intelligence and human learning: What capabilities do learners need for a world with AI? *Computers and Education: Artificial Intelligence, 3*, 100056. DOI: 10.1016/j.caeai.2022.100056

Marsden, E., Craswell, A., Taylor, A., Barnett, A., Wong, P., & Wallis, M. (2022). Translation of the geriatric emergency department intervention into other emergency departments: A post implementation evaluation of outcomes for older adults. *BMC Geriatrics*, *22*(1), 290. Advance online publication. DOI: 10.1186/s12877-022-02999-4 PMID: 35392828

Mart, M., & Kaya, G. (2024). Okul öncesi öğretmen adaylarının yapay zekâya yönelik tutumları ve yapay zekâ okur yazarlığı arasındaki ilişkinin incelenmesi [Investigating the relationship between pre-service preschool teachers' attitudes towards artificial intelligence and artificial intelligence literacy]. *Edutech Research*, *2*(1), 91–109.

Marzano, R. J., Warrick, P. B., Rains, C. L., & DuFour, R. (2018). *Leading a high reliability school: Use data-driven instruction and collaborative teaching strategies to boost academic achievement*. Solution Tree Press.

Marzano, R. J., Waters, T., & McNulty, B. A. (2005). *School leadership that works: From research to results*. Association for Supervision and Curriculum Development.

Ma, S., & Chen, Z. (2024). The development and validation of the artificial intelligence literacy scale for Chinese college students (AILS-CCS). *IEEE Access : Practical Innovations, Open Solutions*, *12*, 146419–146429. DOI: 10.1109/ACCESS.2024.3468378

Maslow, A. H. (1968). *Toward a psychology of being* (2nd ed.). Van Nostrand Reinhold.

Masnan, A., Ismail, N. A., Usop, W., & Zain, D. (2024). Predicting preschool moral development with school climate: Fuzzy Delphi approach. *International Journal of Evaluation and Research in Education*, *13*(3), 1537. DOI: 10.11591/ijere.v13i3.27777

Masten, A. S. (2001). Ordinary magic: Resilience processes in development. *The American Psychologist*, *56*(3), 227–238. DOI: 10.1037/0003-066X.56.3.227 PMID: 11315249

Masud, T., Blundell, A., & Stuck, A. (2022a). Geriatric medicine learning objectives and entrustable professional activities in undergraduate medical curricula: A scoping review. *Age and Ageing*, *51*(5), afac100. Advance online publication. DOI: 10.1093/ageing/afac100 PMID: 35536879

Masud, T., Ogliari, G., Lunt, E., Blundell, A., Gordon, A., Roller-Wirnsberger, R., Vassallo, M., Mari, D., Kotsani, M., Singler, K., Romero-Ortuno, R., Cruz-Jentoft, A. J., & Stuck, A. (2022b). A scoping review of the changing landscape of geriatric medicine in undergraduate medical education: Curricula, topics and teaching methods. *European Geriatric Medicine*, *13*(3), 513–528. DOI: 10.1007/s41999-021-00595-0 PMID: 34973151

Ma, W., Zhang, L., Wang, C., & Xu, X. (2022). Effect of multidisciplinary team collaborative nursing model combined with mind mapping teaching method on postoperative complications and mental health of patients with advanced pancreatic cancer. *Iranian Journal of Public Health*. Advance online publication. DOI: 10.18502/ijph.v51i3.8929 PMID: 35865063

Maxwell, J. A. (2005). *Qualitative research design: An interactive approach* (2nd ed.). Sage Publications.

Mayer, J. D., Salovey, P., & Caruso, D. R. (2008). Emotional intelligence: New ability or eclectic traits? https://cdn2.psychologytoday.com/assets/attachments/1575/rp2008-mayersaloveycarusob.pdf

Mayer, J. D., & Salovey, P. (1997). What is emotional intelligence? In Salovey, P., & Sluyter, D. (Eds.), *Emotional development and emotional intelligence: Educational implications* (pp. 3–31). Basic Books.

Mayer, J. D., Salovey, P., & Caruso, D. R. (2008). Emotional intelligence: New ability or eclectic traits? *The American Psychologist*, *63*(6), 503–517. DOI: 10.1037/0003-066X.63.6.503 PMID: 18793038

McCarthy, C., Lambert, R., Lineback, S., Fitchett, P., & Baddouh, P. (2016). Assessing teacher appraisals and stress in the classroom: Review of the Classroom Appraisal of Resources and Demands. *Educational Psychology Review*, *28*(3), 577–603. DOI: 10.1007/s10648-015-9322-6

McCarthy, J. (2007). From here to human-level AI. *Artificial Intelligence*, *171*(18), 1174–1182. DOI: 10.1016/j.artint.2007.10.009

McCarthy, J., Minsky, M. L., Rochester, N., & Shannon, C. E. (1995). A proposal for the Dartmouth summer research project on artifcial intelligence. *AI Magazine*.

Mclaughlin, M. W., & Marsh, D. D. (1978). Staff development and school change. *Teachers College Record*, *40*, 69–93.

McLean, L., & Connor, C. (2015). Depressive symptoms in third-grade teachers: Relations to classroom quality and student achievement. *Child Development*, *86*(3), 945–954. DOI: 10.1111/cdev.12344 PMID: 25676719

McLeod Saul. (2024). Vygotsky's theory of cognitive intelligence. Simple psychology. https://www.simplypsychology.org-Vygotsky-Theory-of-Cognitive-Development.pdf

Meade, M., Ogundamisi, R., & Tiratelli, M. (2023). Stay calm and educate. *ChangeMaker-funded project investigating SRI students' perspectives on generative AI.*https://www.ucl.ac.uk/teachinglearning/sites/teaching_learning/files/student_perspectives_on_ai_report.docx

Mehrling, P. (1999). The vision of Hyman P. Minsky. *Journal of Economic Behavior & Organization*, *39*(2), 129–158. DOI: 10.1016/S0167-2681(99)00029-3

Meiers, S., Eggenberger, S., & Krumwiede, N. (2018). Development and implementation of a family-focused undergraduate nursing curriculum: Minnesota state university, Mankato. *Journal of Family Nursing*, *24*(3), 307–344. DOI: 10.1177/1074840718787274 PMID: 30101655

Melash, V. D., Molodychenko, V. V., Huz, V. V., Varenychenko, A. B., & Kirsanova, S. S. (2020). Modernization of education programs and formation of digital competences of future primary school teachers. *International Journal of Higher Education*, *9*(7), 377–386. DOI: 10.5430/ijhe.v9n7p377

Melekoğlu, M. A. (2014). Special education today in Turkey. In *Special education international perspectives: Practices across the globe* (pp. 529–557). Emerald Group Publishing Limited. DOI: 10.1108/S0270-401320140000028024

Melekoglu, M. A., Cakiroglu, O., & Malmgren, K. W. (2009). Special education in Turkey. *International Journal of Inclusive Education*, *13*(3), 287–298. DOI: 10.1080/13603110701747769

Merriam, S. B., & Tisdell, E. J. (2015). Dealing with validity, reliability, and ethics. In *Qualitative research: A guide to design and implementation* (4th ed., pp. 237–267). Jossey-Bass.

Metelskaia, I., Ignatyeva, O., Denef, S., & Samsonowa, T. (2018). A business model template for AI solutions. *Proceedings of the international conference on intelligent science and technology* (pp. 35-41). NY: Association for Computing Machinery. DOI: 10.1145/3233740.3233750

Mezirow, J., & Taylor, E. W. (Eds.). (2009). *Transformative learning in practice: Insights from community, workplace, and higher education*. Jossey-Bass.

MGIEP UNESCO. (n.d). SEL for teachers - Course modules. https://mgiep.unesco.org/article/sel-for-teachers-course-modules# Niva Dolev & Shosh Leshem (2016): Developing emotional intelligence competence among teachers, Teacher development. DOI: DOI: 10.1080/13664530.2016.1207093

Miao, F., Holmes, W., Huang, R., & Zhang, H. (2021). *AI and education: A guidance for policymakers*. UNESCO Publishing.

Miguél, M., Gómara, A., Izagirre-Otaegi, A., Elguea-Díaz, F., Badiola-Etxaburu, I., & Gómez-Gastiasoro, A. (2022). Development, application and evaluation of an active learning methodology for health science students, oriented towards equity and cultural diversity in the treatment and care of geriatric patients. *International Journal of Environmental Research and Public Health*, *19*(21), 14573. DOI: 10.3390/ijerph192114573 PMID: 36361451

Miles, M. B., & Huberman, A. M. (1994). *Qualitative data analysis*. Sage Publications.

Miller, J. P. (2000). *Education and the soul: Toward a spiritual curriculum*. State University of New York Press.

Miller, R. (2007). *What are schools for? Holistic education in American culture* (3rd ed.). Holistic Education Press.

Minsky, M. (1988). *The society of mind*. Simon and Schuster.

Mitchell, H., Lucas, C., Charlton, K., & McMahon, A. (2018). Models of nutrition-focused continuing education programs for nurses: A systematic review of the evidence. *Australian Journal of Primary Health*, *24*(2), 101. DOI: 10.1071/PY17088 PMID: 29362024

Monsma, E. V. (2018). *Principles of Effective Goal Setting | Association for Applied Sport Psychology*. Appliedsportpsych.org. https://appliedsportpsych.org/resources/resources-for-athletes/principles-of-effective-goal-setting/

Morgan, T. L. (2014). *Understanding the leadership capacity and practice of assistant principals* (Publication No. 989) [Doctoral dissertation, University of Denver]. Digital Commons @ DU.

Morgan, T. L. (2018). Assistant principals' perceptions of the principalship. *International Journal of Education Policy and Leadership*, *13*(10), 1–20. DOI: 10.22230/ijepl.2018v13n10a743

Morse, J. M. (2015). Critical analysis of strategies for determining rigor in qualitative inquiry. *Qualitative Health Research*, *25*(9), 1212–1222. DOI: 10.1177/1049732315588501 PMID: 26184336

Mudin, N., Ahmad, A., & Kee, P. (2023). Association of stress on well-being among teachers: A systematic literature review. *International Journal of Academic Research in Progressive Education and Development*, *12*(3), 75–85. DOI: 10.6007/IJARPED/v12-i3/17936

Mukattil, N., & Pangandaman, H. (2024). Barriers and strategies in the implementation of virtual learning modalities amid covid-19 among students in nursing schools at sulu province, philippines. *Critical Medical and Surgical Nursing Journal, 13*(1), 20–25. DOI: 10.20473/cmsnj.v13i1.51366

Muñoz, J., & Letouzé, P. (2022). Some considerations on the principles of the cognitive theory of multimedia learning for instructional video design for the elderly. *Research Society and Development, 11*(10), e499111032333. DOI: 10.33448/rsd-v11i10.32333

Munson, J., & Saclarides, E. S. (2022). Getting a foot in the door: Examining content-focused coaches' strategies for gaining access to classrooms. *The Elementary School Journal, 123*(1), 128–154. DOI: 10.1086/720626

Muntinga, M., Leeuwen, K., Schellevis, F., Nijpels, G., & Jansen, A. (2015). From concept to content: Assessing the implementation fidelity of a chronic care model for frail, older people who live at home. *BMC Health Services Research, 15*(1), 18. Advance online publication. DOI: 10.1186/s12913-014-0662-6 PMID: 25608876

Mystakidis, S., & Christopoulos, A. (2022). Teacher Perceptions on Virtual Reality Escape Rooms for STEM Education. *Information (Basel), 13*(3), 1–13. DOI: 10.3390/info13030136

Nakrem, S., Stensvik, G., Skjong, R., & Ostaszkiewicz, J. (2019). Staff experiences with implementing a case conferencing care model in nursing homes: A focus group study. *BMC Health Services Research, 19*(1), 191. Advance online publication. DOI: 10.1186/s12913-019-4034-0 PMID: 30917815

Nascimento, J., Siqueira, T., Oliveira, J., Alves, M., Regino, D., & Dalrí, M. (2021). Development of clinical competence in nursing in simulation: The perspective of bloom's taxonomy. *Revista Brasileira de Enfermagem, 74*(1), e20200135. Advance online publication. DOI: 10.1590/0034-7167-2020-0135 PMID: 33787788

Nashwan, A., Waghmare, J., & Ladd, E. (2020). How will the covid-19 pandemic reshape nursing education globally? *Open Journal of Nursing, 10*(10), 973–976. DOI: 10.4236/ojn.2020.1010068

National Policy Board for Educational Administration. (2015). *Professional Standards for Educational Leaders 2015*.

Neuenschwander, R., Friedman-Krauss, A., Raver, C., & Blair, C. (2017). Teacher stress predicts child executive function: Moderation by school poverty. *Early Education and Development, 28*(7), 880–900. DOI: 10.1080/10409289.2017.1287993

Nguyen, A., Ngo, H. N., Hong, Y., Dang, B., & Nguyen, B. P. T. (2023). Ethical principles for artificial intelligence in education. *Education and Information Technologies, 28*(4), 4221–4241. DOI: 10.1007/s10639-022-11316-w PMID: 36254344

Nguyen, H. T., Tran, V.-T., Nguyen, S. T., & Van Trinh, T. (2023). A Literature Review on the Management of Preschool Teacher's Professional Competence Development in the 4.0 Industrial Revolution. *Journal of Law and Sustainable Development, 11*(9), 1–20. DOI: 10.55908/sdgs.v11i9.1060

Nguyen, L. A. T., & Habók, A. (2024). Tools for assessing teacher digital literacy: A review. *Journal of Computers in Education, 11*(1), 305–346. DOI: 10.1007/s40692-022-00257-5

Nieto, S. (2002). *Affirming diversity: The sociopolitical context of multicultural education* (3rd ed.). Longman.

Niss, M., & Højgaard, T. (2019). Mathematical competencies revisited. *Educational Studies in Mathematics, 102*(1), 9–28. DOI: 10.1007/s10649-019-09903-9

Nong, Y., Cui, J., He, Y., Zhang, P., & Zhang, T. (2024). Development and validation of an AI Literacy Scale. *Journal of Artificial Intelligence Research, 1*(1), 17–26. DOI: 10.70891/JAIR.2024.100029

Norfolk, A. (2021, May 7). The Mental Benefits Of Sport - News. Active Norfolk. https://www.activenorfolk.org/2021/05/mental-benefits-of-sport/

Norris, G., & Norris, H. (2021). Building Resilience Through Sport in Young People With Adverse Childhood Experiences. *Frontiers in Sports and Active Living, 3*, 663587. Advance online publication. DOI: 10.3389/fspor.2021.663587 PMID: 34337400

Northouse, P. G. (2016). *Leadership: Theory and practice* (7th ed.). Sage Publications.

Notarnicola, I., Duka, B., Lommi, M., Prendi, E., Cristofori, E., Mele, T., Ivziku, D., Rocco, G., & Stievano, A. (2024). Empowering nurse health education: Linguistic and cultural validation of the nurse health education competence instrument (NHECI) in the Italian context. *Health Care, 12*(14), 1445. DOI: 10.3390/healthcare12141445 PMID: 39057588

Nunneley, C., Fishman, M., Sundheim, K., Korus, R., Rosen, R., Streater, B., O'Donnell, K. A., Newman, L. R., & Marcus, C. (2020). Leading synchronous virtual teaching sessions. *The Clinical Teacher, 18*(3), 231–235. DOI: 10.1111/tct.13282 PMID: 33063455

Nurhasanah, N., Neviyarni, S., & Effendi, H. (2019). The effectiveness of group counseling with role-playing techniques to increase student empathy. *International Journal of Applied Counseling and Social Sciences, 1*(1), 54–61. DOI: 10.24036/005304ijaccs

Nurse-Clarke, N., & Sockol, L. (2022). An exploration of resiliency among nurse educators during the covid-19 pandemic. *Nursing Education Perspectives, 43*(5), 283–286. DOI: 10.1097/01.NEP.0000000000001014 PMID: 35947139

Nurtanto, M., Sudira, P., Sofyan, H., Kholifah, N., & Triyanto, T. (2022). Professional Identity of Vocational Teachers in the 21 Century in Indonesia. *Journal of Engineering Education Transformations, 35*(3), 30–36. DOI: 10.16920/jeet/2022/v35i3/22085

Nwana, H. S. (1990). Intelligent tutoring systems: An overview. *Artificial Intelligence Review, 4*(4), 251–277. DOI: 10.1007/BF00168958

Nwoko, J., Emeto, T., Malau-Aduli, A., & Malau-Aduli, B. (2023). A systematic review of the factors that influence teachers' occupational well-being. *International Journal of Environmental Research and Public Health, 20*(12), 1–20. DOI: 10.3390/ijerph20126070 PMID: 37372657

Odak, I., Marušić, I., Matić, J., Puzić, S., Bakić, H., Eliasson, N., & Veldin, M. (2023). *Teachers' social and emotional competencies: A lever for social and emotional learning in schools*. Sociologija i Prostor., DOI: 10.5673/sip.61.1.5

OECD. (2015). *Skills for social progress: The power of social and emotional skills.* OECD Publishing.

OECD. (2021). *Beyond academic learning: First results from the Survey on Social and Emotional Skills.* OECD Publishing.

Ohreen, D., Sundararajan, B., Trifts, V., & Comber, S. (2021). Vygotskian business ethics: The influence of peers on moral reasoning in business ethics education. *Journal of Management Education, 46*(1), 70–105. DOI: 10.1177/1052562921996019 PMID: 35013659

Oliver, M. (1990). The Politics of Disablement — New Social Movements. *In: The Politics of Disablement. Critical Texts in Social Work and the Welfare State*. Palgrave, London. https://doi.org/DOI: 10.1007/978-1-349-20895-1_8

Opoku, M. P., Cuskelly, M., Pedersen, S. J., & Rayner, C. S. (2021). Attitudes and self- efficacy as significant predictors of intention of secondary school teachers towards the implementation of inclusive education in Ghana. *European Journal of Psychology of Education, 36*(3), 1–19. DOI: 10.1007/s10212-020-00490-5

Ortíz, J., Argudo, U., López, D., & Mejía, S. (2023). Impact of an educational intervention for healthy eating in older adults: A quasi-experimental study. *International Journal of Environmental Research and Public Health, 20*(19), 6820. DOI: 10.3390/ijerph20196820 PMID: 37835089

Otero, I. Salgado J.F., & Moscoso Silvia (2022). Cognitive Reflection, cognitive intelligence, and cognitive abilities: A meta-analysis. *Intelligence. 90*, 101-614. . https://www.sciencedirect.com/science/article/pii/S0160289621000982DOI: 10.1016/j.intell.2021.101614

Ozturk, M., Wigelsworth, M., & Squires, G. (2024). A systematic review of primary school teachers' wellbeing: Room for a holistic approach. *Frontiers in Psychology, 15*, 1358424. Advance online publication. DOI: 10.3389/fpsyg.2024.1358424 PMID: 38915431

Pagán-Castaño, E., Sánchez-García, J., Garrigós-Simón, F., & Guijarro-García, M. (2021). The influence of management on teacher well-being and the development of sustainable schools. *Sustainability (Basel), 13*(5), 2909. DOI: 10.3390/su13052909

Palmer, P. J. (1998). *The courage to teach: Exploring the inner landscape of a teacher's life*. Jossey-Bass.

Palmer, P. J. (2000). *Let your life speak: Listening for the voice of vocation*. Jossey-Bass.

Pandian, R. (2023, December 24). Why sports needs to be integrated into the academic curriculum. *The Hindu*. https://www.thehindu.com/education/why-sports-needs-to-be-integrated-into-the-academic-curriculum/article67644620.ece

Parasuram, K. (2006). Variables that affect teachers' attitudes towards disability and inclusive education in Mumbai, India. *Disability & Society, 21*(3), 231–242. DOI: 10.1080/09687590600617352

Pargament, K. I. (1997). *The psychology of religion and coping: Theory, research, practice*. Guilford Press.

Pargament, K. I. (2011). *Spiritually integrated psychotherapy: Understanding and addressing the sacred.* Guilford Press.

Paris, D., & Alim, H. S. (2017). *Culturally sustaining pedagogies: Teaching and learning for justice in a changing world*. Teachers College Press.

Park, H., & Lim, S. (2022). Awareness of pain assessment for elderly patients with dementia among geriatric care hospital nurses. *Asia-Pacific Journal of Convergent Research Interchange*, *8*(1), 77–86. DOI: 10.47116/apjcri.2022.01.06

Patel, R. S. (2017). Teacher effectiveness in context to their emotional intelligence. *Voices of Research*, *5*(4), 1–5.

Pathak, V. N., Kaur, R., Jain, M., Chechi, V. K., & Pathak, A. (2024). Coping With Poor Well-Being Issues: Students in Higher Education Institutions. In Aloka, P. (Ed.), *Factors Impacting Student Well-Being and Coping Tactics* (pp. 389–422). IGI Global., DOI: 10.4018/979-8-3693-8860-0.ch016

Pavithra, J., Gangadhar, K. R., & Nanjappan, D. (2023). Evaluate the effectiveness of video assisted teaching on pranayama for reducing stress among geriatric clients at selected old age homes in rural area, Bengaluru. *International Journal of Advanced Psychiatric Nursing*, *5*(1), 73–86. DOI: 10.33545/26641348.2023.v5.i1a.113

Payne, S. C., & Huffman, A. H. (2005). A longitudinal examination of the influence of mentoring on organizational commitment and turnover. *Academy of Management Journal*, *48*(1), 158–168. DOI: 10.5465/amj.2005.15993166

Pekrun, R., & Linnenbrink-Garcia, L. (2014). *International handbook of emotions in education*. Routledge. DOI: 10.4324/9780203148211

Penttinen, V., Pakarinen, E., Von Suchodoletz, A., & Lerkkanen, M. (2020). Relations between kindergarten teachers' occupational well-being and the quality of teacher-child interactions. *Early Education and Development*, *31*(7), 994–1010. DOI: 10.1080/10409289.2020.1785265

Pérez López, S., Montes Peón, J. M., & Vázquez Ordás, C. J. (2004). Managing knowledge: The link between culture and organizational learning. *Journal of Knowledge Management*, *8*(6), 93–104. DOI: 10.1108/13673270410567657

Peters, M., Elasri-Ejjaberi, A., Martínez-Argüelles, M. J., & Fàbregues, S. (2022). Teacher digital competence development in higher education: Overview of systematic reviews. *Australasian Journal of Educational Technology*, *38*(3). Advance online publication. DOI: 10.14742/ajet.7543

Peterson, C., & Seligman, M. E. P. (2004). *Character strengths and virtues: A handbook and classification*. Oxford University Press.

Pham, S. T., & Sampson, P. M. (2022). The development of artificial intelligence in education: A review in context. *Journal of Computer Assisted Learning*, *38*(5), 1408–1421. DOI: 10.1111/jcal.12687

Pimentel, J. L. (2019). Some biases in Likert scaling usage and its correction. [IJSBAR]. *International Journal of Sciences, Basic and Applied Research*, *45*(1), 183–191.

Pombo, L., & Marques, M. M. (2021). Guidelines for teacher training in mobile augmented reality games: Hearing the teachers' voices. *Education Sciences*, *11*(10), 1–12. DOI: 10.3390/educsci11100597

Porter, G. L., & AuCoin, A. (2012). *Strengthening inclusion, strengthening schools: Report of the review of inclusive education programs and practices in New Brunswick Schools: An action plan for growth.* New Brunswick Department of Education and Early Childhood Development.

Presmeg, N. (2014). Contemplating visualization as an epistemological learning tool in mathematics. *ZDM Mathematics Education*, *46*(1), 151–157. DOI: 10.1007/s11858-013-0561-z

Preston, J. P., Jakubiec, B. A. E., & Kooymans, R. (2013). Common challenges faced by rural principals: A review of the literature. *Rural Educator*, *35*(1), 1–12. DOI: 10.35608/ruraled.v35i1.355

psico-smart.com. (2024). *Virtual Reality and Simulation in Leadership Training Programs*. Psico-Smart.com. https://psico-smart.com/en/blogs/blog-virtual-reality-and-simulation-in-leadership-training-programs-177956

Puspandari, L. E. (2023). Exploring pedagogical potentials of technology: Professional development for English teachers of vocational higher education. *Journal of Applied Studies in Language*, *7*(1), 107–116. DOI: 10.31940/jasl.v7i1.107-116

Qin, H. (2024). Stress level of classroom instructors and its influence on their classroom performance. *World Journal of Educational Research (Los Angeles, Calif.)*, *11*(1), p29. Advance online publication. DOI: 10.22158/wjer.v11n1p29

Quennerstedt, M. (2019). Physical education and the art of teaching: Transformative learning and teaching in physical education and sports pedagogy. *Sport Education and Society*, *24*(6), 611–623. DOI: 10.1080/13573322.2019.1574731

Rahiminia, E., Yazdani, S., & Rahiminia, H. (2020). Factors affecting concentration and attendance in the classroom from students' point of view in Qom University of Medical Sciences (2018). *Educational Research in Medical Sciences*, *8*(2). Advance online publication. DOI: 10.5812/erms.93075

Rakap, S., Cig, O., & Parlak-Rakap, A. (2017). Preparing preschool teacher candidates for inclusion: Impact of two special education courses on their perspectives. *Journal of Research in Special Educational Needs*, *17*(2), 98–109. DOI: 10.1111/1471-3802.12116

Rakap, S., & Kaczmarek, L. (2010). Teachers' attitudes towards inclusion in Turkey. *European Journal of Special Needs Education*, *25*(1), 59–75. DOI: 10.1080/08856250903450848

Ralebese, M. D., Jita, L., & Badmus, O. T. (2025). Examining primary school principals' instructional leadership practices: A case study on curriculum reform and implementation. *Education Sciences*, *15*(1), 70. Advance online publication. DOI: 10.3390/educsci15010070

Ramaila, S. (2022). Promoting self-regulated learning in natural sciences teaching through technology integration. https://doi.org/DOI: 10.36315/2022v1end081

Ramnarain, U., Ogegbo, A. A., Penn, M., Ojetunde, S., & Mdlalose, N. (2024). Pre-service science teachers' intention to use generative artificial intelligence in inquiry-based teaching. *Journal of Science Education and Technology*, •••, 1–14. DOI: 10.1007/s10956-024-10159-z

Rance, G., Dowell, R., & Tomlin, D. (2023). The effect of classroom environment on literacy development. *NPJ Science of Learning*, *8*(1), 9. Advance online publication. DOI: 10.1038/s41539-023-00157-y PMID: 37012296

Rashid, M., & Kabir, I. (2024). The role of sports in breaking down cultural barriers: A study. *International Journal of Physical Education. Sports Health*, *11*(4), 198–200. https://www.kheljournal.com/archives/2024/vol11issue4/PartD/11-4-28-407.pdf

Reis da Silva, T. (2024c). The Evolution of Nursing for Older Adult: A Historical Perspective. Associative J Health Sci. 3(3). AJHS. 000561. 2024. DOI: . Link: https://crimsonpublishers.com/ajhs/pdf/AJHS.000561.pdfDOI: 10.31031/AJHS.2024.03.000561

Reis da Silva, T. H (2024b). Oncology and Cancer Medicine: Understanding the complexities in Older Patients. Biomed J Sci & Tech Res 55(3)-2024. DOI: https://doi.org/10.26717/BJSTR.2024.55.008720

Reis da Silva, T. H. (2024d). Death and Its Significance in Nursing Practice. Palliat Med Care Int J. 2024; 4(3): 555640. DOI: https://juniperpublishers.com/pmcij/pdf/PMCIJ.MS.ID.555640.pdf

Reis da Silva, T. H. (2024g). Chapter 7 – Emotional Intelligence in Higher Education: Humanising Technology for Holistic Student Development. In Tikadar, S.; Liu, H.; Bhattacharya, P.; Bhattacharya, S. *Humanizing Technology With Emotional Intelligence*. Pp. 117-140. IGI Global Scientific Publishing. DOI: 10.4018/979-8-3693-7011-7.ch007

Reis da Silva, T. H. (2024h). Chapter 8 – Emotional Intelligence in Teaching Geriatric Nursing: Humanising Technology for Compassionate Care in Higher Education. In Tikadar, S.; Liu, H.; Bhattacharya, P.; Bhattacharya, S. *Humanizing Technology With Emotional Intelligence*. Pp. 141-160. IGI Global Scientific Publishing. DOI: 10.4018/979-8-3693-7011-7.ch008

Reis da Silva, T. H. (2024i). Chapter 12 – Emotional Intelligence in Integrative Medicine: Enhancing Patient-Centred Care Through Humanised Technology. In Tikadar, S.; Liu, H.; Bhattacharya, P.; Bhattacharya, S. *Humanizing Technology With Emotional Intelligence*. Pp. 263-280. IGI Global Scientific Publishing. DOI: 10.4018/979-8-3693-7011-7.ch012

Reis da Silva, T. H. (2023a). Ageing in place: Ageing at home and in the community. *British Journal of Community Nursing*, *28*(5), 213–214. DOI: 10.12968/bjcn.2023.28.5.213

Reis da Silva, T. H. (2023b). Falls assessment and prevention in the nursing home and community. *British Journal of Community Nursing*, *28*(2), 68–72. DOI: 10.12968/bjcn.2023.28.2.68

Reis da Silva, T. H. (2024a). Integrating Effective Strategies for Teaching Nursing Care for Older Adults into the Curriculum. *International Journal of Biomedical Research*, *3*(5). Advance online publication. DOI: 10.31579/2834-5029/063

Reis da Silva, T. H. (2024e). The Nursing Process in Geriatric Care: A Comprehensive Approach. *International Journal of Biomedical Research*, *3*(5). Advance online publication. DOI: 10.31579/2834-5029/064

Reis da Silva, T. H. (2024f). Enhancing Nursing Care Planning Through the Levett-Jones Clinical Reasoning Cycle: A Comprehensive Analysis. *International Journal of Biomedical Research*, *3*(5). Advance online publication. DOI: 10.31579/2834-5029/065

Reis da Silva, T. H. (2024j). Pharmacokinetics in older people: An overview of prescribing practice. *Journal of Prescribing Practice*, *6*(9), 374–381. DOI: 10.12968/jprp.2024.6.9.374

Reis da Silva, T. H. (2024k). Falls prevention in older people and the role of nursing. *British Journal of Community Nursing*, *29*(7), 335–339. DOI: 10.12968/bjcn.2024.0005 PMID: 38963269

Reis da Silva, T. H. (2024l). Chronic kidney disease in older adults: Nursing implications for community nurses. *Journal of Kidney Care*, *9*(4), 174–179. DOI: 10.12968/jokc.2024.9.4.174

Reis da Silva, T. H. (2025b). Promoting Emotional Well-Being in Older Adults: The Role of AI in Supporting Emotional Intelligence and Mental Health. In Jermsittiparsert, K., & Marzo, R. R. (Eds.), *Physical Health, Mental Health, and Human Well-Being in the Age of AI* (pp. 137–167). IGI Global Scientific Publishing., https://doi.org/10.4018/979-8-3693-6190-0.ch8

Reis da Silva, T. H., & Mitchell, A. (2024a). Integrating Digital Transformation in Nursing Education: Best Practices and Challenges in Curriculum Development. In Lytras, M., Serban, A. C., Alkhaldi, A., Malik, S., & Aldosemani, T. (Eds.), *Digital Transformation in Higher Education, Part B Cases, Examples and Good Practices*. Emerald Publishing Limited.

Reis Da Silva, T. H., & Mitchell, A. (2024b). Simulation in nursing: The importance of involving service users. *British Journal of Nursing (Mark Allen Publishing)*, *33*(5), 262–265. DOI: 10.12968/bjon.2024.33.5.262

Reis da Silva, T. M. (2025a). Fostering Emotional Intelligence for Competency Development in Nursing Education. In Aldosemani, T., Lytras, M., & Ordóñez de Pablos, P. (Eds.), *Innovative Educational Frameworks for Future Skills and Competencies* (pp. 39–62). IGI Global Scientific Publishing., DOI: 10.4018/979-8-3693-7555-6.ch002

Reis da Silva, T. M. (2025f). Gamification as a Pedagogical Tool in Geriatric Nursing: Engaging and Empowering Future Nurses. In Marcão, R., & Ribeiro Santos, V. (Eds.), *Enhancing Engagement With Gamification: Education, Business, and Healthcare Perspectives* (pp. 103–136). IGI Global Scientific Publishing., DOI: 10.4018/979-8-3693-8322-3.ch005

Reis da Silva, T. M. H. (2025d). The Role of Emotional Intelligence in Promoting Mental Peace, Healthy Work Environments, and Emotional Well-Being Among Entrepreneurs. In Tunio, M. N. (Ed.), *Supporting Psychological and Emotional Wellbeing Among Entrepreneurs*. IGI Global Scientific Publishing., DOI: 10.4018/979-8-3693-3673-1.ch21

Reis da Silva, T. M., Hammett, R., & Low, G. (2025). Emotional Intelligence in Business: Enhancing Leadership, Collaboration, and Performance. In Sedky, A. (Ed.), *Building Business Knowledge for Complex Modern Business Environments* (pp. 149–178). IGI Global Scientific Publishing., https://doi.org/10.4018/979-8-3693-6447-5.ch006

Reivich, K., & Shatté, A. (2002). *The resilience factor: 7 essential skills for overcoming life's inevitable obstacles*. Broadway Books.

Retelle, E. (2010). Mentoring experiences of first-year assistant principals. *EAF Journal*, *21*(2), 61–87.

Reyes, M. R., Brackett, M. A., Rivers, S. E., White, M., & Salovey, P. (2012). Classroom emotional climate, student engagement, and academic achievement. *Journal of Educational Psychology*, *104*(3), 700–712. DOI: 10.1037/a0027268

Richardson, J., & Milovidov, E. (with Europarat). (2019). *Digital citizenship education handbook: Being online, well-being online, rights online.* Council of Europe.

Riera, V., Moragas-Rovira, M., & Pujadas, X. (2024). Leadership development across the lifespan through sport and physical activity. *Journal of Management Development*, *43*(3), 336–356. Advance online publication. DOI: 10.1108/JMD-10-2022-0253

Rivers, S. E., Brackett, M. A., Reyes, M. R., Elbertson, N. A., & Salovey, P. (2012). Improving the social and emotional climate of classrooms: A clustered randomized controlled trial testing the RULER approach. *Prevention Science*, *14*(1), 77–87. DOI: 10.1007/s11121-012-0305-2 PMID: 23188089

Rodés, V., Porta, M., Garófalo, L., & Enríquez, C. R. (2021). Teacher education in the emergency: A mooc-inspired teacher professional development strategy grounded in critical digital pedagogy and pedagogy of care. *Journal of Interactive Media in Education*, *2021*(1), 1–14. DOI: 10.5334/jime.657

Rodríguez Macías, M., Abad Robles, M. T., & Giménez Fuentes-Guerra, F. J. (2021). Effects of Sport Teaching on Students' Enjoyment and Fun: A Systematic Review and Meta-Analysis. *Frontiers in Psychology*, *12*(708155), 708155. Advance online publication. DOI: 10.3389/fpsyg.2021.708155 PMID: 34421757

Roeser, R. W., Skinner, E., Beers, J., & Jennings, P. A. (2012). Mindfulness training and teachers' professional development: An emerging area of research and practice. *Child Development Perspectives*, *6*(2), 167–173. DOI: 10.1111/j.1750-8606.2012.00238.x

Rogers, C. R. (1969). *Freedom to learn: A view of what education might become*. Charles Merrill.

Roll, I., & Wylie, R. (2016). Evolution and revolution in artificial intelligence in education. *International Journal of Artificial Intelligence in Education*, *26*(2), 582–599. DOI: 10.1007/s40593-016-0110-3

Ross, C. M., Jamieson, L. M., & Mitchell, D. W. (2008). Cross-Cultural Understanding Through Youth Sports: Bridging the Tolerance Gap Through Youth Development. *Journal of Youth Development*, *3*(3), 154–163. DOI: 10.5195/jyd.2008.294

Ross, K., & Tolan, P. (2017). Social and emotional learning in adolescence: Testing the CASEL model in a normative sample. *The Journal of Early Adolescence*, *38*(8), 1170–1199. DOI: 10.1177/0272431617725198

Rowland, T., & Ruthven, K. (Eds.). (2011). *Mathematical knowledge in teaching*. Springer., DOI: 10.1007/978-90-481-9766-8

Rowston, K., Bower, M., & Woodcock, S. (2022). The impact of prior occupations and initial teacher education on post-graduate pre-service teachers' conceptualization and realization of technology integration. In *International Journal of Technology and Design Education* (Vol. 32, Issue 5). Springer Netherlands. DOI: 10.1007/s10798-021-09710-5

Rucinski, C. L., Brown, J. L., & Downer, J. T. (2018). Teacher–child relationships, classroom climate, and children's social-emotional and academic development. *Journal of Educational Psychology, 110*(7), 992–1004. DOI: 10.1037/edu0000240

Rudolph, J., Tan, S., & Tan, S. (2023). ChatGPT: Bullshit spewer or the end of traditional assessments in higher education?. *Journal of applied learning and teaching, 6*(1), 342-363.

Ruggeri, K., Garcia-Garzon, E., Maguire, Á., Matz, S., & Huppert, F. A. (2020). Well-being is more than happiness and life satisfaction: A multidimensional analysis of 21 countries. *Health and Quality of Life Outcomes, 18*(1), 192. Advance online publication. DOI: 10.1186/s12955-020-01423-y PMID: 32560725

Ruijs, N. M., & Peetsma, T. T. (2009). Effects of inclusion on students with and without special educational needs reviewed. *Educational Research Review, 4*(2), 67–79. DOI: 10.1016/j.edurev.2009.02.002

Rus, V., D'Mello, S., Hu, X., & Graesser, A. (2013). Recent advances in conversational intelligent tutoring systems. *AI Magazine, 34*(3), 42–54. DOI: 10.1609/aimag.v34i3.2485

Ryan, R. M., & Deci, E. L. (2000). Self-determination theory and the facilitation of intrinsic motivation, social development, and well-being. *The American Psychologist, 55*(1), 68–78. DOI: 10.1037/0003-066X.55.1.68 PMID: 11392867

Ryff, C. D., & Singer, B. (2000). Interpersonal flourishing: A positive health agenda for the new millennium. *Personality and Social Psychology Review, 4*(1), 30–44. DOI: 10.1207/S15327957PSPR0401_4

Saloviita, T. (2019). Explaining classroom teachers' attitudes towards inclusive education. *Support for Learning, 34*(4), 433–442. DOI: 10.1111/1467-9604.12277

Saloviita, T. (2020a). Attitudes of teachers towards inclusive education in Finland. *Scandinavian Journal of Educational Research, 64*(2), 270–282. DOI: 10.1080/00313831.2018.1541819

Saloviita, T. (2020b). Teachers' changing attitudes and preferences around inclusive education. *International Journal of Disability Development and Education*, •••, 1–18. DOI: 10.1080/1034912X.2020.1828569

Samuel, A.L. (2000). Some studies in machine learning using the game of checkers. *IBM J Res Dev.44(*1.2), 206–226.

Sánchez, V., & García, M. (2014). Socio-mathematical and mathematical norms related to definition in pre-service primary teachers' discourse. *Educational Studies in Mathematics, 85*(2), 305–320. DOI: 10.1007/s10649-013-9516-0

Saripella, A., Wasef, S., Nagappa, M., Riazi, S., Englesakis, M., Wong, J., & Chung, F. (2021). Effects of comprehensive geriatric care models on postoperative outcomes in geriatric surgical patients: A systematic review and meta-analysis. *BMC Anesthesiology, 21*(1), 127. Advance online publication. DOI: 10.1186/s12871-021-01337-2 PMID: 33888071

Sarı, T., Nayir, F., & Kahraman, Ü. (2020). A Study on Inclusive Education in Turkey. *Journal of Education and Future*, (18), 69–82. DOI: 10.30786/jef.642954

Savin-Baden, M., & Howell Major, C. (2012). *Qualitative research: The essential guide to theory and practice*. Routledge.

Saxena, S., & Saxena, S. (2012). Impact of cognitive and emotional intelligence on quality education. [IJMECS]. *International Journal of Modern Education and Computer Science*, *4*(10), 47–58. DOI: 10.5815/ijmecs.2012.10.07

Schatz-Oppenheimer, O., & Goldenberg, J. (2024). The impact of professional mentoring on mentors of novice teachers. *International Journal of Mentoring and Coaching in Education*, *13*(1), 122–141. DOI: 10.1108/IJMCE-08-2022-0070

Schon, D. A. (1987). *Educating the reflective practitioner*. Jossey-Bass.

Schonert-Reichl, K. A. (2017). Social and emotional learning and teachers. *The Future of Children*, *27*(1), 137–155. DOI: 10.1353/foc.2017.0007

Schrum, L. (1999). Technology professional development for teachers. *Educational Technology Research and Development*, *47*(4), 83–90. DOI: 10.1007/BF02299599

Schwab, K. (2017). *The Fourth Industrial Revolution*. Crown Business.

Sekreter, G. (2019). Emotional intelligence as a vital indicator of teacher effectiveness. *International Journal of Social Sciences & Educational Studies*, *5*(3), 286–302.

Seligman, M. E. P. (2011). *Flourish: A visionary new understanding of happiness and well-being*. Free Press.

Seligman, M. E. P., & Csikszentmihalyi, M. (2000). Positive psychology: An introduction. *The American Psychologist*, *55*(1), 5–14. DOI: 10.1037/0003-066X.55.1.5 PMID: 11392865

Semchuk, J., McCullough, S., Lever, N., Gotham, H., Gonzalez, J., & Hoover, S. (2022). Educator-informed development of a mental health literacy course for school staff: Classroom wellbeing information and strategies for educators (Classroom WISE). *International Journal of Environmental Research and Public Health*, *20*(1), 35. DOI: 10.3390/ijerph20010035 PMID: 36612359

Semmar, Y., & Al-Thani, T. (2015). Piagetian and Vygotskian approaches to cognitive development in the kindergarten classroom. *Journal of Educational and Developmental Psychology*, *5*(2). Advance online publication. DOI: 10.5539/jedp.v5n2p1

Shah, M. (2022). Teachers as reflective practitioners: From individualism to Vygotskian social constructivism. *The Alberta Journal of Educational Research*, *68*(3), 297–307. DOI: 10.55016/ojs/ajer.v68i3.68598

Shakir, A., Elue, C., Croft, S. J., & Clegorne, N. (2023). The elementary assistant principal: Exploring mentorship as a development pathway for the principal job role. *Georgia Educational Researcher*, *20*(2), 4. Advance online publication. DOI: 10.20429/ger.2023.200204

Shapiro, S. L., & Carlson, L. E. (2009). *The art and science of mindfulness: Integrating mindfulness into psychology and the helping professions*. American Psychological Association. DOI: 10.1037/11885-000

Sharma, M., & Rush, S. E. (2014). Mindfulness-based stress reduction as a stress management intervention for healthy individuals: A systematic review. *Journal of Evidence-Based Complementary & Alternative Medicine*, *19*(4), 271–286. DOI: 10.1177/2156587214543143 PMID: 25053754

Sharma, U., & Sokal, L. (2016). Can teachers' self-reported efficacy, concerns, and attitudes towards inclusion predict their actual inclusive classroom practices? *Australasian Journal of Special Education, 40*(1), 21–38. DOI: 10.1017/jse.2015.14

Shen, J., Wu, H., Reeves, P., Zheng, Y., Ryan, L., & Anderson, D. (2020). The association between teacher leadership and student achievement: A meta-analysis. *Educational Research Review*, *31*(100357), 1–19. DOI: 10.1016/j.edurev.2020.100357

Shen, Y. (2024). The application of the BOPPPS model in the ward rounds of nurses' standardized training in southwest China: A mixed methods study. *Frontiers in Medicine*, *11*, 1276652. Advance online publication. DOI: 10.3389/fmed.2024.1276652 PMID: 38994347

Shi, J. (2024). Stress level of full-time university classroom instructors affecting their classroom performance. *Journal of Education and Educational Research.*https://doi.org/DOI: 10.54097/q70x5r98

Shinohara, A. (2024). Learning outcomes from preventive home visits practicum in an undergraduate nursing program: a qualitative study. *Journal of International Nursing Research*, 3(1), e2022-0035-e2022-0035. DOI: 10.53044/jinr.2022-0035

Shin, S., Hong, E., Lee, I., & Kim, J. (2024). Effects of governmental support on clinical nurse educators in south Korea: A repeated cross-sectional study. *International Nursing Review*, *71*(4), 933–941. DOI: 10.1111/inr.12942 PMID: 38305725

Shodiyeva, M. (2022). Technologies for Continuous Professional Development of Primary Class Teachers. *The American Journal of Interdisciplinary Innovations and Research*, *04*(11), 40–48. DOI: 10.37547/tajiir/Volume04Issue11-07

Shor, I. (1992). *Empowering education: Critical teaching for social change*. University of Chicago Press. DOI: 10.7208/chicago/9780226147864.001.0001

Shulman, L. S. (1987). Knowledge and teaching: Foundations of the new reform. *Harvard Educational Review*, *57*(1), 1–22. DOI: 10.17763/haer.57.1.j463w79r56455411

Siddique, Ahsaan, Taseer, N. A., & Muhammad Siddique. (2020). Teachers' emotional intelligence and teaching effectiveness: A correlational study. *Ilkogretim Online*, *19*(3), 2411–2417.

Silva, C., Toriyama, A., Claro, H., Borghi, C., Castro, T., & Salvador, P. (2021). Covid-19 pandemic, emergency remote teaching and nursing now: challenges for nursing education. *Revista Gaúcha De Enfermagem*, 42(spe). https://doi.org/DOI: 10.1590/1983-1447.2021.20200248

Silva, D. Y., Gilbert, B., & Nolan, J. (2000). Sliding the doors: Locking and unlocking the possibilities for teacher leadership. *Teachers College Record*, *102*(4), 779–804. DOI: 10.1111/0161-4681.00077

Simran, A. (2024). Lifelong learning for lifelong impact: Professional development for teachers. https://www.suraasa.com/blog/teacher-professional-development?var=as1

Singh, I., & Jha, A. (2012). Teacher effectiveness in relation to emotional intelligence among medical and engineering faculty members. *Europe's Journal of Psychology*, *8*(4), 667–685. DOI: 10.5964/ejop.v8i4.483

Sinvani, L., Kozikowski, A., Pekmezaris, R., Akerman, M., & Wolf-Klein, G. (2016). Delirium: A survey of healthcare professionals' knowledge, beliefs, and practices. *Journal of the American Geriatrics Society*, *64*(12). Advance online publication. DOI: 10.1111/jgs.14544 PMID: 27879983

Siregar, T., Luali, N., Vinalistyosari, R., Hanurawan, F., & Anggraini, A. E. (2024). Implementation of Vygotsky's constructivism learning theory through project-based learning (PJBL) in elementary science education. *Al Qalam Jurnal Ilmiah Keagamaan Dan Kemasyarakatan*, *18*(4), 2586. DOI: 10.35931/aq.v18i4.3620

Siwatu, K. O. (2011). Preservice teachers' self-efficacy beliefs in implementing culturally responsive teaching techniques. *The Urban Review*, *43*(4), 467–487. DOI: 10.1007/s11256-010-0166-3

Skaalvik, E. M., & Skaalvik, S. (2010). Teacher self-efficacy and teacher burnout: A study of relations. *Teaching and Teacher Education*, *26*(4), 1059–1069. DOI: 10.1016/j.tate.2009.11.001

Skrtic, T. M. (1991). *Behind special education: A critical analysis of professional culture and school organization*. Love Publishing.

Smalley, S., Solomonson, J., & Schramm, K. (2020). The role of collaboration throughout the agricultural education student teaching experience. *Journal of Agricultural Education*, *61*(4), 329–342. DOI: 10.5032/jae.2020.04329

Smerek, R. E. (2011). Sensemaking and sensegiving: An exploratory study of the simultaneous 'being and learning' of new college and university presidents. *Journal of Leadership & Organizational Studies*, *18*(1), 80–94. DOI: 10.1177/1548051810384268

Smith, S. (2002). Teacher mentoring and collaboration. *Journal of Special Education Technology*, *17*(1), 47–48. DOI: 10.1177/016264340201700105

Smylie, M. A., Conley, S., & Marks, H. M. (2002). Exploring new approaches to teacher leadership for school improvement. *Yearbook of the National Society for the Study of Education*, *101*(1), 162–188. DOI: 10.1111/j.1744-7984.2002.tb00008.x

Social-Emotional Skills in School. (n.d.). Committee for Children. https://www.cfchildren.org/what-is-social-emotional-learning/schools/

Soldaat, L. (2019). More than a teacher: Understanding the teacher-learner relationship in a public high-school in South Africa. *Journal of Educational and Developmental Psychology*, *9*(1), 77. DOI: 10.5539/jedp.v9n1p77

Song, P., & Tang, W. (2019). The community-based integrated care system in Japan: Health care and nursing care challenges posed by super-aged society. *Bioscience Trends*, *13*(3), 279–281. DOI: 10.5582/bst.2019.01173 PMID: 31327797

Sonmez, N., Alptekin, S., & Bicak, B. 2019. Okul Öncesi Eğitim Öğretmenlerinin Kaynaştırma Eğitiminde Yeterliliklerini Artırmaya Yönelik Geliştirilen Hizmetiçi Eğitim Programının Etkisi [The Effect of the In-Service Training Program Developed to Increase the Preschool Education Teachers' Competence in Inclusive Education]. *İnönü Üniversitesi Eğitim Fakültesi Dergisi, 20*(2), 439-456.

Sorensen, L. C. (2019). “Big data” in educational administration: An application for predicting school dropout risk. *Educational Administration Quarterly, 55*(3), 404–446. DOI: 10.1177/0013161X18799439

Southerland, L., Hunold, K., Fossen, J., Caterino, J., Gulker, P., Bischof, J., & Mion, L. (2021). An implementation science approach to geriatric screening in an emergency department. *Journal of the American Geriatrics Society, 70*(1), 178–187. DOI: 10.1111/jgs.17481 PMID: 34580860

Souza, B. (2023). Regular schools as a straitjacket for zone of proximal development. *Perspectives in Education, 41*(4). Advance online publication. DOI: 10.38140/pie.v41i4.6880

Sperling, K., Stenberg, C. J., McGrath, C., Åkerfeldt, A., Heintz, F., & Stenliden, L. (2024). In search of artificial intelligence (AI) literacy in teacher education: A scoping review. *Computers and Education Open, 6*(100169), 1–13. DOI: 10.1016/j.caeo.2024.100169

Spilker, M., Prinsen, F., & Kalz, M. (2020). Valuing technology-enhanced academic conferences for continuing professional development. A systematic literature review. *Professional Development in Education, 46*(3), 482–499. DOI: 10.1080/19415257.2019.1629614

Sport Values In Every Classroom Teaching respect, equity and inclusion to 8-12 year-old students. (n.d.). https://www.icsspe.org/system/files/Sport%20Values%20in%20Every%20Classroom.pdf

Sports as a Tool for Social Inclusion and Diversity in Schools | Bring it on Sports. (2024, July 9). Bring It on Sports. https://www.bringitonsports.com.au/2024/07/09/sports-as-a-tool-for-social-inclusion-and-diversity-in-schools/

Sports Integration | Holistic Development | LIFE AND CAREER SKILLS. (2024). Dlps.co.in. https://www.dlps.co.in/sportsintegration.aspx

Stahnke, R., Schueler, S., & Roesken-Winter, B. (2016). Teachers' perception, interpretation, and decision-making: A systematic review of empirical mathematics education research. *ZDM Mathematics Education, 48*(1–2), 1–27. DOI: 10.1007/s11858-016-0775-y

Stake, R. E. (2010). *Qualitative research: Studying how things work*. Guilford Press.

States News Service. (2011, August 23). The Wallace Foundation launches major “Principal Pipeline” initiative to help school leaders.

States, N., Karsten, M., Nennig, H., & Cole, R. (2023). Let students work: Analysis of the role of differing facilitation on student engagement in a large stadium-style lecture hall. *Journal of Chemical Education, 100*(11), 4237–4248. DOI: 10.1021/acs.jchemed.3c00750 PMID: 38028750

Steger, M. F., Frazier, P., Oishi, S., & Kaler, M. (2006). The meaning in life questionnaire: Assessing the presence of and search for meaning in life. *Journal of Counseling Psychology, 53*(1), 80–93. DOI: 10.1037/0022-0167.53.1.80

Stembert, N., & Harbers, M. (2019). Accounting for the human when designing with AI: Challenges identified. CHI'19-Extended Abstracts. Glasgow. Scotland.

Stenberg, M., Bengtsson, M., Mangrio, E., & Carlson, E. (2024). Supporting each other towards independence: A narrative analysis of first-year nursing students' collaborative process. *Nursing Inquiry*, *31*(3), e12627. Advance online publication. DOI: 10.1111/nin.12627 PMID: 38351462

Stevanović, N., Hoare, E., McKenzie, V., & O'Connor, M. (2017). Sustaining the use of positive education coping skills to meet the challenges of the emerging adulthood period. *International Journal of Wellbeing*, *7*(3), 39–55. DOI: 10.5502/ijw.v7i3.640

Strahan, D., Geitner, M., & Lodico, M. (2010). Collaborative professional development toward literacy learning in a high school through connected coaching. *Teacher Development*, *14*(4), 519–532. DOI: 10.1080/13664530.2010.533493

Stress. (2022, June 17). *World Health Organization.*https://www.who.int/news-room/questions-and-answers/item/stress

Su, Y. (2023). Delving into EFL teachers' digital literacy and professional identity in the pandemic era: Technological Pedagogical Content Knowledge (TPACK) framework. In *Heliyon* (Vol. 9, Issue 6, pp. 1–7). DOI: 10.1016/j.heliyon.2023.e16361

Su, G. (2018). Unemployment in the AI Age. AI Matters 3(4). 35–43. Tarafdar. M., Beath. C. M., & Ross. J. W. (2019). Using AI to enhance business operations. *MIT Sloan Management Review*, *60*(4), 37–44.

Surbakti, A. M. (2023). *Adaptability and Resilience.* LinkedIn. Www.linkedin.com. https://www.linkedin.com/pulse/adaptability-resilience-ariel-matius-surbakti

Suri, S., & Sharma, A. (2024). Occupational stress and its effect on teaching performance among school teachers. *International Journal of Applied Educational Studies*, *13*(1), 34–45. DOI: 10.5281/zenodo.8237695

Surya, P. S., & Maheshbabu, N. (2023). Emotional Maturity and Resilience among Athletes. *International Journal of Indian Psychology*, *11*(3).

Sutton, R. E., & Wheatley, K. F. (2003). Teachers' Emotions and Teaching: A Review of the Literature and Directions for Future Research. *Educational Psychology Review*, *15*(4), 327–358. DOI: 10.1023/A:1026131715856

Switzer, J., & Schroder, L. (2019). Mobile outreach: An innovative program for older orthopedic patients in care facilities. *Geriatric Orthopaedic Surgery & Rehabilitation*, *10*, 2151459319826476. Advance online publication. DOI: 10.1177/2151459319826476 PMID: 30886761

Tabachnick, B. G., & Fidell, L. S. (2003). *Using multivariate statistics.* Pearson.

Takala, M., Pirttimaa, R., & Törmänen, M. (2009). Inclusive special education: The role of special education teachers in Finland. *British Journal of Special Education*, *36*(3), 162–173. DOI: 10.1111/j.1467-8578.2009.00432.x

Tanaka, F., & Matsuzoe, S. (2012). Children teach a care-receiving robot to promote their learning: Field experiments in a classroom for vocabulary learning. *Journal of Human-Robot Interaction*, *1*(1), 78–95. DOI: 10.5898/JHRI.1.1.Tanaka

Tess, H. (2024). Cognitive intelligence vs emotional Intelligence: Key differences. https://orilearning.com/cognitive-intelligence-vs-emotional-intelligence/#:~:text=IQ%20measures%20abilities%20such%20as,effectively%20in%20oneself%20and%20others

The Importance of Resilience in Leadership: Thrive in Times of Change. (n.d.). Www.utsa.edu. https://www.utsa.edu/pace/news/the-importance-of-resilience-in-leadership-thrive-in-times-of-change.html

The Power of Emotional Intelligence in Sports | Telecoming | MONETIZATION TECHNOLOGY. (2024, May 20). Telecoming | MONETIZATION TECHNOLOGY | Tech for Business Digital Growth. https://www.telecoming.com/blog/the-power-of-emotional-intelligence-in-sports/

Theodorio, A. O. (2024). Examining the support required by educators for successful technology integration in teacher professional development program. *Cogent Education*, *11*(1), 1–18. DOI: 10.1080/2331186X.2023.2298607

Thompson, S., Metcalfe, K., Boncey, K., Merriman, C., Flynn, L., Alg, G., Bothwell, H., Forde-Johnston, C., Puffett, E., Hardy, C., Wright, L., & Beale, J. (2020). Interprofessional education in geriatric medicine: Towards best practice. a controlled before–after study of medical and nursing students. *BMJ Open*, *10*(1), e018041. DOI: 10.1136/bmjopen-2017-018041 PMID: 31964659

Thought, in. (2024). *Effective Decision Making: Adaptive Thinking: Flexibility in Thought: Adaptive Thinking for Effective Decision Making - FasterCapital*. FasterCapital. https://fastercapital.com/content/Effective-Decision-Making--Adaptive-Thinking---Flexibility-in-Thought--Adaptive-Thinking-for-Effective-Decision-Making.html

Timms, M. J. (2016). Letting artificial intelligence in education out of the box: Educational cobots and smart classrooms. *International Journal of Artificial Intelligence in Education*, *26*(2), 701–712. DOI: 10.1007/s40593-016-0095-y

Tisdell, E. J. (2003). *Exploring spirituality and culture in adult and higher education*. John Wiley & Sons.

Tomlinson, C. A., & Imbeau, M. B. (2010). *Leading and managing a differentiated classroom*. ASCD.

Training the cognitive aspects of agility in team sports. (n.d.). Sportsmith. https://www.sportsmith.co/articles/training-cognitive-aspects-of-agility-in-team-sports/

Tsai, S. C., Chen, C. H., Shiao, Y. T., Ciou, J. S., & Wu, T. N. (2020). Precision education with statistical learning and deep learning: A case study in Taiwan. *International Journal of Educational Technology in Higher Education*, *17*(1), 1–13. DOI: 10.1186/s41239-020-00186-2

Tschannen-Moran, M., & Hoy, A. W. (2001). Teacher efficacy: Capturing an elusive construct. *Teaching and Teacher Education*, *17*(7), 783–805. DOI: 10.1016/S0742-051X(01)00036-1

Tulsulkar, G., Mishra, N., Thalmann, N. M., Lim, H. E., Lee, M. P., & Cheng, S. K. (2021). Can a humanoid social robot stimulate the interactivity of cognitively impaired elderly? A thorough study based on computer vision methods. *The Visual Computer*, *37*(12), 3019–3038. DOI: 10.1007/s00371-021-02242-y PMID: 34345091

Turing, A. M. (1950). Computing machinery and intelligence. *Mind*, *59*(236), 433–460. DOI: 10.1093/mind/LIX.236.433

Turnbull, B. J., Anderson, L. M., Riley, D. L., MacFarlane, J. R., & Aladjem, D. K. (2016). *Building a stronger principalship, Volume 5: The Principal Pipeline Initiative in action*. Policy Studies Associates, Inc. DOI: 10.59656/EL-LS5394.001

Turnbull, B. J., Riley, D. L., & MacFarlane, J. R. (2015). *Building a stronger principalship, Volume 3: Districts taking charge of the principal pipeline*. Policy Studies Associates, Inc.

Uka, E. (2024). Exploring differences in Primary School Teachers' Attitudes Towards Inclusive Education in Kosovo. *European Journal of Special Needs Education*, •••, 1–10. DOI: 10.1080/08856257.2024.2334537

Ulliance. (2023). Age Bias at Work? Discover What You Need to Know... *Ulliance.com*. https://doi.org/ DOI: 109050482603/1735937323578

Unal, A., & Unal, Z. (2024). Evaluating the integration of artificial intelligence (AI) in K-12 education. *Journal of Interactive Learning Research, 35*(3), 353–387.

UNESCO. (1994). *The Salamanca Statement and Framework for Action on Special Needs Education*. UNESCO.

United Nations. (2006). *Convention on the Rights of Persons with Disabilities (UNCRPD)*. Retrieved from https://www.un.org/development/desa/disabilities/convention-on-the-rights-of-persons-with-disabilities.html

van der Rijst, R., Baggen, Y., & Sjoer, E. (2019). University teachers' learning paths during technological innovation in education. *The International Journal for Academic Development, 24*(1), 7–20. Advance online publication. DOI: 10.1080/1360144X.2018.1500916

Van der Veer, R., & Valsiner, J. (1991). *Understanding Vygotsky: A quest for synthesis*. Blackwell Publishing.

van Manen, M. (2016). *Phenomenology of practice: Meaning-giving methods in phenomenological research and writing*. Routledge. DOI: 10.4324/9781315422657

Vangrieken, K., & Kyndt, E. (2020). The teacher as an island? A mixed method study on the relationship between autonomy and collaboration. *European Journal of Psychology of Education, 35*(1), 177–204. DOI: 10.1007/s10212-019-00420-0

VanLehn, K. (2011). The relative effectiveness of human tutoring, intelligent tutoring systems, and other tutoring systems. *Educational Psychologist, 46*(4), 197–221. DOI: 10.1080/00461520.2011.611369

Vázquez-Sánchez, M., Casals-Vázquez, A., Sánchez-Ojeda, M., García-Gámez, M., Papadopoulos, I., & Casals, C. (2022). Spanish adaptation and validation of the cultural competence assessment tool (CCATOOL) for undergraduate nursing students. *International Nursing Review, 70*(1), 43–49. DOI: 10.1111/inr.12792 PMID: 35879822

Vaz, S., Wilson, N., Falkmer, M., Sim, A., Scott, M., Cordier, R., & Falkmer, T. (2015). Factors associated with primary school teachers' attitudes towards the inclusion of students with disabilities. *PLoS One, 10*(8), e0137002. DOI: 10.1371/journal.pone.0137002 PMID: 26317862

Verenna, A., Noble, K., Pearson, H., & Miller, S. (2018). Role of comprehension on performance at higher levels of bloom's taxonomy: Findings from assessments of healthcare professional students. *Anatomical Sciences Education*, *11*(5), 433–444. DOI: 10.1002/ase.1768 PMID: 29346708

Villegas, A. M., & Lucas, T. (2002). Preparing culturally responsive teachers: Rethinking the curriculum. *Journal of Teacher Education*, *53*(1), 20–32. DOI: 10.1177/0022487102053001003

von Moltke, N. (2024, October 22). *5 Leadership Development Types For Training Leaders (Plus HR Best Practices)*. AIHR. https://www.aihr.com/blog/leadership-development/

Voorend, C., Joosten, H., Berkhout-Byrne, N., Diepenbroek, A., Franssen, C., Bos, W., Van Buren, M., Mooijaart, S. P., van Alphen, A., Berkhout-Byrne, N., van Breda, F., van Buren, M., Boom, H., Bos, W. J., Diepenbroek, A., Emmelot-Vonk, M., Franssen, C., Gaillard, C. A. J. M., Groeneweg, N., & Stehouwer, C. (2021). Design of a consensus-based geriatric assessment tailored for older chronic kidney disease patients: Results of a pragmatic approach. *European Geriatric Medicine*, *12*(5), 931–942. DOI: 10.1007/s41999-021-00498-0 PMID: 33871790

Vygotsky, L. S. (1978). *Mind in society: The development of higher psychological processes*. Harvard University Press.

Wagner, C., Kawulich, B., & Garner, M. (2019). A Mixed Research Synthesis of Literature on Teaching Qualitative Research Methods. *SAGE Open*, *9*(3), 2158244019861488. Advance online publication. DOI: 10.1177/2158244019861488

Wang, B., Rau, P. L. P., & Yuan, T. (2023). Measuring user competence in using artificial intelligence:validity and reliability of artificial intelligence literacy scale. *Behaviour & Information Technology*, *42*(9), 1324–1337. DOI: 10.1080/0144929X.2022.2072768

Wang, Y. (2019). Is data-driven decision making at odds with moral decision making? A critical review of school leaders' decision making in the era of school accountability. *Values and Ethics in Educational Administration*, *14*(2), 1–9.

Wang, Y. (2021a). When artificial intelligence meets educational leaders' data-informed decision-making: A cautionary tale. *Studies in Educational Evaluation*, *69*, 100872. DOI: 10.1016/j.stueduc.2020.100872

Wang, Y. (2021b). Artificial intelligence in educational leadership: A symbiotic role of human-artificial intelligence decision-making. *Journal of Educational Administration*, *59*(3), 256–270. DOI: 10.1108/JEA-10-2020-0216

Wang, Y. (2023). Artificial Intelligence Technologies in College English Translation Teaching. *Journal of Psycholinguistic Research*, *52*(5), 1525–1544. DOI: 10.1007/s10936-023-09960-5 PMID: 37100967

Watson, A., & Ohtani, M. (Eds.). (2015). *Task design in mathematics education*. Springer., DOI: 10.1007/978-3-319-09629-2

Weick, K. E., Sutcliffe, K. M., & Obstfeld, D. (2005). Organizing and the process of sensemaking. *Organization Science*, *16*(4), 409–421. DOI: 10.1287/orsc.1050.0133

Weick, K. E., & Weick, K. E. (1995). *Sensemaking in organizations* (Vol. 3). Sage publications.

Wettstein, A., Schneider, S., Holtforth, M. G., & La Marca, R. (2021). Teacher stress: A psychobiological approach to stressful interactions in the classroom. *Frontiers in Education*, *6*, 681258. Advance online publication. DOI: 10.3389/feduc.2021.681258

What are the most innovative and cutting-edge mobile learning apps for professional growth? (2024). Linkedin.com. https://www.linkedin.com/advice/1/what-most-innovative-cutting-edge-mobile-learning-frm8c

Wilson, J., Hensley, A., Culp-Roche, A., Hampton, D., Hardin-Fanning, F., & Thaxton-Wiggins, A. (2021). Transitioning to teaching online during the covid-19 pandemic. *SAGE Open Nursing*, *7*, 23779608211026137. Advance online publication. DOI: 10.1177/23779608211026137 PMID: 34222655

Wilson, T. A., & Clayton, J. K. (2020). Novice assistant principals' perceptions of professional learning experiences. *Education Leadership Review of Doctoral Research*, *8*, 20–37.

Winzer, M. A. (1993). *The history of special education: From isolation to integration*. Gallaudet University Press.

Wissanji, T., Forget, M., Muscedere, J., Beaudin, D., Coveney, R., & Wang, H. (2022). Models of care in geriatric intensive care—A scoping review on the optimal structure of care for critically ill older adults admitted in an ICU. *Critical Care Explorations*, *4*(4), e0661. DOI: 10.1097/CCE.0000000000000661 PMID: 35382113

Wong, P. T. P. (2012). Toward a dual-systems model of what makes life worth living. In Wong, P. T. P. (Ed.), *The Human Quest for Meaning* (2nd ed., pp. 3–22). Routledge.

Wood, D., Bruner, J., & Ross, G. (1976). The role of tutoring in problem solving. *Journal of Child Psychology and Psychiatry, and Allied Disciplines*, *17*(2), 89–100. DOI: 10.1111/j.1469-7610.1976.tb00381.x PMID: 932126

Wu, M., Chen, Z., Xu, Y., Zhao, L., Zhao, L., & Xia, L. (2024). A qualitative study of geriatric specialist nurses' experiences to navigate delirium in the elderly. *BMC Nursing*, *23*(1), 426. Advance online publication. DOI: 10.1186/s12912-024-02100-x PMID: 38918789

Wynn, M., Reis da Silva, T. H., & Pearson-Jenkins, J. (2025). Introduction to digital nursing and nursing theory. In Wynn, M. (Ed.), *Digital Nursing: Shaping Practice and Identity in the Age of Informatics*. Routledge. DOI: 10.4324/9781032714547-1

Xie, K., Nelson, M. J., Cheng, S. L., & Jiang, Z. (2023). Examining changes in teachers' perceptions of external and internal barriers in their integration of educational digital resources in K-12 classrooms. *Journal of Research on Technology in Education*, *55*(2), 281–306. DOI: 10.1080/15391523.2021.1951404

Yang, S. J. H., Ogata, H., Matsui, T., & Chen, N.-S. (2021). Human-centered artificial intelligence in education: Seeing the invisible through the visible. *Computers and Education: Artificial Intelligence*, *2*, 100008. DOI: 10.1016/j.caeai.2021.100008

Yang, W., Liu, W., & Liu, X. (2023). Work-related stress among teachers: Interventions and effectiveness. *Educational Psychology*, *43*(4), 783–799. DOI: 10.1080/01443410.2022.2046752

Yellowbrick. (2021, December 24). *How Sports Can Bridge Cultural and Social Differences - Yellowbrick*. Yellowbrick. https://www.yellowbrick.co/blog/sports/how-sports-can-bridge-cultural-and-social-differences

Yeni, G., & Can, T. (2022). The Effect of 21st Century Skills Training on Foreign Language Teachers' Perceptions Regarding Their Educational Technology and Materials Development Competencies. *Bartın Üniversitesi Eğitim Fakültesi Dergisi, 11*(1), 118–136. DOI: 10.14686/buefad.777974

Yıldırım, A., & Şimşek, H. (2021). *Sosyal bilimlerde nitel araştırma yöntemleri* [Qualitative research methods in the social sciences]. 11th ed.). Seçkin Yayıncılık.

Yoffi, N. (2023). The role of team sports in fostering inclusivity and empathy among diverse student groups. *Cross-Cultural Management Journal, XXV*(2). https://seaopenresearch.eu/Journals/articles/CMJ2023_I2_1.pdf

York-Barr, J., & Duke, K. (2004). What do we know about teacher leadership? Findings from two decades of scholarship. *Review of Educational Research, 74*(3), 255–316. DOI: 10.3102/00346543074003255

Yureva, O. V., Burganova, L. A., Kukushkina, O. Y., Myagkov, G. P., & Syradoev, D. V. (2020). Digital transformation and its risks in higher education: Students' and teachers' attitude. *Universal Journal of Educational Research, 8*(11 B), 5965–5971. DOI: 10.13189/ujer.2020.082232

Zajonc, A. (2013). *Meditation as contemplative inquiry: When knowing becomes love*. Lindisfarne Books.

Zaki, J., & Cikara, M. (2015). Addressing empathic failures. *Current Directions in Psychological Science, 24*(6), 471–476. DOI: 10.1177/0963721415599978

Zambrotta, M., Aylward, P., Roy, C., Piper-Vallillo, E., Pelletier, S., Honan, J., Heller, N., Ramani, S., & Shields, H. (2021). Nurse-doctor co-teaching: A pilot study of the design, development, and implementation of structured interprofessional co-teaching sessions. *Advances in Medical Education and Practice, 12*, 339–348. DOI: 10.2147/AMEP.S300231 PMID: 33889044

Zaslow, M., & Martinez-Beck, I. (2006). *Self-efficacy and teachers' beliefs about their ability to teach diverse students*. Teachers College Press.

Zdravkova, K., Krasniqi, V., Dalipi, F., & Ferati, M. (2022). Cutting-edge communication and learning assistive technologies for disabled children: An artificial intelligence perspective. *Frontiers in Artificial Intelligence, 5*, 970430. DOI: 10.3389/frai.2022.970430 PMID: 36388402

Zembylas, M., & Papanastasiou, E. (2006). Sources of teacher job satisfaction and dissatisfaction in Cyprus. *Compare: A Journal of Comparative Education, 36*(2), 229–247. DOI: 10.1080/03057920600741289

Zhang, W. (2022). The Role of Technology-Based Education and Teacher Professional Development in English as a Foreign Language Classes. In *Frontiers in Psychology* (Vol. 13, pp. 1–7). DOI: 10.3389/fpsyg.2022.910315

Zhang, C., Schießl, J., Plößl, L., Hofmann, F., & Gläser-Zikuda, M. (2023). Acceptance of artificial intelligence among pre-service teachers: A multigroup analysis. *International Journal of Educational Technology in Higher Education, 20*(49), 1–22. DOI: 10.1186/s41239-023-00420-7

Zhang, L., & Zhang, B. A. (1999). Geometrical representation of McCulloch-Pitts neural model and its applications. *IEEE Transactions on Neural Networks*, *10*(4), 925–949. DOI: 10.1109/72.774263 PMID: 18252589

Zhao, Y., & Li, X. (2023). Mindfulness and teacher resilience: An intervention study in China. *International Journal of Psychology*, *58*(2), 229–241. DOI: 10.1002/ijop.12956

Zhong, X., Liu, X., & Sheng, Y. (2021). The effect of the humanistic care teaching model on nurse patient conflict and nurse turnover intention in a pediatric outpatient department: Results of a randomized trial. *Translational Pediatrics*, *10*(8), 2016–2023. DOI: 10.21037/tp-21-214 PMID: 34584871

Zhou, D. (2024). The influence of smart classrooms on students' learning experience in the digital age. *Advances in Educational Technology and Psychology*, *8*(2). Advance online publication. DOI: 10.23977/aetp.2024.080210

Zhou, X., Padron, Y., & Waxman, H. (2022). Exploring the Relationship Between Professional Development Experience and Skills in Educational Technology Integration Among Primary EFL Teacher. *Contemporary Educational Technology*, *14*(1), 1–13. DOI: 10.30935/cedtech/11365

Zins, J. E., Weissberg, R. P., Wang, M. C., & Walberg, H. J. (2004). *Building academic success on social and emotional learning: What does the research say?* Teachers College Press.

Zins, J. E., Weissberg, R. P., Wang, M. C., & Walberg, H. J. (Eds.). (2004). *Building Academic Success on Social and Emotional Learning: What Does the Research Say?* Teachers College Press.

Zohuri, B., & Mossavar-Rahmani, F. (2024). Revolutionizing education: The dynamic synergy of personalized learning and artificial intelligence. *International Journal of Advanced Engineering and Management Research*, *9*(1), 143–153. DOI: 10.51505/ijaemr.2024.9111

Zúñiga, F., Geest, S., Guerbaaï, R., Basinska, K., Nicca, D., Kressig, R., & Simon, M. (2019). Strengthening geriatric expertise in Swiss nursing homes: InterCare implementation study protocol. *Journal of the American Geriatrics Society*, *67*(10), 2145–2150. DOI: 10.1111/jgs.16074 PMID: 31317544

Zúñiga, F., Guerbaaï, R., Geest, S., Popejoy, L., Bartáková, J., Denhaerynck, K., & Simon, M. (2022). Positive effect of the InterCare nurse-led model on reducing nursing home transfers: A nonrandomized stepped-wedge design. *Journal of the American Geriatrics Society*, *70*(5), 1546–1557. DOI: 10.1111/jgs.17677 PMID: 35122238

About the Contributors

Bahaddin Demirdiş is an Assistant Professor at Çankırı Karatekin University. He earned his PhD in Educational Psychology from University College London in 2021. He also holds a Master's degree in Educational Psychology from University College London, completed in 2015, and a Bachelor's degree from Ondokuz Mayıs University in 2012.

Alper Uslukaya is an Assistant Professor in the Department of Educational Administration, Supervision, Planning, and Economics at Çankırı Karatekin University. He earned his PhD in Educational Administration, Supervision, Planning, and Economics from Fırat University in 2023. He also holds a Master's degree in Philosophy and Religious Studies from Fırat University, awarded in 2018, and a Bachelor's degree from Çukurova University, completed in 2005. Dr. Uslukaya has made significant academic contributions through his research on teacher professionalism, educational leadership, presenteeism, and work engagement in schools. His work has been published in numerous international refereed journals, where he has explored topics such as the relationship between school leadership and teacher well-being, teacher burnout, and the impact of social support in educational settings. His research uses advanced methodologies such as multilevel moderated mediation analyses, which contribute to the understanding of complex dynamics within educational institutions. In addition to his academic career, Dr. Uslukaya has extensive experience as a teacher and school administrator within Turkey's Ministry of National Education, where he worked for over 15 years before joining academia. His practical experience in educational leadership enriches his research and teaching.

Mehmet Tufan Yalçın is an Associate Professor in the Department of Educational Administration, Supervision, Planning, and Economics at Çankırı Karatekin University. He holds a PhD in Educational Administration and Supervision from Gazi University, awarded in 2018. He also completed his Master's degree in the same field at Gazi University in 2013, following his Bachelor's degree from Dokuz Eylul University in 2007. Dr. Yalçın has made significant contributions to the field of educational leadership, teacher professionalism, and instructional practices. He has published numerous articles in internationally refereed journals, focusing on topics such as the impact of school leadership on student achievement, social capital in education, and instructional capacity in schools. His research often explores the dynamics of leadership and its role in educational improvement and innovation. In addition to his academic publications, Dr. Yalçın has co-authored book chapters and presented papers at prestigious international conferences. His work emphasizes the role of distributed leadership and organizational innovation in enhancing professional development and teacher collaboration. Dr. Yalçın's research continues to shape the understanding of educational leadership and management, making him a key contributor to the field of educational sciences.

***

Monisha A K is working as a Assistant Professor at Department of Psychology, in Jain University School of Allied Healthcare & Sciences Bangalore. Currently she is pursuing her Ph.D in Psychology.

Veysel Altunel is a Ph.D. student and Graduate Research Assistant in Curriculum & Instruction: English Education at LSU's Lutrill & Pearl Payne School of Education. He is interested in Learner Autonomy, Data-Driven Learning, Individual Differences, and Educational Technologies. Veysel's research explores innovative approaches to enhance education, fostering personalized and technology-driven instruction.

Elizabeth Kardze Annan is a Ghanaian final-year Clinical Psychology master's degree international student at Lovely Professional University, India. She graduated from the University of Ghana in Legon, Ghana, with a Bachelor of Science in Family and Consumer Sciences. Being able to make a difference in people's lives is her ultimate life aim. She became interested in psychology in order to learn more about human behavior and mental processes. Her hobby lies in engaging in sporting activities especially soccer, and served as an assistant sports prefect in both junior high school and senior high school. Since teachers are important members of society who can influence a child's development, it is imperative that they have certain skills and competences to meet the needs of their students. For this reason, she wrote a chapter in this book titled "Holistic Approaches to Teacher Development: Leadership, Pedagogical Practices, and Cognitive Insights" to draw the awareness of investing in teacher development for the betterment of students' learning and achievements, and teachers' job fulfillment and satisfaction.

Ramazan Atasoy. He received his MA in Girne American University and Ph.D. in educational sciences from Gazi University. He currently works as an assistant professor at Harran University. His research interests include leadership, quality of education, education policy, PIAAC adult competency, literacy skills, numeracy skills and problem solving skills in technology rich environments.

Abdullah Ciftci is a lecturer at Akdeniz University. He earned an MA in Special Education from the University of Nottingham and a PhD in Special Education from University College London.

Ömür Çoban worked as an English teacher in public schools affiliated to the Ministry of National Education between 1999-2018. At the beginning of 2018, he started to work as a doctoral faculty member at Karamanoğlu Mehmetbey University, Faculty of Education, Department of Educational Sciences, Department of Educational Administration and is still working in Karaman. Ömür ÇOBAN conducts courses in thesis and non-thesis master's programmes within the department. His academic studies are generally on leadership, educational technologies, organisational behaviour, school effectiveness and school development.

Mehmet Durnali, Ph.D.. He is an Associate Professor of Educational Administration & Leadership, an experienced educator, quantitative researcher resorting to (SPSS, AMOS, MPlus, LISREL, STATA), project developer/executer/advisor/supervisor with a demonstrated history of working in the education industry. Skilled in Idea/Theory Development, Analyzing & Reporting Data, Analytical Skills, Writing Skills, C# Programming Language, Lecturing, and + Supervising Nonprofit Organizations and

Arboriculture. Published papers regarding higher education, organizational goals, organizational behaviors, and the use of ICT in education, knowledge management, technological leadership, lifelong learning, and project management. His current researches are on the internationalization of higher education, the professionalization of higher education, the relationship between Turkey's economic growth and international students, sustainable higher education, and inclusion diversity equity & access (IDEA) in higher education. Well-rounded, hardworking resilient, and disciplined professional with a Doctor of Philosophy - Ph.D. focused on Educational Leadership and Administration, from Hacettepe University.

Tiago Horta Reis da Silva is a Lecturer in Nursing Education (AEP) at King's College London, with a range of qualifications including BSc in Nursing, BSc in Traditional Chinese Medicine, MSc in Emotional Intelligence, Counselling and NLP, a Master of Business Administration, and MSc in Traditional Chinese Medicine. He has experience in other Higher Education Institutions and NHS and is a member of the Adult Nursing department, assisting with the running, development, and administration of BSc and MSc programmes. He also assists with teaching other modules' content and assists with interviews and assessment matters for other courses. Tiago is a Senior Fellow in Higher Education, a Fellow of the Royal Society of Medicine, and a Fellow for Faculty Nursing and Midwifery at the Royal College of Southern Ireland. .

Pallavi Kumari holds degrees in Economics, English, and Management, with a Ph.D. in Economics from Ranchi University. She has worked as an Assistant HR Manager in the corporate sector and is currently an Associate Professor, leading research and publications at the university. Dr. Kumari serves as the Associate Editor for the university's publications and Editor of its newsletter. She has published research in prestigious journals, presented at national and international conferences, and is a regular reviewer for various journals. Recognized with multiple awards, including for research and publication, her areas of expertise include Entrepreneurship, Population and Health Economics, and Industrial Relations. She is also an Editorial member of Science Publishing Group, New York.

Henderson Lewis, Jr., Ph.D. is an Assistant Professor of Educational Leadership within the School of Education, College of Human Sciences and Education, Louisiana State University, Baton Rouge. His research interest includes mentoring first-generation scholars of color, instructional leadership in urban environs, and decentralization. Dr. Lewis has served as an educator in Louisiana for over 25 years with services extending across the southern region of the state. Dr. Lewis began his career as an educator in his hometown of St. Bernard Parish. He served his community in the role of a mathematics teacher, curriculum coordinator, and later as an 11-year elected school board member. Dr. Lewis continued his career journey by the assuming role of a charter school and district leader in New Orleans. His career journey afforded him the opportunity to serve the residents of East Feliciana Parish as the Superintendent of Schools. Dr. Lewis has served his last 7 years as the Superintendent of NOLA Public Schools, where he oversaw the historic 2018 unification of public schools in New Orleans with those schools run by the state of Louisiana and created the first decentralized school district in America.

Amitava Pal is the Associate Dean at the International Institute of Sports Management (IISM), which received the prestigious President's Award in 2020. A seasoned academic and industry expert, he has guided pivotal industry reports in collaboration with organizations like FICCI, ASSOCHAM, and IMC. As Training and Research Head, he guided SOPs for the 37th National Games, Khelo India

Youth Games, and Khelo India University Games. His research has been featured in leading publications, including Bonn University, Germany, and the University of California, and he received the Best Paper Award for Sustainability in Sports. Additionally, Amitava has mentored IIC fellows for the University of Chicago Harris School of Public Policy and spearheaded training, operation and implementation of the computer science curriculum researched and developed at IIT, Bombay for InOpen, SINE, IIT Bombay.

Vijendra Nath Pathak is a distinguished Associate Professor in the Department of Psychology at the Parul Institute of Liberal Arts, Parul University, India. With a robust academic and research background, Dr. Pathak has made significant contributions to psychology, including health psychology, psychometrics, military psychology, and Indian psychology. He has also held the position of Research Associate Fellow at DRDO-Defence Institute of Psychological Research, Delhi, and is a recipient of a Post-Doctoral Fellowship from the Indian Council of Social Science Research in Delhi.

Sandhya Reddy S is working as a Assistant Professor at Department of Psychology, PG, Bishop Cotton Women's Christian College, Bangalore. Currently she is pursuing her Ph.D in Psychology.

Rocío Rodríguez-Padín was born in Pontevedra and is an outstanding academic who is part of the teaching staff of the Faculty of Education Sciences at the University of Coruña, where she teaches in the Department of Pedagogy and Didactics.

Kavya Shree S is working as a Assistant Professor at Department of Psychology, PG, Bishop Cotton Women's Christian College, Bangalore.

Preksha Yadav is an accomplished academician with over 16 years of experience in teaching, research, and corporate training. She is currently an Associate Professor at Krupanidhi College of Management, Bangalore. A UGC-NET-qualified professional, she holds a Ph.D. in Management with a focus on Green HRM and sustainability. She has also completed a prestigious Faculty Development Program (FDP) at the Indian Institute of Management (IIM) Indore, enhancing her expertise in management education, research, and innovative pedagogy. Dr. Yadav has authored and edited multiple books, including Human Resource Strategies in the Era of Artificial Intelligence (IGI Global) and Sustainable Pathways: Aligning Business with the SDGs (Nova Science Publishers). Her research interests span Green HRM, sustainability, digital well-being, and the role of AI in HR. She has published over 20 Scopus-indexed research papers and actively reviews for esteemed journals. Beyond academia, she has extensive experience as a corporate trainer, delivering impactful sessions on stress management, leadership, and interpersonal skills.

Seema Yadav has academic qualifications of M. Sc. (Chemistry), M Ed (RIE), UGC-NET, and PhD in Education. Presently she is working as an Assistant Professor, in the Department of Education at The Bhopal School of Education, Bhopal. She has almost 18 years of teaching experience. Her areas of specialization are educational psychology, environmental education, digital learning, and curriculum development.

Index

Symbols

21st Century Skills 306

A

Adaptability 8, 10, 30, 32, 41, 53, 54, 60, 61, 63, 70, 72, 76, 135, 149, 150, 157, 158, 159, 160, 161, 163, 165, 167, 169, 172, 176, 183, 184, 192, 205, 206, 207, 210, 212, 213, 214, 215, 216, 218, 220, 221, 261, 262, 264, 266, 274, 280
Affective Skills 69, 70, 71, 72, 75, 77, 78, 79, 80, 81, 82, 83, 85, 86, 87, 89, 91
AI in Education 6, 9, 25, 26, 27, 34, 38, 39, 41, 44
AI Literacy 1, 7, 8, 9, 11, 13, 14, 16, 17, 18, 19, 23, 38, 41, 43
Artificial Intelligence 1, 2, 4, 5, 6, 7, 9, 12, 14, 15, 16, 17, 18, 19, 20, 21, 22, 23, 24, 25, 26, 27, 28, 31, 33, 36, 40, 42, 44, 45, 46, 125, 145, 161, 223, 225, 287, 290, 291, 292, 293, 298, 299, 306, 307
Artificial Intelligence Literacy 1, 2, 7, 12, 14, 15, 16, 17, 18, 19, 20, 22, 24
Assistant Principals 121, 122, 123, 124, 125, 126, 127, 128, 129, 130, 132, 133, 134, 135, 136, 137, 138, 139, 140, 142, 143, 144, 145
Attitudes Toward Inclusion 102, 103, 105, 106, 107, 111, 113, 117

C

Classroom Performance 235, 237, 239, 242, 243, 244, 245, 246, 247, 248, 249, 250, 253, 255, 256, 260
Classroom Practices 10, 18, 60, 74, 75, 76, 77, 78, 89, 119, 179, 294, 297
Cognitive Agility 148, 150, 151, 152, 156, 158, 163, 166
Cognitive Intelligence 48, 49, 50, 51, 52, 53, 56, 59, 60, 61, 62, 63, 64, 67, 68
Cognitive Strategies 58, 63, 206
Critical Thinking 7, 9, 11, 27, 31, 36, 41, 42, 48, 56, 59, 75, 91, 160, 171, 172, 173, 174, 176, 177, 179, 182, 183, 184, 185, 187, 188, 189, 192, 201, 205, 206, 207, 208, 210, 211, 212, 213, 215, 216, 217, 220, 221, 236, 300
Cultural Awareness 213, 261, 271, 274, 279
Cultural Rcsponsivcncss 155, 157, 178, 272, 273

D

Development 1, 2, 3, 4, 5, 6, 7, 9, 10, 18, 19, 20, 21, 22, 23, 25, 26, 27, 31, 33, 34, 35, 36, 37, 38, 40, 41, 43, 47, 48, 49, 50, 51, 52, 54, 55, 56, 57, 58, 59, 60, 61, 62, 63, 64, 65, 66, 67, 68, 69, 70, 71, 72, 73, 74, 75, 76, 77, 78, 80, 81, 82, 83, 84, 85, 86, 87, 88, 89, 90, 91, 92, 94, 95, 97, 99, 100, 101, 102, 103, 106, 113, 114, 116, 117, 118, 119, 121, 122, 124, 125, 126, 127, 128, 129, 130, 132, 133, 134, 135, 137, 138, 139, 140, 142, 143, 144, 147, 148, 149, 150, 152, 156, 157, 158, 159, 160, 161, 162, 163, 164, 165, 166, 167, 168, 169, 171, 172, 173, 174, 175, 176, 177, 178, 179, 181, 184, 186, 189, 190, 191, 194, 195, 196, 197, 198, 199, 200, 201, 202, 203, 207, 208, 209, 210, 213, 214, 215, 218, 219, 221, 224, 227, 228, 229, 230, 231, 233, 236, 237, 240, 241, 242, 245, 247, 248, 250, 253, 254, 255, 256, 259, 260, 264, 265, 267, 268, 269, 270, 271, 273, 274, 275, 276, 277, 278, 279, 281, 284, 285, 287, 288, 289, 290, 291, 292, 293, 294, 295, 296, 297, 298, 299, 300, 301, 302, 303, 304, 305, 306, 307
Digital Pedagogy 43, 306

E

Education 4.0 1, 2
Educational Administration 79, 123, 143, 144, 145
Educational Equity 98, 100, 114, 125, 281, 282
Educational Technology 19, 20, 24, 44, 45, 96, 287, 288, 289, 296, 298, 299, 300, 301, 302, 303, 305, 306, 307
Education Faculties 97, 98, 107, 108, 109, 111, 113
Emotional Intelligence 9, 42, 47, 48, 49, 52, 53, 54, 55, 56, 57, 59, 60, 61, 62, 63, 65, 66, 67, 68, 72, 73, 75, 76, 90, 92, 93, 94, 141, 147, 148, 150, 151, 152, 156, 158, 161, 163, 169, 210, 223, 229, 230, 268, 269
Emotional Regulation 54, 58, 59, 69, 70, 75, 81, 82, 89, 156, 162, 175, 193, 194, 210, 244, 247, 248, 253, 274, 281
Emotional Resilience 58, 75, 89, 148, 150, 152, 154, 158, 160, 163, 164, 172, 173, 175, 191, 193, 194, 196, 197, 198, 200, 247, 250, 251, 253, 256, 261, 262, 267, 275, 276, 278, 280
Empathy Development 113
Equity in Education 40, 116, 165, 171
Evidence-Based Approaches 100, 205

F

Flexible Teaching 205, 214, 218, 221, 270

G

Gamification 152, 153, 154, 157, 162, 165, 230, 292, 293, 297, 300, 301
Geriatric Nursing 205, 206, 207, 208, 209, 210, 211, 212, 213, 214, 215, 216, 217, 218, 219, 220, 221, 223, 224, 226, 229, 230

I

Inclusive Education 26, 97, 98, 99, 100, 101, 102, 103, 104, 105, 106, 107, 108, 109, 110, 111, 112, 113, 114, 115, 116, 117, 118, 119, 281
Inclusive Teaching 98, 100, 102, 103, 114, 267, 270, 271, 274, 275, 276, 279
Inquiry-Based Learning 183, 184, 188, 197, 198, 201
Instructional Leadership 9, 121, 124, 125, 130, 132, 133, 134, 135, 136, 138, 139, 140, 141, 144
Intelligent Tutoring Systems 26, 28, 30, 32, 37, 44, 45, 46

L

Leadership 1, 2, 3, 4, 9, 10, 18, 20, 21, 23, 24, 25, 36, 53, 64, 66, 71, 88, 89, 114, 121, 122, 123, 124, 125, 126, 127, 128, 129, 130, 132, 133, 134, 135, 136, 137, 138, 139, 140, 141, 142, 143, 144, 145, 147, 148, 149, 150, 152, 153, 154, 156, 157, 158, 159, 160, 161, 162, 163, 164, 165, 166, 167, 168, 169, 171, 173, 174, 175, 176, 178, 179, 180, 181, 182, 183, 196, 197, 198, 201, 230, 238, 240, 242, 245, 246, 247, 248, 252, 253, 255, 256, 277, 278, 281, 296
Leadership in Education 66, 148, 167, 169

M

Manipulatives 171, 176, 178, 183, 184, 185, 186, 187, 188, 189, 196, 197, 198, 201
Microlearning 287, 300, 301, 302, 307
Mindfulness Practices 63, 175, 182, 193, 194, 235, 243, 247, 248, 274, 276

N

Nursing Education 205, 206, 207, 208, 209, 210, 211, 212, 213, 214, 215, 217, 218, 219, 220, 221, 223, 226, 228, 230, 231

O

Onto-Semiotic Approach 173, 177, 203

P

Pedagogical Content Knowledge 172, 173, 176, 186, 203, 289, 290, 303, 306
Pedagogical Formation 97, 98, 107, 108, 109, 111, 113, 117
Pedagogical Practices 7, 18, 33, 91, 152, 153, 156, 159, 198, 200, 211, 218
Personalized Learning 21, 27, 28, 30, 31, 34, 35, 36, 37, 41, 44, 46, 100, 184, 186, 265, 299, 300, 301, 307
Problem-Solving 5, 28, 32, 36, 40, 50, 52, 54, 55, 58, 59, 61, 62, 75, 84, 91, 124, 127, 148, 151, 156, 161, 163, 171, 172, 173, 174, 175, 176, 177, 178, 179, 180, 182, 183, 184, 185, 187, 189, 191, 192, 197, 199, 201, 205, 206, 207, 211, 215, 216, 217, 220, 266, 268, 299
Professional Growth 9, 36, 40, 47, 58, 121, 122, 129, 130, 133, 134, 135, 136, 137, 138, 139, 147, 152, 159, 162, 164, 169, 171, 173, 175, 190, 194, 196, 197, 198, 215, 262, 269, 273, 275, 276, 277, 278, 279, 288, 289, 291, 293, 294, 295, 296, 300, 302
Professional Learning Communities 91, 121, 127, 128, 129, 134, 137, 195, 196, 198, 277, 289

Q

Qualitative Inquiry 77, 92, 94, 141

R

Reflective Practice 58, 91, 113, 158, 174, 181, 199, 208, 211, 216, 280
Resilience 57, 58, 63, 70, 71, 72, 74, 75, 87, 89, 91, 147, 148, 149, 150, 152, 154, 158, 159, 160, 161, 162, 163, 164, 165, 166, 168, 169, 171, 172, 173, 175, 176, 180, 181, 182, 184, 191, 192, 193, 194, 196, 197, 198, 200, 210, 247, 250, 251, 253, 256, 260, 261, 262, 263, 264, 265, 266, 267, 268, 269, 271, 273, 274, 275, 276, 278, 280, 281, 282, 284, 285, 295

S

School Leadership 21, 121, 122, 123, 127, 128, 129, 130, 132, 133, 134, 140, 141, 142, 143, 196,

240, 247, 281
Secondary Education 69, 92, 241
Smart Classrooms 37, 40, 46, 96, 289, 291, 293
Socio-Emotional Support 171, 194, 195, 196, 197, 198, 200
Special Educational Needs 97, 98, 116, 118, 200
Sports-Based Learning 157, 158, 159, 160, 161, 162, 163, 164
Student Engagement 27, 31, 32, 33, 34, 38, 39, 52, 59, 61, 63, 74, 95, 126, 148, 153, 157, 160, 175, 182, 186, 188, 189, 192, 193, 199, 200, 206, 237, 240, 242, 246, 247, 248, 249, 250, 253, 270, 278, 279, 280, 291, 293
Student Outcomes 25, 60, 73, 74, 126, 128, 129, 134, 137, 140, 142, 150, 172, 174, 179, 181, 183, 185, 195, 198, 199, 200, 213, 235, 239, 240, 241, 243, 246, 247, 249, 251, 271, 273, 277, 278, 279, 294, 298
Students 1, 2, 3, 4, 6, 7, 9, 10, 11, 17, 18, 20, 22, 26, 27, 28, 29, 30, 31, 32, 33, 34, 35, 36, 37, 38, 39, 40, 41, 42, 43, 44, 45, 47, 48, 49, 50, 52, 54, 55, 56, 57, 58, 59, 60, 61, 62, 63, 64, 65, 67, 70, 71, 72, 73, 74, 75, 76, 77, 78, 79, 80, 81, 82, 83, 84, 85, 86, 87, 88, 89, 90, 91, 92, 93, 94, 95, 96, 98, 99, 100, 101, 102, 103, 104, 105, 106, 107, 108, 112, 113, 114, 116, 118, 119, 123, 124, 125, 127, 128, 133, 136, 138, 140, 148, 150, 151, 152, 153, 154, 155, 156, 157, 158, 160, 161, 163, 166, 167, 168, 171, 172, 173, 174, 175, 176, 177, 178, 179, 180, 181, 182, 183, 184, 185, 186, 187, 188, 189, 190, 191, 192, 193, 194, 195, 197, 199, 200, 201, 203, 205, 206, 207, 208, 209, 210, 211, 212, 213, 214, 215, 216, 217, 218, 219, 220, 221, 222, 223, 224, 226, 227, 232, 236, 237, 238, 239, 240, 241, 244, 246, 247, 248, 249, 250, 251, 253, 254, 256, 258, 259, 261, 262, 264, 265, 266, 267, 268, 269, 270, 271, 272, 273, 274, 275, 276, 277, 278, 279, 280, 281, 282, 283, 286, 288, 289, 291, 292, 295, 296, 299, 301, 304, 307

T

Teacher 1, 2, 3, 4, 7, 9, 10, 11, 13, 14, 18, 20, 21, 23, 24, 25, 26, 32, 36, 38, 39, 40, 41, 42, 43, 47, 48, 49, 50, 51, 52, 54, 55, 56, 57, 58, 59, 60, 61, 62, 63, 64, 65, 66, 67, 68, 71, 72, 73, 74, 77, 81, 82, 84, 85, 87, 89, 91, 92, 93, 95, 97, 98, 99, 100, 102, 103, 104, 105, 106, 107, 108, 109, 110, 111, 112, 113, 114, 115, 116, 117, 118, 121, 122, 124, 125, 126, 127, 129, 132, 133, 135, 136, 137, 138, 139, 140, 141, 142, 143, 144, 145, 147, 148, 150, 151, 152, 153, 154, 155, 156, 157, 158, 159, 160, 161, 162, 163, 164, 165, 167, 171, 172, 173, 174, 175, 176, 178, 179, 188, 191, 192, 193, 195, 196, 197, 198, 199, 200, 201, 203, 228, 235, 236, 237, 238, 239, 240, 241, 242, 243, 244, 245, 246, 247, 248, 249, 250, 251, 252, 253, 254, 255, 256, 257, 258, 259, 260, 264, 266, 267, 269, 270, 271, 272, 274, 276, 278, 279, 280, 281, 282, 283, 284, 285, 286, 287, 288, 289, 293, 294, 295, 296, 297, 298, 299, 300, 301, 302, 303, 304, 305, 306, 307
Teacher Candidates 97, 98, 103, 105, 106, 107, 108, 109, 110, 111, 112, 113, 114, 116, 118
Teacher Collaboration 99, 121, 122, 127, 129, 133, 135, 138, 139, 279
Teacher Development 25, 47, 48, 49, 51, 52, 54, 55, 56, 57, 58, 60, 61, 62, 63, 64, 66, 67, 72, 127, 144, 147, 148, 150, 152, 156, 157, 158, 159, 160, 161, 162, 163, 164, 167, 172, 173, 174, 175, 176, 197, 198, 199, 200, 201, 279, 299, 303
Teacher Engagement 243
Teacher Identity 113, 281
Teacher Leadership 1, 2, 3, 4, 9, 10, 18, 20, 21, 23, 24, 25, 141
Teacher Perceptions 303, 305
Teacher Resilience 165, 260
Teacher Roles 25, 26, 39
Teachers 1, 2, 3, 4, 7, 9, 10, 11, 12, 13, 14, 15, 16, 17, 18, 19, 21, 22, 23, 24, 25, 26, 27, 28, 31, 32, 33, 34, 35, 36, 37, 38, 39, 40, 41, 42, 43, 45, 47, 48, 49, 50, 52, 54, 55, 56, 57, 58, 59, 60, 61, 62, 63, 64, 65, 66, 67, 68, 69, 71, 72, 73, 74, 75, 76, 77, 78, 79, 80, 81, 82, 83, 84, 85, 86, 87, 88, 89, 90, 91, 92, 93, 94, 95, 96, 100, 101, 102, 103, 104, 105, 106, 107, 109, 113, 114, 115, 116, 117, 118, 119, 121, 122, 123, 124, 125, 126, 127, 128, 129, 132, 133, 134, 135, 136, 137, 138, 139, 140, 141, 142, 144, 148, 149, 150, 151, 152, 153, 154, 155, 156, 157, 158, 159, 160, 161, 162, 163, 164, 165, 166, 171, 172, 173, 174, 175, 176, 177, 178, 179, 180, 181, 182, 183, 184, 185, 186, 187, 188, 189, 190, 191, 192, 193, 194, 195, 196, 197, 198, 199, 200, 201, 202, 203, 204, 231, 235, 236, 237, 238, 239, 240, 241, 242, 243, 244, 245, 246, 247, 248, 249, 250, 251, 252, 253, 254, 255, 256, 257, 258, 259, 260, 261, 262, 264, 265, 266, 267, 268, 269, 270, 271, 272, 273, 274, 275, 276, 277, 278, 279, 280, 281, 282, 283, 284, 285, 286, 288, 289, 290, 291, 292, 293, 294, 295, 296, 297, 298, 299, 300, 301, 303, 304, 305, 306, 307
Teacher Stress 200, 235, 239, 241, 242, 243, 244, 247, 248, 249, 250, 251, 253, 255, 256, 257, 258, 260

Teacher's Well-Being 235, 239, 241, 255, 256
Teacher Training 7, 32, 47, 48, 56, 58, 59, 60, 62, 63, 66, 89, 91, 97, 98, 99, 100, 102, 103, 104, 105, 107, 108, 111, 112, 113, 115, 158, 163, 172, 176, 195, 199, 288, 296, 297, 305
Teacher Well-Being 59, 63, 91, 102, 171, 173, 175, 191, 192, 193, 195, 196, 197, 198, 199, 200, 201, 237, 238, 240, 241, 243, 244, 245, 247, 248, 249, 250, 251, 252, 253, 254, 255, 258, 260, 279, 281
Türkiye 10, 11, 13, 69, 78, 79, 97, 98, 100, 101, 102, 103, 104, 105, 106, 107, 108, 109, 110, 111, 112, 113, 114

W

Work-life Balance 175, 194, 197, 235, 237, 238, 243, 246, 247, 252, 254, 256, 258, 259, 276

www.ingramcontent.com/pod-product-compliance
Lightning Source LLC
LaVergne TN
LVHW081251011026
849162LV00006BA/1402